Wireless Signal Processing and Radio Engineering

Wireless Signal Processing and Radio Engineering

Edited by **Adrian Franel**

Willford Press
New York

Published by Willford Press,
118-35 Queens Blvd., Suite 400,
Forest Hills, NY 11375, USA
www.willfordpress.com

Wireless Signal Processing and Radio Engineering
Edited by Adrian Franel

International Standard Book Number: 978-1-68285-067-1 (Hardback)

Printed in the United States of America.

Contents

Preface

This book has been an outcome of determined endeavour from a group of educationists in the field. The primary objective was to involve a broad spectrum of professionals from diverse cultural background involved in the field for developing new researches. The book not only targets students but also scholars pursuing higher research for further enhancement of the theoretical and practical applications of the subject.

Wireless signal processing forms the core of most of the devices and technologies that are used in the present era. The aim of this book is to present researches that have transformed this discipline and aided its advancement. Included in this book are detailed discussions on areas such as antennas, microwaves, circuits, optics, applications of wireless technologies, etc. With state-of-the-art inputs by acclaimed experts of this field, this book targets students and professionals. A number of latest case studies have been included to keep the readers up-to-date with the global concepts in this area of study.

It was an honour to edit such a profound book and also a challenging task to compile and examine all the relevant data for accuracy and originality. I wish to acknowledge the efforts of the contributors for submitting such brilliant and diverse chapters in the field and for endlessly working for the completion of the book. Last, but not the least; I thank my family for being a constant source of support in all my research endeavours.

Editor

Three-Pole Tunable Filters with High Rejection using Mixed $\lambda/4$ Resonators and Asymmetric $\lambda/2$ Resonators

Zhiyuan ZHAO[1,2], *Jiang CHEN*[2], *Lin YANG*[2], *Kunhe CHEN*[2]

[1] Institute of Communications Engineering, PLA University of Science and Technology, Nanjing 210007, China
[2] Nanjing Telecommunication Technology Institute, Nanjing 210007, China

zhaozhiyuan1986@sina.com, chenjiang999@sina.com

Abstract. *A novel three-pole tunable bandpass filter using varactor-loaded quarter-wavelength combline and asymmetric half-wavelength resonators is proposed in this paper. A nearly constant 3-dB absolute bandwidth is 150 ± 13 MHz (10.7%~6.9% fractional bandwidth) within the tuning range of 1.4-2.0 GHz (42.8%). The filter is designed on a Rogers substrate with $\varepsilon_r = 2.2$ and h = 1 mm with its insertion loss varying from 3.6 dB to 2.8 dB and return loss better than 10 dB over the entire tuning range. The creation of two transmission zeros near the passband edges is analyzed by the even-odd-method. By using dissimilar resonators, the proposed tunable filter could obtain > 33 dB rejection levels at the second harmonics. The measured results show good agreement with the simulated ones.*

Keywords

Three-pole tunable filters, combline resonators, modified combline resonators, constant absolute bandwidth (CABW), transmission zeros (TZs), varactor-tuned.

1. Introduction

Electronically tunable or reconfigurable pre-select microwave filters significantly improve the overall performance of multi-bands and multi-functions wireless communication systems, which bring the reduction of the system size, complexity and cost. Planar tunable bandpass filters (BPFs) can be realized by varactor diodes [1], ferroelectric diodes [2], or RF micro-electro-mechanical systems (RF-MEMS) devices [3]. However, the varactor-diode filters are widely studied due to the advantages of continuous and high-speed tuning and economical fabrication.

Due to compactness and ease of integration, combline and modified combline resonators loaded by varactor diodes, commonly used in small-size tunable filters for the RF front end, have been studied in numerous literatures [4]-[11]. However, the combination of quarter-wavelength ($\lambda/4$) combline and asymmetric half-length ($\lambda/2$) resonators, especially for realizing a constant absolute bandwidth (CABW) of three-pole tunable filter with high rejection, has not been reported. Hunter and Rhodes firstly described that the optimum electrical length is approximately 53° at the center frequency of the tuning range for achieving the constant bandwidth of the stripline tunable filter [4]. For microstrip-line tunable filter, a systematic approach was presented to choose the tunable filter design parameters [5]. In [6], a tunable combline filter with plural transmission zeros originating from source-load and multi-resonators coupling was realized. Note that most microstrip combline or modified combline filters only have transmission zeros in the lower stopband or upper stopband, and this leads to an asymmetrical transmission response and poor rejection in the stopband without transmission zeros. Yi-Chyun Chiou et. al. proposed a non-adjacent resonator coupling to generate an extra zero at the lower band, except the intrinsic zero at the upper stopband [7]. In [8], there were two transmission zeros at both sides of the passband, respectively, however the source-load coupling structure was complicated. In [9], there were two transmission zeros near the passband due to the tap connections of input and output port, however the rejection level in the stopband was not high enough without adding the bandpass network to the two ends. Therefore, it remains difficult to realize a high-rejection tunable filter with CABW and two transmission zeros within a wide tuning range.

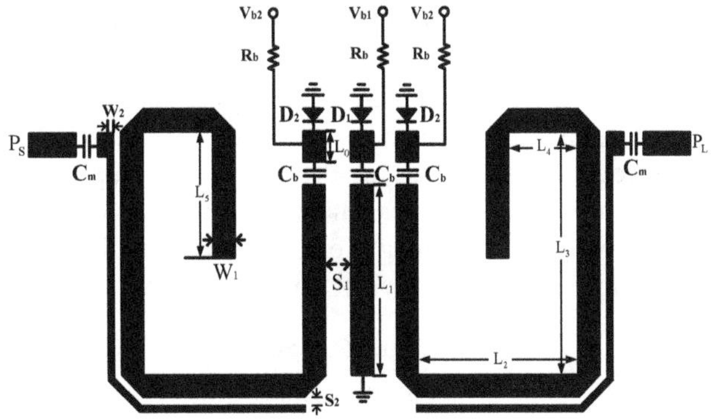

Fig. 1. Proposed three-pole tunable BPF using mixed $\lambda/4$ combline and asymmetric $\lambda/2$ resonators.

This paper proposes a new three-pole tunable filter with two transmission zeros using mixed $\lambda/4$ combline and asymmetric $\lambda/2$ resonators, as shown in Fig. 1. The design aim is to maintain the absolute bandwidth constant and obtain high rejection. It consists of an asynchronously varactor-tuned $\lambda/4$ combline resonator between the two varactor-loaded asymmetric $\lambda/2$ resonators. Parallel-coupled-lines are selected as the external coupling structure to achieve impedance matching across a wide tuning range without any other tuning components [10]. By properly controlling the bias voltages, a high-rejection tunable filter with nearly constant absolute bandwidth within a wide frequency tuning range can be realized.

This paper is organized as follows. The characteristic of the proposed filter topology is introduced and design theory is analyzed in Section 2. In Section 3, experimental results of a built prototype are shown. Finally, the conclusions of this work are drawn in Section 4.

2. Tunable Filter Theory

2.1 Bandwidth Characteristic of the Coupled Resonators

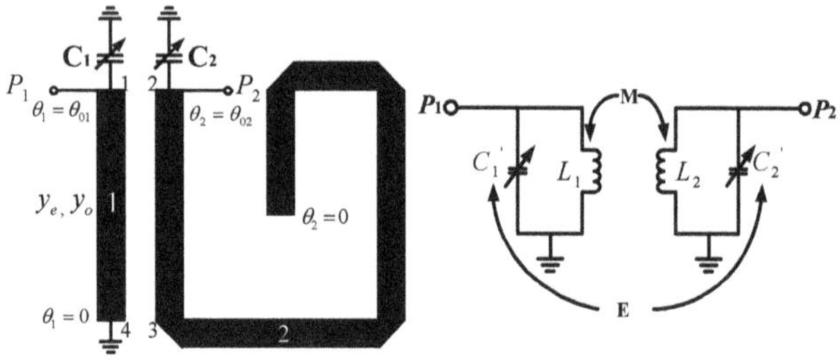

(a) (b)

Fig. 2. (a) Coupled varactor-loaded combline and modified-combline resonators, and (b) its equivalent circuit.

The design is based on varactor-loaded $\lambda/4$ combline resonator and varactor-loaded asymmetric $\lambda/2$ modified-combline resonators [9]. Fig. 2 presents a coupled-resonator section of the proposed tunable filter and its simplified equivalent circuit. The lumped-element capacitances are $C_1^{'} = C_c^{'} + C_1$ and $C_2^{'} = C_a^{'} // C_2$, and the inductances are L_1 and L_2, where L_1 and $C_c^{'}$ are the equivalent inductor and capacitor of the $\lambda/4$ combline resonator, respectively, L_2 and $C_a^{'}$ are the equivalent inductor and capacitor of the asymmetric $\lambda/2$ modified-combline resonator, respectively [11]. The coupling coefficient of the equivalent coupled resonators model in Fig. 2 is calculated by energetic coupling approach [12]. Assuming weak coupling condition ($k_M k_E << 1$), the coupling coefficient is determined as

$$k \approx k_M + k_E = \frac{(k_L^{'} - k_C^{'})[\cos\Delta\theta - \cos(\theta_{02} + \theta_{01})]}{\sqrt{(2\theta_{01} + \sin 2\theta_{01})(2\theta_{02} - \sin 2\theta_{02})}} + \frac{2(k_L^{'} + k_C^{'})\theta_{01}\sin\Delta\theta}{\sqrt{(2\theta_{01} + \sin 2\theta_{01})(2\theta_{02} - \sin 2\theta_{02})}} \quad (1)$$

$$\Delta\theta = \theta_{02} - \theta_{01} \quad (2)$$

where, $k_L^{'}$ and $k_C^{'}$ are the inductive and capacitive coupling coefficients per unit length in the coupling region. θ_{01} ($\theta_{01} < \pi/4$) and θ_{02} ($\theta_{02} > \pi/2$) indicate electrical lengths of the $\lambda/4$ combline resonator and the asymmetric $\lambda/2$ resonator, respectively. Using the standard differentiation procedure, given $\theta_{02} = 4\theta_{01}$, it can be calculated that the optimum electrical length of θ_{01} in the midband of the tuning range. The crude optimum electrical length of the $\lambda/4$ combline resonator corresponds to the frequency at which the bandwidth is maximized.

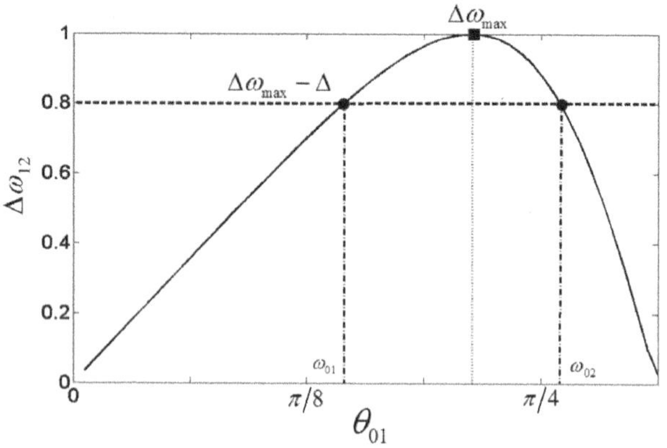

Fig. 3. Normalized bandwidth of the proposed architecture using quasi-TEM resonators.

Fig. 3 is a plot of normalized coupling bandwidth $\Delta\omega_{12}$ ($\Delta\omega_{12} = k\omega_0$) versus θ_{01} according to (1) when quasi-TEM resonators are used, with $\omega_0 = \theta_{01}$. If the allowed percentage change of bandwidth is $\pm 10\%$, a wide center frequency tuning range (ω_{01}, ω_{02}) can be obtained to fulfill the bandwidth requirement [$\Delta\omega_{max} - \Delta$, $\Delta\omega_{max}$]. Due to the linear relationship between electrical length and frequency, the plot shows the bandwidth dependence on the tuning frequency. The electrical length corresponding to $\Delta\omega_{max}$ is the optimum resonator length (θ_{01} = 39°) at the midband.

2.2 Analysis of the Coupled Resonators

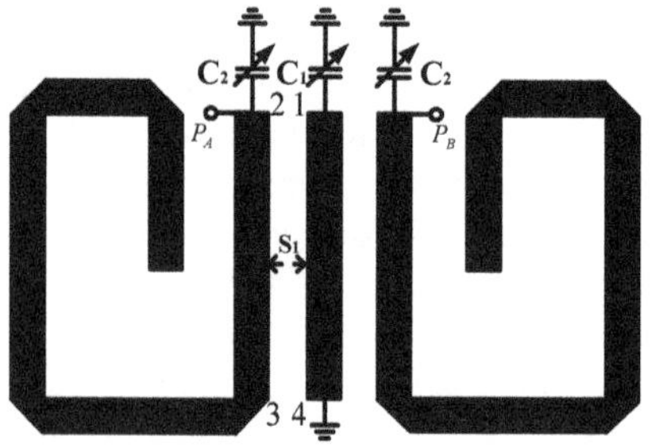

Fig. 4. The three coupled-resonators without I/O coupling structure.

In the analysis of the three coupled-resonators of the proposed tunable filter, the *Y*-parameter is adopted. Fig. 4 is three coupled-resonators without I/O coupling structure. The reference ports (P_A and P_B) are added for deriving the admittance matrix, the *Y*-parameter matrix of the two asynchronously coupled-resonators is calculated as

$$Y = \begin{bmatrix} y_{11} - \dfrac{y_{13}y_{31}}{y_{33} + jY_0 \tan\Delta\theta} & y_{12} - \dfrac{y_{13}y_{32}}{y_{33} + jY_0 \tan\Delta\theta} \\ y_{21} - \dfrac{y_{13}y_{32}}{y_{33} + jY_0 \tan\Delta\theta} & y_{22} - \dfrac{y_{23}y_{32}}{y_{33} + jY_0 \tan\Delta\theta} \end{bmatrix} \quad (3)$$

And the four-port Y parameters of the parallel coupled-line are [13],

$$y_{11} = y_{22} = y_{33} = -\frac{j}{2}(Y_{ro} + Y_{re})\cot\theta_{01}\ , \quad (4a)$$

$$y_{12} = y_{21} = -\frac{j}{2}(Y_{ro} - Y_{re})\cot\theta_{01}\ , \quad (4b)$$

$$y_{13} = y_{31} = -\frac{j}{2}(Y_{ro} - Y_{re})\csc\theta_{01}\ , \quad (4c)$$

$$y_{23} = y_{32} = -\frac{j}{2}(Y_{ro} + Y_{re})\csc\theta_{01} \quad (4d)$$

where Y_{re}, Y_{ro} are the even- and odd-mode characteristic admittances of the parallel coupled-line, respectively.

The resonant angular frequency ω_0 can be found by the following equation, ($i = 1, 2$).

$$\mathrm{Im}[Y_{ii}(\omega_0)] + \omega_0 C_i = 0 \quad (5)$$

For the filter composed of two varactor-loaded asymmetric $\lambda/2$ resonators, their fundamental resonant frequency is f_0, and the second-order spurious response is at the frequency higher than $2f_0$. Furthermore, that composed of a varactor-loaded $\lambda/2$ combline resonator has spurious passband far from $3f_0$. Therefore, the rejection level of the second harmonics of the proposed filter in Fig. 1 can be improved significantly.

The coupling coefficient of the coupled resonators is [11]

$$k_{12} = \frac{\mathrm{Im}[Y_{12}(\omega_0)]}{\sqrt{b_1 b_2}} = \frac{BW}{f_0\sqrt{g_1 g_2}} \quad (6)$$

where BW is the 3-dB absolute bandwidth, g_1, g_2 are the element values of low-pass prototype filter, respectively. And the slope parameter b_1 and b_2 are derived as [13]

$$b_i = \frac{\omega_0}{2}\frac{\partial\,\mathrm{Im}[Y_{ii}]}{\partial\omega}\bigg|_{\omega=\omega_0} + \frac{\omega_0 C_i}{2}\ . \quad (7)$$

2.3 Transmission Zeros Creation

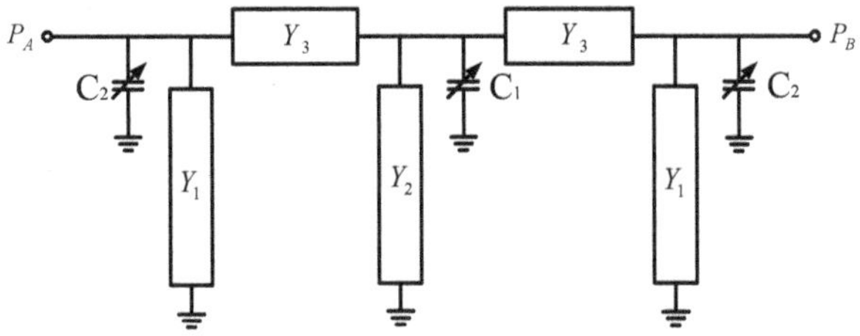

Fig. 5. Equivalent circuit model of the three coupled-resonators.

As was explained in the past literatures, adding I/O coupling structure shifts the transmission zero at infinite frequency downwards, but it does not really add a zero. For facilitating the analysis of the TZs' location, the I/O matching networks are not calculated. Since the topology is symmetrical to the center plane, the S-parameter of the three coupled resonators can be calculated by using the even-odd-mode method [13]. It is assumed that the even- and odd-mode electrical lengths of the parallel symmetrical pair of coupled-lines are identical to the value of θ_{01}, and the parasitic effects of the grounded via-holes and the line discontinuity due to the DC block capacitor C_b are ignored. Fig. 5 presents the exact equivalent circuit model of the three coupled-resonators in Fig. 4. In this circuit, $Y_1 = Y_{22} - Y_3$, $Y_2 = Y_{11} - Y_3$, $Y_3 = -Y_{12}$, where Y_{ij} is the element of the $\boldsymbol{Y}$ matrix in (3).

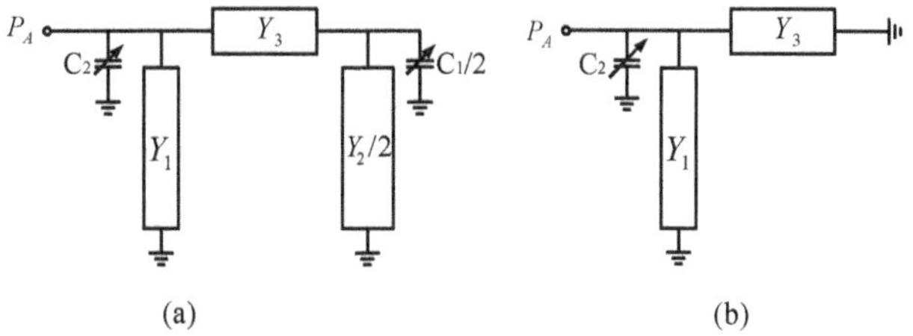

Fig. 6. (a) Its equivalent even-mode circuit, and (b) its equivalent odd-mode circuit in Fig. 5.

Fig. 6 is the equivalent even- and odd-mode circuit obtained by placing a short or open circuit at the symmetric plane. The even- and odd-mode input admittance Y_{Ae} and Y_{Ao} seen from port P_A can be respectively solved by (8) and (9)

$$Y_{Ae} = j\omega C_2 + Y_1 + \frac{Y_3(Y_2 + j\omega C_1)}{Y_2 + 2Y_3 + j\omega C_1}\ , \quad (8)$$

$$Y_{Ao} = j\omega C_2 + Y_1 + Y_3\ . \quad (9)$$

By superposition, the transfer function S_{21} can then be written as [13]

$$S_{21} = \frac{Y_0(Y_{Ae} - Y_{Ao})}{(Y_0 + Y_{Ae})(Y_0 + Y_{Ao})} \quad (10)$$

where Y_0 is terminal admittance. Enforcing $S_{21} = 0$ or $Y_{Ae} - Y_{Ao} = 0$, the two transmission zero frequencies can be solved. As can be seen in the above transcendental equations, it is not easy to obtain the explicit solution of the TZs' location.

The transmission response is obtained by feeding the three coupled resonators at P_A and P_B, and the S_{21} response indicates that the filter possesses three poles (f_{p1} through f_{p3}) and two transmission zeros (f_{z1} and f_{z2}), as shown in Fig. 7. One transmission zero appears at the lower stopband, and the other appears at the upper stopband. The TZs creation is usually due to the multiple coupling paths, which means several signals cancel each others at a certain frequency. According to (8)-(10), the transmission zero is a function of the loading capacitance, so they move with the center frequency of the filter varied by controlling the biasing voltages.

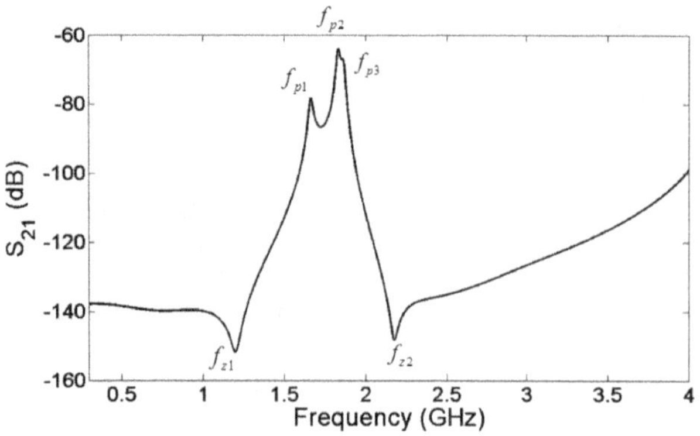

Fig. 7. S_{21} response of Fig. 4 with weakly capacitive coupling at its two ports.

2.4 Extraction of k_{12} and Q_e of the Tunable Filter

As the center frequency of the filter is varied, all of the filter parameters change. Thus, the coupling coefficients between resonators are always changed. To assure CABW, the desired coupling coefficient must satisfy the conditions described in [3]. The gap S_1 is chosen so that the two asynchronously tuned resonators are coupled through a suitable coupling coefficient k_{12}. The coupling coefficient k_{12} of the coupled resonators can be calculated by [13]

$$k_{12} = \pm\frac{1}{2}\left(\frac{f_{02}}{f_{01}} + \frac{f_{01}}{f_{02}}\right)\sqrt{\left(\frac{f_{p2}^2 - f_{p1}^2}{f_{p2}^2 + f_{p1}^2}\right)^2 - \left(\frac{f_{02}^2 - f_{01}^2}{f_{02}^2 + f_{01}^2}\right)^2} \quad (11)$$

where f_{01}, f_{02} are resonant frequencies of each single resonator, f_{p1} and f_{p2} are frequencies of the two peaks. The choice of the sign in (4) depends on the definition of electric and magnetic coupling.

To maintain nearly constant absolute bandwidth, the external quality factor Q_e should increase as the frequency shifts upward. This can be realized by using the parallel-line structure as shown in Fig. 1. The geometrical dimensions and capacitance of the matching capacitor C_m of the I/O structure can be determined by matching the singly loaded Q. The required Q_e for a 3-pole filter is given by [13]

$$Q_e = \frac{f_0}{\Delta f_{\pm 90^\circ}} = \frac{g_0 g_1 f_0}{BW} \quad (12)$$

where $\Delta f_{\pm 90^\circ}$ is determined from the frequency at which the phase shifts ±90° with respect to the absolute phase at f_0.

3. Design and Experimental Verification of Tunable Filter

A nearly CABW three-pole tunable filter with high rejection using $\lambda/4$ combline and asymmetric $\lambda/2$ resonators is presented in this section to verify the analysis above. A full-wave electromagnetic simulator, Computer Simulation Technology (CST2009) Microwave Studio and Design Studio software packages are employed for the circuit dimensions of final filter simulation. The designed filter is simulated and fabricated in microstrip technology with the following specifications:

Frequency tuning range: 1.4-2.0 GHz;

Passband bandwidth: 150 ± 10 MHz;

Number of poles: three;

Type: 0.04-dB ripple Chebyshev at 1.7 GHz.

3.1 Design Procedure

Step 1). Subjecting to the specifications, select the three-pole low-pass prototype with elements g_i, i =0,...4. Then calculate the external quality factors and coupling coefficient according to (6) and (12).

Step 2). Determine the dimensions of the proposed tunable resonators. The electrical length of the combline resonator is set to be 39° (θ_{01} = 39°, θ_{02} = $4\theta_{01}$ = 156°) at 1.7 GHz. The characteristic impedance of combline filter is traditionally 71 or 72 Ω to obtain the optimum unloaded Q [11]. For minimizing the size of the circuit, especially the asymmetric half-wavelength resonator, we choose a little higher characteristic impedance (W_1 = 1 mm). The gap S_1 between $\lambda/4$ combline resonator and asymmetric $\lambda/2$ resonators can then be determined to satisfy k_{12}. The parameters of the I/O structure are typically chosen to be a high impedance line for tight coupling. The matching capacitor C_m, the gap S_2, and the width W_2 of the line are then chosen to satisfy Q_e.

Step 3). Finally, the optimization of circuit dimensions of filter is employed CST2009 EM simulator. The electrical length of the comb-line resonator is optimized to be 8 mm (27°) at 1.7 GHz because of the additional equivalent electrical lengths of grounding via-holes, grounding pads and components of DC blocking capacitor and varactor.

According to the extraction method of coupling coefficient and external quality factor in [13], the k_{12} and Q_e versus different center frequencies of the tunable filter is extracted and plotted in Fig. 8 with the desired values for the constant absolute bandwidth. Due to the inconsistency between the simulated k_{12}, Q_e and the desired ones, the ripple of the filter will change and the absolute bandwidth will vary as the frequency shifts downward or upward from the mid-band.

For the optimized resonators, the capacitance of the loading varactors C_1 and C_2 corresponding to the resonant frequency is plotted in Fig. 9 for comparison. The simulated results agree well with the calculated ones by (5).

The filter is designed on Rogers RT5880 substrate (ε_r = 2.2, h = 1 mm, and tanδ = 0.0009) with overall size of ~40 × 36 mm², as shown in Fig. 10.

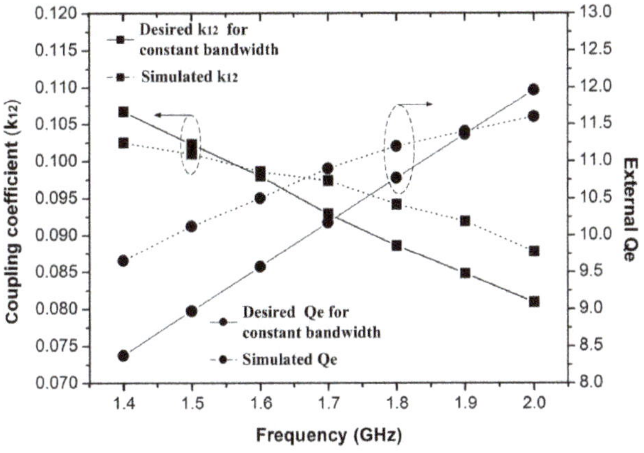

Fig. 8. Desired and simulated coupling coefficients and Q_e of the proposed filter.

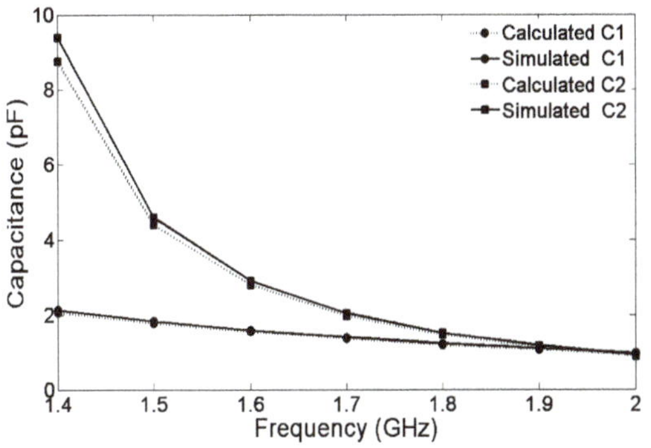

Fig. 9. Calculated and simulated capacitance of the loading varactors for the optimized resonators.

Fig. 10. Photograph of the fabricated 3-pole tunable filter including biasing circuits.

W_1 / W_2	L_0 / L_1 / L_2 / L_3 / L_4 / L_5	S_1 / S_2	C_m (pF) / C_b (pF)
1 / 0.1	2 / 8 / 6/10.6 / 2.5 / 5.5	2 / 0.1	5.6 / 15

Tab. 1. Dimensions for the filter (Dimensions are in millimeters).

The physical dimensions of the filter are summarized in Tab. 1. The matching capacitor C_m and DC blocking capacitor C_b are realized by ATC 600S series capacitors. The DC blocking capacitor has a high self-resonant frequency of over 2.5 GHz, which works as open circuit at DC frequency and short circuit at RF frequency. The capacitors C_1 and C_2 are implemented by MA46H202-1088 GaAs diodes. MA46H202 varactor diodes (C_1 or C_2 = 0.6-11 pF, R_s = 0.2-0.9 Ω, over a 22 V reverse bias range) are selected as D_1 and D_2 for frequency tuning, respectively. The biasing circuits are realized using two 100-kΩ resistors to minimize RF signal leakage.

3.2 Measurements

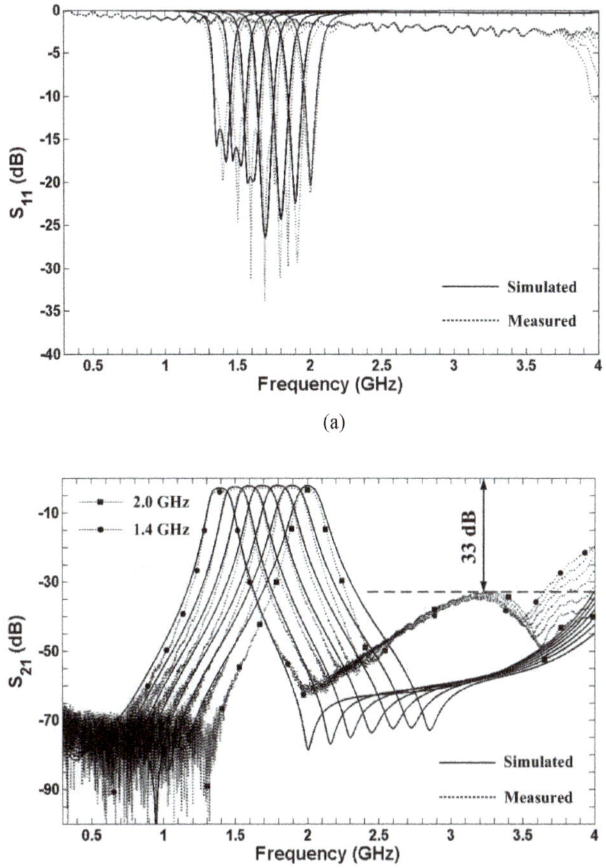

(a)

(b)

Fig. 11. Measured and simulated *S*-parameters of the proposed tunable BPF. (a) S_{11}. (b) S_{21}.

The measured frequency responses of the designed filter with the center frequency tuning obtained by using Agilent N5230A network analyzer are in good agreement with the simulation results, as shown in Fig. 11. The measured return loss ($|S_{11}|$) is worse than the simulation results due to the inconsistency of the varactor diodes biased by only one bias voltage for the asymmetric λ/2 resonators, however better than 10 dB for all states. The 3-dB absolute bandwidth is maintained nearly constant about 150 MHz (150 ± 13 MHz) over the entire center frequency tuning range from 1.4 GHz to 2.0 GHz (a wide tuning range of 42.8%). The bandwidth slightly increases at the midband frequency and decreases at the both ends of the tuning range. The characteristic of bandwidth variation is consistent with that of Fig. 3. Because of the effect of the package inductances of the varactor diodes and parasitic inductance coupling among the via-holes, the passband bandwidth has slight variation between simulation and measurement. Its insertion loss ($|S_{21}|$) varies from 3.6 dB to 2.8 dB due to the low overall unloaded Q of the resonators by using varactors [14]. The insertion loss improves as the center frequency is tuned to the higher end of the tuning range, and it is related to the characteristic of the varactor diodes under different bias condition and the variation of fractional BW (10.7% at lower end and 6.9% at higher end). The discrepancies mainly result from fabrication tolerance in the implementation and the simulation error between SPICE models and actual varactor diodes.

Note that the two transmission zeros near to passband are shifted with the center frequency varied, and the rejection of the lower and upper stopbands remains high with the tuning. Furthermore, the extra zero appears at the higher side due to the package inductance of the varactor diodes, which can enhance the rejection of the second harmonics (> 33 dB). By using dissimilar resonators [15], the rejection level is > 20 dB in the stopband from $1.07f_0$ to $2.85\ f_0$. The nonlinear distortion performance is another important figure of merit due to the involvement of the varactor. The measured 1-dB compression point at the filter input is higher than 8 dBm over the tuning frequency. The measured IIP3 is ranging from 33 dBm to 38 dBm for $\Delta f = 1$ MHz, meeting most of the requirements for software-defined radio applications. The summary of the measured results is shown in Tab. 2.

f_0 (GHz)	I.L. (dB)	BW (MHz)	P1 (dBm)	IIP3 (dBm)	Vb_1 (V)	Vb_2 (V)
1.4	3.6	150	8	33	5.45	0.5
1.7	3.0	163	9	38	8.65	6.0
2.0	2.8	137	8.5	36	13.2	14.9

Tab. 2. Measured results of the tunable filter.

4. Conclusions

In this paper, we proposed a novel three-pole tunable filter topology with high rejection by using mixed varactor-loaded $\lambda/4$ combline and asymmetric $\lambda/2$ resonators. A 1.4-2.0 GHz three-pole microstrip-line tunable filter with a nearly constant 3-dB absolute bandwidth of 150 ± 13 MHz and the rejection level > 33 dB at the second harmonics is simulated, fabricated, and measured. The measured results agree well with the simulated ones.

Acknowledgements

This work was supported by Key Pre-research Foundation of PLA university of Science and Technology under grant. KY63ZLXY1301. The authors would like to thank M/A-COM Corporation for the high-performance varactor diodes and Rogers Corporation for the low-loss substrate.

References

[1] WONG, P. W., HUNTER, I. C. Electronically tunable filters. *IEEE Microwave Magazine,* 2009, vol. 10, no. 6, p. 46–54.

[2] HONG, J. S. Reconfigurable planar filters. *IEEE Microwave Magazine*, 2009, vol. 10, no. 6, p. 73–83.

[3] REBEIZ, G. M., REINES, I. C., EL-TANANI, M. A., ET AL. Tuning in to RF MEMS. *IEEE Microwave Magazine*, 2009, vol. 10, no. 6, p. 55–72.

[4] HUNTER, I. C., RHODES, J. D. Electronically tunable microwave bandpass filters. *IEEE Transactions on Microwave Theory and Techniques*, 1982, vol. 30, no. 9, p. 1354–1360.

[5] TORREGROSA-PENALVA, G., LOPEZ-RISUENO, G., ALONSO, J. I. A simple method to design wide-band electronically tunable combline filters. *IEEE Transactions on Microwave Theory and Techniques,* 2002, vol. 50, no. 1, p. 172–177.

[6] SANCHEZ-RENEDO, M. High-selectivity tunable planar combline filter with source/load-multiresonator Coupling. *IEEE Microwave Wireless Components Letter*, 2007, vol. 17, no. 7.

[7] CHIOU, Y. C., REBEIZ, G. M. A quasi elliptic function 1.75-2.25 GHz 3-pole bandpass filter with bandwidth control. *IEEE Transactions on Microwave Theory and Techniques*, 2012, vol. 60, no. 2, p. 244–249.

[8] EL-TANANI, M. A., REBEIZ, G. M. A two-pole two-zero tunable filter with improved linearity. *IEEE Transactions on Microwave Theory and Techniques*, 2009, vol. 57, no. 4, p. 830–839.

[9] ZHANG, X. Y., XUE, Q., CHAN, C. H., ET AL. Low-loss frequency-agile bandpass filters with controllable bandwidth and suppressed second harmonic. *IEEE Transactions on Microwave Theory and Techniques*, 2010, vol. 58, no. 6, p. 1557–1564.

[10] PARK, S. J., REBEIZ, G. M. Low-loss two-pole tunable filters with three different predefined bandwidth characteristics. *IEEE Transactions on Microwave Theory and Techniques*, 2008, vol. 56, no. 5, p. 1137–1148.

[11] MATTHAEI, G. L., YOUNG, L., JONES, E. M. T. *Microwave Filters Impedance-Matching Networks, and Coupling Structures.* Norwood, MA: Artech House, 1980.

[12] GUYETTE, A. C. Alternative architectures for narrowband varactor-tuned bandpass filters. In *Proceedings of the 39th European Microwave Conference.* Rome (Italy), 2009, p. 1828–1831.

[13] HONG, J. S., LANCASTER, M. J. *Microstrip Filters for RF/Microwave Applications.* New York: Wiley, 2001.

[14] BROWN, A. R., REBEIZ, G. M. A varactor tuned RF filter. *IEEE Transactions on Microwave Theory and Techniques*, 2000, vol. 48, no. 7, p. 1157–1160.

[15] LI, Y. C., ZHANG, X. Y., XUE, Q. Bandpass filter using discriminating coupling for extended out-of-band suppression. *IEEE Microwave Wireless Components Letter*, 2010, vol. 20, no. 7, p. 369–371.

About Authors ...

Zhiyuan ZHAO was born in 1986. He received his bachelor's degree from Lanzhou University in 2008, and received his master's degree from PLA University of Science and Technology in 2011. He is now a Ph. D candidate in the Institute of Communications Engineering, PLA University of Science and Technology, Nanjing, China. His research interests include microwave and RF tunable filters.

Jiang CHEN was born in 1965 and he is now a professor in Nanjing Telecommunication Technology Institute, Nanjing, China. His research interests include RF power amplifiers and RF filters.

Lin YANG was born in 1974 and he is now an engineer in Nanjing Telecommunication Technology Institute, Nanjing, China. His research interests include RF power amplifiers.

Kunhe CHEN was born in 1975 and he is now an engineer in Nanjing Telecommunication Technology Institute, China. His research interests include RF tunable filters.

2

Near Sea-Surface Mobile Radiowave Propagation at 5 GHz: Measurements and Modeling

Yee Hui LEE[1], *Feng DONG*[1], *Yu Song MENG*[2]

[1]School of Electrical and Electronic Engineering, Nanyang Technological University, 50 Nanyang Avenue, Singapore 639798, Singapore
[2] National Metrology Centre, Agency for Science, Technology and Research (A*STAR) 1 Science Park Drive, Singapore 118221, Singapore

eyhlee@ntu.edu.sg, ysmeng@ieee.org, meng_yusong@nmc.a-star.edu.sg

Abstract. *Near sea-surface line-of-sight (LoS) radiowave propagation at 5 GHz was investigated through narrowband measurements in this paper. Results of the received signal strength with a transmission distance of up to 10 km were examined against free space loss model and 2-ray path loss model. The experimental results have good agreement with the predicted values using the 2-ray model. However, the prediction ability of 2-ray model becomes poor when the propagation distance increases. Our results and analysis show that an evaporation duct layer exists and therefore, a 3-ray path loss model, taking into consideration both the reflection from sea surface and the refraction caused by evaporation duct could predict well the trend of LoS signal strength variations at relatively large propagation distances in a tropical maritime environment.*

Keywords

Evaporation duct, maritime mobile, modeling, path loss prediction, sea reflection.

1. Introduction

Radiowave propagation in maritime environments has been of great interest to many researchers over the years [1–9]. The understanding of over-sea radiowave propagation is very important for system designers when planning to establish a reliable radio link between a sea vessel and an onshore base station (BS). Over the years, different application scenarios have been investigated; in [2–4], over-the-horizon fixed links were examined; in [6], [7], mobile channels were studied; and in [8], [9], slant-path propagation for aeronautical applications was analyzed.

Recently, there has been growing interest in implementations of wireless systems in 5 GHz band for maritime environments, mainly in the applications of WiMAX at seaports [10] and for military off-shore anti-terrorist surveillance [11]. These applications require the 5 GHz systems to be operated in a near sea-surface line-of-sight (LoS) mobile environment (e.g., between a BS and a moving vessel). The radio channel for this scenario differs from those reported in [1–4], [8], [9]. Moreover, our previous investigations [9], [12] indicate that ducting is significant for radiowave propagation over the tropical ocean of Singapore in 5 GHz band. Therefore, although similar scenarios [5–7], [13] were reported, their maritime conditions and frequencies are different and may not be applicable to the 5 GHz radio link over a tropical ocean where the occurrence of an evaporation duct is known to be of a higher probability and more predominant [5].

From the literature, it is well-recognized that radiowave propagation over a sea surface is affected by the ducts such as surface ducts, elevated ducts, and evaporation ducts [14], [15]. For near sea-surface LoS propagation, the effects of evaporation duct are obvious and dominant amongst all the ducts. This is because the evaporation duct exists over the ocean almost all of the time [14], since it is a result of the rapid decrease in vapor pressure from a saturation condition at the sea surface to an ambient vapor pressure at levels several tens of meters above the sea surface. This decrease in vapor pressure generally results in a decrease in the modified refractivity and thus, creates a ducting layer adjacent to the sea surface. Research work in [2] indicated that an evaporation duct above the sea surface can result in a substantial increase in the received signal strength at frequencies above 3 GHz. Measurement results in [3] also showed an enhancement in signal strength of more than 10 dB, observed 48% of the time along a 27.7 km over-sea path at 5.6 GHz.

Therefore, it is important to study the effect of evaporation duct in characterizing and modeling of near sea-surface radiowave propagation at the frequency of 5 GHz. As a continuation of our previous work [16] where channel characteristics of a non-line-of-sight (NLoS) sea-surface link at 5 GHz for Singapore maritime environment was reported, the main objective of this paper is to perform a detailed investigation on near sea-surface LoS propagation through path loss modeling with the considerations the evaporation duct and the sea-surface reflection.

In the following, measurement campaign is described in Section 2. In Section 3, free space loss model and 2-ray path loss model are examined against the experimental results first. Based on the observed discrepancy, a 3-ray path loss model is introduced to take into consideration the refracted wave by evaporation duct. Section 4 then gives a discussion on empirical estimations of the effective duct height. Finally, conclusions of this paper are given in Section 5.

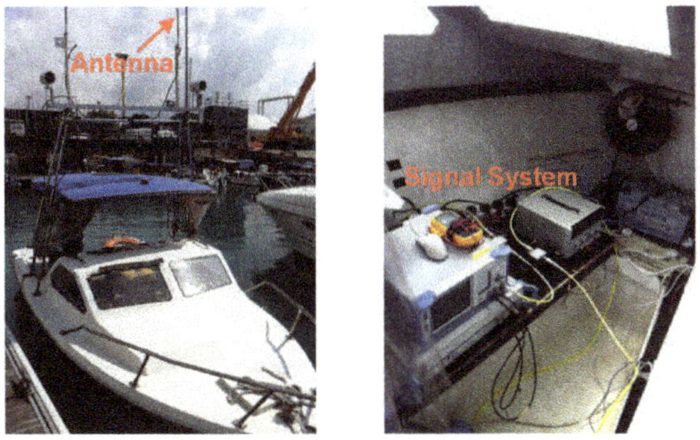

(a) Transmitter on a speed boat

(b) Receiver on shore

Fig. 1. Setup of transmission and data-logging systems for near sea-surface LoS measurements.

2. Measurement Campaign

2.1 Measurement System

Fig. 1 shows the main equipment and setup used for narrowband measurements. It is noted that 5.15 GHz was chosen and approved for continuous-wave transmission where no interfering signal was detected. As shown in Fig. 1a, an omni-directional antenna was mounted on top of a speed boat at a height of approximately 3.5 m above sea level. It was connected to a signal generator and a high-power amplifier housed within the boat cabin, forming a mobile transmitter with an output power of 30 dBm. During the measurements, GPS data was continuously logged so as to obtain the instantaneous time, altitude, longitude and latitude coordinates. The pitch and roll of the moving boat was logged using a fluxgate compass.

The BS was on shore, close to the sea. In order to ensure a LoS condition between the transmitter and the receiver, the receiver consisted of an identical omni-directional antenna installed on the roof of a building as shown in Fig. 1b with different heights above sea level (in order to study the effect of antenna height), and connected to a spectrum analyzer. The span of the spectrum analyzer was set to 20 kHz around its center frequency to reduce the noise bandwidth. Peak marker readings of the received signal were recorded at intervals of 1 second using a Labview program. All the data recorded was time-stamped for synchronization with the location, heading and orientation of the mobile transmitter.

The whole system was carefully calibrated on-site before the sea trials, and checked again after the measurements. The system effect was minimized through the removal of the antenna gains and the measurement of a back-to-back connection between the transmitter and the receiver. Weather conditions were also recorded using a weather meter at the transmitter side and a weather station at the receiver side respectively. However due to the lack of more suitable meteorological instruments such as a weather balloon, the weather information was restricted at the transmitter and receiver altitude levels.

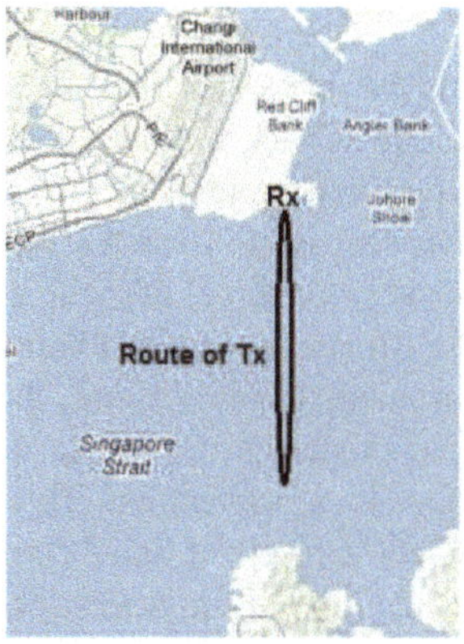

Fig. 2. Measurement route and receiver location, taken from Google Earth.

Parameter	Value
Carrier frequency	5.15 GHz
Transmitted power	30 dBm
Transmitter height	3 to 4 m
Receiver height	20 m, 10 m and 7.6 m
Maximum route distance	10 km
Antennas type	Omni-directional

Tab. 1. Summary of measurement parameters.

2.2 Measurement Routes

Measurements were carried out over an open sea area off the southeast coast of Singapore. BS was located at a yacht club (N 01°19′09″, E 104°1′22″). The mobile transmitter was on board a speed boat with a maximum speed of 30 knots. During the trials, the boat traveled along a 10 km route as shown in Fig. 2 with a speed of approximately 6 knots. LoS was maintained throughout the measurements except for some occasional passing-by ships for a short period of time. For each receiver height, multiple trials on different days have been carried out in order to compare and verify the results. It is noted that all the measurement campaign was conducted under similar sea status; calm or near calm. The main measurement parameters are summarized in Tab. 1.

3. Near Sea-Surface Path Loss Model

ITU-R P.1546 model [13] provides a set of curves for prediction of propagation loss over a sea path at a frequency from 30 MHz to 3 GHz, which limits its application in our study at 5.15 GHz. However, a general trend can be observed from those field strength versus distance curves in [13] that radiowave propagation over sea paths (path distance less than 10 km) approaches free space propagation at 50% of time when the frequency increases. Since the frequency of operation in this study is 5.15 GHz, free space loss (FSL) model could therefore be used. Moreover, works in [5] also reveals that propagation loss of a 5-GHz over sea-water radio link can be predicted using the FSL model when the transmission range is less than 10 nautical miles (18.52 km) but with some interference nulls.

These nulls could be due to the interference between the direct ray and the sea-surface reflected ray. Our previous study [9] also found that around 86% of all the trials for over-sea radiowave propagation can be represented by a 2-ray multipath model (i.e., a direct ray with a sea-surface reflected ray) over the tropical ocean. Although the information in [9] is for aeronautical applications, there is a very high probability for the existence of a sea-surface reflected wave for near sea-surface LoS propagation. Therefore in this section, we will examine the received signal strength against the predicted results using the FSL model and the 2-ray path loss model.

3.1 Propagation Loss Models

For the radiowave propagation in free space, the propagation loss can be predicted by the FSL model [17],

$$L_{FSL} = -27.56 + 20\log_{10}(f) + 20\log_{10}(d) \qquad (1)$$

where L_{FSL} is the free space loss in dB, f is the frequency in MHz, and d is the propagation distance in meters.

When a reflected ray from the sea surface exists besides the LoS path (direct ray), the propagation loss could be predicted by a 2-ray path loss model. Since there is a near-grazing incidence on the sea surface in our measurements, the reflection coefficient for a vertically polarized wave approaches -1. Therefore, the 2-ray path loss model can be simplified as [17]

$$L_{2-ray} = -10\log_{10}\left\{\left(\frac{\lambda}{4\pi d}\right)^2\left[2\sin\left(\frac{2\pi h_t h_r}{\lambda d}\right)\right]^2\right\} \qquad (2)$$

where L_{2-ray} is the 2-ray propagation loss in dB, λ is the wavelength in meters, and h_t, h_r are the heights of a transmitter and a receiver in meters. In the following, both the FSL model and the 2-ray model will be used to predict the received signal strength under ideal conditions for near sea-surface radiowave propagation at 5.15 GHz.

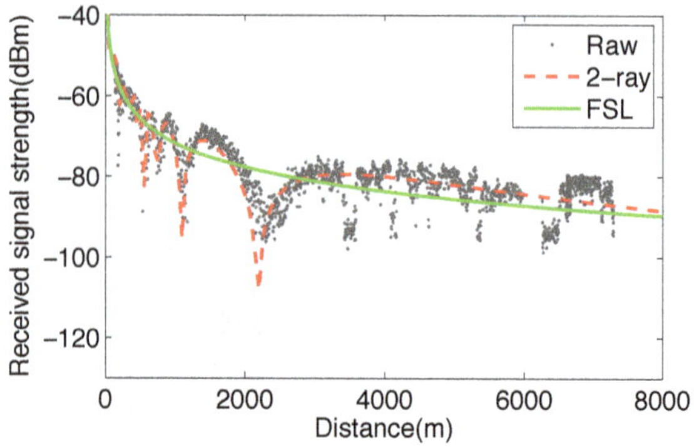

(a) Results at h_r = 20 m conducted on 23 November 2011

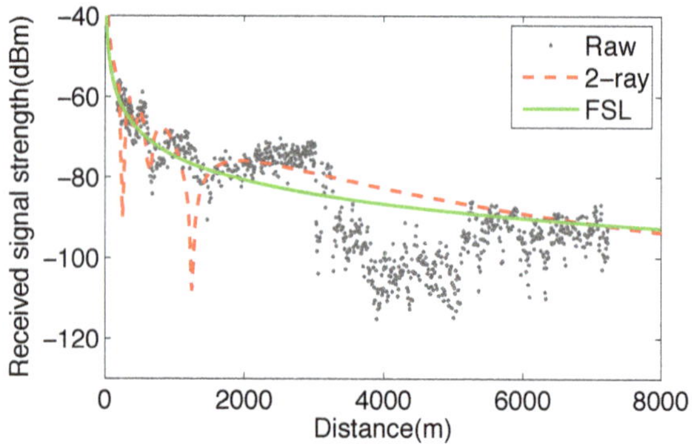

(b) Results at h_r = 10 m conducted on 28 September 2011

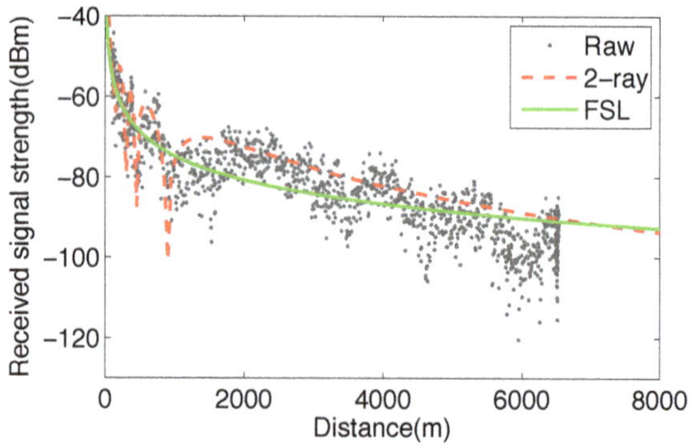

(c) Results at h_r = 7.6 m conducted on 3 April 2012

Fig. 3. Received signal strengths versus distance at different receiver heights; measured and predicted.

3.2 Comparison with the Predicted Results

Typical measurement results at the receiver height h_r of 7.6 m, 10 m, and 20 m are shown in Fig. 3, together with the predicted signal strengths using the FSL model and the 2-ray model (assuming a perfect sea-surface reflection). From Fig. 3, it can be observed that the FSL model is able to predict the exponential decreasing strength trend for all the 3 receiver heights. The FSL model is well-suited for prediction of the local mean (large scale) propagation loss. These observations are consistent with those reported in [5], [13]. More interestingly, when the propagation distance is less than about 2000 m to 3000 m (from Fig. 3), the measured results show a similar trend as the predicted results using the 2-ray model for all the measurement scenarios. That is, there are some interference nulls which are due to the destructive summation of the radiowaves (the direct wave and the sea-surface reflected wave from modeling process of the 2-ray model) arriving at the receiver. The slight misalignments of the predicted nulls with respect to the measured ones in Fig. 3 are due to the sea-surface roughness [18], [19] and the refraction of the propagating waves mainly. Both of them can weaken the applicability of 2-ray path loss model which is assumed for a perfect reflection and straight rays.

However, it is observed from Fig. 3 that as the propagation distance increases beyond about 2000 m to 3000 m (known hereby as the break point d_{break}), the prediction abilities of both the FSL model and the 2-ray model become poor for near sea-surface LoS environments. There are some interference nulls beyond d_{break} that cannot be predicted using both the models. As seen in Fig. 3, both the models tend to approach a stable signal level with an exponential decay.

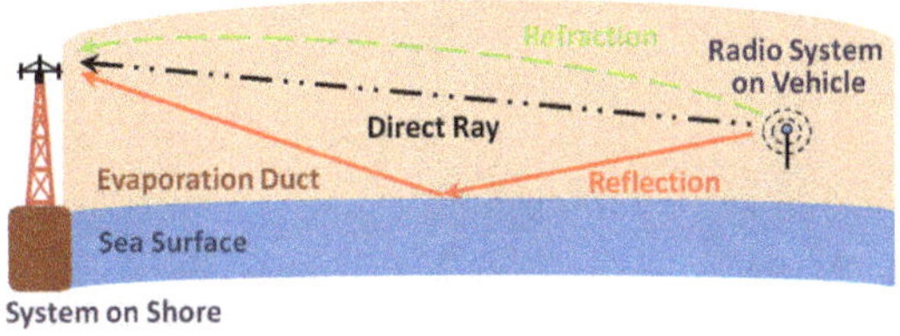

(a) Main propagation mechanisms in an evaporation duct

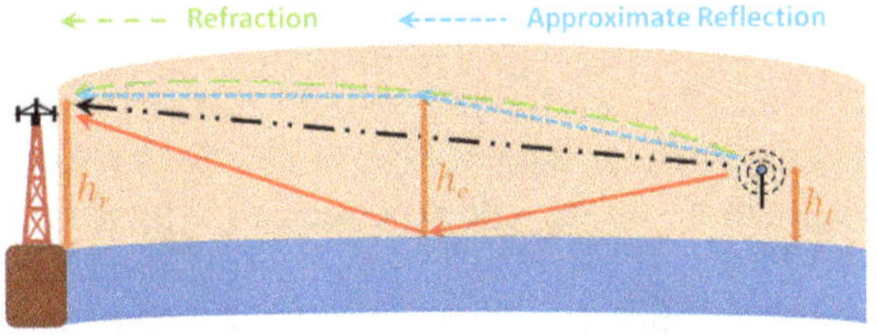

(b) Approximate representation of the refracted wave

Fig. 4. Near sea-surface radiowave propagation where a refracted wave is approximately represented as a reflected wave.

These additional interference nulls beyond d_{break} could be due to an additional ray which appears as the propagation distance increases. It can be a refracted ray (as shown in Fig. 4) caused by the evaporation duct which exists over the tropical sea surface almost all of the time [14]. From the reported works, the evaporation duct can trap the over-water propagating signals [20]. Therefore, a multi-ray path loss model which takes into consideration the signals trapped in an evaporation duct should be considered beyond d_{break}.

In order to estimate this break point d_{break}, we reviewed the trends of the FSL model and the 2-ray model. As the distance increases beyond the last null predicted by 2-ray model, both the models tend to level out slowly. Therefore, it could be concluded that the 2-ray path loss model will provide accurate predictions roughly up to a distance of its last predictable null (or more precisely the first maximum after the last predictable null). This break distance d_{break} can be roughly estimated as

$$d_{break} = \frac{4h_t h_r}{\lambda}. \quad (3)$$

In our measurement campaign, d_{break} is around 4000 m, 2000 m and 1500 m when h_r is 20 m, 10 m and 7.6 m respectively. For path loss modeling beyond d_{break}, a multi-ray path loss model is considered. In the following, a 3-ray path loss model, taking into consideration the refraction due to the evaporation duct as shown in Fig. 4 is proposed.

3.3 Modeling with the Ducting Effect

As discussed above, 2-ray path loss model will lose its prediction ability when the propagation distance exceeds d_{break} roughly. For example, interference nulls observed in Fig. 3a at around 900 m, 1100 m and 2100 m ($< d_{break}$, $d_{break} \approx 4000$ m) can be predicted using the 2-ray model. While the nulls at 4100 m, 5300 m and 6200 m ($> d_{break}$) cannot be predicted by the 2-ray model. The interference nulls could be caused by the trapped wave within the evaporation duct as illustrated in Fig. 4.

Although a distance of about 5000 m is reported as the one after which the ducting should be accounted for in [21], there is a higher probability for a shorter ducting distance (e.g., < 4500 m) in the tropical ocean where our measurements were carried out. This is because the evaporation duct heights in tropical waters are typically larger than those reported for temperate cooler waters [22]. It therefore could start to trap the radiowaves at a shorter distance. The trapped wave in the evaporation duct is more likely due to the radiowave refraction as shown in Fig. 4a. This is because the upper boundary of the evaporation duct layer is not homogenous [20], the radiowaves incident onto the upper boundary would be diffused immediately and make the reflection from the boundary insignificantly in the received field. Moreover for the refraction, the refracted ray (3rd ray) will not appear at short distances, but would appear as the distance increases. Especially when the distance increases much further, more additional rays could also appear although there is minor probability for them to happen in our measurements.

Therefore in this study, a 3-ray path loss model (including a direct LoS ray, a reflected ray from sea surface, and also a refracted ray by evaporation duct) is used for modeling and predicting near sea-surface LoS propagation preliminarily. Although there are other methods which may be better for modeling the radiowave propagation in a duct (e.g., parabolic equation approximation method [19]), ray-tracing is preferred in this study due to the simplicity of its final mathematical expression that describes the scenario and hence, its straightforward application in radio planning. As shown in Fig. 4b, the refracted ray is approximately represented by a near-grazing reflected wave to simplify the process of ray-tracing modeling since the antenna heights are much smaller than the propagation distance.

For 3-ray path loss modeling, the evaporation duct layer is assumed to be horizontally homogeneous. Similar to the 2-ray path loss model described earlier, a near-grazing incidence on the sea surface is assumed and finally, the reflection coefficient for a vertically polarized wave approaches to -1. With these assumptions, the 3-ray path loss model [23] can be simplified into (4),

$$L_{3-ray} = -10\log_{10}\left\{\left(\frac{\lambda}{4\pi d}\right)^2 [2(1+\Delta)]^2\right\}, \qquad (4)$$

with

$$\Delta = 2\sin\left(\frac{2\pi h_t h_r}{\lambda d}\right)\sin\left(\frac{2\pi(h_e-h_t)(h_e-h_r)}{\lambda d}\right) \qquad (5)$$

where h_t and h_r are the heights of the transmitter and the receiver in meters, and h_e is the effective duct height as shown in Fig. 4b. h_e is approximately equal to (or slightly less than) the height of evaporation duct layer which is used as a reference in 3-ray path loss modeling. A preliminary investigation of 3-ray path loss modeling is then performed through evaluating the measured signal strengths against the estimated ones using (4) with different h_e assumed. Fig. 5 shows an example of results with h_e = 25 m and 35 m respectively.

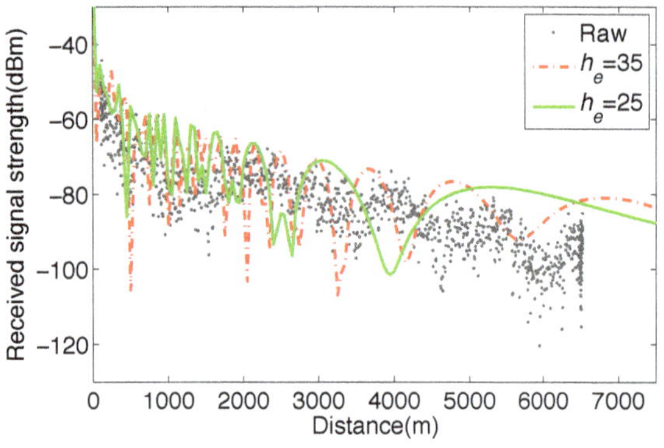

Fig. 5. Example of the received signal strength versus distance with the estimated values using 3-ray model with h_e = 25 m and 35 m.

From Fig. 5, it can be observed that the signal nulls beyond d_{break} of 1500 m in the measured results which cannot be predicted by the 2-ray model previously could be estimated using the 3-ray model. The misalignments of the measured nulls with the predicted ones are due to the improper values assigned to h_e. Therefore for modeling of radiowave propagation over a tropical sea surface, 3-ray path loss model that not only considers the direct ray and the sea-surface reflected ray, but also the refracted ray by an evaporation duct, should be considered.

4. Estimation of Effective Duct Height

In order to know the effective duct height h_e as shown in Fig. 4b, the height h of a typical evaporation duct which is approximately equal to (or slightly higher than) h_e is used as a reference. For deriving h, weather information such as humidity, temperature, wind speed and pressure is required at regular vertical intervals above the sea surface up to a height of 40 m or more. However, due to the limitation of meteorological instruments (e.g., lack of a weather balloon as we mentioned above), it is difficult to get the vertical weather information on site which restricts an accurate estimation of h. Therefore, the single point weather information collected at both the transmitter and receiver has to be used to estimate the duct height h for all the 21 measurement campaign, using a modified P-J formulation given in [24]. The P-J formulation was developed for an open ocean and therefore it may lose the prediction accuracy when applied to this coastal environment. It is also noted that the P-J formulation is very sensitive to the weather information.

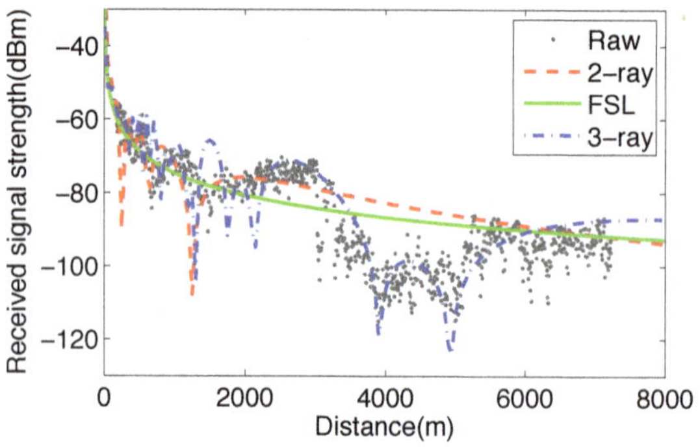

Fig. 6. The received signal strength versus distance with the predicted results using the FSL, 2-ray and 3-ray models: an example with h_r = 10 m.

An empirical method of determining the effective duct height h_e is therefore performed through the curve fitting/regression technique on the measured data in this study. The method is similar to the one reported in [25] where h_e was obtained based on a split-step Fourier solution of the parabolic equation approximation to the wave equation. Empirical values of h_e in (5) are then estimated by performing curve fitting onto our measurement data to align the predicted nulls with the measured ones at larger distances for each trial. An example of the curve fitting of 3-ray path loss model to the experimental results for determining h_e is shown in Fig. 6. For completeness, both the FSL model and the 2-ray model are also shown in Fig. 6.

From Fig. 6, it can be observed that the fitted 3-ray model shows a good prediction ability when the propagation distance is beyond d_{break}, especially for the sudden drop of the received signal level at distances between 4 km and 5 km. The sudden signal drop is due to the destructive summation of additional refracted wave by evaporation duct with the LoS ray and the sea-surface reflected wave. The effective evaporation duct height h_e obtained from the curve fittings of (4) to the measurement data is able to predict the signal variations very well. The observations hold for all the 21 measurement campaign performed over the tropical sea environment, and correspondingly h_e are estimated and summarized in Fig. 7 with the calculated h using the P-J formulation [24].

The observations from Fig. 6 also indicate that the 3-ray path loss model taking into consideration the contribution of the refracted wave by evaporation duct is more appropriate for near sea-surface propagation loss prediction as compared to the popular 2-ray path loss model particularly at a distance beyond d_{break}. This is because the ducting effect usually becomes significant for long-range over-sea radiowave propagation. When the propagation distance is below d_{break}, the ducting effect is almost negligible where the reflection from the sea surface and the direct LoS ray dominate. Thus, the 2-ray model should be only used for short-range ($< d_{break}$) near sea-surface propagation.

Moreover from the literature, it is found that the occurrence probability of an evaporation duct around nearby marine environments such as the South China Sea is around 80%, and the annual average height h of the evaporation duct is around 7 m to 15 m [26]. These values are similar to those calculated h using the P-J formulation as shown in Fig. 7a.

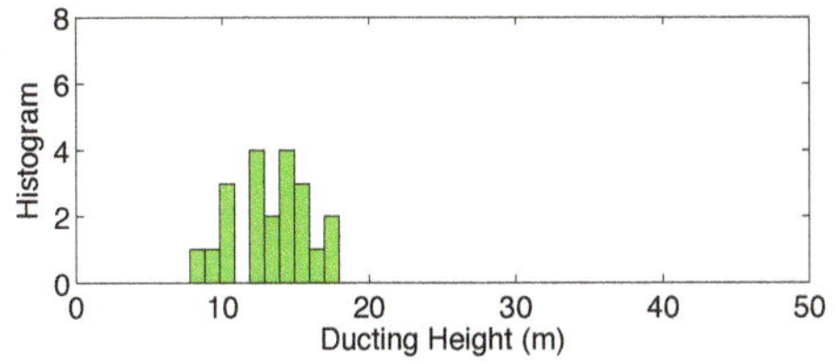

(a) Calculated ducting height h

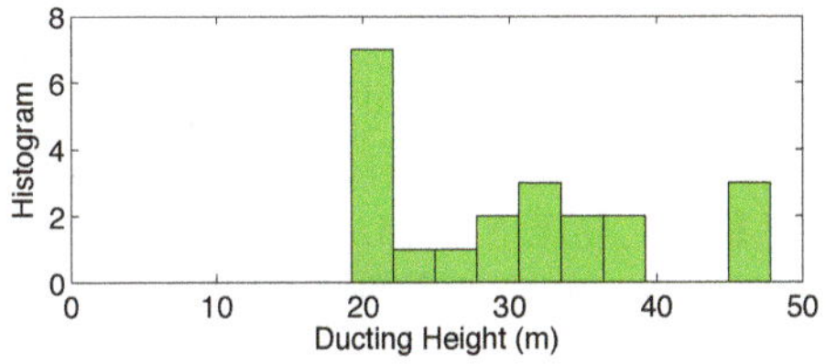

(b) Effective ducting height h_e

Fig. 7. Histograms of the calculated duct height h and the empirical estimated effective duct height h_e.

However, the results in Fig. 7b show that most (around 86%) of the estimated h_e (supposed to be approximately equal to/slightly less than h) in this study falls within the range of 20 m to 40 m, with a median value of 30.5 m. This statistic is found to be close to those reported in [26] where the averaged evaporation duct height h at nearby marine environment is between 25 m to 40 m during similar months of March, April, September and November. Another example for the empirical h at a similar environment is reported in [22], which was based on the data measured between the Palm Islands and the Australian mainland in Northern Queensland which is also with a tropical climate. The duct height h is found to increase as the wind speed increases, and can be up to 25 m at a wind speed of 10 knots roughly. Referring to the reported information in [22], the estimated effective duct height h_e in Fig. 7b which has a measured wind speed within the range of 10 knots to 15 knots looks reasonable.

The values of h_e in Fig. 7b are then more reliable than the calculated h using the P-J formulation which mainly falls into the range of 8 m to 18 m as shown in Fig. 7a. That is, the P-J formulation tends to underestimate the actual duct height in this coastal environment. A possible reason for this observed discrepancy may be because: the measurements were made in a coastal environment where stable atmospheric conditions could exist, while the P-J formulation is for an open ocean which may lead to underestimation of the evaporation duct height as discussed [4].

5. Conclusions

This paper reported an experimental investigation of near sea-surface LoS radiowave propagation at 5 GHz through narrowband measurements. Good agreement has been observed between the measured results and the predicted values using a 2-ray path loss model when the propagation distance is less than d_{break}. However when the propagation distance increases beyond d_{break}, its prediction ability becomes poor.

Our results and analysis indicated that a 3-ray path loss model taking into consideration the refracted wave by an evaporation duct and the reflection from the sea surface could well predict the trend of the signal strength variations in the tropical marine environments when the propagation distance increases beyond d_{break}. The results also show that most (around 86%) of the estimated effective duct height h_e falls into the range of 20 m to 40 m with a median value of 30.5 m, which is close to the reported average height h of evaporation duct at a nearby marine environment.

Therefore for radio planning, near sea-surface LoS radiowave propagation loss L in dB could be estimated generally with the following,

$$L = \begin{cases} -10\log_{10}\left\{\left(\frac{\lambda}{4\pi d}\right)^2\left[2\sin\left(\frac{2\pi h_t h_r}{\lambda d}\right)\right]^2\right\}, & d \le d_{break}, \\ -10\log_{10}\left\{\left(\frac{\lambda}{4\pi d}\right)^2\left[2(1+\Delta)\right]^2\right\}, & d > d_{break}. \end{cases}$$

Here, $\Delta = 2\sin\left(\frac{2\pi h_t h_r}{\lambda d}\right)\sin\left(\frac{2\pi(h_e-h_t)(h_e-h_r)}{\lambda d}\right)$ as introduced previously. Furthermore, although the break point $d_{break} = \frac{4h_t h_r}{\lambda}$ has been defined based on the antenna heights and the signal wavelength, the accuracy of the path loss model above could vary around this point depending on the sea status.

Acknowledgements

This work was supported in part by the Defence Science and Technology Agency, Singapore.

References

[1] INOUE, T., AKIYAMA, T. Propagation characteristics on line-of-sight over-sea paths in Japan. *IEEE Transactions on Antennas and Propagation*, 1974, vol. AP-22, no. 4, p. 557 - 565.

[2] HITNEY, H. V., HITNEY, L. R. Frequency diversity effects of evaporation duct propagation. *IEEE Transactions on Antennas and Propagation*, 1990, vol. 38, no. 10, p. 1694 - 1700.

[3] HEEMSKERK, H. J. M., BOEKEMA, R. B. The influence of evaporation duct on the propagation of electromagnetic waves low above the sea surface at 3-94 GHz. In *Proceedings of the Eighth International Conference on Antennas and Propagation.* Edinburgh (UK), 1993, p. 348 - 351.

[4] GUNASHEKAR, S. D., WARRINGTON, E. M., SIDDLE, D. R., VALTR, P. Signal strength variations at 2 GHz for three sea paths in the British Channel Islands: Detailed discussion and propagation modeling. *Radio Science*, 2007, vol. 42, p. 1 - 13.

[5] HITNEY, H. V., RICHTER, J. H., PAPPERT, R. A., ANDERSON, K. D., BAUMGARTNER, G. B. Tropospheric radio propagation assessment. *Proceedings of the IEEE*, 1985, vol. 73, no. 2, p. 265 - 283.

[6] MALIATSOS, K., CONSTANTINOU, P., DALLAS, P., IKONOMOU, M. Measuring and modeling the wideband mobile channel for above the sea propagation paths. In *Proceedings of the 2006 First European Conference on Antennas and Propagation.* Nice (France), 2006.

[7] YANG, K., ROSTE, T., BEKKADAL, F., HUSBY, K., TRANDEM, O. Long-distance propagation measurements of mobile radio channel over sea at 2 GHz. In *Proceedings of the 2011 IEEE Vehicular Technology Conference.* San Francisco (USA), 2011.

[8] LEI, Q., RICE, M. Multipath channel model for over-water aeronautical telemetry. *IEEE Transactions on Aerospace and Electronic Systems*, 2009, vol. 45, no. 2, p. 735 - 742.

[9] MENG, Y. S., LEE, Y. H. Measurements and characterizations of air-to-ground channel over sea surface at C-band with low airborne altitudes. *IEEE Transactions on Vehicular Technology*, 2011, vol. 60, no. 4, p. 1943 - 1948.

[10] JOE, J., HAZRA, S. K., TOH, S. H., TAN, W. M., SHANKAR, J., HOANG, V. D., FUJISE, M. Path loss measurements in sea port for WiMAX. In *Proceedings of the 2007 IEEE Wireless Communications and Networking Conference.* Kowloon, 2007, p. 1871 - 1876.

[11] DONG, F., CHAN, C. W., LEE, Y. H. Channel modeling in maritime environment for USV. In *Defence Technology Asia 2011.* Singapore, 2011.

[12] LEE, Y. H., MENG, Y. S. Empirical modeling of ducting effects on a mobile microwave link over a sea surface. *Radioengineering*, 2012, vol. 21, no. 4, p. 1054 - 1059.

[13] ITU-R P.1546-4. *Method for Point-To-Area Predictions for Terrestrial Services in the Frequency Range 30 MHz to 3000 MHz.* Geneva (Switzerland): International Telecommunication Union, 2009.

[14] TETI, Jr. J. G. Wide-band airborne radar operating considerations for low altitude surveillance in the presence of specular multipath. *IEEE Transactions on Antennas and Propagation*, 2000, vol. 48, no. 2, p. 176 - 191.

[15] ITU-R P.453-10. *The Radio Refractive Index: Its Formula and Refractivity Data.* Geneva (Switzerland): International Telecommunication Union, 2012.

[16] LEE, Y. H., DONG, F., MENG, Y. S. Stand-off distances for non-line-of-sight maritime mobile applications in 5 GHz band. *Progress in Electromagnetics Research B*, 2013, vol. 54, p. 321 - 336.

[17] PARSONS, J. D. *The Mobile Radio Propagation Channel.* 2nd ed. Chichester (UK): Wiley, 2000.

[18] TIMMINS, I. J., O'YOUNG, S. Marine communications channel modeling using the finite-difference time domain method. *IEEE Transactions on Vehicular Technology*, 2009, vol. 58, no. 6, p. 2626 - 2637.

[19] ZHAO, X., HUANG, S. Influence of sea surface roughness on the electromagnetic wave propagation in the duct environment. *Radioengineering*, 2010, vol. 19, no. 4, p. 601 - 605.

[20] GUNASHEKAR, S. D., SIDDLE, D. R., WARRINGTON, E. M. Transhorizon radiowave propagation due to evaporation ducting. *Resonance*, 2006, vol. 11, no. 1, p. 51 - 62.

[21] ITU-R P.452-14. *Prediction Procedure for the Evaluation of Interference between Stations on the Surface of the Earth at Frequencies above About 0.1 GHz.* Geneva (Switzerland): International Telecommunication Union, 2009.

[22] KERANS, A., KULESSA, A. S., LENSSON, E., FRENCH, G., WOODS, G. S. Implications of the evaporation duct for microwave radio path design over tropical oceans in Northern Australia. In *Proceedings of the 2002 Workshop on the Applications of Radio Science.* Leura (Australia), 2002.

[23] LEE, W. C. Y. *Mobile Communications Engineering: Theory and Applications.* 2nd ed. McGraw-Hill, 1997.

[24] PAULUS, R. A. Practical application of an evaporation duct model. *Radio Science*, 1985, vol. 20, p. 887 - 896.

[25] LEVADNYI, I., IVANOV, V., SHALYAPIN, V. Assessment of evaporation duct propagation simulation. In *Proceedings of the XXXth URSI General Assembly and Scientific Symposium.* Istanbul (Turkey), 2011.

[26] ZHAO, X. L., HUANG, J. Y., GONG, S. H. Statistical analysis of an over-the-sea experimental transhorizon communication at X-band in China. *Journal of Electromagnetic Waves and Applications*, 2008, vol. 22, no. 10, p. 1430 - 1439.

About Authors . . .

Yee Hui LEE received the B.Eng. (Hons.) and M.Eng. degrees in Electrical and Electronics Engineering from Nanyang Technological University, Singapore, in 1996 and 1998, respectively, and the Ph.D. degree from the University of York, York, U.K., in 2002. Since July 2002, she has been with the School of Electrical and Electronic Engineering, Nanyang Technological University where she is currently an Associate Professor. Concurrently, she is also appointed as Assistant Chair (Student) for the School of Electrical and Electronic Engineering. Her interest is in channel characterization, rain propagation, antenna design, electromagnetic bandgap structures, and evolutionary techniques.

Feng DONG received the B.Eng. (Hons.) degree in Electrical and Electronics Engineering from Nanyang Technological University, Singapore, in 2010, where he is currently working toward the M.Eng degree. His research interest is in wireless channel characterizations and modeling.

Yu Song MENG received the B.Eng. (Hons.) and Ph.D. degrees in Electrical and Electronic Engineering from Nanyang Technological University, Singapore, in June 2005 and February 2010, respectively. From May 2008 to June 2009, he was a Research Engineer with the School of Electrical and Electronic Engineering, Nanyang Technological University, Singapore. Since July 2009, he has been with the Agency for Science, Technology and Research (A*STAR), Singapore. He was firstly with A*STAR's Institute for Infocomm Research as a Research Fellow, and then a Scientist I. In September 2011, he was transferred to A*STAR's National Metrology Centre where he is currently a Scientist II. His research interests include electromagnetic metrology, electromagnetic measurements and standards, and radiowave propagation.

3

Space-time Characteristics and Experimental Analysis of Broadening First-order Sea Clutter in HF Hybrid Sky-surface Wave Radar

Yajun LI, Yinsheng WEI, Rongqing XU, Tianqi CHU, Zhuoqun WANG

School of Electronics and Information Engineering, Harbin Institute of Technology, Harbin, 150001, China

liyajun1985happy@163.com, weiys@hit.edu.cn, xurongqing@hit.edu.cn, ctqhit@126.com, wangzhuoqun_HIT@126.com

Abstract. *In high frequency (HF) hybrid sky-surface wave radar, the first-order sea clutter broadening is very complex and serious under the influence of ionosphere and bistatic angle, which affects the detection of ship target. This paper analyzes the space-time characteristics based on the HF sky-surface wave experimental system. We first introduce the basic structure, working principle and position principle based on our experimental system. Also analyzed is the influence of ionosphere and bistatic angle on the space-time coupling characteristics of broadening first-order sea clutter and the performance of space-time adaptive processing (STAP). Finally, the results of theoretic analysis are examined with the experimental data. Simulation results show that the results of experiment consist with that of theoretic analysis.*

Keywords

Space-time characteristics of first-order sea clutter, bistatic angle, ionosphere, STAP, HF hybrid sky-surface wave radar.

1. Introduction

HF sky-wave or surface-wave radar has some natural advantages, such as the capability of detecting low-flying target and stealth target, avoiding the attack of anti-radiation missile, as well as providing wide area surveillance with longer early warning time than some normal HF radar. Therefore, HF radar plays a more and more important role in modern air defense system and has potential significant contribution to military applications. Based on the propagation mode associated with HF sky-surface wave, HF hybrid sky-surface wave system is one new detection technique, which attains the information of the target and ocean surface. This technique can improve target Over-The-Horizon (OTH) detection and ocean surface dynamical environment surveillance capability. Because of the special working system and the influence of bistatic system layout, its sea echo spectrum property is more complicated than that of merely sky-wave reflection or ground-wave diffraction mode. Due to the combined influences of ionosphere and bistatic angle, the broadening of first-order sea clutter spectrum is very serious, which seriously affects the effective detection of slowly moving targets like ships.

The sea clutter spectrum is mainly composed of first-order scatter and the second-order scatter which is a continuum spectrum with amplitude less than the first-order returns about 20-45 dB. In general, the first-order returns can be denoted by two distinct Bragg lines. The first-order sea clutter characteristics of HF hybrid sky-surface wave radar are not only related to oceanic dynamics, beam width and operating frequency, but also influenced by bistatic angle and ionosphere. The time-varying characteristic and hierarchical structure of the ionosphere will cause the sea clutter Doppler frequency shift and broadening. Thus the sea clutter characteristics in the radar largely depend on the ionospheric state. In addition, the HF hybrid sky-surface wave radar is actually a bistatic radar system. Harbin Institute of Technology carried out an integrated HF sky-surface wave experimental system, but the scale of receiving arrays is so limited that the width of the received beam is quite broad, and in this condition, the bistatic angle might cause different broadening characteristics in the different resolution cells. What's worse, ionospheric contamination further contributes to the broadening characteristic of sea clutter.

P. A. Melyanovski et al. mentioned this new concept radar in 1997 [1]. G. J. Frazer described an experimental multi-static HF radar in 2007 [2]. R. J. Riddolls respectively discussed theories of the experimental configuration and the influence of ionosphere in this radar system in 2007 and 2008 [3], [4]. Jiao et al. investigated the possible application of the new mode in 2007 [5]. Jiang et al. analyzed the phenomenon in this radar system that the first-order sea clutter varies with azimuth in 2011 [6]. Lou et al. studied the frequency selection problem in 2011 [7]. Zhao et al. analyzed the detection performance of hybrid sky-surface wave propagation mode based on DRM digital AM broadcasting in 2013 [8]. Harbin Institute of Technology analyzed the detection principle and the characteristics of sea clutter by experimental system in recent years [6], [9]. However, experimental results show that broadening of some sea clutter is serious, and difficult to effectively detect. Therefore, it requires deep study on the space-time characteristic of sea clutter in order to effectively suppress the broadening sea clutter and detect the ship target.

The paper conducts study on the space-time characteristics of first-order sea clutter in HF hybrid sky-surface wave radar. Firstly, the basic structure, working principle, radar equation and positioning principle of HF sky-surface wave experimental system were introduced respectively in Section 2. Secondly, this paper conducts detailed study on space-time characteristics of sea clutter in HF hybrid sky-surface wave radar in Section 3, and the space-time distribution and spreading model of first-order sea clutter under the action of bistatic angle and ionosphere were given. Then, space-time characteristics of sea clutter were analyzed through the simulation in Section 4, the experimental data for the space-time spreading spectrum of sea clutter in real HF sky-surface wave system were given, which verified the results of theoretical analysis. The theoretical analysis of space-time distribution for spreading sea clutter was also in agreement with experimental results. Finally, the influence of ionospheric disturbance on the performance of STAP algorithm was analyzed based on experimental data. It has been proved that serious ionospheric disturbance will influence the space-time coupling characteristics of first-order sea clutter, and lead to performance reduction of STAP. The results are useful for practical application of STAP in HF hybrid sky-surface wave radar.

2. Summarization of HF Hybrid Sky-surface Wave Experimental System

2.1 Basic Structure and Working Principle

The system layout for the HF hybrid sky-surface wave radar is shown in Fig. 1. In the HF sky-surface wave radar, skywave radar station behind the coast is used as

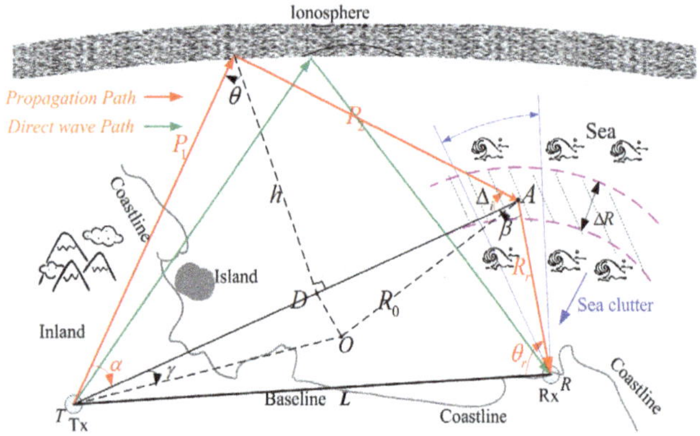

Fig. 1. The layout of HF sky-surface wave experimental system.

transmitting station, and the energy radiated by the transmitting station is reflected by ionosphere to monitor the sea area. Then the scattering energy of sea wave and target in the monitoring sea area arrives at the ground-wave radar receiving station by ground-wave diffraction on the sea surface. Thus it realizes the acquisition of sea information and the detection of targets on the sea. In Fig. 1, L is the baseline distance between transmitting and receiving stations, R_r is the distance between target and receiving station, θ_r is the angle between R_r and baseline, h is the ionospheric reflection height, D is the distance between transmitting station and target, R_0 is the earth radius, α is the elevation angle of the emitted electromagnetic wave, θ is the reflected angle of ionosphere.

2.2 Positioning Principle

Combined with the geometric relationships in Fig. 1, the following two relationships can be obtained by the law of cosines:

$$P_1^2 = h^2 + (D/2)^2 \ , \tag{1}$$

$$D^2 = L^2 + R_r^2 - 2LR_r \cos\theta_r \ . \tag{2}$$

Distance R is obtained by the conventional time delay estimation. Fig. 1 shows that $R = P_1 + P_2 + R_r$. Assuming the ionosphere is not inclined, i.e., $P_1 = P_2$. Then, we can obtain the following formula:

$$P_1 = (R - R_r)/2 \ . \tag{3}$$

Considering equations (1) to (3), the positioning relationship of the radar system can be obtained as follows:

$$R_r = \frac{R^2 - 4h^2 - L^2}{2(R - L\cos\theta_r)} \ . \tag{4}$$

According to (4), in order to realize the target location, namely measure the distance from target to receiving station, details need to be known including the distance obtained from time-delay ranging, the azimuth of target, the ionospheric height and baseline length.

3. Analysis of Space-time Characteristics of the First-order Sea Clutter in HF Hybrid Sky-surface Wave Radar

3.1 The Space-time Characteristics of the First-order Sea Clutter with Bistatic Angle

HF radar sea clutter consists of the first, second and higher order components, of which the first-order sea clutter consists of two symmetrical peaks, namely the negative and positive first-order Bragg peaks. Its mechanism can be explained by the Bragg resonant scattering process [10]. The interaction of incident radio waves and sea waves can cause the radio wave scattering. Because the scattering echo is caused by the first effect of HF radio waves and sea waves, this process is called the first-order effect of HF radio waves and sea waves. When the wavelength of radio and sea wave satisfies the Bragg resonance condition, Bragg resonant scattering occurs.

The Bragg frequency of first-order sea clutter in HF bistatic radar can be calculated by the following formula [10-12]:

$$f_B = \pm\sqrt{\frac{gf_0}{\pi c}}\sqrt{\cos(\beta/2)} \quad (5)$$

where f_0 is radar operating frequency, β is the bistatic angle.

In (5), the positive and negative sign corresponds to toward and away from the radar. In fact, the Bragg resonance condition of bistatic radar is shown in Fig. 2.

$$L_w(\cos\Delta_i + \cos\Delta_s)\cos(\beta/2) = m\lambda \qquad m = 1,2,3,\ldots \quad (6)$$

where λ is radio wavelength, Δ_i is incident grazing angle, Δ_s is the angle of reflection of radio waves, L_w is sea wave length.

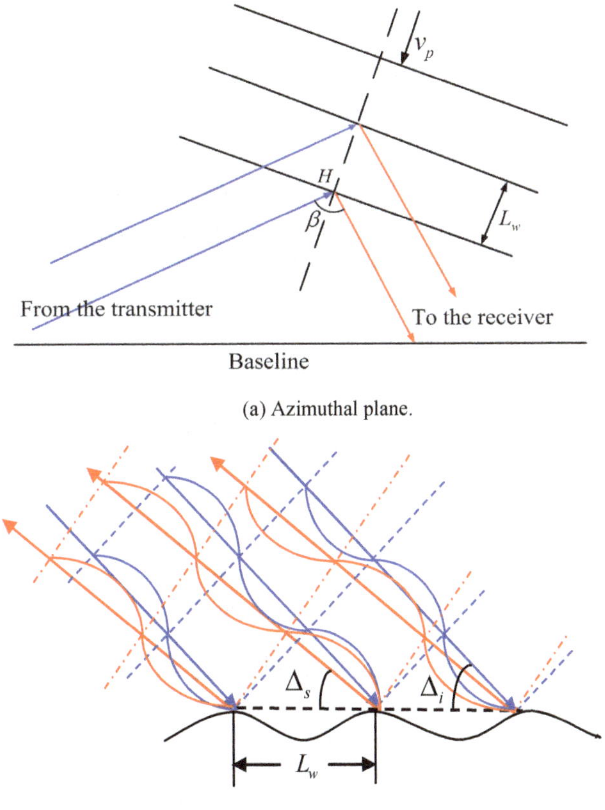

(a) Azimuthal plane.

(b) Elevation-angle plane.

Fig. 2. The Bragg scattering geometry of bistatic radar: (a) azimuthal plane and (b) elevation-angle plane.

When $m = 1$, corresponding to the first-order sea clutter. Considering the hybrid system radar's geometrical relation and configuration, we have the incident angle $\Delta_i = \alpha$ and the reflection angle $\Delta_s = 0$. Thus we can get the resonance condition of the first-order sea clutter under the configuration of hybrid propagation mode:

$$L = \frac{\lambda}{\cos(\beta/2)(\cos\alpha + 1)} \,. \quad (7)$$

Waves with other wavelength will not produce coherent scattering, which echoes can be ignored. In the spectrum, the Doppler frequency shift caused by the moving velocity is:

$$f_B = \pm v_p \cos(\beta/2)(\cos\alpha + 1)/\lambda \quad (8)$$

where v_p is the moving velocity of the sea wave's phase.

Known from the theory of hydraulics, for gravity wave, its v_p and its wavelength has the following relation [12]:

$$v_p = \sqrt{gL/2\pi} \,. \quad (9)$$

Thus we can derive the formula of first-order sea clutter Bragg frequency of hybrid system HF OTH radar:

$$f_B = \pm\sqrt{\frac{gf_0}{\pi c}}\sqrt{\cos(\beta/2)}\sqrt{\frac{\cos\alpha + 1}{2}} \,. \quad (10)$$

The bistatic angle β can be obtained by geometric relation in Fig. 1:

$$R_t^2 = R_r^2 + L^2 - 2R_r L\cos\theta_r \,, \quad (11)$$

$$L^2 = R_t^2 + R_r^2 - 2R_r R_t \cos\beta \,. \quad (12)$$

Combining with (10)-(12), we can derive the space-time distribution of first-order sea clutter with angle-Doppler in HF hybrid sky-surface wave radar:

$$f_B = \pm\sqrt{\frac{gf}{\pi c}}\sqrt{\frac{\cos\alpha + 1}{2}} \cdot \sqrt{\sqrt{\frac{1}{2} + \frac{R_r - L\cos\theta_r}{2\sqrt{R_r^2 + L^2 - 2R_r L\cos\theta_r}}}} \quad (13)$$

From the above formula (13), we can see the Bragg frequency of first-order sea clutter in HF hybrid sky-surface wave is a function of the bistatic angle, range cell, and also the radar looking direction. In distance R_r, the Bragg frequency f_B of first-order sea clutter varies with azimuth angle θ_r (the angle between beam pointing and baseline) changes, and shows the obvious angle-Doppler coupling characteristics.

Next, we need to analyze the spreading range of the first-order Bragg frequency for the HF hybrid propagation mode. Harbin Institute of Technology developed an integrated HF sky-surface wave radar experimental system in recent years. Given the size and cost of this radar system, the array aperture of the receiving array for our experimental system is small. Thus the width of the received beam is quite broad, the scattering cell E corresponding to a certain range gate cannot be equivalent to a point but a small area, leading to the corresponding bistatic angle is not a single value β, but a range of $\beta \in [\beta_{\min}, \beta_{\max}]$. Therefore, the

Bragg frequency of first-order sea clutter is also a range of $f_B \in \left[f_{B_{\min}}, f_{B_{\max}} \right]$. Thus, the theoretical broadening value Δf_B caused by bistatic angle is calculated by (14):

$$\begin{aligned}\Delta f_B &= f_{B_{\max}} - f_{B_{\min}} \\ &= \sqrt{\frac{g}{\pi\lambda}}\left(\sqrt{\cos\left(\frac{\beta_{\min}}{2}\right)} - \sqrt{\cos\left(\frac{\beta_{\max}}{2}\right)}\right) .\end{aligned} \tag{14}$$

3.2 The Space-time Characteristics of First-order Sea Clutter with Ionosphere Disturbance

HF sky-surface wave radar echo is also influenced by the ionospheric disturbance. Ionospheric phase path approximately moves with the variable motion for a long CIT, and the phase path variation of radio wave is also nonlinear variation, resulting in sea clutter spectral broadening [14]-[17]. In this paper, two typical ionosphere contamination functions (corresponding to two different ionospheric motion state) are used to simulate the space-time spectrum of contaminated sea clutter with ionospheric disturbance: (1) sine phase contamination mode: $\phi_1(t) = A_m \sin(\omega_m t)$, and (2) polynomial phase contamination mode: $\phi_2(t) = 2\pi(a_0 + a_1 t + a_2 t^2)$.

The space-time distribution of first-order sea clutter under the action of bistatic angle and ionospheric disturbance for HF hybrid sky-surface wave radar can be expressed as:

$$\begin{aligned} f_B &= \sqrt{\frac{g}{\pi\lambda}\cdot\frac{\cos\alpha+1}{2}}\cdot\sqrt{\sqrt{\frac{1}{2}+\frac{R_r - L\cos(\theta_i)}{2\sqrt{R_r^2+L^2-2R_r L\cos\theta_i}}}} + \frac{1}{2\pi}\frac{d(\phi_i(t))}{dt} \\ &= \pm 0.102\times10^{-3}\sqrt{f_0(MHz)\cdot\frac{\cos\alpha+1}{2}} \cdot\sqrt{\sqrt{\frac{1}{2}+\frac{R_r - L\cos(\theta_i)}{2\sqrt{R_r^2+L^2-2R_r L\cos\theta_i}}}} + \frac{1}{2\pi}\frac{d(\phi_i(t))}{dt}\end{aligned} \tag{15}$$

Ionospheric disturbance also cause first-order sea clutter spectrum broadening. Here

$$\Delta f_B = \left[\max\left(d(\phi_i(t))/dt\right) - \min\left(d(\phi_i(t))/dt\right)\right]/2\pi$$

is used to represent the sea clutter broadening caused by ionosphere. Combining with (14), we can derive and obtain the biggest possible broadening value Δf_B of first-order sea clutter Bragg frequency under the action of bistatic angle and ionospheric disturbance:

$$\left[\begin{aligned}&-0.102\times10^{-3}\sqrt{f_0(MHz)\cdot\cos\left(\frac{\beta_{\min}}{2}\right)} + f_{ion\_\min}, \\ &-0.102\times10^{-3}\sqrt{f_0(MHz)\cdot\cos\left(\frac{\beta_{\max}}{2}\right)} + f_{ion\_\max}\end{aligned}\right] \tag{16}$$

$$\left[\begin{aligned}&+0.102\times10^{-3}\sqrt{f_0(MHz)\cdot\cos\left(\frac{\beta_{\max}}{2}\right)} + f_{ion\_\min}, \\ &+0.102\times10^{-3}\sqrt{f_0(MHz)\cdot\cos\left(\frac{\beta_{\min}}{2}\right)} + f_{ion\_\max}\end{aligned}\right] \tag{17}$$

where $f_{ion\_\min}$ and $f_{ion\_\max}$ represent the maximum and minimum instantaneous frequency caused by ionosphere, respectively.

The following uses the formula (16) and (17) to quantitatively represent the degree of sea clutter broadening. Here f_{dbin} represents the Doppler resolution cell.

$$\varepsilon = \frac{\Delta f_B}{f_{dbin}} . \tag{18}$$

The sea clutter signal within range cell l can be expressed by the superposition of K scatter point return signals with each angle-range resolution cell. Therefore, the broadening sea clutter signal model with combined actions of bistatic angle and ionosphere can be written as the following form:

$$\begin{aligned} s(t) = &\sum_{i=1}^{K} \mathrm{a}_i(t)\cdot\left[\exp(j2\pi f_B t) + \exp(-j2\pi f_B t)\right] \\ &\cdot\exp(j\phi_i(t))\end{aligned} \tag{19}$$

where $\mathrm{a}_i(t)$ is the amplitude of signal component, meets Gaussian distribution; $f_B (i = 1...K)$ is K complex frequency component within Bragg frequency band, and $f_{B\_\min} \le f_B \le f_{B\_\max}$; $\phi_i(t)$ is the ionospheric phase contamination function.

3.3 Analysis of Space-time Coupling Characteristics of the First-order Sea Clutter

The bistatic angle characteristics of HF hybrid sky-surface wave radar make the first-order sea clutter has obvious space-time coupling characteristics. In order to quantitatively analyze the influence of ionosphere contamination on space-time coupling characteristics of sea clutter, this paper defines the average derivative ξ. Assume that positive and negative Bragg peak is affected by the ionospheric disturbance, ξ is defined as the mean of Doppler frequency derivative on the sine of angle. Here, the above Doppler frequency refers to the Doppler frequency at the position of the power of positive Bragg peak at each beam pointing within range cell l (Nth beam pointing).

$$\xi_l = \frac{1}{N}\sum_{i=1}^{N}\frac{d\left(f_{B_{\theta_i},\max}\right)}{d(\sin\theta_i)} . \tag{20}$$

Doppler frequency of ground clutter has a linear relation with the sine of angle for airborne side-looking radar, therefore $\xi = 1$, which has strong space-time coupling. The Doppler frequency of first-order sea clutter is constant value and independent of angle in HF surface wave radar (HFSWR), therefore $\xi \to 0$, which almost do not have the space-time coupling (the coupling relationship between angle and Doppler). From (13), we know that the Doppler frequency of first-order sea clutter changes with angles for HF hybrid sky-surface wave radar, and its space-time coupling characteristics are between airborne radar and HFSWR, and the space-time distribution is relatively smooth, so using the defined average derivative ξ in this paper can better analyze the space-time coupling characteristics.

4. Simulation and Experimental Analysis of Space-time Distribution of First-order Sea Clutter

4.1 The Space-time Distribution of First-order Sea Clutter with Bistatic Angle

Fig. 3 shows the array layout diagram of HF sky-surface wave experimental system. System parameters: baseline length $TR = L = 800$ km; normal direction is 50° counterclockwise from north; operating frequency $f_0 = 10$ MHz; uniform linear receiving array element $N = 8$, array element spacing $d = 20$ m; digital beam forming (DBF) DBF =-30°-30°, CIT = 40 s.

(a) Actual receiving antenna array of HF sky-surface wave system.

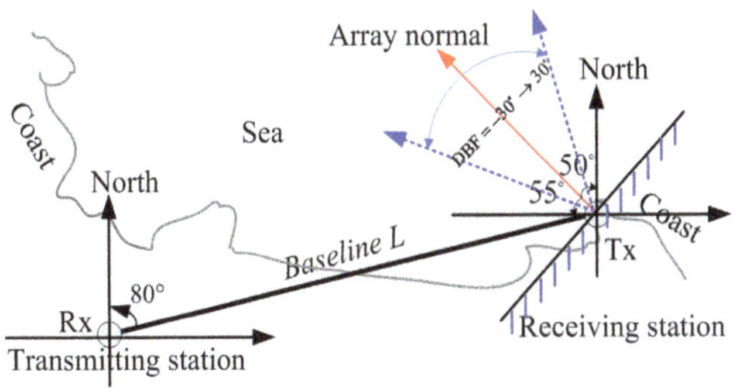

(b) Array geometric diagram of the transmitting station and the receiving station.

Fig. 3. Array layout diagram of HF sky-surface wave radar.

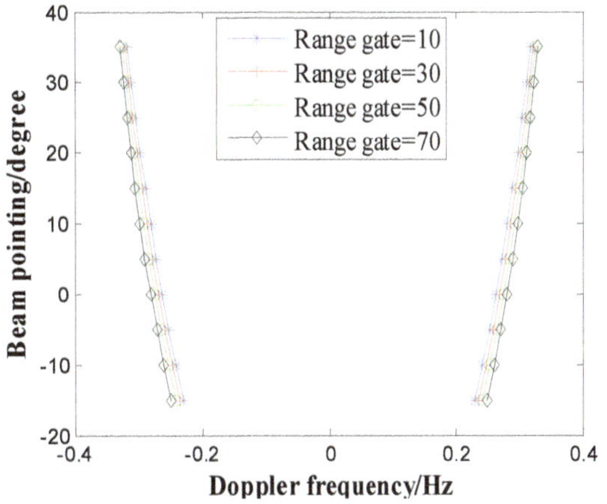

(a) Theoretical space-time distribution of sea clutter.

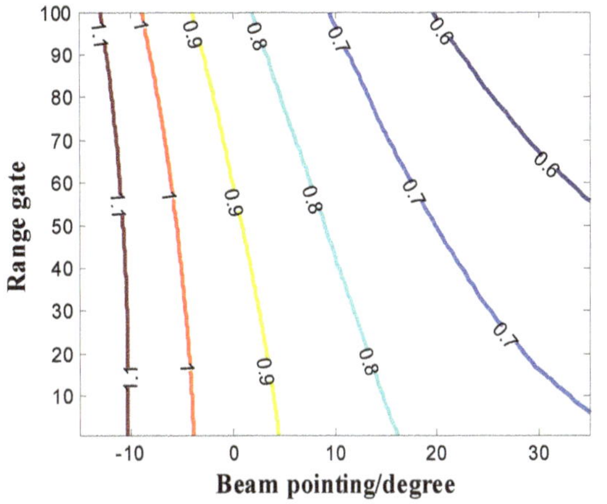

(b) Theoretical broadening characteristics of sea clutter.

Fig. 4. The theoretical distribution characteristics of first-order sea clutter spectrum without the influence of ionosphere.

The space-time distribution characteristics of first-order sea clutter with different ranges are simulated and analyzed. It is noted that there is no consider the effect of the ionosphere. From Fig. 4(a), we can see that it is consistent with the space-time distribution characteristics of first-order sea clutter at different range cells. For the same azimuth angle, the corresponding bistatic angle decreases and Doppler frequency increases with range increasing, so there are Doppler frequency shifts between range cells. From (15), we can see that the Doppler frequency of first-order sea clutter increases with the increase of the azimuth angle at the same range cell, showing the obvious space-time coupling characteristics.

According to (16) and (17), calculate and simulate the broadening characteristics of first-order sea clutter. Here, formula (18) is used as the sea clutter broadening measurement. Fig. 4(b) shows the theoretical broadening characteristics (unit: Hz) of the first-order sea clutter Bragg frequency with different ranges and azimuths. From Fig. 4(b), we can see that: (1) when the azimuth angle θ_r is little, corresponding to large bistatic angle, sea clutter broadening is serious within some receiving beam width; (2) when the azimuth angle θ_r increases, the broadening extent is small which is similar to the monostatic HFSWR.

4.2 The Space-time Distribution of First-order Sea Clutter with Ionospheric Disturbance

Combining with (19) and different ionospheric phase contamination functions, the space-time distribution of first-order sea clutter with angle-Doppler at range gate = 30 was simulated and analyzed, as shown in Fig. 5. The black line represents the theoretical space-time distribution of first-order sea clutter without the influence of ionosphere. Compared with the characteristics of HF bistatic radar, sine contamination function of ionospheric phase path makes the space-time spectrum broadening obviously, polynomial contamination function of ionospheric phase path makes the space-time spectrum produced obvious Doppler frequency shift and broadening.

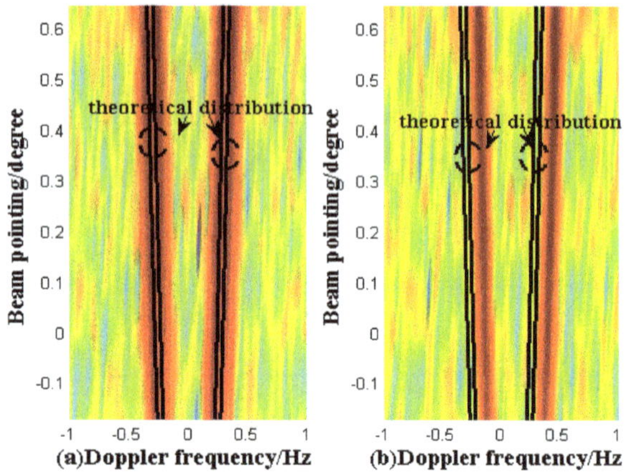

Fig. 5. The space-time distribution characteristics of first-order sea clutter with the influence of ionosphere: (a) space-time distribution of sea clutter with sine phase contamination mode; (b) space-time distribution of sea clutter with polynomial phase contamination mode.

4.3 Analysis of Space-time Coupling Characteristics of First-order Sea Clutter with Ionospheric Disturbance

Combining with the previous analysis, the influence of the Doppler frequency shift and broadening caused by ionospheric disturbance on space-time coupling of first-order sea clutter was analyzed. The ionospheric phase contamination function is set to [17]:

$$\phi(t) = \omega_{m0}t + A_m \sin(\omega_m t) \tag{21}$$

where, ω_{m0} is the translation frequency, A_m and ω_m is respectively the amplitude and frequency of periodic item. $A_m \sin(\omega_m t)$ was used to control the degree of sea clutter broadening caused by ionosphere; ω_{m0} was used to control the degree of the Doppler frequency shift of sea clutter. From Fig. 6(a), we can see that sea clutter has a certain space-time coupling characteristics under the array configuration, $\xi \approx 0.1$. When there is not Doppler frequency shift disturbance ω_{m0} between beam forming, single Doppler spectrum broadening $A_m \sin(\omega_m t)$ will not cause serious influence on space-time coupling of first-order sea clutter, approaching the ideal first-order sea clutter without the influence of ionosphere. When there is Doppler frequency shift disturbance, space-time coupling characteristics of sea clutter significantly deviate from the ideal first-order sea clutter without the influence of ionosphere, and the more serious the frequency shift disturbance, the more obvious the space-time coupling characteristics of sea clutter change.

Through the simulation and analysis, we can get the same conclusion by polynomial phase contamination function. Fig. 6(b) shows the comparison of space-time coupling characteristics of ideal and measured sea clutter under different array layouts. Here, array layout 1 refers to the angle between normal direction of receiving array and baseline is 150°; array layout 2 refers to the angle is 50°. From Fig. 6(b), we can see that actual space-time coupling of sea clutter was affected by the ionosphere. The space-time coupling from ideal sea clutter without the influence of ionosphere is identical to the actuality of experimental system in array layout 1.

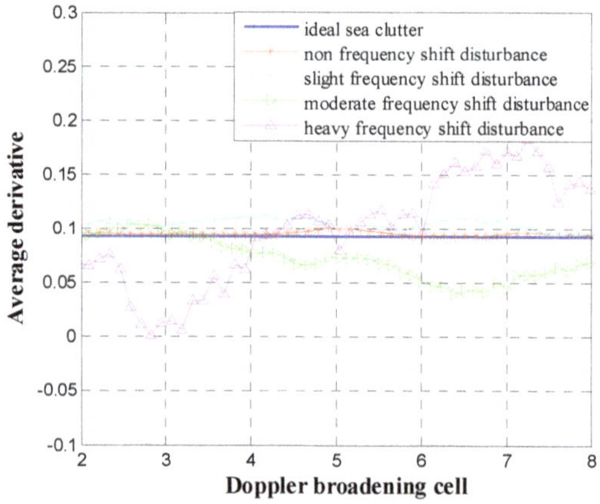

(a) Simulation of influence of ionospheric disturbance on space-time coupling characteristics of sea clutter.

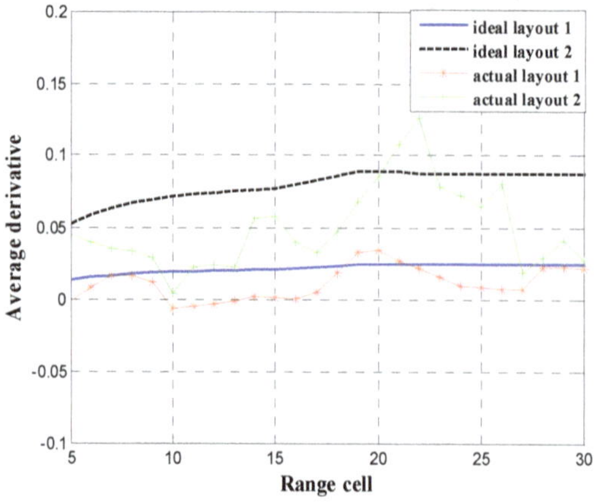

(b) Comparison of space-time coupling characteristics of ideal and measured sea clutter.

Fig. 6. Analysis of space-time coupling characteristics of first-order sea clutter.

4.4 Comparative Analysis of Theoretical Simulation and Experimental Results

With the support of the developed integrated HF sky-surface wave radar experimental system, experimental data were analyzed with the theoretical simulation. System parameters: Baseline length $TR = 800$ km, operating frequency $f_0 = 13$ MHz, CIT = 43 s, receiving array element $N = 8$, array element interval d=20m, repetition interval is 20 ms. The angle between normal direction of receiving array and baseline respectively is 150° (array layout 1) and 50° (array layout 2). Processing of experimental data includes pulse compression, beam forming, coherent accumulation, etc.

In Fig. 7, the black line represents the space-time distribution of first-order sea clutter under the action of bistatic angle and ionospheric phase contamination (obtained by direct wave) by (16) and (17). It is noted that the direct wave path in Fig. 1 represents that from the sky-wave station to the ionosphere and then directly reflects back to ground-wave stations, therefore, the direct echoes are only under the influence of the ionosphere, while not affected by the bistatic layout. For this reason, we can take advantage of the direct wave data to analyze the influence of ionosphere on sea clutter. As can be seen from Fig.7, the theoretical and measured results are basically met. The angle between normal direction of receiving array and baseline of array layout 2 is much smaller than array layout 1. Therefore, the change of bistatic angle is very obvious and Bragg frequency of first-order sea clutter has obvious differences at different azimuths in array layout 2, and λ is large, space-time coupling is strong.

From Fig. 7, we can see that there is frequency shift disturbance caused by ionosphere at different ranges and azimuths in real-world environments, causing the obvious changes of space-time coupling of sea clutter between theoretical and practical, and the difference between different ranges is larger, space-time coupling characteristics of sea clutter at most of range cells is less than ideal situation.

In the following, the influence of ionospheric disturbance on sea clutter spectrum was analyzed by comparing the simulated and measured Doppler spectrum of sea clutter, as shown in Fig. 8. Through the oblique ionogram corresponding the batch of measured data at that moment [18], we know that the transmitting electromagnetic wave reflects at layer Es when $f_0 = 13$ MHz, which means no multimode effects. The mean of ideal sea clutter broadening without the influence of ionosphere at 3 dB below the two

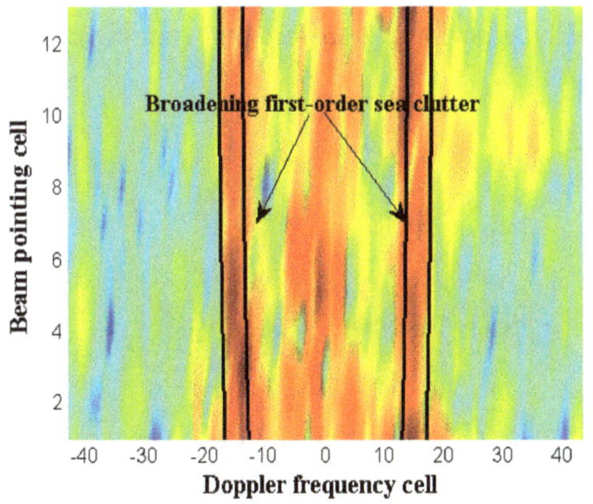

(a) Array layout 1, range cell = 37.

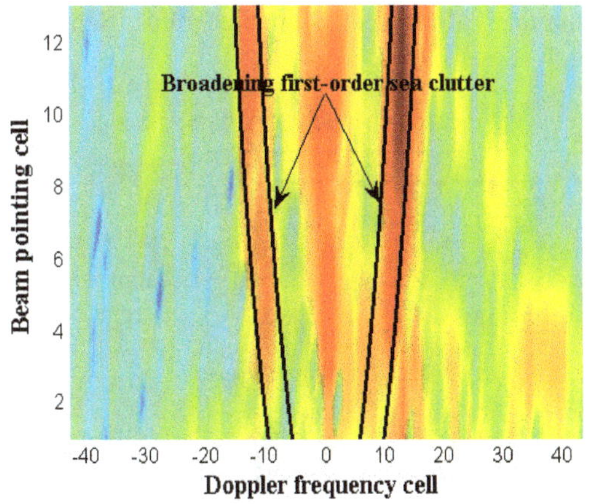

(b) Array layout 2, range cell = 8.

Fig. 7. Analysis of angle-Doppler distribution of measured first-order sea clutter.

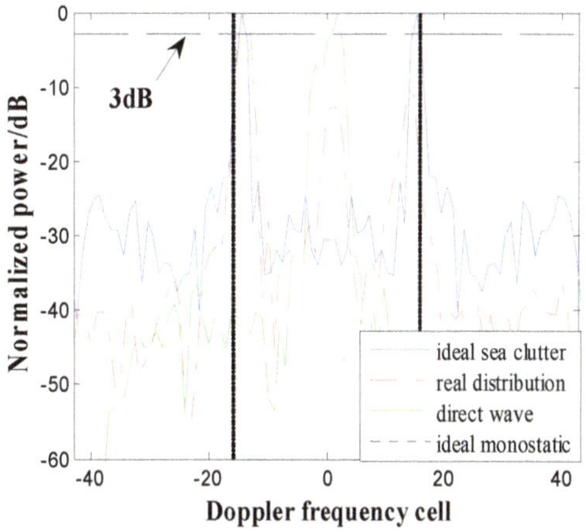

(a) Range cell = 27, azimuth angle =-25°.

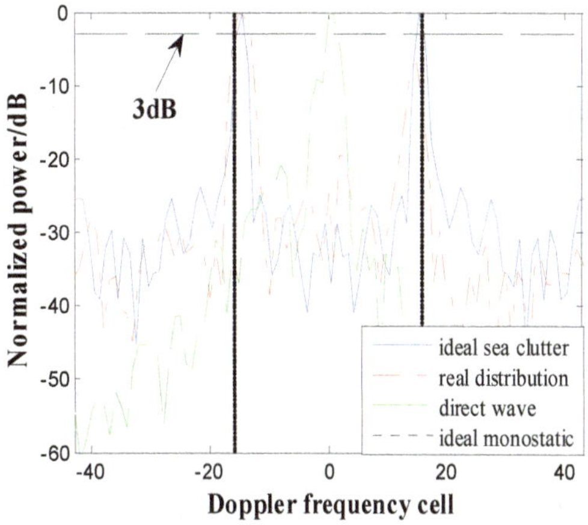

(b) Range cell = 27, azimuth angle =-15°.

Fig. 8. Comparison of ideal and measured sea clutter Doppler spectrum.

Bragg peak $(\Delta f^{+}+\Delta f^{-})/2$ is used as spectrum broadening measurement. Fig. 8 shows the Doppler spectrum of first-order sea clutter at the same range cell = 27 with two different azimuth angles. The Doppler spectrum of the real direct wave signal has significant broadening, and there is obvious broadening by comparison of ideal and measured sea clutter at different azimuths.

Tab. 1 shows the degree of Doppler broadening of first-order sea clutter at range cells = 27-30 (R27-R30) with different azimuth angles DBF = -25°-25° (interval 5°, B1-B11). From Tab. 1, we can see that there is obvious broadening at different azimuths. We can obtain the Doppler spectrum broadening of direct wave is about 0.04 Hz by the method of literature [14], which is consistent with the actual broadening value.

It should be noted that the broadening value of sea clutter at different azimuths is not identical; this is due to spatial difference of ionospheric propagation path at different azimuths. Through the analysis of measured data we find that the spatial difference of ionospheric propagation path is small.

Broadening /Hz	B1	B3	B5	B7	B9	B11
R27	0.035	0.045	0.017	0.017	0.033	0.034
R30	0.025	0.040	0.034	0.023	0.018	0.017

Tab. 1. First-order sea clutter broadening at different azimuths.

5. Influence of Ionospheric Disturbance on the Performance of STAP

STAP is a kind of low complexity and effective way for using training samples to suppress clutter, having become an important direction of research scholars from various countries and has been used in airborne radar [19]-[21]. Similar to HF ground clutter [19], [20], there are angle-Doppler coupling characteristics of sea clutter in HF hybrid sky-surface wave radar; so, using STAP to suppress broadening sea clutter becomes possible. The Joint Domain Localized (JDL) algorithm mainly takes advantage of the transformation vector **T** to transform the space-time data to the angle-Doppler domain and select a small local area for adaptive processing [19]. It is a kind of dimension reduced algorithm of the STAP processing, solving the lack of training samples and excessive computation load.

5.1 JDL Algorithm Principle

Space Time Adaptive Processing is using the training samples close to the range bin to be detected to estimate clutter and noise covariance matrix **R**, and according to linearly constrained minimum variance criteria (LCMV) to estimate adaptive weights w, and then weight the received data to maximize SNR. Covariance matrix is given by:

$$\mathbf{R}=\frac{1}{K}\sum_{i=1}^{K}\mathbf{X}_i\cdot\mathbf{X}_i^H \tag{22}$$

wherein, X_i is a training sample data.

Weight vector can be given by:

$$\mathbf{w}=\mu\mathbf{R}^{-1}\mathbf{v}=\mu\left(\mathbf{R}^{-1/2}\right)\left(\mathbf{R}^{-1/2}\mathbf{v}\right) \tag{23}$$

wherein, μ is a complex normalized constant.

The JDL algorithm mainly takes advantage of the transformation vector **T** to transform the space-time data to the angle-Doppler domain, and select a small local area for adaptive processing. It is a kind of dimension reduced algorithm of the STAP processing, solving the lack of training samples and excessive computation load.

S_s and S_t respectively are space steering vector and time steering vector. Elements of them are the discrete Fourier transform coefficients. Therefore, the inner product of the spatial and temporal steering vector is equivalent to 2-D DFT , and then the process of transforming the received data from space-time domain to the *i*th angle bin w_{si} , the *j*th Doppler bin we concern can be expressed as:

$$\mathbf{X}=\left(\mathbf{S_s}\left(w_{si}\right)\right)\otimes\left(\mathbf{S_t}\left(w_{tj}\right)\right)^H\mathbf{X}\,. \tag{24}$$

The JDL algorithm transforming matrix can be represented as:

$$\mathbf{T}=\begin{pmatrix}\left[\mathbf{S_s}(w_{si})\otimes\mathbf{S_t}(w_{tj})\right]^T\\ \left[\mathbf{S_s}(w_{si})\otimes\mathbf{S_t}(w_{tj}+w_k)\right]^T\\ \vdots\\ \left[\mathbf{S_s}(w_{si})\otimes\mathbf{S_t}(w_{tj}+(q-1)w_k)\right]^T\\ \vdots\\ \left[\mathbf{S_s}(w_{si}+(p-1)w_n)\otimes\mathbf{S_t}(w_{tj}+(q-1)w_k)\right]^T\end{pmatrix}^T \tag{25}$$

w_n and w_k respectively represent angle and Doppler interval, p represents the number of adjacent angle bins, q represents the number of adjacent Doppler bins. Schematic diagram Fig. 9 is as follows:

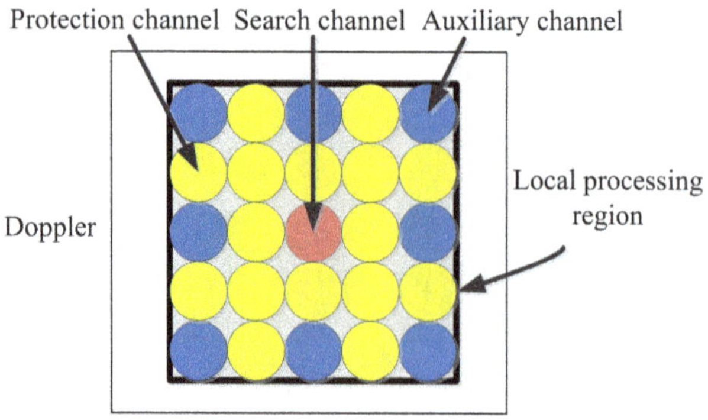

Fig. 9. JDL schematic diagram.

In Fig. 9, the red channel is the search channel, the blue channel is auxiliary channel, and the yellow channel is protection channel. The selection of a protection channel in angle-Doppler domain may be effective in preventing the spread of the target signal to clutter covariance matrix, affecting the actual detection performance.

When select the 3×3 local shown in Fig. 9, space-time transformation matrix can be expressed as:

$$\mathbf{T}=\left[\mathbf{S}_{\mathbf{t}}\left(w_{t,j-2}\right);\ \mathbf{S}_{\mathbf{t}}\left(w_{t,j}\right);\mathbf{S}_{\mathbf{t}}\left(w_{t,j+2}\right)\right]\otimes\left[\mathbf{S}_{\mathbf{s}}\left(w_{s,i-2}\right);\ \mathbf{S}_{\mathbf{s}}\left(w_{s,i}\right);\ \mathbf{S}_{\mathbf{s}}\left(w_{s,i+2}\right)\right] \tag{26}$$

The transformed space-time steering vector is:

$$\tilde{\mathbf{v}}=\mathbf{T}^{H}\cdot\mathbf{v}\ . \tag{27}$$

The receiving data vector is:

$$\tilde{X}=T^{H}\cdot X\ . \tag{28}$$

The corresponding adaptive weight vector is:

$$\tilde{w}=\mu\tilde{R}^{-1}\tilde{v}\ . \tag{29}$$

5.2 Experimental Results

The target is injected into the position of range cell = 50 and azimuth angle = 0° of measured data under array layout 2, signal to clutter ratio (SCR) is SCR = -5 dB. The performance of JDL algorithm with contaminated sea clutter is analyzed by measured data. Improvement factor *IF* is used as the performance measurement:

$$IF=\frac{SCR_{out}}{SCR_{in}}\ . \tag{30}$$

Fig. 10(a)-(b) show the performance of JDL algorithm with ideal and contaminated point target (because of target signal is also contaminated by ionosphere in practice) based on the measured data. The ideal point target without the influence of ionospheric disturbance is analyzed in Fig. 10(a); the contaminated point target with the influence of ionospheric disturbance is analyzed in Fig. 10(b). In Fig. 10(a)-(b), the blue solid line represents the performance of JDL with contaminated sea clutter; the red dotted line represents the performance of JDL after decontamination, where phase gradient algorithm (PGA) was used to ionospheric phase decontamination.

As can be seen from Fig. 6 in Section 4.3 and Fig. 10(a)-(b): (1) space-time coupling characteristics of sea clutter is stable when there is no ionospheric disturbance; even if the Doppler spectrum of sea clutter gradually broadening and submerged target, JDL algorithm can effectively suppress the sea clutter and find target, improvement factor is above 10 dB; (2) when there is ionospheric disturbance, space-time coupling of sea clutter is significantly affected, and clutter distribution between the training sample is no longer even, the improvement factor of JDL algorithm decreased; (3) the more severe iono-

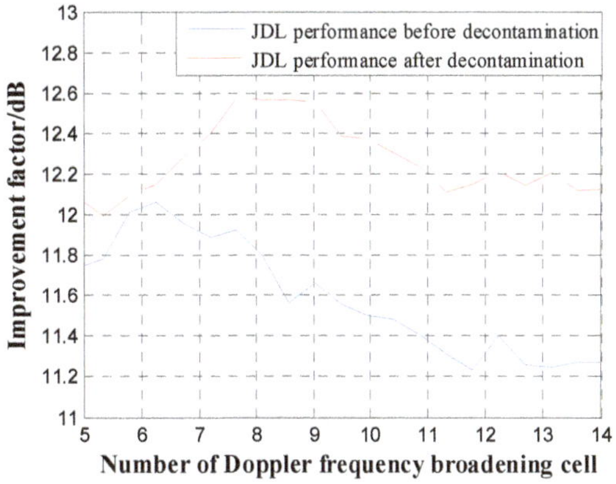

(a) Ideal point target without ionospheric disturbance is injected.

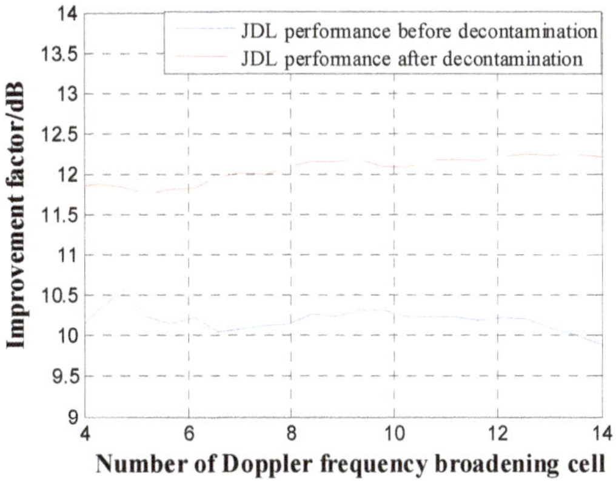

(b) Contaminated point target with ionospheric disturbance is injected.

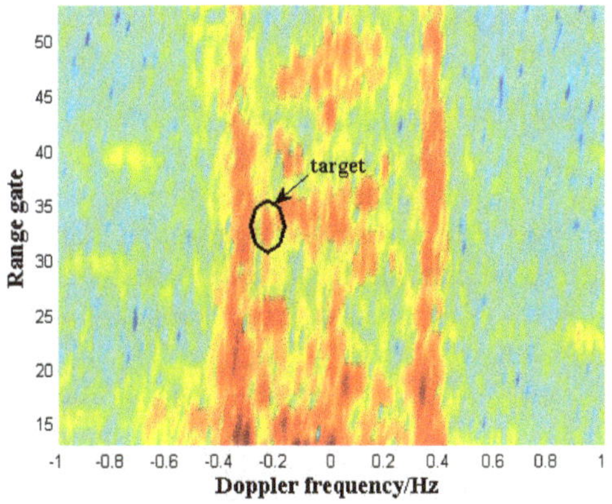

(c) Measured sea clutter spectrum with range-Doppler distribution.

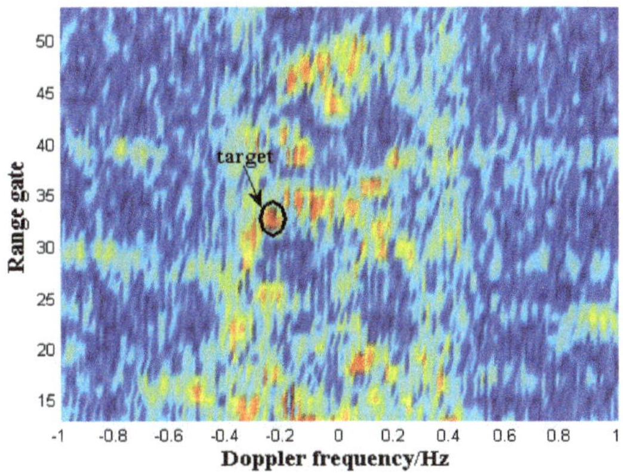

(d) Measured sea clutter suppression by PGA&JDL algorithm.

Fig. 10. Analysis of performance of JDL algorithm with measured data.

spheric disturbance, the more obvious the decrease of the improvement factor; (4) the JDL performance after decontamination is superior to the JDL performance before decontamination.

Fig. 10(c)-(d) show the measured sea clutter spectrum with range-Doppler distribution (DBF = -15°) and the sea clutter suppression results after PGA&JDL algorithm. From Fig. 10(c)-(d), we can see that the measured sea clutter is significantly reduced by PGA&JDL processing, thus the PGA&JDL algorithm can achieve good suppression performance for broadening sea clutter.

6. Conclusions

In this paper, the basic working principle of HF sky-surface wave experimental system was firstly introduced. Secondly, the space-time characteristics of sea clutter were analyzed, and space-time distribution and spreading model of the first-order sea clutter under the action of bistatic angle and ionosphere were given. Then, space-time coupling characteristics of sea clutter were analyzed through the simulation, and the results of theoretical analysis were examined with the experimental results based on our developed integrated HF sky-surface wave radar system. The theoretical analysis of space-time distribution for spreading sea clutter is also in agreement with experimental results. Finally, the influence of ionospheric disturbance on the performance of STAP algorithm was analyzed based on experimental data.

It must be pointed out that these are just preliminary results and there is still a need for more studies and improvements. More details about the broadening sea clutter suppression for HF hybrid sky-surface wave radar need to be investigated to improve the detection performance. In addition, target detection and localization are also key areas of our future work.

Acknowledgements

At the point of finishing this paper, we would like to express our sincere thanks to the National Natural Science Foundation Key Project (No.:61032011), the support of the measured data and members of the School of Electronics and Information Engineering, Research Center, Harbin Institute of Technology for technical support.

References

[1] MELYANOVSKI, P. A., TOURGENEV, I. S. Bistatic HF radar for oceanography applications with the use of both ground and space waves. *Telecommunications and Radio Engineering*, 1997, vol. 51, no. 2, p. 73–79.

[2] FRAZER, G. J. Forward-based receiver augmentation for OTHR. In *2007 IEEE Radar Conference*. Boston (MA, USA), April 2007, p. 373–378.

[3] RIDDOLLS, R. J. Ship detection performance of a high frequency hybrid sky-surface wave radar. *Defense R&D Canada-Ottawa*, 2007, p. 1–42.

[4] RIDDOLLS, R. J. Limits on the detection of low-Doppler targets by a high frequency hybrid sky-surface wave radar system. In *IEEE Radar Conference 2008*. Rome (Italy), May 2008, p. 1–4.

[5] JIAO PEINAN, YANG LONGQUAN, FAN JUNMEI New propagation mode associating with HF sky-and-surface wave and its application. *Chinese Journal of Radio Science*, 2007, vol. 22, no. 5, p. 746–750.

[6] WEI JIANG. WEI-BO, D., QIANG, Y. Analyses of sea clutter for HF over the horizon hybrid sky-surface wave radar. *Journal of Electronic & Information Technology*, 2011, vol. 33, no. 8, p. 1786–1791.

[7] LOU PENG, FAN JUNMEI, JIAO PEINAN, YANG LONGQUAN The operating frequency selection in HF hybrid sky-surface wave radar. In *9th International Symposium on Antennas, Propagation and EM Theory (ISAPE)*. 2010, p. 513–516.

[8] ZHIXIN ZHAO, XIANRONG WAN, DELEI ZHANG, FENG CHENG An experimental study of HF passive bistatic radar via hybrid sky-surface wave mode. *IEEE Transactions on Antennas and Propagation*, 2013, vol. 61, no. 1, p. 415-424.

[9] YAJUN LI, YINSHENG WEI Analysis of first-order sea clutter spectrum characteristics for HF sky-surface wave radar. In *2013 International Conference on Radar*. Adelaide (Australia), 9-12 September 2013, p. 368–373.

[10] TRIZNA, D., GORDON, J. Results of a bistatic HF radar surface wave sea scatter experiment. In *IEEE International Geoscience and Remote Sensing Symposium*. 2002, vol. 3, p. 1902–1904.

[11] BARRICK, D. Remote sensing of sea state by radar. In *IEEE International Conference on Engineering in the Ocean Environment*. 1972, p. 186–192.

[12] GILL, E. The scattering of high frequency electromagnetic radiation from the ocean surface: An analysis based on a bistatic ground wave radar configuration. *PhD Thesis*. Faculty of Engineering and Applied Science, Memorial University of Newfoundland, 1999, p. 1–178.

[13] WALSH, J., HUANG, W., GILL, E. The first-order HF radar ocean surface cross section for an antenna on a floating platform. *IEEE Transactions on Antennas and Propagation*, 2010, vol. 58, p. 2994–3003.

[14] GEORGES, T. M. The effects of space and time resolution on the quality of sea echo Doppler spectra measured with HF sky wave radar. *Radio Science*, 1979, vol. 14, p. 455–469.

[15] DAVIES, K., BARKER, D. M. On frequency variations of ionospherically propagated HF radio signals. *Radio Science*, 1966, p. 545–556.

[16] BOURDILLON, A., GAUTHIER, F., PARENT, J. Use of maximum entropy spectral analysis to improve ship detection by over-the horizon radar. *Radio Science*, 1987, vol. 22, no. 2, p. 313–320.

[17] PARENT, J., BOURDILLON, A. A method to correct HF skywave ionosphere frequency modulation. *IEEE Transactions on Antennas and Propagation*, 1988, vol. 36, no. 1, p. 127–135.

[18] REINISH, B. W. Ionospheric sounding in support of over-the-horizon radar. *Radio Science*, 1997, vol. 32, no. 4, p. 1689–1694.

[19] SALEH, O., ADVE, R. S., RIDDOLLS, R. J., RAVAN, M., PLATANIOTIS, K. Adaptive processing in high frequency surface wave radar. In *Proceeding of the IEEE Radar Conference, RADAR 2008*. Rome (Italy), 2008, p. 1–6.

[20] RAVAN, M., ADVE, R. S. Robust STAP for HFSWR in dense target scenarios with nonhomogeneous clutter. In *2012 IEEE Radar Conference*. Atlanta (GA, USA), May 2012, p. 0028–0033.

[21] FABRIZIO, G. A., FRAZER, G. J., TURLEY, M. D. STAP for clutter and interference cancellation in a HF radar system. In *2006 IEEE International Conference on Acoustics, Speech and Signal Processing*. Toulouse (France), vol. 4, p. 14–19.

About Authors ...

Yajun LI was born in 1983. He received the M.S. degree in Information and Communication Engineering from Harbin Engineering University in 2011. He is now a PhD student of the School of Electronics and Information Engineering at Harbin Institute of Technology, China. His current research interests include space-time adaptive processing, suppression of sea clutter, and HF radar system simulation.

Yinsheng WEI was born in 1974. He received his M.S. and Ph.D. degrees in Communication and Information Systems from Harbin Institute of Technology (HIT) in 1998 and 2002, respectively. And then, he joined the Department of Electronics Engineering in HIT as a lecturer, and became a professor in 2011. He is a member of IEEE AES, and a senior member of CIE. His main researches include radar signal processing and radar system analysis and simulation.

Rongqing XU was born in 1958. He received his M.S. and Ph.D. degrees in Communication and Information Systems from Harbin Institute of Technology (HIT) in 1984 and 1990, respectively. He is a professor at the Department of Electronics Engineering in HIT. His research interests are in the fields of radar signal processing, multi-target tracking and data processing.

Tianqi CHU was born in Heilongjiang, China, in 1989. He received the B.S. degree in Electronic Information Engineering from Harbin Institute of Technology in 2012. He is currently proceeding the M.S. degree in Electronics and Communication Engineering in HIT. His researches mainly focus on electromagnetic waves transmission and signal processing in HF OTH radar.

Zhuoqun WANG was born in 1985. She received the M.S. degree in Information and Communication Engineering from Harbin Engineering University in 2011. She is now a PhD student of the School of Electronics and Information Engineering at Harbin Institute of Technology, Harbin, China. Her current research interests include SAR imaging processing and radar signal processing.

4

An Extended Virtual Aperture Imaging Model for Through-the-wall Sensing and Its Environmental Parameters Estimation

Yongping SONG, Tian JIN, Biying LU, Jun HU, Zhimin ZHOU

College of Electronics Science and Engineering, National University of Defense Technology, Changsha, 410073, China

sypopqjkl@163.com, tianjin@nudt.edu.cn

Abstract. *Through-the-wall imaging (TWI) radar has been given increasing attention in recent years. However, prior knowledge about environmental parameters, such as wall thickness and dielectric constant, and the standoff distance between an array and a wall, is generally unavailable in real applications. Thus, targets behind the wall suffer from defocusing and displacement under the conventional imaging operations. To solve this problem, in this paper, we first set up an extended imaging model of a virtual aperture obtained by a multiple-input-multiple-output array, which considers the array position to the wall and thus is more applicable for real situations. Then, we present a method to estimate the environmental parameters to calibrate the TWI, without multiple measurements or dominant scatterers behind-the-wall to assist. Simulation and field experiments were performed to illustrate the validity of the proposed imaging model and the environmental parameters estimation method.*

Keywords

Through-the-wall radar, environmental parameter estimation, virtual aperture, multiple-input multiple-output (MIMO).

1. Introduction

Through-the-wall sensing is highly desired in many civilian and military applications. Through-the-wall imaging radar (TWIR) achieves good wall penetration by transmitting low-frequency electromagnetic waves and provides imaging description of targets of interest behind walls [1-3] or the inside structure and layout of buildings [4-5]. Therefore, it has attracted more and more attention [6-10]. Generally speaking, two techniques, namely synthetic apertures and virtual apertures, are adopted in TWIR. Compared with the synthetic aperture formed by moving radar antennas, the virtual aperture obtained by the multiple-input multiple-output (MIMO) array can collect imaging data in a much shorter time and is thus more applicable for real-time implementation. In this paper, we focus on the MIMO array based TWIR.

When electromagnetic waves propagate in a layered medium composed of the air and wall, reflection and refraction will occur at the air-wall interface. This requires through-the-wall imaging (TWI) to consider the non-linear propagation path of electromagnetic waves, where some wall parameters, i.e., wall thickness and dielectric constant, are required to determine the propagation path. Currently, most through-the-wall image formations are based on the prior knowledge of the aforementioned wall parameters [11-13]. However, prior knowledge about these wall parameters is usually unavailable in practical applications, and the measured or estimated error of the wall parameters will greatly affect imaging quality of behind-the-wall targets [14]. Therefore, accurate estimation of wall parameters is an important technique in TWI.

The emerging techniques to estimate wall parameters can be categorized into three types. First, the image autofocusing method searches the reasonable wall parameters by assessing the image focusing quality [15]. It is effective but affords heavy computing burden. To improve the efficiency, an estimation method by minimizing the cross-range resolution of a special dominant scatterer rather than assessing the whole image was introduced in [16]. The method is invalid if there are no dominant scatterers behind the wall. Second, the wall parameters can be estimated by adjusting the array structure [17] or stand-off distance [18], both of which involve extra measurements. Third, echoes reflected by the wall are utilized to estimate the wall parameters. The dielectric constant can be provided by the back and forth propagation time if the wall thickness is known [19]. Moreover, searching the maximum of the correlation coefficient between the measured return and the corresponding estimating return in different wall parameters is also valid [20], and the greatest difficulty is the indecisive searching direction. Instead, using some special information extracted from echoes caused by the front surface and the rear surface of the wall is more promising, e.g., the dielectric constant can be estimated using the amplitude information of the front surface reflection [21]. Because the amplitude estimation is easily influenced by noise, a more practical method is presented by performing time-delay-only measurements in [22]; however, to achieve adequately high accuracy, it requires adjusting the transceiver-receiver separation repeatedly.

Furthermore, all of the above methods assume that the TWIR parallels the wall during data collection in the imaging model. When the array is placed close to the wall, the assumption can be easily met. However, in some applications, e.g., fire rescue, standoff operation is required. Therefore, it is necessary to setup a more reasonable model that considers the distance and inclination angle between the radar and wall. In such an extended imaging model for through-the-wall sensing, the unknown parameters include not only the traditional wall parameters, i.e., thickness and dielectric constant, but also the distance and the inclination angle. These four unknown parameters, denoted as the environmental parameters in this paper, are all required to be estimated in practical applications.

In the next section, an extended imaging model is proposed to fit the detection situation where a linear MIMO array sets in front of a wall with unknown distance and inclination angle. Then, in Section 3, we will show a novel environmental parameters estimation algorithm without any extra measurements or dominant scatters behind-the-wall to assist. To improve the image quality, an effective compensation imaging method for the proposed imaging model is introduced in Section 4. Section 5 & 6 show the corresponding processing results of the simulation and field measured data, which verify the imaging model and environmental parameters estimation algorithm. Conclusions will be drawn in Section 7.

2. Extend Virtual Aperture Imaging Model for TWI

The conventional linear MIMO array through-the-wall imaging model assumes an antenna array strictly parallel to the wall, as shown in Fig. 1. At this point, the relative position of the antenna array and the wall can be described with only one distance value R, which reduces complexity in image processing. However, in practice, limiting factors, such as the probe scene, make it difficult to ensure that the antenna array is strictly parallel to the wall, so it is necessary to extend the conventional model and take into consideration the case in which the wall inclines the antenna array.

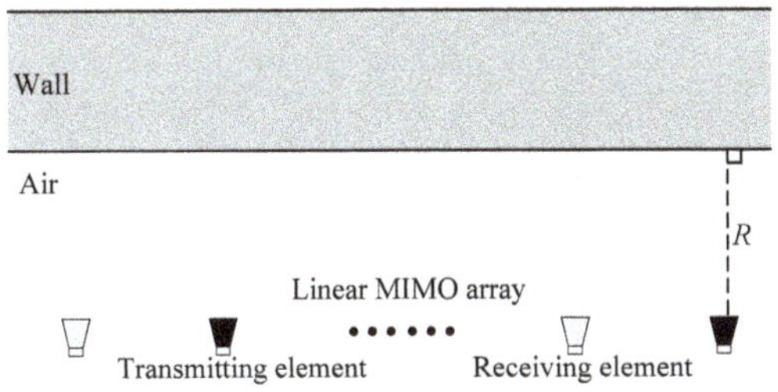

Fig. 1. Conventional linear MIMO array TWI model.

When the antenna array is inclined to the wall, the relative position of array and the wall needs to be described by both the distance and the inclination angle. For a linear MIMO array, the inclination angle can be described by one inclination angle θ. However, the array elements are at different distances to the wall. Because the array structure is known, we can consider anywhere in the array to be the reference position and use the reference position's distance to the wall to describe the distance information R from the array to the wall. Here we use the center of the array as the reference position.

Suppose we have a linear MIMO array TWIR with M transmitting elements and N receiving elements. To simplify our problem, we assume that the array is only inclined to the wall in the horizontal plane. Considering reflection and refraction caused by the wall, the imaging model can be expressed as Fig. 2.

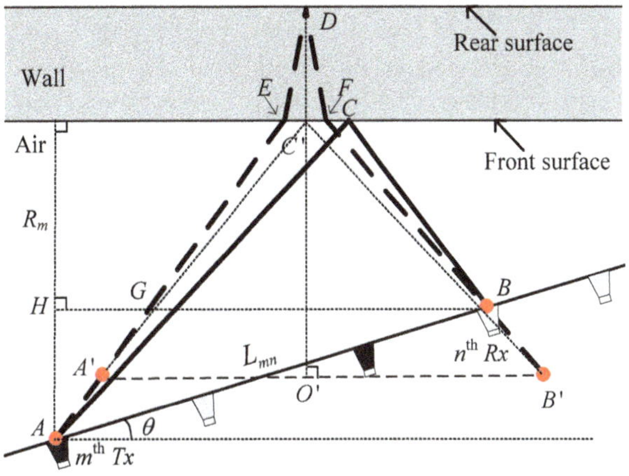

Fig. 2. TWI model of inclined linear MIMO array.

Fig. 2 shows a tilted linear MIMO array set in front of a wall with an inclination angle θ. When the left end point of the array is further from the wall than the right one, we define $\theta > 0$, otherwise $\theta < 0$. Obviously, $-90^\circ < \theta < 90^\circ$. Donate the mth transmitting element as m^{th} *Tx*, and the nth receiving element as n^{th} *Rx*. Then, the distance from m^{th} *Tx* to the wall is R_m.

Without consideration of the direct wave, the first two echoes are caused by the front surface and the rear surface of the wall, and the intersections of their propagation paths would be m^{th} *Tx* and n^{th} *Rx*. Donate A to be the location of mth *Tx*, B to be the location of nth *Rx*, and L_{mn} to be their spacing. C is the front surface echo reflection point, and the corresponding rear surface echo reflection point is D. Their respective refraction points are E and F. BH is parallel to the wall. G is the intersection of BH & AE. A' is the midpoint of AG. BB' is the extension line of FB, and $A'B' // BH$. $DO' \perp A'B'$, and C' is the intersection of DO' and the front surface.

It can be proved that for the echo propagation path of the front surface:

$$|AC|+|CB|=|A'C'|+|C'B'| \tag{1}$$

and for the echo propagation path of the rear surface:

$$|AE|+|ED|+|FD|+|BF|=|A'E|+|ED|+|FD|+|B'F| \tag{2}$$

Furthermore, the spacing of $A'B'$ is $L_{mn}\cos\theta$, and the distance to the wall is $R_m + 0.5\alpha_{mn} L_{mn} \sin|\theta|$. When m^{th} Tx is further than n^{th} Rx from the wall, $\alpha_{mn} = -1$, otherwise $\alpha_{mn} = 1$ (See Appendix).

Thus, for the echo propagation paths of the front surface and the rear surface, the tilted linear MIMO array can be equivalent to several virtual element pairs whose connections are parallel to the wall.

3. Environmental Parameters Estimation of the Linear MIMO Array TWI Model

Already mentioned in the introduction, for our proposed linear MIMO array TWI model, the environmental parameters should include the distance of the array to the wall R, the inclination angle θ, the wall thickness d and the dielectric constant ε_r. We will use the conclusions in Section 2 to show the estimation method of the environmental parameters.

3.1 Estimation of R and θ

By the geometric relationship in Fig. 2 we get:

$$(ct_f(m,n))^2 - (L_{mn}\cos\theta)^2 = (2R_m + \alpha_{mn} L_{mn} \sin|\theta|)^2 \quad (3)$$

where, c is the light velocity in air, and $t_f(m,n)$ is the front surface echo delay caused by m^{th} Tx and n^{th} Rx. It can be further converted to:

$$c^2 t_f^2(m,n) - L_{mn}^2 = 4R_m^2 + 4\alpha_{mn} L_{mn} R_m \sin|\theta| \quad (4)$$

set

$$g_{mn} = c^2 t_f^2(m,n) - L_{mn}^2 , \quad (5)$$

$$E_m = (R_m^2, R_m \sin|\theta|)^T , \quad (6)$$

$$h_{mn} = (4, 4\alpha_{mn} L_{mn}) , \quad (7)$$

then

$$g_{mn} = h_{mn} E_m \quad (8)$$

Considering all the echoes generated by m^{th} Tx, we get

$$G_m = H_m E_m \quad (9)$$

where:

$$G_m = (g_{m1}, g_{m2}, \ldots, g_{mN})^T , \quad (10)$$

$$H_m = (h_{m1}^T, h_{m2}^T, \ldots, h_{mN}^T)^T . \quad (11)$$

Accordingly,

$$E_m = (H_m^T H_m)^{-1} H_m^T G_m . \quad (12)$$

That is, from

$$E_m = \begin{bmatrix} e_{m1} \\ e_{m2} \end{bmatrix} = \begin{bmatrix} R_m^2 \\ R_m \sin|\theta| \end{bmatrix} \quad (13)$$

we obtain the estimation of the distance R and the inclination angle θ on m^{th} Tx:

$$\begin{bmatrix} \hat{R}_m \\ |\hat{\theta}_m| \end{bmatrix} = \begin{bmatrix} \sqrt{e_{m1}} \\ \arcsin(e_{m2} / \sqrt{e_{m1}}) \end{bmatrix} . \quad (14)$$

By performing the same processing on the echo data of the remaining $M-1$ transmitting elements we will get

$$\hat{R}_E = (\hat{R}_1, \hat{R}_2, \ldots, \hat{R}_M)^T , \quad (15)$$

$$\hat{\theta}_E = (|\hat{\theta}_1|, |\hat{\theta}_2|, \ldots, |\hat{\theta}_M|)^T . \quad (16)$$

Taking the average of elements in vector $\hat{\theta}_E$ as the final estimation of $|\hat{\theta}|$:

$$|\hat{\theta}| = \frac{1}{M} \sum_{m=1}^{M} |\hat{\theta}_m| . \quad (17)$$

Then, if the left end point of the array is further from the wall than the right end point:

$$\hat{\theta} = -|\hat{\theta}| , \quad (18)$$

else

$$\hat{\theta} = |\hat{\theta}| . \quad (19)$$

Accordingly, $\hat{R}$, which represents the distance between the array and wall, can be estimated as:

$$\hat{R} = \sum_{m=1}^{M} w_m \hat{R}_m \quad (20)$$

wherein w_m is the weighting coefficient, which depends on the geometry of the antenna array.

3.2 Estimation of d and ε_r

It has been proved in Section 2 that m^{th} Tx and n^{th} Rx can be equivalent to a virtual element pair for the echo paths of the front surface and the rear surface. Denoting the virtual element pair as $m'n'$, with the estimation $\hat{R}_m$ and $\hat{\theta}$ in Section 3.1, the spacing of $m'n'$ can be expressed as:

$$\hat{L}_{m'n'} = L_{mn} \cos\hat{\theta} \quad (21)$$

and the distance of $m'n'$ to the wall $\hat{R}_{m'n'}$ is:

$$\hat{R}_{m'n'} = \hat{R}_m + 0.5\alpha_{mn} L_{mn} \sin|\hat{\theta}| . \quad (22)$$

So that the echo path model of the virtual element pair $m'n'$ can be showed as Fig. 3.

Because $m'n'$ is parallel to the wall, we have the following equation [23]:

$$d^2 \varepsilon_r - \frac{d^2 \hat{L}_{m'n'}^2 \cos^2\hat{\theta}}{\hat{L}_{m'n'}^2 \cos^2\hat{\theta} + 4\hat{R}_{m'n'}^2} = 0.25 c^2 t_d^2(m,n) \quad (23)$$

where

$$t_d(m,n) = t_r(m,n) - t_f(m,n) \cdot \quad (24)$$

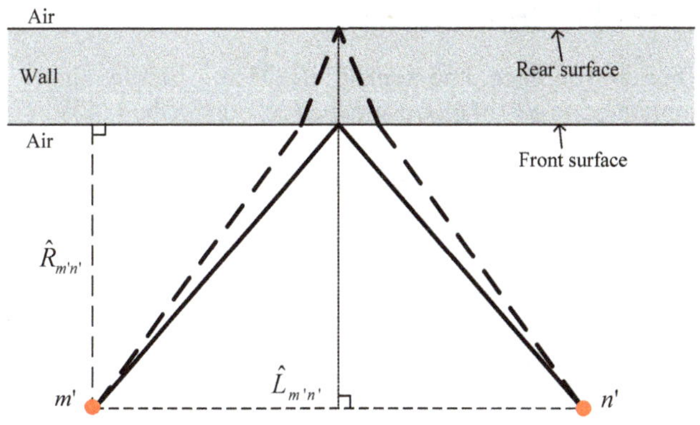

Fig. 3. The echo path model of a virtual element pair.

$t_r(m, n)$ is the rear surface echo delay caused by m^{th} *Tx* and n^{th} *Rx*. Then, we get:

$$Ap = b \tag{25}$$

where:

$$A = \begin{bmatrix} 1 & -\dfrac{\hat{L}^2_{1'1'}\cos^2\hat{\theta}}{\hat{L}^2_{1'1'}\cos^2\hat{\theta}+4\hat{R}^2_{1'1'}} \\ 1 & -\dfrac{\hat{L}^2_{1'2'}\cos^2\hat{\theta}}{\hat{L}^2_{1'2'}\cos^2\hat{\theta}+4\hat{R}^2_{1'2'}} \\ \vdots & \vdots \\ 1 & -\dfrac{\hat{L}^2_{M'N'}\cos^2\hat{\theta}}{\hat{L}^2_{M'N'}\cos^2\hat{\theta}+4\hat{R}^2_{M'N'}} \end{bmatrix}_{MN\times 2} \tag{26}$$

$$p = \begin{bmatrix} d^2\varepsilon_r \\ d^2 \end{bmatrix}_{2\times 1} = \begin{bmatrix} p_1 \\ p_2 \end{bmatrix}_{2\times 1} \tag{27}$$

$$b = \begin{bmatrix} 0.25c^2t_d^2(1,1) \\ 0.25c^2t_d^2(1,2) \\ \vdots \\ 0.25c^2t_d^2(M,N) \end{bmatrix}_{MN\times 1} \tag{28}$$

then:

$$p = (A^TA)^{-1}A^Tb. \tag{29}$$

Finally, we can obtain the estimates of the left two environmental parameters:

$$\hat{d} = \sqrt{p_2}, \tag{30}$$

$$\hat{\varepsilon}_r = p_1 / p_2. \tag{31}$$

4. Compensation Image Formation Based on the Extended Virtual Aperture Imaging Model

Virtual aperture radar, such as MIMO radar using multi-elements, greatly improves detection performance. Accordingly, the imaging model becomes more complicated, which makes many traditional synthetic aperture radar imaging algorithms no longer applicable [24]. Due to the absence of any restrictions on the antenna array, the BP algorithm is widely used in virtual aperture radar imaging systems. We will use the environmental parameters estimated in Section 3 to show the compensation BP algorithm for the extended virtual aperture imaging model.

The BP algorithm replaces phase compensation by calculating the exact propagation delay of the target to the antenna elements [25]. In traditional through-the-wall BP imaging algorithm, the antenna array center is usually set as the origin, whereas the array itself is the abscissa. The direction perpendicular to the linear MIMO array is the ordinate. We name this coordinate system the antenna-coordinate-system. Correspondingly, the coordinate system that sets the front surface of the wall as the horizontal axis and the wall perpendicular as the longitudinal axis is called the wall-coordinate-system.

The purpose of TWI is usually to describe targets behind the wall. When the targets' position information is relative to the wall, interpretation of targets is easier. Therefore, using the wall-coordinate-system to show the results of TWI is more suitable. Assuming the array center is located in the center on the connection of mth *Tx* and n^{th} *Rx*, we can obtain the TWI model in the wall-coordinate-system:

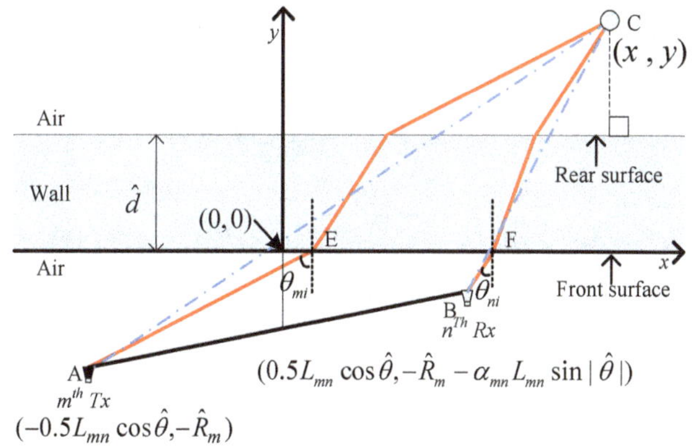

Fig. 4. TWI model in wall-coordinate-system.

According to the BP algorithm, the image value $I(x,y)$ of a point target (x,y) behind the wall could be calculated as follows:

$$I(x,y) = \sum_{m=1}^{M}\sum_{n=1}^{N} s_{mn}(t_{TE}(m,n)) \tag{32}$$

where, $s_{mn}(t_{TE}(m,n))$ represents the sampling at time $t_{TE}(m,n)$ in the echo generated by m^{th} *Tx* and n^{th} *Rx*. $t_{TE}(m,n)$ is the propagation delay from the point target to *mn*.

As shown in Fig. 4, $t_{TE}(m,n)$ can be obtained as:

$$t_{TE}(m,n) = \frac{l_{AC} + l_{BC}}{c} \tag{33}$$

l_{AC} is the propagation path from the target to m^{th} Tx, and l_{BC} is the propagation path from the target to n^{th} Rx.

The key to obtaining l_{AC} and l_{BC} is to determine the position of refraction points E and F in Fig. 4. Using the estimated thickness and dielectric constant of the wall, a quartic equation with one unknown quantity can be created to obtain the analytical solutions of the refraction point according to Snell's law [26] or to search for the location of refraction points by the minimum time criteria [27]. Both algorithms have high accuracy but costly calculations.

To avoid calculation resource depletion in solving the exact solutions, l_{AC} can be approximated by the expression [23]:

$$l_{AC} \approx r_{AC} + \hat{d}(\sqrt{\hat{\varepsilon}_r - \sin^2\theta_{mi}} - \cos\theta_{mi}) \tag{34}$$

where r_{AC} is the straight-line distance from the target to m^{th} Tx, and θ_{mi} is the incident angle of l_{AC}. θ_{mi} can be approximately obtained by:

$$\theta_{mi} \approx \arctan(\frac{x + 0.5L_{mn}\cos\hat{\theta}}{y + \hat{R}_m}). \tag{35}$$

Similarly:

$$l_{BC} \approx r_{BC} + \hat{d}(\sqrt{\hat{\varepsilon}_r - \sin^2\theta_{ni}} - \cos\theta_{ni}) \tag{36}$$

where r_{BC} is the straight-line distance from the target to n^{th} Rx, and θ_{ni} is the incident angle of l_{BC}.

In summary, for the point target (x, y) in the imaging scene, set:

$$\Delta r_m = \hat{d}(\sqrt{\hat{\varepsilon}_r - \sin^2\theta_{mi}} - \cos\theta_{mi}), \tag{37}$$

$$\Delta r_n = \hat{d}(\sqrt{\hat{\varepsilon}_r - \sin^2\theta_{ni}} - \cos\theta_{ni}), \tag{38}$$

then the propagation delay in the echo generated by m^{th} Tx and n^{th} Rx is calculated as follows:

$$t_{TE}(m,n) = \frac{r_{AC} + \beta\Delta r_m + r_{BC} + \beta\Delta r_n}{c} \tag{39}$$

where β is a scale factor to distinguish the positional relationship between the point target and the wall:

$$\beta = \begin{cases} 0 & , \quad y \le 0 \\ y/\hat{d} & , \quad 0 < y \le \hat{d} \\ 1 & , \quad y > \hat{d} \end{cases} \tag{40}$$

Combining formula (32), the pixel value on (x,y) can be obtained. By traversing every point in the imaging scene we can achieve the compensated image for the entire scene. Compared with conventional algorithms, the computation greatly reduces.

5. FDTD Simulations

Finite-difference-time-domain (FDTD) simulations are conducted to test the performance of our method. The MIMO array can be regarded as several associated single-input multiple-output (SIMO) arrays. To simplify the simulating complexity, a SIMO array is set in front of the wall:

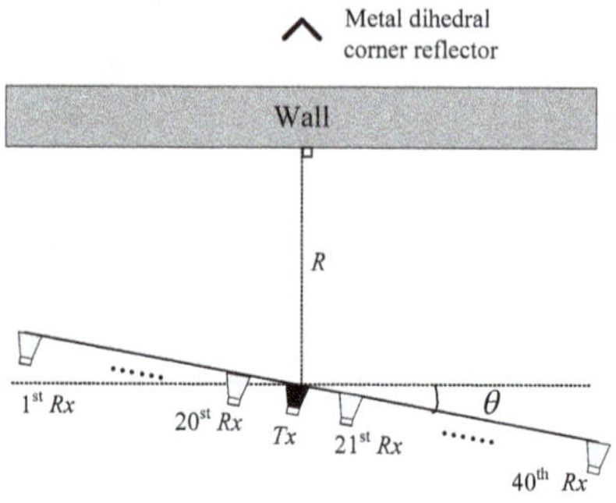

Fig. 5. The simulating scene.

It is excited by a Gaussian derivative pulse with 0.6 ns width. The white noise is added to the simulating echoes. The signal to noise ratio (SNR) is 15 dB. The estimation results are shown in Tab. 1 and Tab. 2.

$\theta(^\circ)$	$\hat{\theta}(^\circ)$	$\hat{R}$(m)	$\hat{d}$(m)	$\hat{\varepsilon}_r$
-5	-5.0708	1.4033	0.2002	4.2121
-10	-10.0998	1.4034	0.1936	4.5025
-15	-15.1011	1.4034	0.1915	4.6020
-25	-25.1440	1.4035	0.1875	4.7965
True value		1.4000	0.2000	4.5000

Tab. 1. Estimation results of the simulation data when $R = 1.4$ m.

R (m)	$\hat{R}$(m)	$\hat{\theta}(^\circ)$	$\hat{d}$(m)	$\hat{\varepsilon}_r$
1	1.0018	-15.1372	0.1994	4.2456
1.4	1.4034	-15.1011	0.1915	4.6020
1.8	1.8047	-15.0857	0.1944	4.4645
2.2	2.2058	-15.0642	0.1862	4.8620
True value		-15.0000	0.2000	4.5000

Tab. 2. Estimation results of the simulation data when $\theta = -15°$.

The results are satisfactory. To indicate the benefits of our virtual aperture imaging model, an imaging process is performed on the simulation data of $R = 1.4$ m and $\theta = -15°$:

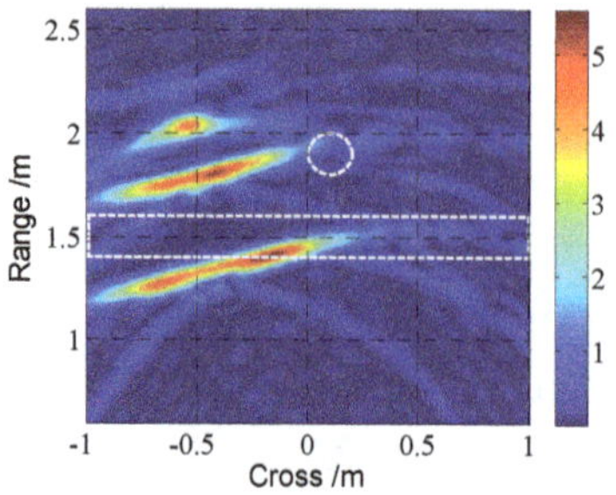

Fig. 6. BP imaging under the antenna-coordinate-system.

Without considering the array position to the wall, direct BP imaging has to be performed on the antenna-coordinate-system, resulting in whole scene bias. To make the scene easier to understand, the wall-coordinate-system should be built with the environment parameters $\hat{R}$ and $\hat{\theta}$:

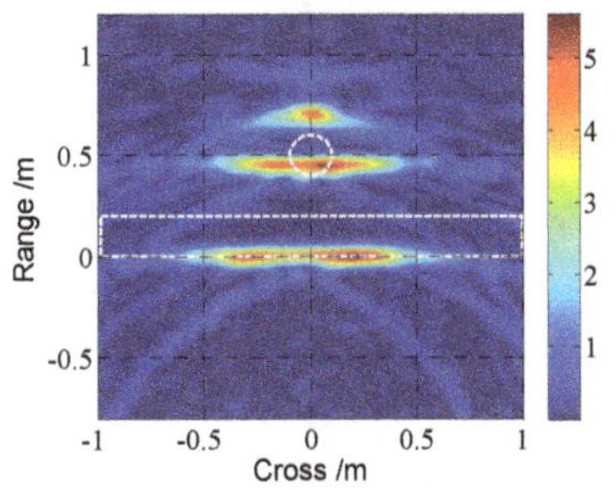

Fig. 7. BP imaging under the wall-coordinate-system.

Then, the defocusing and displacement of the target behind the wall can be fixed by the left two environment parameters $\hat{d}$ and $\hat{\varepsilon}_r$:

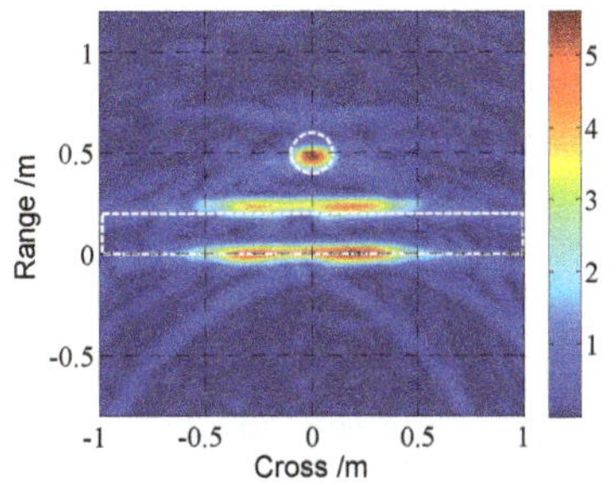

Fig. 8. BP imaging based on the extended virtual aperture imaging model.

Finally, benefiting from the extended virtual aperture imaging model, the quality of TWI is successfully enhanced.

6. Measurement Results

To validate the proposed imaging model and environmental parameter estimation algorithm, we have designed a vehicle virtual aperture radar system with 2 transmitting elements and 11 receiving elements. The transmitted signal is a stepping-frequency signal from 0.5 GHz to 3 GHz, and the step frequency is 2 MHz. The experiment scene and corresponding layout are shown in Fig. 9 and Fig. 10.

Fig. 9. The vehicle virtual aperture radar system and experiment scene.

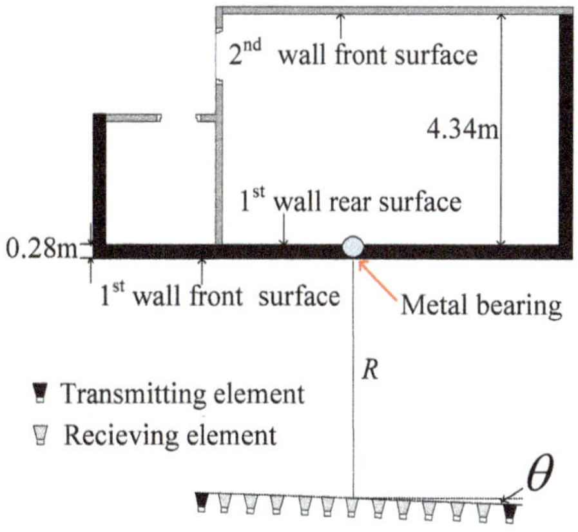

Fig. 10. Experiment layout.

The antenna array and the 1st wall are center-aligned. The distance from the array center to the wall is 11 m, and the measured inclination angle is –3.22°. The thickness of the 1st wall is 0.28 m, but the dielectric constant is unknown. In addition, there is a metal bearing in the center of the 1st wall, and we will see its shadowing effect in the later imaging result.

The derivation in Section 3 shows that the environmental parameters estimation depends on the echo time delay estimation of the front surface and rear surface. For the front surface echo, there are only reflections on the air-wall interface, so the dispersion effect is not serious. Therefore, by extracting the peak position of the echo processed by the matched-filter, we can obtain an effective echo time delay estimation of the front surface.

After the echo time delay estimation of the front surface is finished, we can obtain the relative position of the array and the 1st wall by $\hat{R}$ and $\hat{\theta}$. Using these two environmental parameters to build the wall-coordinate-system, we can achieve the BP imaging result without compensation, as shown in Fig. 11.

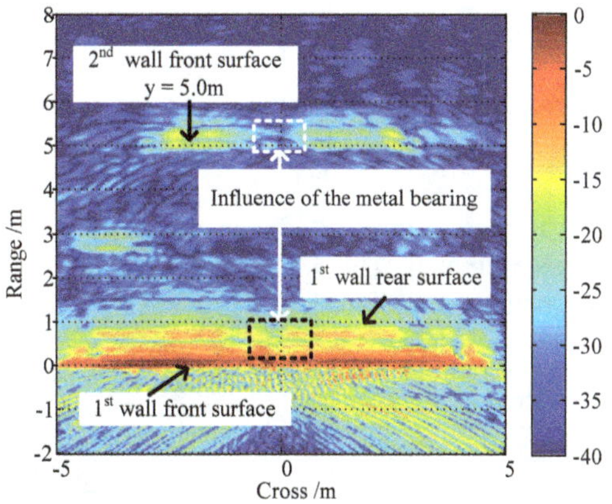

Fig. 11. BP imaging result without compensation under the wall-coordinate-system.

In Fig. 11, we obtain a relatively clear image of the building's inner structure. To show the proposed environmental parameters in this paper, we pay close attention to the 1st wall and the 2nd wall. By the shielding effect of the metal bearing, we find that the 1st wall rear surface and the

2nd wall have suffered different degrees of fault. Then, the range position of the 2nd wall front surface under the wall-coordinate-system should be 4.62 m from Fig. 6. However, it is approximately 5.0 m in Fig. 11. That is to say, the 2nd wall front surface lags 0.38 m because of the 1st wall. The 1st wall thickness is 0.28 m; thus, the deduced dielectric constant is 5.5561. This value can be used as a reference value to assess the final estimation result.

When estimating the echo time delay of the 1st wall rear surface, we have to consider the dispersion effect caused by the wall. Dispersion reduces the correlation between the echo and transmitting signal greatly, and this means that we cannot obtain an effective estimation of the 1st wall rear surface echo time delay from the traditional matched-filtering echo [28]. Protiva et al. proposed that the echo time-delay can be estimated by subspace decomposition on the deconvoluted echo data [29]. Based on this idea and referencing the frequency domain deconvolution methods, which are widely used in ultrasonic detection [30], we adopt the frequency domain Wiener inverse filter to complete deconvolution of the original echo data. Then, the maximum entropy (ME) power spectrum [31] is used to estimate the echo time delay of the 1st wall rear surface.

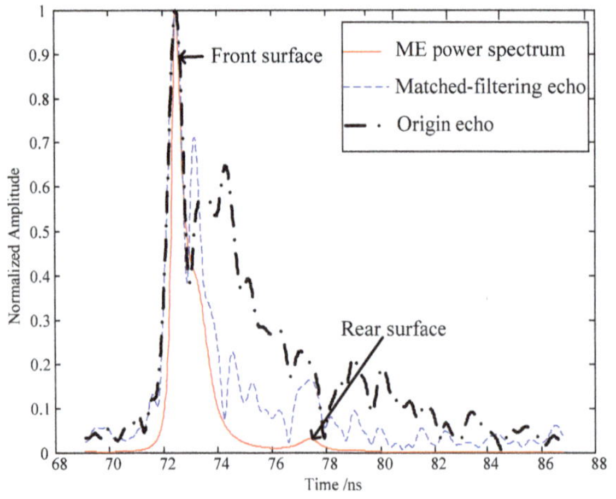

Fig. 12. Time delay estimation.

Fig. 12 shows the results of time delay estimation. With the help of deconvolution, the 1st wall rear surface echo becomes easy to identify in the ME power spectrum. Then, from Section 3.2, the wall thickness and dielectric constant can be estimated. The whole environmental estimation results are shown in Tab. 3.

	R (m)	θ (°)	d (m)	ε_r
Estimates	10.9256	-3.5409	0.2884	5.2273
True value	11.0000	-3.22	0.2800	5.5561
Relative error	-0.68%	9.97%	3.00%	-5.92%

Tab. 3. Estimation results of the environmental parameters.

Finally, we use these environmental parameters to obtain the compensation BP imaging for the echo data:

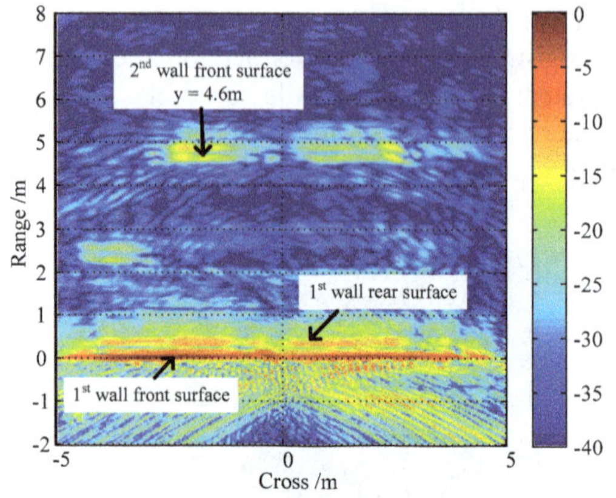

Fig. 13. Compensation BP imaging result.

The image entropy is commonly used to evaluate the image quality [32], and its definition is as follows:

$$H = \frac{\left(\sum_x \sum_y |I(x,y)|^2\right)^2}{\sum_x \sum_y |I(x,y)|^4}. \tag{41}$$

Usually, the smaller H is, the better the image quality is. Then, we use image entropy to evaluate the image with/without compensation, and the results are shown in Tab. 4.

	Image without compensation	Image with compensation
Image entropy	12161	9145

Tab. 4. The comparison of the image entropy in the image with/without compensation.

Combining Fig. 13 and Tab. 4, we find that by the compensation of environmental parameters, the position deviation of the scene behind the outer wall has been effectively corrected, and the image quality has also improved.

7. Conclusion

In this paper, an extended through-the-wall imaging model and its associated environmental parameters estimation algorithm are presented for the virtual aperture radar system. Because it has no special requirements for the array attitude, this imaging model is more suitable for the actual situation. Simultaneously, without any extra measurements or behind-the-wall dominant scatterers to assist, the environmental parameters estimation algorithm is of low computational complexity and easy to implement. The processing results of the measured data show the improvements to the through-the-wall imaging.

Furthermore, the environmental parameters estimation algorithm is tested on a single wall. In fact, it has the potential to estimate the parameters of multi walls, and this will be our future work.

Acknowledgements

This work is supported in part by the National Natural Science Foundation of China under Grant 61271441 and 61372161, and the research project of National University of Defense Technology under Grant CJ12-04-02.

Appendix

We mentioned in Section 2 that the element pair *AB*, which inclines the wall, can be equivalent to a virtual element pair $A'B'$, which parallels the wall. Here, we will prove this. Before proving, we reaffirm the relevant definitions and assumptions:

- m^{th} *Tx* is located at *A*, and its distance to the wall is R_m.
- n^{th} *Rx* is located at *B*, and the length of *AB* is L_{mn}.
- To simplify the proving process, we define that $\alpha_{mn} = -1$ when m^{th} *Tx* is further than n^{th} *Rx* from the wall. Otherwise, $\alpha_{mn} = 1$.

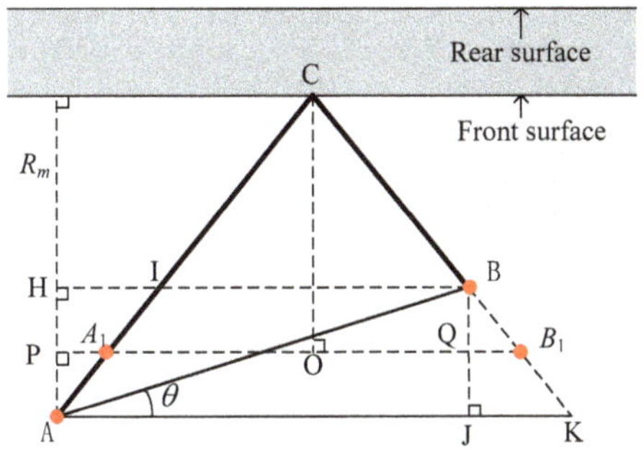

Fig. 14. The front surface echo path.

Fig. 14 shows the echo path of the front surface AK, and BH is parallel to the wall. *K* is the intersection of *CB*'s extension line and *AK*. *I* is the intersection of *BH* and *AC*. *P* is the midpoint of *AH*, and B_1 is the midpoint of *BK*. $BJ \perp AK$. A_1 is the intersection of B_1P and *AI*, and *Q* is the intersection of B_1P and *BJ*. $CO \perp B_1P$.

$\because$ reflection law

$\therefore |AC| = |KC| \,\&\, |IC| = |BC|$

$\therefore |AI| = |KB|$

$\because$ *P* & B_1 are the midpoints of *AH* & *BK*, respectively.

$\therefore PB_1 \,//\, AK \,//\, BH$

$\therefore A_1$ & *Q* are the midpoints of *AI* & *BJ*, respectively.

$\therefore |AA_1| = |BB_1|$

Then, we have:

$$|AC| + |BC| = |A_1C| + |B_1C| \tag{42}$$

Therefore, *AB* is equivalent to A_1B_1 for the length of the rear surface echo path. Furthermore:

$$\Delta APA_1 \cong \Delta BQB_1 . \tag{43}$$

Then, for the isosceles ΔA_1CB_1

$$|A_1B_1| = |PQ| = L_{mn}\cos\theta , \tag{44}$$

$$|OC| = R_m + 0.5\alpha_{mn}L_{mn}\sin|\theta| . \tag{45}$$

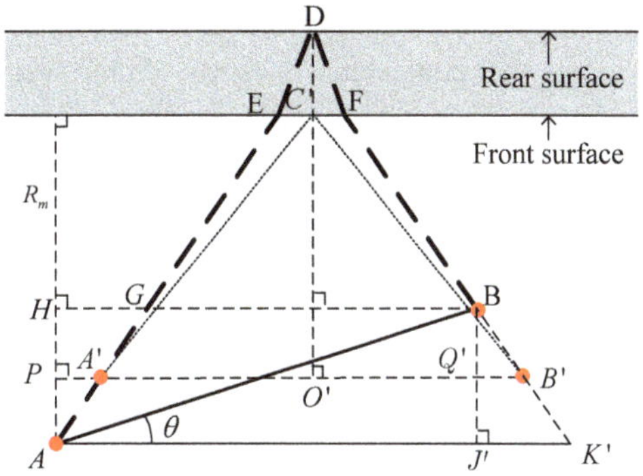

Fig. 15. The rear surface echo path.

Fig. 15 shows the echo path of the rear surface. AK' & *BH* are parallel to the wall. K' is the intersection of *AK* and the extension line of *FB*. *G* is the intersection of *BH* and *AE*. $BJ' \perp AK$, P & B' are the midpoints of *AH* & BK', respectively. A' is the intersection of $B'P$ and *AG*, and Q' is the intersection of $B'P$ and BJ'. *D* is the reflection point on the rear surface, $DO' \perp BH$, and C' is the intersection point of DO' and the front surface.

$\because$ reflection law & refraction law

$\therefore |AE| = |K'F| \,\&\, |GE| = |BF|$

$\therefore |AG| = |BK'|$

$\because$ *P* & *B'* are the midpoints of *AH* & BK', respectively.

$\therefore PB' \,//\, A'K \,//$ BH

$\therefore A'$ & Q' are the midpoints of AG & BJ', respectively.

$\therefore |AA'| = |BB'|$

Then, we have:

$$|AE| + |ED| + |FD| + |BF| = |A'E| + |ED| + |FD| + |B'F| \tag{46}$$

Therefore, *AB* is equivalent to *A'B'* for the length of the rear surface echo path. Furthermore:

$$\Delta APA' \cong \Delta BQB' . \tag{47}$$

Then, for the isosceles $\Delta A'C'B'$:

$$|A'B'| = |PQ'| = L_{mn}\cos\theta , \tag{48}$$

$$|O'C'| = R_m + 0.5\alpha_{mn}L_{mn}\sin|\theta| . \tag{49}$$

According to (44) and (45), we get:

$$\Delta A'C'B' \cong \Delta A_1CB_1 . \quad (50)$$

Consequently:

$$|AC|+|BC|=|A_1C|+|B_1C|=|A'C'|+|B'C'| . \quad (51)$$

So, AB is equivalent to $A'B'$ for the length of the front surface echo path.

In summary, for the path length of the echo from front and rear surface, m^{th} Tx and n^{th} Rx, whose connection AB inclines the wall, can be equivalent to a virtual element pair $A'B'$, which parallels the wall. The equivalent spacing is $L_{mn}\cos\theta$, and the equivalent distance to the wall is $R_m + 0.5\alpha_{mn}L_{mn}\sin|\theta|$ while $\alpha_{mn} = -1$ when m^{th} Tx is further than n^{th} Rx away from the wall. Otherwise, $\alpha_{mn} = 1$.

References

[1] CHETTY, K., SMITH, G. E., WOODBRIDGE, K. Through-the-wall sensing of personnel using passive bistatic wifi radar at standoff distances. *IEEE Transactions on Geoscience and Remote Sensing,* 2012, vol. 50, no. 4, p. 1218–1226.

[2] SU, Y. J. The research on receiver technology of the through-the-wall surveillance radar. *Journal of China Academy of Electronics and Information Technology,* 2011, vol. 6, no. 6, p. 648–651.

[3] LI, J., ZENG Z., SUN, J., LIU, F. Through-wall detection of human being's movement by UWB radar. *IEEE Geoscience and Remote Sensing Letters,* 2012, vol. 9, no. 6, p. 1079–1083.

[4] AFTANAS, M., DRUTAROVSKY, M. Imaging of the building contours with through the wall UWB radar system. *Radioengineering,* 2009, vol. 18, no. 3, p. 258–264.

[5] CHANG, P. C. Physics-based inverse processing and multi-path exploitation for through-wall radar imaging. *Doctor of Philosophy Thesis.* USA, Ohio State University, 2011.

[6] BARANOSKI, E. J. Through-wall imaging historical perspective and future directions. In *Proceedings of the IEEE International Conference on Acoustics, Speech and Signal Processing.* Las Vegas (USA), 2008, p. 5173–5176.

[7] SISMA, O., GAUGUE, A., LIEBE, CH., OGIER, J. M. UWB radar: vision through the wall. *Telecommun. Syst.,* 2008, vol. 38, no. 1 - 2, p. 53–59.

[8] SONG, L. P., YU, C., LIU, Q. H. Through wall imaging (TWI) by radar: 2-D tomo graphic results and analyses. *IEEE Transactions on Geoscience and Remote Sensing,* 2005, vol. 43, no. 12, p. 2793–2798.

[9] DEBES, C., AMIN, M. G., ZOUBIR, A. M. Target detection in single- and multiple-view through-the-wall radar imaging. *IEEE Transactions on Geoscience and Remote Sensing,* 2009, vol. 47, no. 5, p. 1349–1361.

[10] SHOUHEI, K., TAKUYA S., TORU, S. High resolution 3-D imaging algorithm with an envelope of modified spheres for UWB through-the-wall radars. *IEEE Transactions on Antennas and Propagation,* 2009, vol. 57, no. 11, p. 3520–3529.

[11] BROWNE, K. E., BURKHOLDER, R. J., VOLAKIS, J. L. Fast optimization of through-wall radar images via the method of Lagrange multipliers. *IEEE Transactions on Antennas and Propagation,* 2013, vol. 61, no. 1, p. 320–328.

[12] CHEN, P. H., NARAYANAN, R. M. Shifted pixel method for through-wall radar imaging. *IEEE Transactions on Antennas and Propagation,* 2012, vol. 60, no. 8, p. 3706–3716.

[13] WANG, Y., FATHY, A. E. Advanced system level simulation platform for three-dimensional UWB through-wall imaging SAR using time-domain approach. *IEEE Transactions on Geoscience and Remote Sensing,* 2012, vol. 50, no. 5, p. 1986–2000.

[14] LIU, X., LEUNG, H., LAMPROPOULOS, G. A. Effect of wall parameters on ultra-wideband synthetic aperture through-the-wall radar imaging. *IEEE Transactions on Aerospace and Electronic Systems,* 2012, vol. 48, no. 4, p. 3435–3449.

[15] LI, L., ZHANG, W., LI, F. A novel autofocusing approach for real-time through-wall imaging under unknown wall characteristics. *IEEE Transactions on Geoscience and Remote Sensing,* 2010, vol. 48, no. 1, p. 423–431.

[16] JIN, T., CHEN, B., ZHOU, Z. Image-domain estimation of wall parameters for autofocusing of through-the-wall SAR imagery. *IEEE Transactions on Geoscience and Remote Sensing,* 2013, vol. 51, no. 3, p. 1836–1843.

[17] WANG, G. Y., AMIN, M. G. ZHANG, Y. M. New approach for target locations in the presence of wall ambiguities. *IEEE Transactions on Aerospace and Electronic System,* 2006, vol. 42, no. 1, p. 301–315.

[18] WANG, G. Y., AMIN, M. G. Imaging through unknown walls using different standoff distances. *IEEE Transactions on Signal Processing,* 2006, vol. 54, no. 10, p. 4015–4025.

[19] SAGNARD, F., ZEIN, G. E. In situ characterization of building materials for propagation modeling: frequency and time responses. *IEEE Transactions on Antennas and Propagation,* 2005, vol. 53, no. 10, p. 3166–3173.

[20] LI, X., HUANG, X., JIN, T. Estimation of wall parameters by exploiting correlation of echoes in time domain. *Electronics Letters,* 2010, vol. 46, no. 23, p. 1563–1564.

[21] AFTANAS, M., SACHS, J., DRUTAROVSKY, M., KOCUR, D. Efficient and fast method of wall parameter estimation by using UWB radar system. *Frequenz Journal,* 2009, vol. 63, no. 11 - 12, p. 231–235.

[22] PROTIVA, P., MRKVICA, J., MACHAC, J. Estimation of wall parameters from time-delay-only through-wall radar measurements. *IEEE Transactions on Antennas and Propagation,* 2011, vol. 59, no. 11, p. 4268–4278.

[23] JIN, T., CHEN, B., ZHOU, Z. Estimation of wall parameters for cognitive imaging in through-the-wall radar. In *Proceedings of the 2012 IEEE 11th International Conference on Signal Processing.* Beijing (China), 2012, p. 1936–1939.

[24] MCCORKLE, J. W. Focusing of synthetic aperture ultra wideband data. In *Proceedings of the IEEE International Conference on Systems Engineering.* Dayton (USA), 1991, p. 1–5.

[25] WANG, H. J., HUANG, C. L., LU, M., SU, Y. Back projection imaging algorithm for MIMO radar. *Systems Engineering and Electronics (China),* 2010, vol. 32, no. 8, p. 1567 - 1673.

[26] AHMAD, F., AMIN, M. G., KASSAM, S. A. Synthetic aperture beamformer for imaging through a dielectric wall. *IEEE Transactions on Aerospace and Electronic Systems,* 2005, vol. 41, no. 1, p. 271–283.

[27] JIA, Y., KONG, L., YANG, X. Improved cross-correlated back-projection algorithm for through-wall-radar imaging. In *Proceedings of the 2013 IEEE Radar Conference.* Ottawa (Canada), 2013, p. 1–3.

[28] WEISS, L. G. Wavelets and wideband correlation processing. *IEEE Signal Processing Magazine,* 1994, vol. 11, no. 1, p. 13–32.

[29] PROTIVA, P., MRKVICA, J., MACHAC, J. Time delay estimation of UWB radar signals backscattered from a wall. *Microwave & Optical Technology Lett.,* 2011, vol. 53, no. 6, p. 1444 to 1450.

[30] ALI, M. G. S., ELSAYED, N. Z., EBEID, M. R. Signal processing of ultrasonic data by frequency domain deconvolution. *Walailak*

Journal of Science and Technology, 2013, vol. 10, no. 3, p. 297 to 304.

[31] QIU, T. S., WANG, H. Y. A high time delay estimation based on the maximum entropy power spectrum estimation. *Journal of Electronics (China),* 1997, vol. 14, no. 3, p. 279–284.

[32] LI, L., ZHANG, W., LI, F. A novel autofocusing approach for real-time through-wall imaging under unknown wall characteristics. *IEEE Transactions on Geoscience and Remote Sensing,* 2010, vol. 48, no. 1, p. 423–431.

About Authors ...

YONGPING SONG received his B.S. degree in Electronic Engineering from the National University of Defense Technology, Changsha, China in 2012, and now he is studying for the M.S. degree in Information and Communication Engineering. His research interests include radar imaging and automatic target detection.

TIAN JIN received B.S., M.S. and Ph.D. degrees in Information and Communication Engineering from the National University of Defense Technology, Changsha, China, in 2002, 2003, and 2007, respectively. He is currently an associate professor of the National University of Defense Technology. His Ph.D. dissertation was awarded as the National Excellent Doctoral dissertation of China in 2009. His fields of interest include radar imaging, automatic target detection, and machine learning.

5

Analysis of Energy Consumption Performance towards Optimal Radioplanning of Wireless Sensor Networks in Heterogeneous Indoor Environments

Peio LÓPEZ ITURRI[1], *Leire AZPILICUETA*[1], *Juan Antonio NAZABAL*[1], *Carlos FERNÁNDEZ-VALDIVIELSO*[1], *Jesús SORET*[2], *Francisco FALCONE*[1]

[1] Dept. of Electrical and Electronic Engineering, Public Univ. of Navarre, Campus de Arrosadía, 31006, Pamplona, Spain
[2] Dept. of Electrical and Electronic Engineering, University of Valencia, Burjassot, Valencia, Spain

{peio.lopez, leyre.azpilicueta, juanantonio.nazabal, carlos.fernandez, francisco.falcone} @unavarra.es, jesus.soret@uv.es

Abstract. *In this paper the impact of complex indoor environment in the deployment and energy consumption of a wireless sensor network infrastructure is analyzed. The variable nature of the radio channel is analyzed by means of deterministic in-house 3D ray launching simulation of an indoor scenario, in which wireless sensors, based on an in-house CyFi implementation, typically used for environmental monitoring, are located. Received signal power and current consumption measurement results of the in-house designed wireless motes have been obtained, stating that adequate consideration of the network topology and morphology lead to optimal performance and power consumption reduction. The use of radioplanning techniques therefore aid in the deployment of more energy efficient elements, optimizing the overall performance of the variety of deployed wireless systems within the indoor scenario.*

Keywords

Radioplanning, wireless sensor networks, energy consumption, ray launching, CyFi.

1. Introduction

The use of wireless sensor networks is growing rapidly into a large number of fields of application, such as industrial monitoring, farming and agriculture, structural monitoring, health assistance, location and guiding or security and defense, among others [1-8]. The use of these wireless sensor networks within domestic environments is leading towards the fast paced development of the so called smart homes, linked with the more global concept of Internet of Things. One of the key issues is to reduce energy consumption of the individual elements of these wireless sensor networks, due to the fact that in the near future, a great deal of these devices will be operating within a conventional indoor environment. This is in line with ambitious energy reduction strategies, such as those stated in Europe 20/20 strategy.

Typically, the deployment of wireless systems is performed by initial coverage estimations (usually by empirical based models) which can later on be validated by field measurements. These field measurements, in the case of WLAN/WPAN systems are performed by using sniffers or protocol analyzers in order to obtain estimations of RSSI values and initial metrics of link level quality, such as Packet Error Ratio levels. In the case of mobile networks, performance analysis is conducted typically by means of test drives and Key Performance Indicator validation, based on mobile terminal tracing as well as by correlation to network statistics managed by the radio subsystem. Even though these approaches give an initial point to validate network operation, issues such as intra-system or inter-system interference are not considered, as well as the large variability in signal strength and quality due to the strong multipath characteristics inherent to indoor scenarios. Furthermore, not only coverage should be considered but also dynamic variation due to changes in traffic demands (and hence, in overall interference values) should be taken into account, leading to coverage-capacity relations. This is a relevant issue that is gaining importance as a larger amount of wireless networks are coexisting, leading to a heterogeneous wireless environment.

Moreover, minimizing energy consumption has become one of the main goals, driven by international Green policies. In this context, radio channel features of indoor environments pose a challenge to energy consumption, due to the fact that the complexity of the scenario increases losses due to strong multipath propagation and multi-screen diffraction, as well as absorption due to lossy dispersive media, as it is shown in previous works [9]. The existence of interference sources in these complex environments also affects the deployment strategies and overall power consumption [10].

In this paper, the topological influence of a layout of in-house developed CyFi based wireless sensors will be analyzed in terms of power consumption and radio coverage. For that purpose, the characteristics of the CyFi motes are presented in Section 2. Then, in Section 3, the analysis of the considered scenario by means of an in-house developed 3D ray tracing simulation tool is presented, showing radioplanning results as received power planes, power delay profiles or current consumption planes. Finally, in Section 4, received power and consumption measurements are presented for different test cases, showing the dependence between network topology and power consumption. In conclusion, the application of deterministic radioplanning approaches, like the method presented in this work, lead to an optimal network configuration, minimizing energy consumption and achieving desired quality of service.

2. CyFi Wireless Devices

For the purpose of this work, a system based on a set of wireless motes has been designed in house by the University of Valencia. Each mote includes sensor/actuator elements, a PSoC processor core, expansion ports and power, and a CyFi transceiver. Depending on the role that the nodes play in the protocol, the motes can be configured as a master or as a slave.

The basic network topology is a star configuration, in which a master node manages a certain number of peripheral slave nodes. Following a hierarchical scheme, different master nodes can be connected at slave-type stages to form a second layer around a master node that is responsible for monitoring the platform. Each mote has two parts: a main card which contains the microcontroller, and an additional one that contains the radio frequency part. The main board consists of different blocks, as shown schematically in Fig. 1. The real implementation can be seen in Fig. 2.

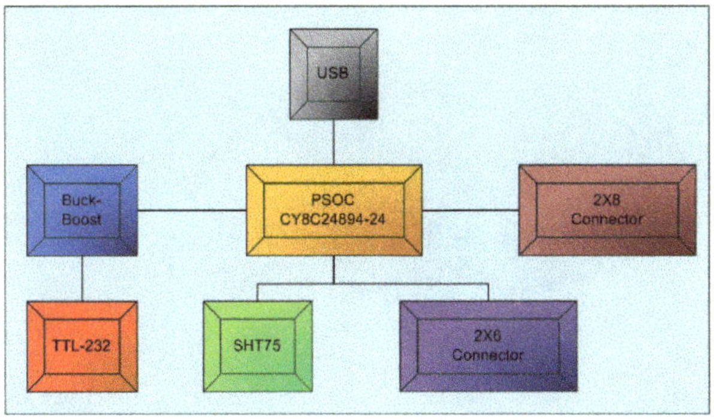

Fig. 1. Block diagram for the implemented mote device.

Fig. 2. Image of the main board (left) and the radio frequency module (right) of the mote device.

The microcontroller is the Cypress CY8C24894 PSoC. It incorporates an 8-bit microcontroller M8C and up to 4 MIPS. As reconfigurable elements, it contains 4 digital blocks and 6 analog blocks, in addition to 1KB of SRAM and 16KB of Flash. Also, the device has the possibility to communicate via USB without requiring any additional items. The humidity and temperature sensor used is an SHT75 model. This is calibrated at the factory, controlled digitally, has high resolution and accuracy, with an operating range of 0-100% relative humidity and a temperature range between -40ºC and 123ºC.

The used CyFi technology is a cost effective low-power wireless solution developed by Cypress Semiconductor that operates in the unlicensed 2.4 GHz ISM band, with active link and power management features. The network topology consists in a simple star network controlled by a central hub. Due to the lightweight network protocol stack of CyFi nodes, a bidirectional communication to up to 250 nodes is provided. CyFi output maximum power is 4 dBm and the receiver sensitivity -97 dBm, with a typical range in a line of sight, interference-free environment between 50 m and 70 m.

The mentioned active link and power management provide interesting dynamic functionalities. For more robustness, the transmitter changes the modulation and data rates dynamically between 1 Mbps and 250 Kbps depending on the environment's interference. If the interference increases, then the power output level is dynamically increased to overcome that interference. Besides, if a node detects that its power output is excessive, it will dynamically reduce it, reducing power consumption. Also, to ensure that the central hub of the network receives packets correctly, an interference free channel is selected whenever possible. If the hub detects a noisy channel, a CyFi network will look for a clean channel and settle there. The dynamic power handling capability, in terms of coverage/capacity estimation, will lead to smaller coverage radius as overall transmission speed increases, as will be shown in the following sections.

3. Indoor Scenario Analysis

In order to perform estimations of the influence of the indoor environment in a wireless sensor network, radioplanning simulation results have been obtained. For that purpose, radiopropagation analysis can be performed by means of empirical methods (such as COST-231, Walfish-Bertoni, Okumura Hata, etc.) [11-13], based on statistical approaches and non-linear regression techniques. They give rapid results but require calibration based on measurements to give an adequate fit of the results. On the other hand, deterministic methods [14-20] are based on numerical approaches to the resolution of Maxwell's equations, such as ray launching and ray tracing (based on geometrical approximations) or full-wave simulation techniques (method of moment (MoM), finite difference time domain (FDTD) [21], FITD, etc.). These methods are precise but are time-consuming to inherent computational complexity.

As a midpoint, methods based on geometrical optics, for radioplanning calculations with strong diffractive elements, offer a reasonable trade-off between precision and required calculation time [22]. The Ray Tracing method combined with uniform theory of diffraction (UTD) [23] is most frequently applied to radio coverage prediction [24-27]. The Ray Tracing models, including modifications as reception sphere technique [28], potentially represent the most accurate and versatile methods for urban and indoor multipath propagation characterization or prediction [29-33].

3.1 Simulation Technique

This work presents an in-house developed 3D Ray Launching algorithm to analyze the influence of the indoor environment in the propagation of electromagnetic signals, validated in previous works [34-38]. The novelty of the proposed method is that it takes into account all the obstacles within the scenario, with their different shapes and material properties. It is important to emphasize that a grid is defined in the space to save the parameters of each ray. Accordingly, the environment is divided into a number of cuboids of a fixed size. When a ray enters a specific hexahedron, its parameters are saved in a matrix. Electromagnetic phenomena such as reflection, refraction and diffraction are taken into account, based on Geometrical Optics and Geometrical Theory of Diffraction. Firstly, an indoor scenario has been created, taking into account the material parameters of all of the elements within it (e.g., furniture, walls, windows, etc.), in terms of conductivity and dielectric permittivity. Electromagnetic phenomena such as reflection, refraction and diffraction have been taken into consideration.

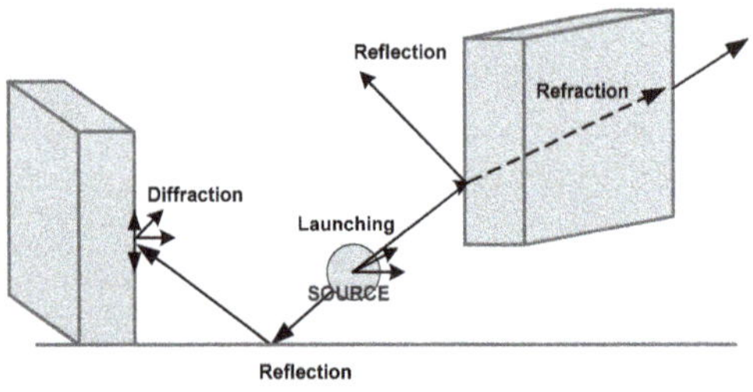

Fig. 3. Principle of Ray Launching method.

Fig. 3 shows the principle of ray launching method. The equivalent transmitter antenna located in the simulation scenario launches rays in different directions following the radiation pattern of the antenna. The reflection and refraction coefficients are calculated using the well-known Fresnel's equations by

$$T^{\perp}=\frac{E_t^{\perp}}{E_i^{\perp}}=\frac{2\eta_2\cos(\Psi_i)}{\eta_2\cos(\Psi_i)+\eta_1\cos(\Psi_t)}, \tag{1}$$

$$R^{\perp}=\frac{E_r^{\perp}}{E_i^{\perp}}=\frac{\eta_2\cos(\Psi_i)-\eta_1\cos(\Psi_t)}{\eta_2\cos(\Psi_i)+\eta_1\cos(\Psi_t)}, \tag{2}$$

$$R^{\parallel}=\frac{E_r^{\parallel}}{E_i^{\parallel}}=\frac{\eta_1\cos(\Psi_i)-\eta_2\cos(\Psi_t)}{\eta_1\cos(\Psi_i)+\eta_2\cos(\Psi_t)}, \tag{3}$$

$$T^{\parallel}=\frac{E_t^{\parallel}}{E_i^{\parallel}}=\frac{2\eta_2\cos(\Psi_i)}{\eta_1\cos(\Psi_i)+\eta_2\cos(\Psi_t)} \tag{4}$$

where $\eta_1=120\pi/\sqrt{\varepsilon_{r1}}$, $\eta_2=120\pi/\sqrt{\varepsilon_{r2}}$ and Ψ_i,Ψ_r,Ψ_t are the incident, reflected and transmitted angles respectively.

The diffraction coefficients are considered by the Uniform Theory of Diffraction (UTD) [39-40] as follows

$$D^{\parallel\perp}=\frac{-e^{(-j\pi/4)}}{2n\sqrt{2\pi k}}\left\{\begin{array}{l}\mathrm{cotg}\left(\frac{\pi+(\Phi_2-\Phi_1)}{2n}\right)F\left(kLa^{+}(\Phi_2-\Phi_1)\right)\\+\mathrm{cotg}\left(\frac{\pi-(\Phi_2-\Phi_1)}{2n}\right)F\left(kLa^{-}(\Phi_2-\Phi_1)\right)\\+R_0^{\parallel\perp}\mathrm{cotg}\left(\frac{\pi-(\Phi_2+\Phi_1)}{2n}\right)F\left(kLa^{-}(\Phi_2+\Phi_1)\right)\\+R_n^{\parallel\perp}\mathrm{cotg}\left(\frac{\pi+(\Phi_2+\Phi_1)}{2n}\right)F\left(kLa^{+}(\Phi_2+\Phi_1)\right)\end{array}\right\} \tag{5}$$

where $n\pi$ is the wedge angle, Φ_2 and Φ_1 angles, F, L and $a\pm$ are defined in [39], $R_{0,n}$ are the reflection coefficients for the appropriate polarization for the 0 face or n face, respectively.

The commitment between accuracy and computational time is acquired with the number of launching rays and the cuboids size of the considered scenario. Several transmitters can be placed within an indoor scenario. Parameters such as frequency of operation, radiation patterns of the antennas, number of multipath reflections, separation angle between rays and cuboids dimension are introduced. Fig. 4 depicts ray launching method within a defined indoor scenario.

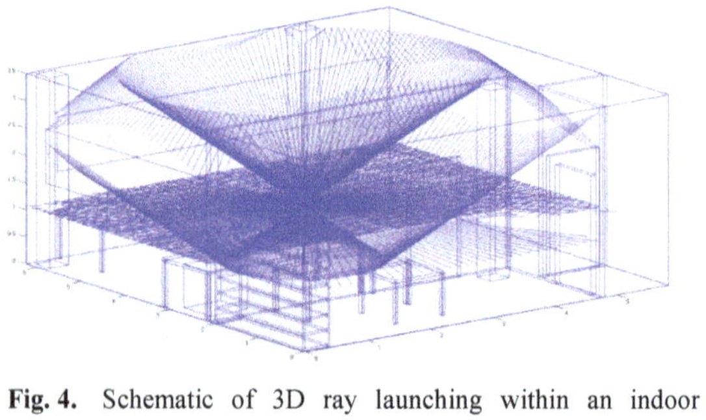

Fig. 4. Schematic of 3D ray launching within an indoor scenario.

The scenario that has been analyzed in this paper is the Radiocommunication Laboratory, placed in the Electric and Electronic Engineering Department of the Public University of Navarra. The scenario has the inherent complexity of an indoor scenario, as it has interior columns, many furniture elements, different types of instruments and walls made of different materials (wood, concrete, bricks, metal and glass). The scenario can be seen in Fig. 5a and its schematic representation for the ray launching software can be seen in Fig. 5b. Red points in Fig. 5b represent the different points where the wireless motes have been placed. These positions have been chosen in order to simulate a possible morphology of a real wireless network. For that reason, the motes have been placed at different heights. The exact coordinates for the motes are shown in Tab. 1.

(a)

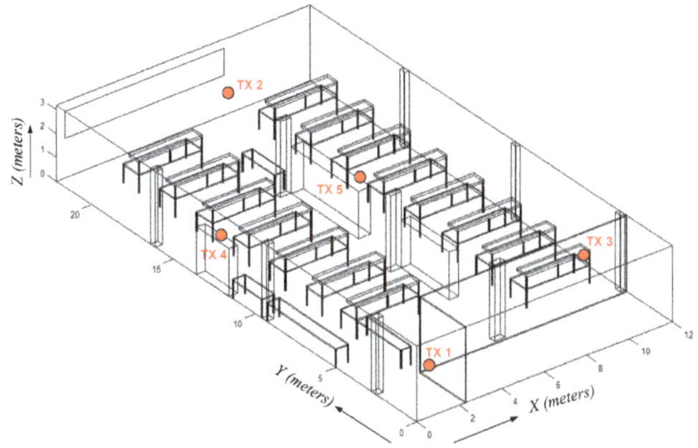

(b)

Fig. 5. (a) Indoor scenario under analysis, corresponding to Radiocommunication Laboratory, UPNA. (b) Schematic description of the scenario in the 3D ray launching software.

Transmitter	*Coordinates*
TX1	(2, 2, 0.81) m
TX2	(6.5, 21, 2.7) m
TX3	(11, 4, 1.5) m
TX4	(0.3, 12.5, 2.1) m
TX5	(7, 12.5, 2.1) m

Tab. 1. Coordinates where the wireless motes are placed in the indoor scenario.

3.2 Simulation Results

3D ray launching simulation results have been obtained for the whole volume of the simulation scenario. The parameters used in the simulation are the following: uniform cuboids resolution of 20 cm, vertical plane angle resolution $\Delta\theta = \pi/180$, horizontal plane angle resolution $\Delta\Phi = \pi/180$, maximum number of tolerated reflections $N = 5$, frequency of operation 2.4 GHz and power transmission of 4 dBm. The considered parameters are equivalent to those of a conventional ZigBee system. Fig. 6 shows the obtained received power levels for the same bidimensional plane at height 0.81 m (the same height as the tables within the laboratory) for different number of transmitters. For each of the represented planes (from Fig. 6a to Fig. 6e), a new transmitter has been added consecutively, starting with a single transmitter (Fig. 6a) and finishing with a wireless network composed by five transmitters (Fig. 6e).

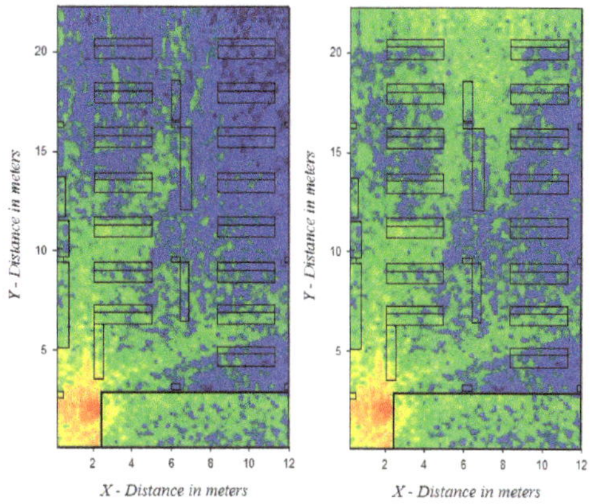

(a) (b)

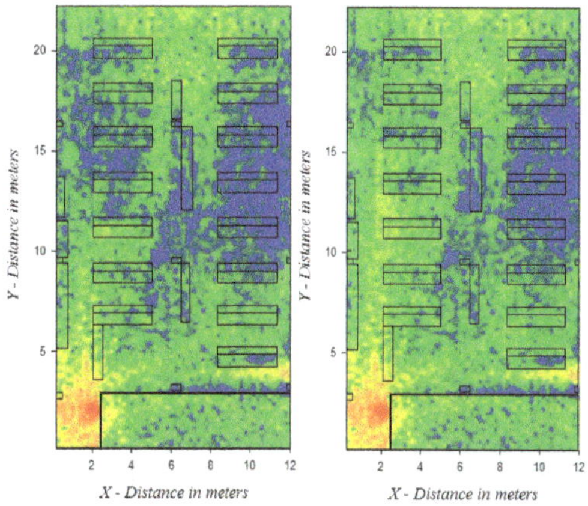

(c) (d)

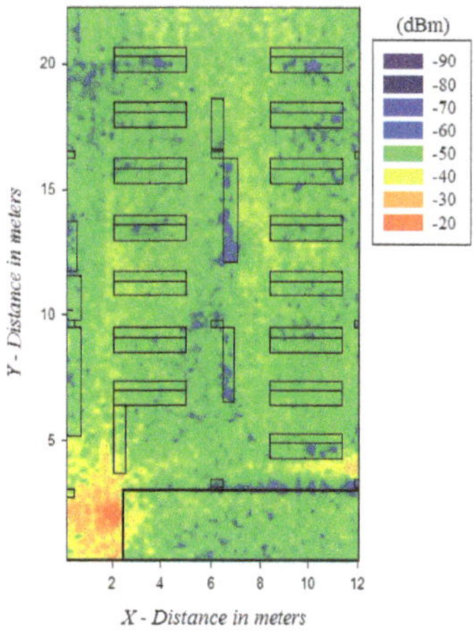

(e)

Fig. 6. RSSI 3D ray launching simulation results obtained at a bidimensional plane at a height of 0.81 m for different number of transmitters: (a) TX 1, (b) TX1 and TX2, (c) TX1, TX2 and TX 3, (d) TX1, TX2, TX3 and TX4, (e) TX1, TX2, TX3, TX4 and TX5.

As it can be seen from the previous figures, received power level is strongly dependent on the position of the potential receiver element and the morphology of the wireless network. Variations can be in order of 10 dB within 1 meter when the number of transmitters is low, which has a strong impact on the performance of the sensors, not only in terms of receiver sensitivity limits but also on overall system capacity, which is dependent on signal level as well as on signal to noise ratio. As it is shown in Fig. 6, this received power variations can be strongly mitigated changing the morphology (e.g. adding wireless motes) of the wireless network, thus obtaining a reasonable received power level for every position of the complete indoor scenario.

The multipath propagation is the strongest phenomenon in this type of complex indoor environments, hence, to appreciate the variability of estimated received power level more precisely, Fig. 7 shows this variability within a given line path for a fixed value of X for two different heights in the indoor scenario. The X value has been set to 3.5 m randomly, since this phenomenon happens all alike within the whole scenario. As stated above, the signal variation is driven by strong multipath components, as can be seen from the short term variation component within the received power level. For a more thorough analysis of the impact of the multipath propagation in the scenario, time domain results are shown in Fig. 8 and Fig. 9. Specifically, power delay profiles are presented for the locations of TX2 and TX3 respectively, when TX1 transmits. A red line has been depicted in both graphics to delimit the sensitivity level of the CyFi motes. As expected, a lot of components reached the TX2 and TX3 points due to the multipath propagation, but in Fig. 8, which corresponds to the farthest mode of the network (from TX1), there are a lot of components under the sensitivity level, due mainly the distance. On the other hand, in Fig. 9 most of the components are above the sensitivity level, as it corresponds to the nearest node of the network. In order to complete the time domain results, the delay spread for a plane of 0.81 m height (the height of the tables and TX1) is presented in Fig. 10.

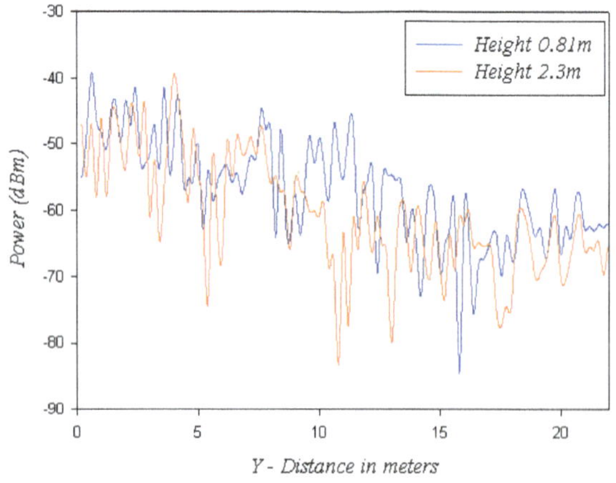

Fig. 7. Simulation results for height 0.81 m and height 2.3 m, for $X = 3.5$ m, along the Y-axis of the indoor scenario under analysis.

The obtained simulation results and mainly the estimated values of received power can lead to the analysis of the performance of the wireless system. As an example, Fig. 12 represents the signal to noise ratio (SNR) for two different heights in the same scenario, which could be used to consider the most adequate deployment strategy of a set

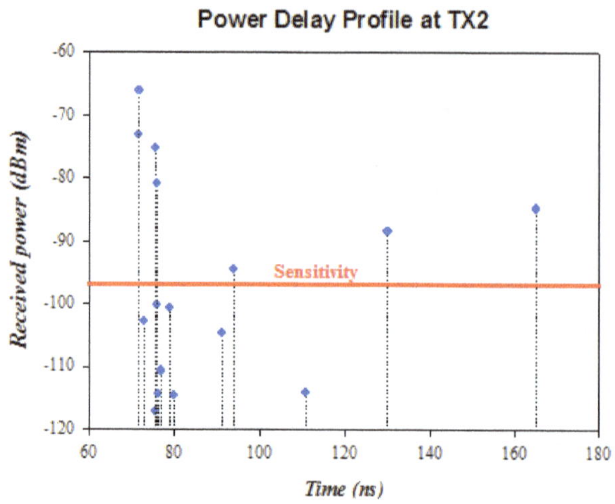

Fig. 8. Power Delay Profile at location of TX2 (the farthest node), while TX1 is transmitting.

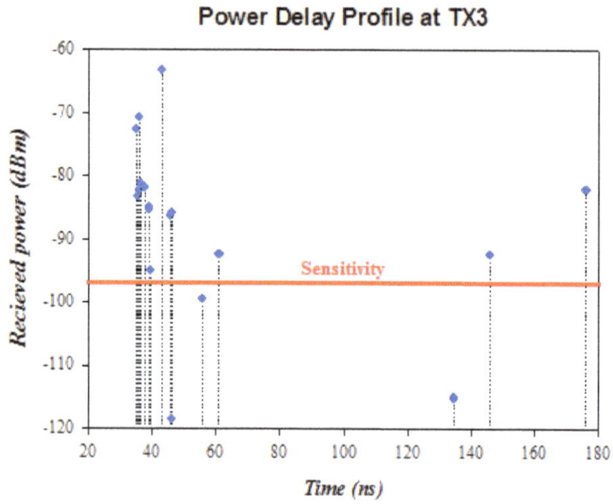

Fig. 9. Power Delay Profile at location of TX3 (the nearest node), while TX1 is transmitting.

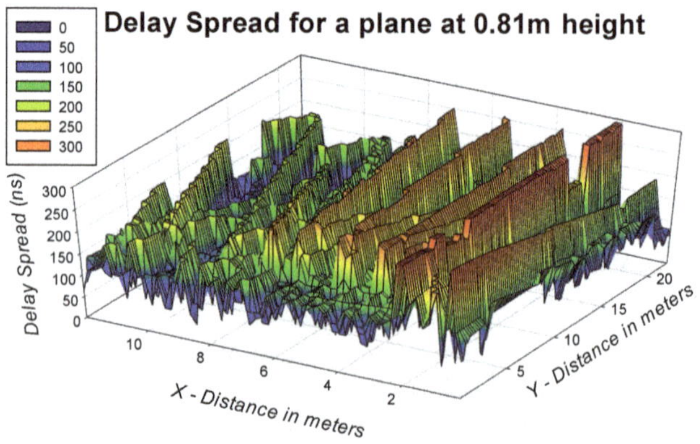

Fig. 10. Delay Spread values for a plane at 0.81 meters height (i.e. the height of TX1 within the scenario).

of wireless sensor networks within the indoor scenario. Specifically, SNR planes depicted in Fig. 12 have been calculated taking into account that the whole CyFi network is deployed (see Fig. 6e) and as noise sources, an interfering WiFi network (an access point and 3 laptops) operating at the same frequency band of the CyFi motes has been simulated. The simulation parameters have been the same than those used for the simulation of the CyFi motes, but the transmitted power of WiFi nodes has been set to 20 dBm, which corresponds to a typical maximum transmitted power of a commercial device. In Fig. 11 the schematic configuration of the wireless networks within the scenario is shown. Tab. 2 shows the position within the scenario of the WiFi nodes.

Device	*Coordinates*
WiFi access point	(3, 3, 2.5) m
Laptop1	(4.5, 6.9, 0.9) m
Laptop2	(2, 13.5, 0.9) m
Laptop3	(8.5, 17.7, 0.9) m

Tab. 2. Coordinates where WiFi devices have been placed within the scenario.

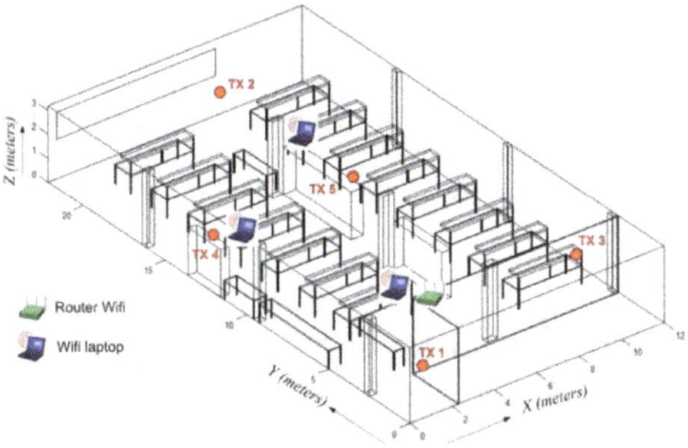

Fig. 11. Schematic representation of the scenario used to calculate the SNR planes of Fig. 12.

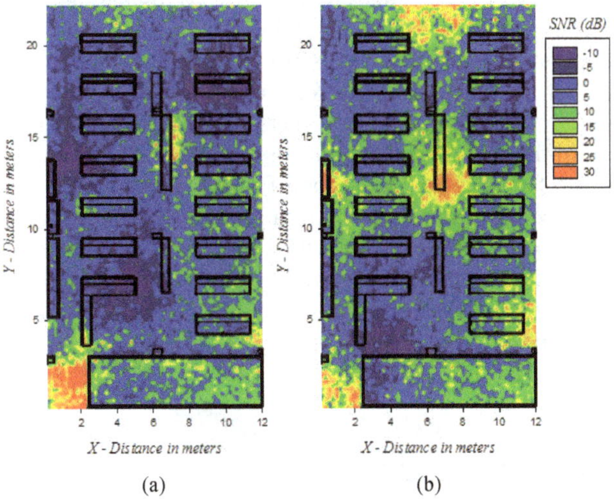

Fig. 12. Spatial distribution of signal to noise ratio in the indoor scenario of Fig. 5 for two different heights: (a) 0.81 meters, (b) 2.3 meters.

Thus, the SNR value is obtained for each point within the room, giving valuable information about the zones and points where the placement of a mote will be better in terms of received signal quality, whilst maintaining the optimal wireless power transmission (and hence energy consumption) of the system. As can be clearly seen in Fig. 12, the zones where interfering devices have been placed are the zones with lower SNR, as expected. For a more in-depth analysis of the proposed CyFi network in terms of SNR, in Fig. 13, the SNR at each receiver CyFi mote is depicted, when a single mote is transmitting. In Tab. 3 the preset transmission power levels for the CyFi motes are shown, which have been used for the calculation of the SNR, in order to show how it affects the SNR at the receiver motes. As the CyFi motes can change dynamically the transmission rate between 1 Mbps and 250 Kbps, the minimum SNR needed has been calculated for both data rates, and these limits have been depicted by red dashed lines (0 dB for 1 Mbps, and -7.23 dB for 250 Kbps). Fig. 13a shows the results for the worst noise case, i.e. when the WiFi devices transmit 20 dBm, whilst Fig. 13b shows the results for WiFi devices transmitting 0 dBm. These results show that the presented method can aid in an adequate deployment strategy within a harsh indoor scenario, which has a direct impact on the network efficiency.

Defined internal level	*Transmitted power* (dBm)
7	4
6	0
5	-5
4	-13
3	-18
2	-24
1	-30
0	-35

Tab. 3. CyFi motes' preset levels and their correspondent transmission power level.

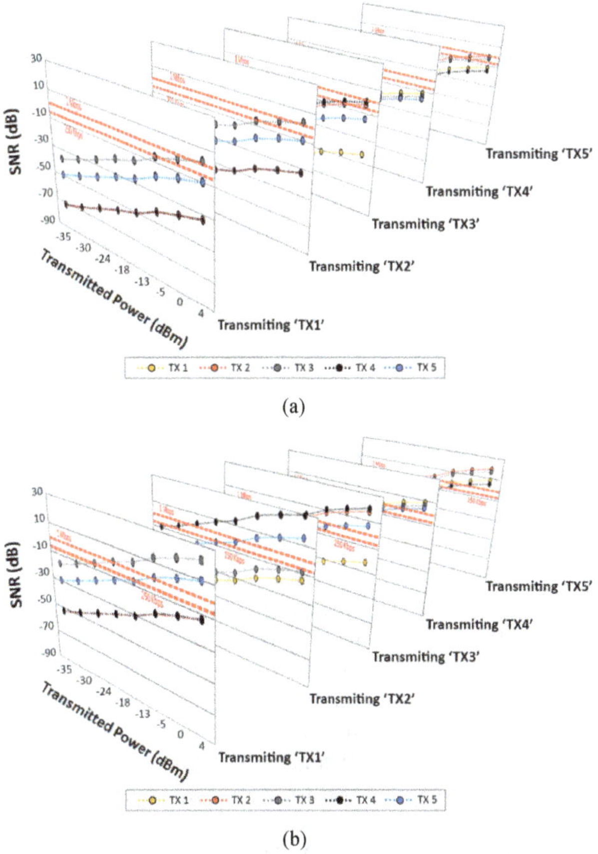

Fig. 13. SNR values at each CyFi mote position for a single transmitting mote, while WiFi-interference sources are transmitting (a) 20 dBm and (b) 0 dBm.

Considering the overall power consumption of the deployed motes is also a highly important issue in radio planning strategies. As it has been shown previously, the location of the transceiver has a significant role in terms of the variations of the expected value of received power within the scenario, which has a great impact on the power consumption of transmitting motes. This is given by the fact that as the received power varies, the link balance within the sensitivity threshold limit also varies, modifying the required current for the transceiver to operate. Therefore, it is possible to estimate energy consumption of the transmitter as a function of the receiver location. In order to gain insight on the effect of topology and morphology on energy consumption in the previous scenario, Fig. 14 shows the consumption increase maps for two cases: first when only two motes are operating (TX1 and TX2) and afterwards when the whole wireless network is operating (5 transmitters), which corresponds to the optimal configuration of the network for the presented indoor scenario.

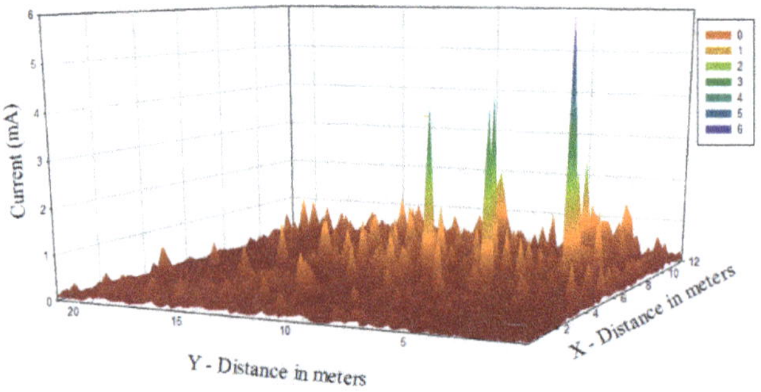

(a)

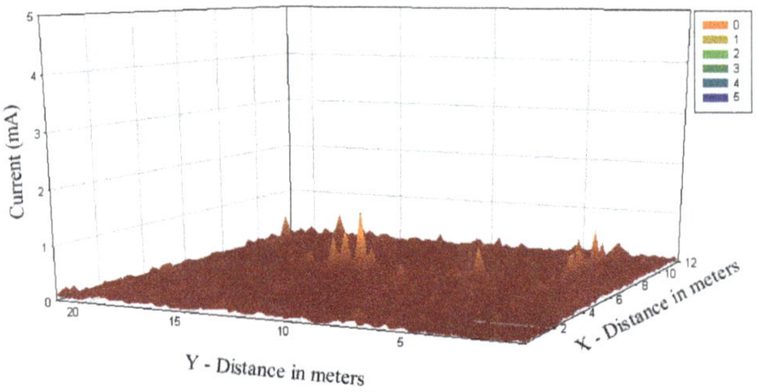

(b)

Fig. 14. Estimation of energy consumption in terms of current values in mA for different locations of the scenario depicted in Fig. 5.

These maps represent the overall increase of current consumption of the transmitting motes placed at the scenario for each possible receiver location. From the real measurements explained in the next section, it is shown that the lowest measured consumption of a mote when transmitting is 40.5 mA (distance between transmitter and receiver of 5 cm). So, as it can be seen in the consumption maps for the first case (Fig. 14a), a maximum current consumption increase of 6.03 mA is reached for a specific location (this maximum peak seems not to reach that value, but it is due to the perspective of the graph). This corresponds to the worst location for a receiver to be placed in terms of current consumption. This means that for that receiver location, the overall current consumption for the transmitters will increase in 6.03 mA due to the power level received at that location, which is equal to 14.8% more consumption. On the other hand, for the optimal case of 5 transmitters deployed (Fig. 14b), the worst receiver location implies an increase of 3.11% of the total consumption of the five transmitting motes of the network. This lower increase of consumption is expected as the received power level for the whole scenario is higher due to the higher amount of deployed transmitting nodes.

These results can be really useful in order to plan the design of the optimal network, taking into account the number of employed nodes, the required transmission bandwidth and the sensitivity level. Moreover, as it is shown in Fig. 14, the density of the nodes within the network has a clear impact on energy consumption, due to the fact that link balance limitations are lower when the whole network is operating. Nevertheless, it is important to achieve a commitment between the density of nodes and interference levels, because a larger number of nodes could lead to increased interference levels, which could degrade system performance.

4. Measurement Results of Deployed Wireless Sensors

In order to validate the previously obtained simulation estimations, in which the morphological dependence of the network performance in terms of received signal is observed, wireless CyFi motes have been configured and measured. For that purpose, power distribution and current consumption measurement results are presented. In Fig. 15, a layout of the tested setup is shown. The laboratory has two zones, separated by several metallic shelves. For the purpose of the study, the left hand zone has been measured, due to the fact that this is a zone of interaction with students and collaborators, leading to a realistic situation for the deployment and use of a wireless sensor network. The measurements have been performed by programming a test setup among the motes, given by a coordinator element and a wireless sensor.

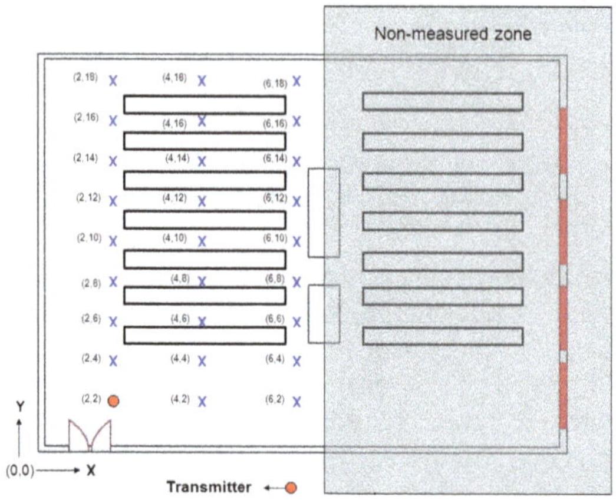

Fig. 15. Schematic of the indoor scenario (Public University of Navarre).

Initially, the RSSI values in different points of the scenario (Fig. 15) have been measured: The transmitter is located in coordinates (2, 2) and it is represented by a red point. The receiver has been placed in different points, which are represented by blue X marks in the figure. Both the transmitter and receiver have been placed at the same height of 81 cm, which is the height of the tables located in the scenario. The same antenna orientation in all the measurements has been carefully maintained in order to minimize variations due to the radiation pattern of the receiving antenna. The motes have been programmed to transmit at low data rate of 1 packet every 20 seconds, emulating a possible wireless sensor network application linked to Ambient Intelligence or Smart Homes. The RSSI values have been read directly from the data provided by the motes, by means of protocol analysis of the air interface. The obtained values for different positions in the laboratory are shown in Fig. 16. The scale has been set up to -100 dBm in order to account for the sensitivity value of the motes (-97 dBm). As expected, due to the CyFi's active link and power management, signal level does not clearly decrease with the distance as it happens in a common radiowave transmission (with a transmitter transmitting a fixed power level). The signal level is maintained quite well within the scenario, although variations on received power can be seen due to the multipath radiopropagation effects (mainly diffraction and reflection), very significant in an indoor complex scenario like this.

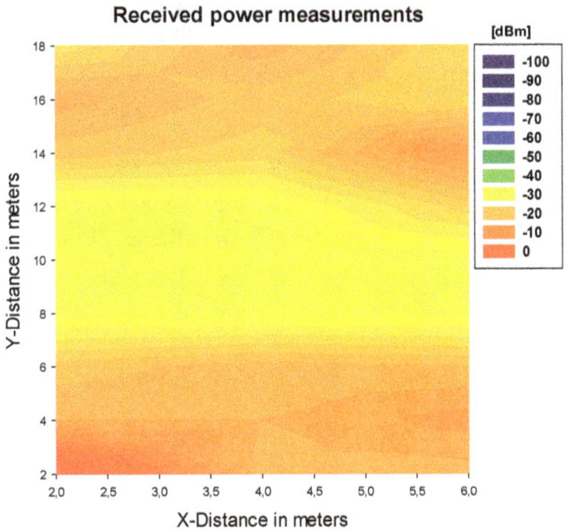

Fig. 16. Measured power levels in dBm for a pair of mote coordinator/sensor in the indoor scenario.

To gain more insight in the operation of the sensor motes and the influence of the topology and morphology of the scenario, current consumption has been measured for several positions. For that purpose, the motes have been programmed to transmit at their highest packet transmission rate (1 packet per 15 ms) and the highest transmission power level (4 dBm) in order to increase the current demand of the motes. As in the previous case, both the transmitter and receiver have been placed at the same height of 81 cm. With this new approach, unlike the previous case, at certain distance from the coordinator no packets are received. The RSSI values shown in Fig. 17 are the mean values of the RSSI data of the packets received in a 2-second duration time slot, which correspond approximately to 130 packets as long as the communication has been correctly done.

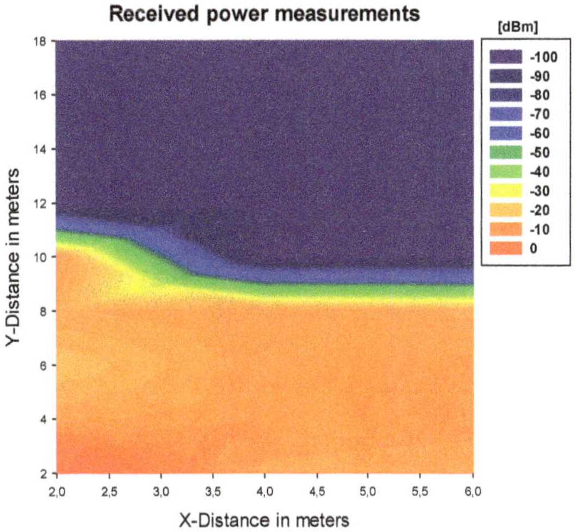

Fig. 17. Measured power levels in dBm for a pair of mote coordinator/sensor in the indoor scenario at highest transmission data rate.

When the distance is higher, the communication has problems and the number of received packets decreases up to 10 packets in 2 seconds due to the approach of the received power to the sensitivity level of the motes. Fig. 17 shows how the received power level near the transmitter is quite constant as well as the sensitivity level zone, in which no packets are received. This is due, once again, to the automatic regulation of transmit power that is embedded in the motes. But, despite of that, variations on received power level could be measured (on a smaller extent than in a case without transmitter power regulation), once again, due to the particularities of radiopropagation within a heterogeneous scenario with a complex morphology like this (a lot of furniture composed of different materials). As an example, although coordinates (6,6) are located further from the transmitter than coordinates (2,6), and the transmitter tries to maintain the received power level throughout the scenario, the zone corresponding to coordinates (2,6) has lower received power level. It is worth noting that due to the increase in the overall transmission rate, the sensitivity of the motes is decreased, leading to lower coverage zones, as clearly observable from Fig. 17. This again is given by the auto regulation power function embedded within the motes, reducing the available transmission power in order to handle a decreased sensitivity value given by a higher transmission speed. As an example, a PER (Packet Error Rate) measurement has been made between a transmitter in (2,2) and a receiver in (2,6). As mentioned previously, a low received power zone is detected surrounding the coordinate (2,6). This zone has the same characteristics as the zone near the sensitivity level, in which the number of received packets decreases abruptly. This is clearly shown in the PER value obtained for the transmission of 100,000 packets: only 1,363 packets arrived (PER = 98.637 %).

In order to see the evolution of the current consumption of the transmitter in this scenario, the receiver has been placed at different distances from the transmitter. A Tektronix DPO 3014 oscilloscope has been used to obtain the current consumption measurements. For this purpose, a 1 ohm resistor has been introduced in series in the feeding circuit of the mote. In this way, by measurement of the voltage difference in the resistor, an estimation of the current value is obtained. The obtained results are shown in Fig. 18 and Fig. 19.

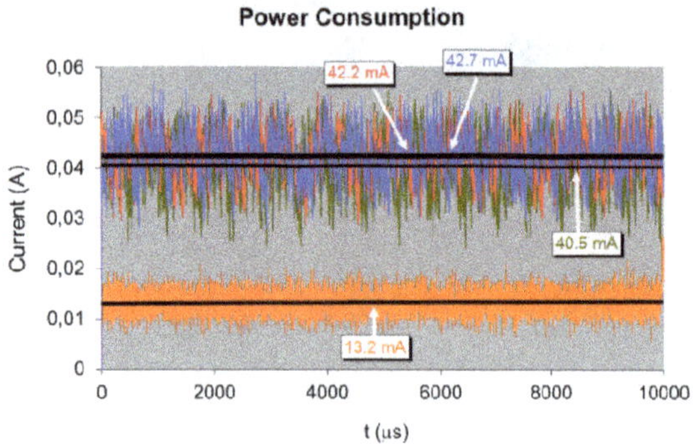

Fig. 18. Power consumption variation as a function of time for different positions. The bottom curve (13.2 mA) is for standby, whereas as the rest of the curves span from the closest to the farthest mote within the measurement scenario.

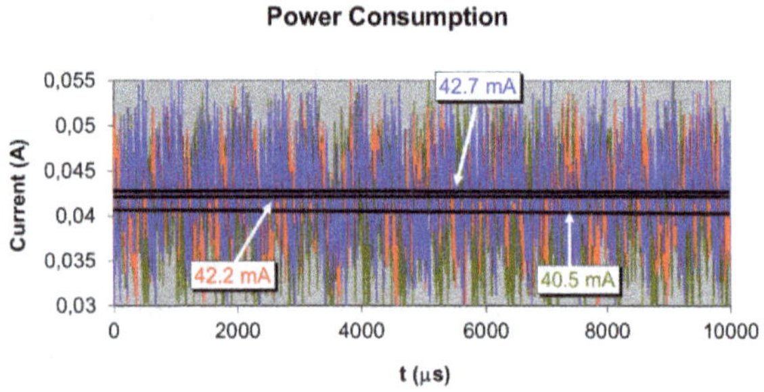

Fig. 19. Detail on the power consumption for the designed mote device within the indoor scenario. The mean values for different currents are proportional to the separation between motes.

As it can be seen in Fig. 18, a clear difference exists between the standby (orange curve) and the transmit mode (the rest of the curves), which is expected due to the normal operational procedure in the wireless transceiver. In Fig. 19, a detail of the different transmit mode current values can be seen, given for different positions of the sensor within the scenario. These distances have been 0.05 m, 2 m and 4.2 m, respectively. The last distance corresponds to the point in which the sensitivity level has been almost reached, in which few packets are received. This sensitivity point varies between 4 and 6 m depending on the environment and the objects surrounding the motes. From the measured values of the current consumption from the motes in operating mode at different distances, an increase in current consumption in the order of 4.2% in the case of 2 m and 5.4% in the case of 4.2 m is observed. Therefore, by considering the pre-existent levels of interference as well as the expected fading losses of the scenario, the optimal location of the motes can be planned prior to real network deployment.

As the distance increases, the power consumption level also increases, which is in accordance to the operation of the power management of CyFi motes: By increasing the distance between motes, the receiver power decreases. But due to the power management features of the motes, the transmitted power level increases in order to maintain the received power level in each position, leading to higher power demands of the transmitter.

5. Conclusions

In this work, the topological and morphological influence in the operation of a wireless sensor network is described. An indoor scenario has been analyzed by means of deterministic 3D ray launching in-house algorithm as well as by measurements with an in-house developed wireless sensor platform. The results show that the radiopropagation characteristic of indoor scenarios is complex, leading to strong topological dependences in the overall received signal power, which affects other parameters such as capacity of the wireless sensor network. The results show that by considering radioplanning issues in the deployment of the wireless sensor networks, power consumption as well overall system performance can be strongly optimized, due to direct impact on energy consumption of the wireless transceivers. In the future, these results can aid deployment and planning of complex indoor sensor networks, optimizing the overall power consumption without degrading system performance. With the advent of LTE and Internet of Things, the use of precise radioplanning techniques to aid in wireless transceiver deployment can be a determining factor for successful adoption of these emerging technologies.

Acknowledgements

This work has been supported by projects TEC2013-45585-C2-1-R, funded by the Ministry of Economy and Competiveness, Government of Spain..

References

[1] BOSE, R. Sensor networks-motes, smart spaces and beyond. *IEEE Pervasive Computing*, 2009, vol. 8, no. 3, p. 84–90.

[2] GROSSE, C.U., GLASER, S.D., KNUGER, M. Initial development of wireless acoustic emission sensor motes for civil infrastructure state monitoring. *Smart Structures and Systems*, 2010, vol. 6, no. 3, p. 197–209.

[3] BUCKNER, B.D., MARKOV, V., LAI, L.C., EARTHMAN, J.C. Laser-scanning structural health monitoring with wireless sensor motes. *Optical Engineering*, 2008, vol. 47, no. 5, Art. No. 054402.

[4] BERISHA, V., KWON, H., SPANIAS, A. Real-time acoustic monitoring using wireless sensor motes. In *Proceedings of the IEEE International Symposium on Circuits and Systems*. Island of Kos (Greece), 2006, p. 847–850.

[5] TRUBILOWICZ, J., CAI, K., WEILER, M. Viability of motes for hydrological measurements. *Water Resources Research*, 2009, vol. 45, Art. No. W00D22.

[6] KUANG, K.S.C., QUEK, S.T., MAALEJ, M. Remote flood monitoring system based on plastic optical fibers and wireless motes. *Sensors and Actuators A – Physical*, 2008, vol. 147, no. 2, p. 449–455.

[7] YUNSEOP, K., EVANS, R.G., IVERSEN, W.M., Remote sensing and control of an irrigation system using a distributed wireless sensor network. *IEEE Transactions on Instrumentation and Measurement*, 2008, vol. 57, no. 7, p. 1379–1387.

[8] RUIZ-GARCÍA, L., BARREIRO, P., ROBLA, J.I., LUNADEI, L. Testing ZigBee motes for monitoring refrigerated vegetable transportation under real conditions. *Sensors*, 2010, vol. 10, no. 5, p. 4968–4982.

[9] NAZABAL, J.A., LOPEZ ITURRI, P., AZPILICUETA, L., FALCONE, F., FERNÁNDEZ-VALDIVIELSO, C. Performance analysis of IEEE 802.15.4 compliant wireless devices for heterogeneous indoor home automation environments. *International Journal of Antennas and Propagation*, 2012, article number 176383.

[10] LOPEZ ITURRI, P., NAZABAL, J.A., AZPILICUETA, L., RODRIGUEZ, P., BERUETE, M., FERNÁNDEZ-VALDIVIELSO, C., FALCONE, F. Impact of high power interference sources in planning and deployment of wireless sensor networks and devices in the 2.4 GHz frequency band in heterogeneous environments. *Sensors*, 2012, vol. 12, no. 11, p. 15689–15708.

[11] HATA, M. Empirical formula for propagation loss in land mobile radio services. *IEEE Transactions on Antennas and Propagation*, 1980, vol. 29, no. 3, p. 317–325.

[12] IKEGAMI, F., YOSHIDA, S., TAKEUCHI, T., UMEHIRA, M. Propagation factors controlling mean field strength on urban streets. *IEEE Transactions on Antennas and Propagation*, 1984, vol. 32, no. 8, p. 822–829.

[13] PHAIBOON, S., PHOKHARATKUL, P. Path loss prediction for low-rise buildings with image classification on 2-D aerial photographs. *Progress in Electromagnetics Research*, 2009, vol. 95, p. 135–152.

[14] LEE, S. H. A photon modeling method for the characterization of indoor optical wireless communication. *Progress in Electromagnetics Research*, 2009, vol. 92, p. 121–136.

[15] LEE, D. J. Y., LEE, W. C. Y. Propagation prediction in and through buildings. *IEEE Transactions on Vehicular Technology*, 2000, vol. 49, no. 5, p. 1529–1533.

[16] TAN, S. Y., TAN, H. S. A microcellular communications propagation model based on the uniform theory of diffraction and multiple image theory. *IEEE Transactions on Antennas and Propagation*, 1996, vol. 44, no. 10, p. 1317–1326.

[17] KANATAS, A. G., KOUNTOURIS, I. D., KOSTARAS, G. B., CONSTANTINOU, P. A UTD propagation model in urban microcellular environments. *IEEE Transactions on Vehicular Technology*, 1997, vol. 46, no. 1, p. 185–193.

[18] DIMITRIOU, A. G., SERGIADIS, G. D. Architectural features and urban propagation. *IEEE Transactions on Antennas and Propagation*, 2006, vol. 54, no. 3, p. 774–784.

[19] FRANCESCHETTI, M., BRUCK, J., SCHULMAN, L. J. A random walk model of wave propagation. *IEEE Transactions on Antennas and Propagation*, 2004, vol. 52, no. 5, p. 1304–1317.

[20] BLAS PRIETO, J.,LORENZO TOLEDO, R. M., FERNÁNDEZ REGUERO, P., ABRIL, E. J., BAHILLO MARTÍNEZ, A., MAZUELAS FRANCO, S., BULLIDO, D. A new metric to analyze propagation models. *Progress In Electromagnetics Research*, 2009, vol. 91, p. 101–121.

[21] SCHUSTER, J. W., LUEBBERS, R. J. Comparison of GTD and FDTD predictions for UHF radio wave propagation in a simple outdoor urban environment. In *IEEE Antennas and Propagation Society International Symposium*. 1997, vol. 3, p. 2022–2025.

[22] ISKANDER, M. F., YUN, Z. Propagation prediction models for wireless communications systems. *IEEE Transactions on Microwave Theory and Techniques*, 2002, vol. 50, p. 662–673.

[23] KOUYOUMJIAN, R. G., PATHAK, P. H. A uniform theory of diffraction for an edge in a perfectly conducting surface. *Proc. IEEE*, 1974, vol. 62, no. 4, p. 1448–1462.

[24] GENNARELLI, G., RICCIO, G. A UAPO-based model for propagation prediction in microcellular environments. *Progress in Electromagnetics Research*, 2009, vol. 17, p. 101–116.

[25] SON, H. W., MYUNG, N. H. A deterministic ray tube method for microcellular wave propagation prediction model. IEEE *Transact. on Antennas and Propagation*, 1999, vol. 47, no. 8, p. 1344–1350.

[26] TAYEBI, A., GÓMEZ J., DE ADANA, F. S., GUTIERREZ, O. The application of arrival and received signal strength in multipath indoor environments. *Progress In Electromagnetics Research*, 2009, vol. 91, p. 1–15.

[27] SONG, H. B., WANG, H. G., HONG, K., WANG, L. A novel source localization scheme based on unitary esprit and city electronic maps in urban environments. *Progress In Electromagnetics Research*, 2009, vol. 94, p. 243–262.

[28] LU, W., CHAN, K. T. Advanced 3D ray tracing method for indoor propagation prediction. *Electronics Letters*, 1998, vol. 54, no. 12, p. 1259–1260.

[29] SEIDEL, S. Y., RAPPAPORT, T. S. Site-specific propagation prediction for wireless in-building personal communication system design. *IEEE Transactions on Vehicular Technology*, 1994, vol. 43,no. 4, p. 879–891.

[30] DURGIN, G., PATWARI, N., RAPPAPORT, T. S. An advanced 3D ray launching method for wireless propagation prediction. In *IEEE 47th Vehicular Technology Conference*. Phoenix (AZ, USA), 4-7 May 1997.

[31] CHANG-FA YANG, BOAU-CHENG WU, CHUEN-JYI KO. A ray-tracing method for modeling indoor wave propagation and penetration. *IEEE Transactions on Antennas and Propagation*, 1998, vol. 46, no. 6, p. 907–919.

[32] LOTT, M. On the performance of an advanced 3D ray tracing method. In *Proc. of European Wireless & ITG Mobile Communication*. Munich (Germany), 6-8 Oct. 1999.

[33] ROSSI, J.-P., GABILLET, Y. A mixed ray launching/tracing method for full 3-D UHF propagation modeling and comparison with wide-band measurements. *IEEE Transactions on Antennas and Propagation*, 2002, vol. 50, no. 4, p. 517–523.

[34] AZPILICUETA, L., FALCONE, F., ASTRÁIN, J. J., VILLADANGOS, J., GARCÍA ZUAZOLA, I. J., LANDALUCE, H., ANGULO, I., PERALLOS, A. Measurement and modeling of a UHF-RFID system in a metallic closed vehicle. *Microwave and Optical Technology Letters*, 2012, vol. 54, no. 9, p. 2126–2130.

[35] AGUIRRE, E., ARPÓN, J., DE MIGUEL, S., RAMOS, V., FALCONE, F. Evaluation of electromagnetic dosimetry of wireless systems in complex indoor scenarios with human body interaction. *Progress In Electromagnetic Research B*, 2012, vol. 43, p. 189 to 209.

[36] LED, S., AZPILICUETA, L., AGUIRRE, E., MARTÍNEZ DE ESPRONCEDA, M., SERRANO, L., FALCONE, F. Analysis and description of HOLTIN service provision for AECG monitoring in complex indoor environments. *Sensors*, 2013, vol. 13, no. 4, p. 4947–4960.

[37] AGUIRRE, E., ARPON, J., AZPILICUETA, L., LOPEZ, P., DE MIGUEL, S., RAMOS, V., FALCONE, F. Estimation of electromagnetic dosimetric values from non-ionizing radiofrequency fields in an indoor commercial airplane environment. *Electromagnetic Biology and Medicine*, published online in Aug. 2013.

[38] AGUIRRE, E., LOPEZ ITURRI, P., AZPILICUETA, L., DE MIGUEL-BILBAO, S., RAMOS, V., GARATE, U., FALCONE, F. Analysis of estimation of electromagnetic dosimetric values from non-ionizing radiofrequency fields in conventional road vehicle environments. *Electromagnetic Biology and Medicine*, published online in Jan. 2014.

[39] LUEBBERS, J. R. A heuristic UTD slope diffraction coefficient for rough lossy wedges. *IEEE Transactions on Antennas and Propagation*, 1989, vol. 37, no. 2, p. 206–211.

[40] LUEBBERS, J. R. Comparison of lossy wedge diffraction coefficients with application to mixed path propagation loss prediction. *IEEE Transactions on Antennas and Propagation*, 1988, vol. 36, no. 7, p. 1031–1034.

About Authors ...

Peio LÓPEZ ITURRI received his Telecommunications Engineering Degree from the Public University of Navarre (UPNA), Pamplona, Navarre, in 2011. Since then he has been working in the 'FASTER' research project at UPNA. He obtained a Master of Communications in 2012, held by the UPNA and he is currently pursuing the Ph.D degree in Telecommunication Engineering. His research interests include radio propagation, modeling of radio interference sources and mobile radio systems.

Leire AZPILICUETA received her Telecommunications Engineering Degree from the Public University of Navarre (UPNa), Pamplona, Spain, in 2009. In 2010 she worked in the R&D department of RFID Osés as radio engineer. In 2011, she obtained a Master of Communications held by the Public University of Navarre. She is currently pursuing the Ph.D. degree in telecommunication engineering. Her research interests are on radio propagation, mobile radio systems, ray tracing and channel modeling.

Juan Antonio NAZABAL was born in Pamplona, Spain, in 1977. He received his B.S. degree in Telecommunications Engineering from the Public University of Navarre (UPNA), Pamplona, Spain, in 2003. In 2010, he obtained a Master of Communications held by the Public University of Navarre. He is currently pursuing the Ph.D. degree in Telecommunication Engineering. His research interests are system integration, building automation and mobile radio systems.

Carlos FERNÁNDEZ-VALDIVIELSO received his Telecommunications Engineering Degree in 1998 and in 2003 his PhD in Communications, both from the Universidad Pública de Navarra, Navarra, Spain. In 1998 he co-founded Ingeniería Domótica, a company devoted to smart buildings and home automation systems. In 2005 he became an Associate Professor at UPNA. Since 2012 he is Director of SODENA, a venture capital company.

Jesús SORET received his Telecommunications Engineering Degree in 1998 and in 2003 his PhD in Communications, both from the Universidad de Valencia (UV), Spain. Since 2005 he is an Associate Professor at UV, working on wireless sensor systems.

Francisco FALCONE received his Telecommunications Engineering Degree in 1999 and his PhD in Communications in 2005, both from the Universidad Pública de Navarra, Navarra, Spain. From 1999 to 2000 he worked in Siemens-Italtel as a Microwave Engineer. From 2000 to 2008 he was a Radio Network Engineer in Telefónica Móviles. In 2009 he co-founded Tafco Metawireless, a spin-off company devoted to complex EM media. In parallel, he was an Assistant Professor at UPNA and since 2009, an Associate Professor at UPNA.

6

On the Comparative Performance Analysis of Turbo-Coded Non-Ideal Single-Carrier and Multi-Carrier Waveforms over Wideband Vogler-Hoffmeyer HF Channels

Fatih GENÇ[1], *M. Anıl REŞAT*[2], *Asuman SAVAŞÇIHABEŞ*[3], *Özgür ERTUĞ*[4]

[1] CU OPEN Lab, Dept. of Electronic and Communication Engineering, Çankaya University,Yukarıyurtçu Mahallesi Mimar Sinan Caddesi No:4, 06790, Etimesgut, Ankara, Turkey
[2] Dept. of Electrical and Electronics Engineering, Yildırım Beyazıt University, Çankırı Caddesi Çiçek sok. No:3 , Altındağ, Ankara, Turkey
[3] Dept. of Electrical and Electronics Engineering, Nuh Naci Yazgan University, Erkilet Dere Mah., Kocasinan, Kayseri/Turkey
[4] Telecommunication and Signal Processing Laboratory, Dept. of Electrical and Electronics Engineering, Gazi University, Yükseliş Sk. No:5, Maltepe, Ankara/Turkey

c1182604@student.cankaya.edu.tr, anilresat@yahoo.com, asavascihabes@gazi.edu.tr, ertug@gazi.edu.tr

Abstract. *The purpose of this paper is to compare the turbo-coded Orthogonal Frequency Division Multiplexing (OFDM) and turbo-coded Single Carrier Frequency Domain Equalization (SC-FDE) systems under the effects of Carrier Frequency Offset (CFO), Symbol Timing Offset (STO) and phase noise in wide-band Vogler-Hoffmeyer HF channel model. In mobile communication systems multipath propagation occurs. Therefore channel estimation and equalization is additionally necessary. Furthermore a non-ideal local oscillator generally is misaligned with the operating frequency at the receiver. This causes carrier frequency offset. Hence in coded SC-FDE and coded OFDM systems; a very efficient, low complex frequency domain channel estimation and equalization is implemented in this paper. Also Cyclic Prefix (CP) based synchronization synchronizes the clock and carrier frequency offset. The simulations show that non-ideal turbo-coded OFDM has better performance with greater diversity than non-ideal turbo-coded SC-FDE system in HF channel.*

Keywords

SC-FDE, OFDM, FFT-based channel estimation, ML synchronization, HF channel.

1. Introduction

Spectral and power efficiency of terminal in the limited bandwidth and transmit power have been developing continuously for the new generation of wireless communication systems. To meet the new user demands new air interfaces are needed to be enhanced. Orthogonal Frequency-Division Multiplexing (OFDM) is a popular modulation technique to satisfy these requirements, adopted to broadcast systems, such as Digital Video Broadcasting (DVB) [13], Digital Audio Broadcasting (DAB), Wireless Local Area Networks (WLAN) and Asymmetric Digital Subscriber Line (ADSL) for wired systems. In OFDM systems, one Inverse Fast-Fourier Transform (IFFT) block is used at the transmitter and also one FFT block is used at the receiver sides of the link. In the IFFT block, OFDM transmitter multiplexes the information into many low-rate streams which are transmitted parallelly instead of sending the information as a single stream [1],[16]. The modulated signals in an OFDM system have high peak values in time domain since many sub-carriers are added via an IFFT operation. Therefore, OFDM systems are known to have high Peak-to-Average Power Ratio (PAPR). Due to the limited battery life in mobile terminals, the PAPR problem is a main disadvantage of the OFDM system for the up-link [14].

On the other hand, Single-Carrier Frequency-Domain Equalization (SC-FDE) is a desirable alternative to OFDM systems. In the case of SC-FDE technique, no IFFT and FFT blocks exist at the transmit side while FFT and IFFT operators are performed at the receiver side of the link. SC-FDE experiences lower PAPR levels than OFDM because no IFFT is performed at the transmitter to precode the signal. In order to mitigate the PAPR problem, Single-Carrier (SC) transmission uses single-carrier modulation instead of many sub-carriers [1]-[3].

Low-complexity channel equalization and estimation in the frequency domain are used to mitigate the inter-symbol interference (ISI) [4]. For this purpose, frequency domain MMSE equalizer is generally used to minimize the attenuations of the fading channel. For wide-band channels, conventional time domain equalizers are impractical because of the very long channel impulse response in the time domain. Frequency domain equalization is more practical for such channels because the FFT size does not grow linearly with the length of the channel response and the complexity of the FDE is much lower than the time domain equalizer.

At the same time, frequency domain MMSE estimation is preferred with the comb-type pilot tone arrangement to predict the multi-path channel coefficients [9]-[11]. In order to estimate the channel characteristics, the comb-type pilot symbols are placed as periodically as possible in coherence time. The coherence time is the inverse of the Doppler spread in the channel.

An additional way to eliminate ISI almost completely, is to use a guard interval which is called cyclic-prefix (CP). The CP is the replica of the last L symbols of the block as shown in Fig. 4. The guard time L must be larger than the expected channel delay spread. At the receiver, the received CP is discarded before processing the block. By doing so CP also prevents inter-block interference.

CP is also used in CP-based channel synchronization to compensate the inter-carrier interference (ICI) caused by the Doppler effect [15],[16]. CP-based synchronization enables CFO estimation without need of additional redundant pilots. In fact, the key point is that CP already contains sufficient information to perform synchronization. Without CFO, the sub-channels do not interfere with one another. The impact of frequency offset is loss of orthogonality between the tones. Hence, the received signal is not a white process because of its probabilistic structure and it contains information about the timing offset and carrier frequency offset [12]. Estimations of timing offset θ and frequency offset $\hat{\varepsilon}$ are achieved by the relation of the CPs of consecutive frames.

In this paper, the performances of the turbo-coded SC-FDE and OFDM systems in wide-band Vogler-Hoffmeyer HF channels are compared. In practice, a wide-band radio channel has time-variant, frequency-selective and noisy properties. Most commonly used HF channel model is recommended by CCIR and ITU-R that is called as Watterson HF channel [5],[6]. The main restriction of the Watterson model is that the model is designed and tested for narrow-band channels but not for ones having more than 12 kHz bandwidth. In the design of high data speed wide-band HF communication systems, exact modeling and simulation of HF channel are needed. Therefore, Vogler-Hoffmeyer HF channel model is used in this paper [7],[8].

The remainder of this paper is organized as follows. Section II gives an overview of the wide-band Vogler-Hoffmeyer HF channel model. Section III overviews OFDM and SC-FDE structures. In Section IV, channel equalization/estimation and synchronization methods are defined in detail. Numerical results and discussions are given in Section V and, finally, conclusions are drawn from the results in Section VI.

2. Wide-Band HF Channel Model

HF channel characteristics are directly shaped by the ionosphere behavior because HF channels utilize the ionospheric reflections in order to provide long-distance communications.

The wide-band HF channel can be modeled as a FIR filter where the taps are time-variant and have complex values. This model can be described by the following equation:

$$y_t = \sum_{i=0}^{L-1} h_i x_t + n_i \tag{1}$$

where y_t is the complex output of the channel, L is the length of the channel, h_i is one of the L taps of the time-varying transversal filter, x_t is the complex input to the channel and n_i is Additive White Gaussian Noise (AWGN).

This type of a complex-valued FIR filter can be formed easily by convolving the input signal with the channel impulse response. Thus, the coefficients of the filter can be defined as the samples of the HF channel impulse response which is given as:

$$h(t,\tau) = \sqrt{P(\tau)}D(t,\tau)\psi(t,\tau) \tag{2}$$

where $P(\tau)$ is the delay power profile and its square root $\sqrt{P(\tau)}$ describes the shape in delay dimension; $D(t,\tau)$ is the deterministic phase function showing each path's Doppler shift, and $\psi(t,\tau)$ is the stochastic modulation function which describes the fading value of the impulse response.

The Doppler Effect can also be given with the formula:

$$D(t,\tau) = e^{j2\pi f_D t} \tag{3}$$

where f_D is the Doppler shift value. The stochastic modulation function $\psi(t,\tau)$ can be stated as random variables with an auto-correlation function that possesses a Gaussian shape.

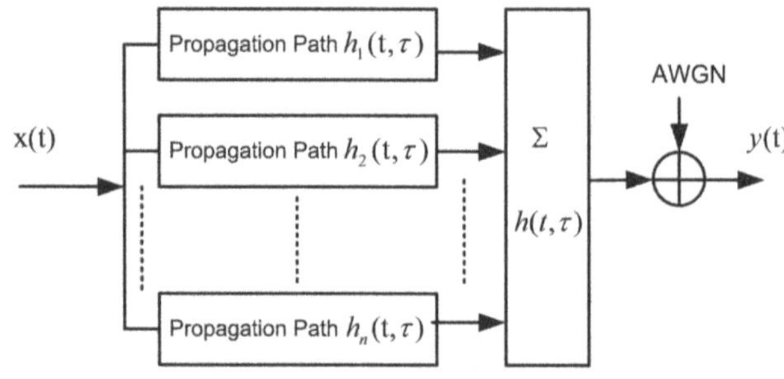

Fig. 1. HF channel model.

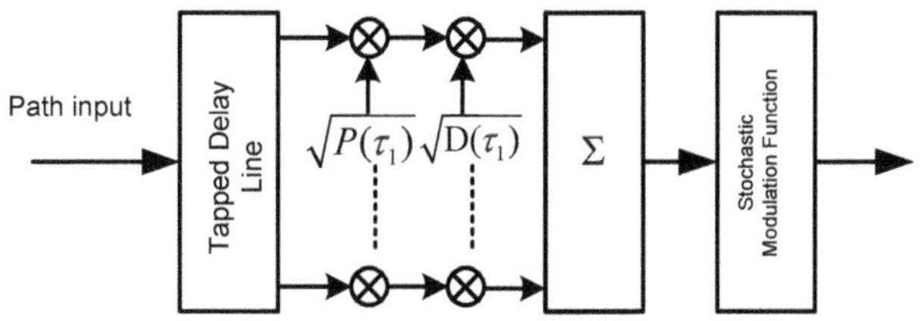

Fig. 2. Single propagation path.

Whilst Fig. 1 shows the structure of the wide-band HF channel model with propagation paths, Fig. 2 shows the model of a single propagation path [8]. It is important to specify the main difference between the narrow-band Watterson model and the wide-band channel model here. In the Watterson model time delay spread is neglected and the time delay of each path has a single value. On the other hand, the wide-band model has a delay power profile symbolized with $P(\tau)$ and relates Doppler effect with the time delay of each path.

3. System Models

3.1 OFDM System

In this section OFDM system model is described. The system structure is illustrated in Fig. 3. First, each binary source data are encoded by non-punctured, $R = 1/3$ code-rate turbo encoder and Log-Map algorithm is chosen for best decoding performance with low-complexity for turbo decoder. In this time, N sub-carriers X_k for $k = 0, 1, \ldots, N-1$ are modulated by a signal alphabet $A = \{\pm 1, \pm 3, \pm j, \pm 3j\}$ used for transmitting the information for 16-QAM. After base-band modulation, pilot tone symbol insertion is applied for the channel estimation where the pilot pattern is shown in Fig. 4.

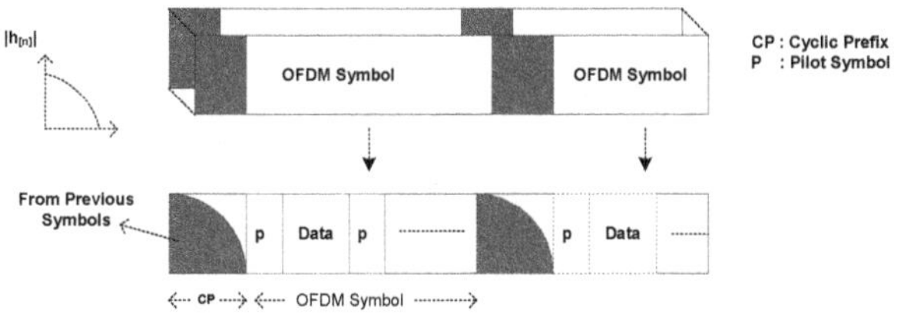

Fig. 4. Frame structure.

Pilot arrangement in OFDM system issue is discussed in more detail under Section IV. Here, output of the IFFT after the serial to parallel conversion can be expressed as:

$$x(n) = \frac{1}{\sqrt{N}} \sum_{k=0}^{N-1} X(k) e^{j2\pi kn/N} \tag{4}$$

where the constant $\frac{1}{\sqrt{N}}$ normalizes the power, N is the sub-carrier number and $X(k)$ are the modulated input symbols.

Cyclic prefix (CP) of length N_c is added at the beginning of the frame which must be greater than the maximum channel delay spread then the composite symbols are transmitted through the HF channel. In order to eliminate inter-carrier interference (ICI), this guard time includes the cyclic extended part of the OFDM symbol. Next, Root-Raised-Cosine (RRC) pulse shaping filtering is used to reconstruct on the data symbols.

After FFT is applied at the receiver, the received signal is given by:

$$\begin{aligned}
Y[k] &= \sum_{n=0}^{N-1} y[n] e^{-2\pi kn/N} \\
&= \sum_{n=0}^{N-1} \left\{ \sum_{m=0}^{\infty} h[n] x[n-m] + z[n] \right\} e^{-j2\pi kn/N} \\
&= \sum_{n=0}^{N-1} \left\{ \sum_{m=0}^{\infty} h[m] \left\{ \frac{1}{N} \sum_{i=0}^{N-1} X[i] e^{j2\pi i(n-m)/N} \right\} \right\} e^{-j2\pi kn/N} + Z[k] \\
&= \frac{1}{N} \sum_{i=0}^{N-1} \left\{ \left\{ \sum_{m=0}^{\infty} h[m] e^{-j2\pi im/N} \right\} X[i] \sum_{n=0}^{\infty} e^{-j2\pi(k-i)n/N} \right\} e^{-j2\pi kn/N} \\
&\quad + Z[k],
\end{aligned}$$

$$Y[k] = \mathrm{H}[k]\mathrm{X}[k] + \mathrm{Z}[k] \tag{5}$$

where $\mathrm{X}[k]$ denotes the kth sub-carrier frequency components transmitted symbol, $Y(k)$ is received symbol, $\mathrm{H}[k]$ is channel frequency response and $\mathrm{Z}[k]$ is noise in frequency domain, respectively.

At the receiver, after passing to discrete-time domain through A/D converter and pulse shaping filter, CP-based ML synchronization is applied to compensate the Carrier Frequency Offset (CFO) which is mentioned in Section IV, then guard time is removed:

$$y[n] = \{y_g(n) \quad M < n < N\} \tag{6}$$

where N is the sub-carrier, M is CP length, y_g received signal that have guard interval insertion.

Then y_n is received to FFT block for the following operation:

$$\begin{aligned}
Y[k] &= FFT\{y(n)\},\ k, n = 0, 1, \ldots N-1 \\
&= \frac{1}{N} \sum_{n=0}^{N-1} y(n) e^{-j2\pi kn/N}.
\end{aligned} \tag{7}$$

Next *Least Square* estimated $\hat{H}_{LS}^{p}[k] = \frac{Y^p[k]}{X^p[k]}$ is obtained by extracting the pilot signals $Y^p[k]$. The interpolated $\hat{H}[k]$ for all data sub-carriers is obtained in MMSE channel estimation. Then, in the Frequency domain equalization (FDE) block the transmitted data is equalized by MMSE equalizer as

$$\hat{X}_n = IFFT\{Y_k C_k\} = y_n \otimes c_n \tag{8}$$

where C_k represents the equalizer correction term, which is computed according to the FDE as follows:

- MMSE Equalizer:

$$C_k = \frac{\hat{H}^*_{FFT}}{\left|\hat{H}_{FFT}\right|^2 + (E_b/N_o)^{-1}} \tag{9}$$

where $(.)^*$ denotes conjugate. MMSE equalizer in (9) makes an optimum trade-off between noise enhancement and channel correction term, while using the signal-to-noise ratio (SNR) value [4]. Finally, the binary information data is obtained back in 16-QAM modulation and turbo decoding respectively.

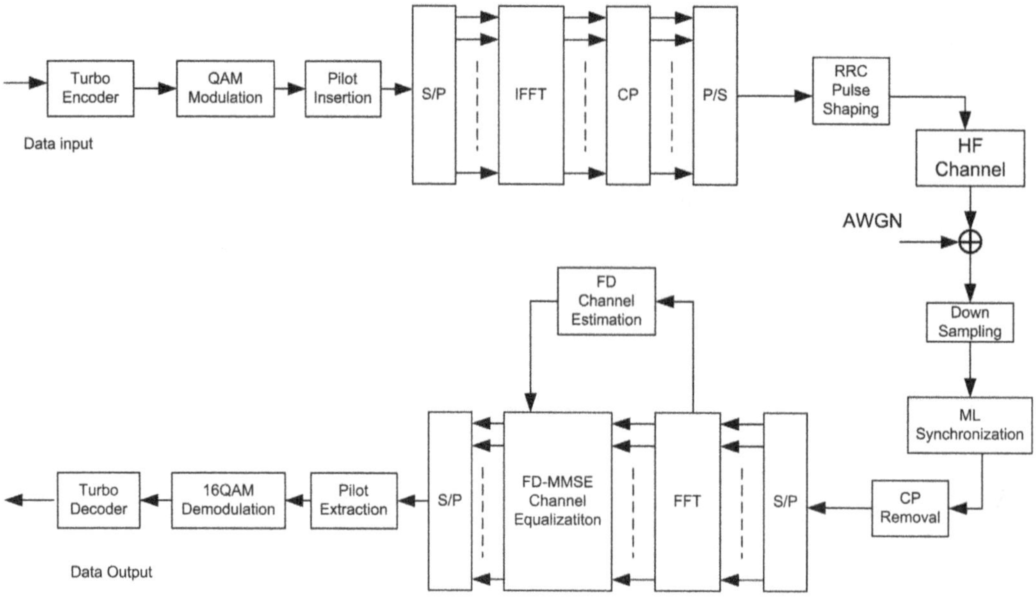

Fig. 3. OFDM structure.

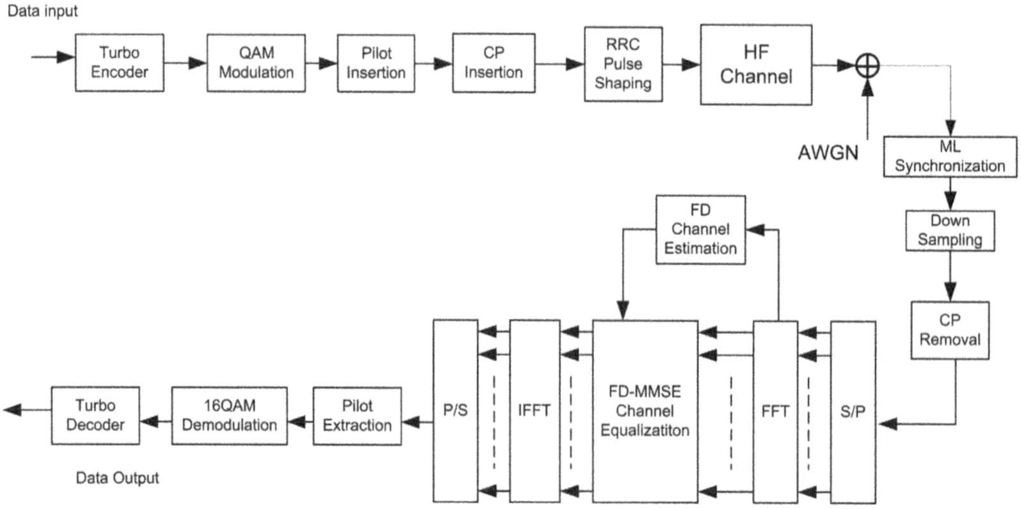

Fig. 5. SC-FDE structure.

3.2 SC-FDE System

OFDM and SC-FDE are similar in many ways. However there are explicit differences that make the two systems perform differently. As shown in Fig. 5, the main difference between OFDM and SC systems is the placement of the IFFT block. In SC systems, it is placed at the receiver side to transform the frequency domain equalized signals, thus compensating for channel distortion, bringing back to the time domain [3]. All the other blocks are formed with the same manner like OFDM system at both sides of the transmission.

In the OFDM system, symbols are exposed to an additional transformation by using the IFFT, $x(n) = IFFT\{X[k]\}$, but in the SC-FDE system no transformation is used. The frame of SC-FDE is transmitted during the time instant after the Turbo encoder, 16-QAM modulation, pilot insertion and CP insertion are applied respectively and the receiver maps received data into the frequency domain in order to equalize. When the channel delay spread is large it is more efficient computationally to equalize in the frequency domain. In addition, SC-FDE has better behavior when used with non-linear power amplifiers.

4. Channel Estimation & Synchronization

4.1 Channel Estimation

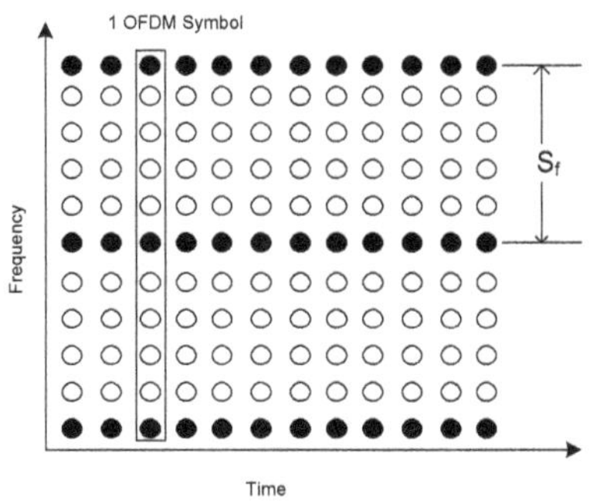

Fig. 6. Comb-type pilot arrangement.

Comb-type pilot arrangement, shown in Fig. 6, is used for frequency domain interpolation to estimate channel frequency response that is the Fourier transform of the channel impulse response [9],[10]. In Comb-type pilot arrangement, every OFDM and SC symbol has pilot tones which are periodically located at the sub-carriers. Notice that S_f the periods of pilot tones in frequency domain must be placed in the coherence bandwidth. The coherence bandwidth is determined by an inverse of the *maximum delay spread* $\sigma_{\max}$. The pilot symbol period is shown as following inequality:

$$S_f = \frac{1}{\sigma_{\max}}. \tag{10}$$

Let us consider the $\hat{H}_{LS} = X^{-1}Y \triangleq \tilde{H}$, using the weight matrix W channel estimate $\hat{H} \triangleq W\tilde{H}$ is defined and MSE of the channel estimate is calculated as below :

$$J(\hat{H}) = E\left\{\| e\|^2\right\} = E\left\{\| H - \hat{H}\|^2\right\}. \tag{11}$$

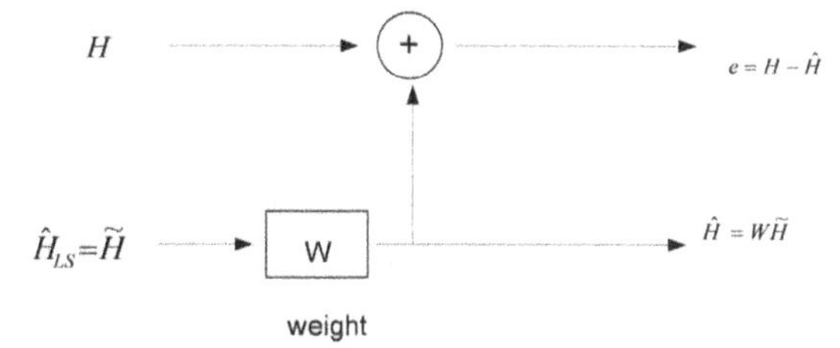

Fig. 7. MMSE channel estimation.

MMSE estimation method shown in Fig. 7 finds a better estimate that minimizes the MSE in (11). For the derivation of MMSE channel estimation, the cross-correlation $R_{e\tilde{H}}$, error vector e with channel estimate $\tilde{H}$ is forced to zero.

$$\begin{aligned} R_{e\tilde{H}} &= E\left\{e\tilde{H}^H\right\} \\ &= E\left\{(H-\hat{H})\tilde{H}^H\right\} \\ &= E\left\{(H-W\tilde{H})\tilde{H}^H\right\} \\ &= E\left\{H\tilde{H}^H\right\} - WE\left\{\tilde{H}\tilde{H}^H\right\} \\ &= R_{H\tilde{H}} - WR_{\tilde{H}\tilde{H}} = 0 \end{aligned} \tag{12}$$

where $(.)^H$ is the *Hermitian* operator. Solving (12) for W yields

$$W = R_{H\tilde{H}}R_{\tilde{H}\tilde{H}}^{-1}. \tag{13}$$

Using (13) the MMSE channel estimate follows as:

$$\hat{H} = W\tilde{H} = R_{H\tilde{H}}R_{\tilde{H}\tilde{H}}\tilde{H}. \tag{14}$$

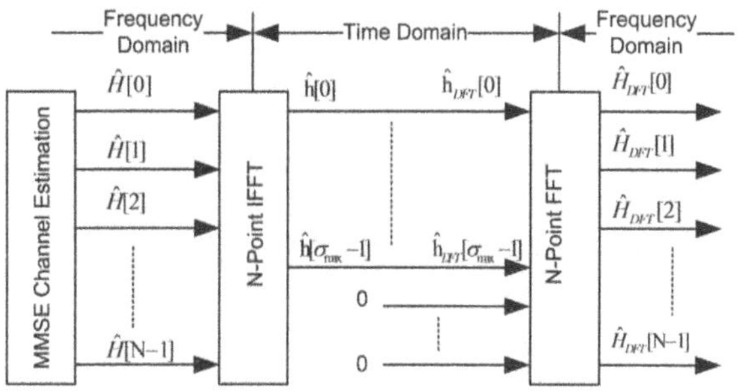

Fig. 8. FFT-based channel estimation.

Fig. 8 shows the block diagram of FFT-based channel estimation, given the MMSE channel estimation. An important point is that $\sigma_{\max}$ must be known formerly to remove the effect of noise outside the channel delay. Taking the IFFT of the MMSE channel estimate $\hat{H}$ to get in the time domain, that the coefficients contain the noise are ignored with zero padding and then transform the remaining $\sigma_{\max}$ elements back to the frequency domain to achieve $\hat{H}_{FFT}$. Finally $\hat{H}_{FFT}$ is used in (9) at the Frequency Domain MMSE Channel Equalizer block.

4.2 Channel Synchronization

In general, there are two types of distortion related to the carrier signal. One is the Phase Noise due to the Voltage Control Oscillator (VCO) and the other is Carrier Frequency

Offset (CFO) caused by Doppler Frequency shift f_d. Let us define the normalized CFO, ε, as a ratio of the CFO to subcarrier spacing Δ_f, i.e.,

$$\varepsilon = \frac{f_d}{\Delta_f} \tag{15}$$

where f_d is the Doppler Frequency. Here Δ_f is the ratio of the bandwidth to subcarrier number $(BW/N_{\Delta f}) = \frac{24000}{256} = 93.75$ Hz.

CP-based channel synchronization estimates the time and carrier-frequency offset. This algorithm exploits the cyclic prefix preceding the OFDM and SC symbols, thus reducing the need for pilots. The received data in the time domain are represented by $e^{j2\pi\varepsilon k/N}$, where ε denotes the difference in the transmitter and receiver oscillators as a fraction of the inter-carrier spacing, that is calculated in (15). Notice that all sub-carriers are affected by the same shift ε, shown as:

$$r(k) = s(k-\theta)e^{j2\pi\varepsilon k/N} + n(k) \tag{16}$$

where $r(k)$ is the received data, $s(k-\theta)$ is the unknown arrival time transmitted signal and $n(k)$ is the AWGN. Hence $r(k)$ contains information about the time offset θ and carrier offset ε. From the observation shown in Fig. 9 the estimation of frequency offset and the estimation of timing offset are calculated as:

$$\begin{aligned}
\gamma(m) &\triangleq \sum_{k=m}^{m+L-1} r(k)r^*(k+N), \\
\Phi(m) &\triangleq \frac{1}{2}\sum_{k=m}^{m+L-1} |r(k)|^2 + |r(k+N)|^2, \\
\hat{\theta}_{ML} &= \arg\max_{\theta}\left\{|\gamma(\theta)| - \tfrac{SNR}{SNR+1}\Phi(\theta)\right\}, \\
\hat{\varepsilon}_{ML} &= -\tfrac{1}{2}\angle\gamma(\hat{\theta}_{ML})
\end{aligned} \tag{17}$$

where L is the CP length, m is the index of samples, $\gamma(m)$ is the correlation coefficient and $\Phi(m)$ is an energy term [11], [12]. Fig. 9 shows the estimation of timing offsets and frequency offsets. Notice that peaks of the timing estimate, six frame are obtained from the observation interval and the index values of the peaks of the timing estimate gives the estimates of carrier frequency offsets $\hat{\varepsilon}$ values:

$$\hat{\varepsilon} = \hat{\gamma}(maxindexvalues(\hat{\Phi}(m))). \tag{18}$$

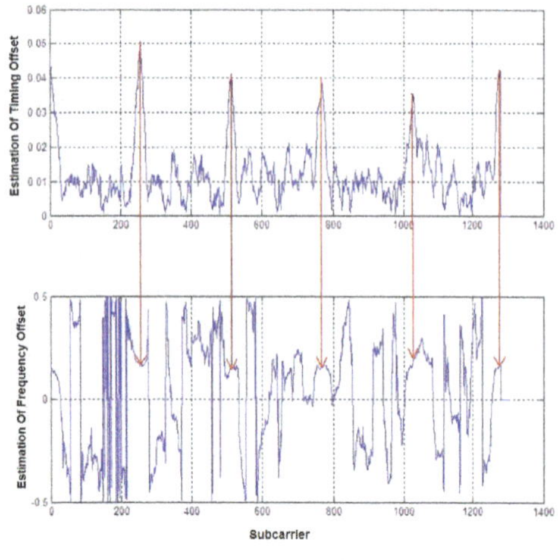

Fig. 9. STO and CFO estimates.

Finally, these estimates are used in channel synchronization block to compensate the carrier frequency offset as:

$$\hat{s}(k) = r(k)\cdot e^{-j2\pi\hat{\varepsilon}k/N} \tag{19}$$

where $\hat{s}(k)$ is the synchronized signal.

5. Simulation Results

In this section, BER performance of the proposed systems for CFO, STO and phase noise are shown. The simulation parameters are compliant to the wide-band HF channel model: 24 kHz bandwidth, 16 QAM constellation, 256 subcarriers, 210 occupied sub-carriers, 16 cyclic prefix length and pilot tone number is equal to 30. The code rate of the turbo code is $1/3$, the interleaver is 512 block interleaver and 10-iteration log-MAP decoding is used.

In all of the simulations, normalized frequency offset of each system is a constant value between 0.1 and 0.5. For the channel model, multipath Rayleigh fading channel is used which can be modeled as a tapped-delay line with $L_{ch} = 3$ delay taps of [3 7 10] ms. Furthermore the channel gains of the taps are [0 -3 -8] dB .

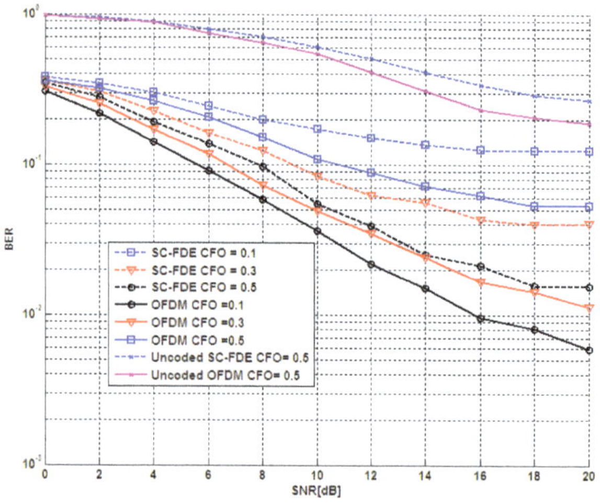

Fig. 10. BER performance versus SNR parametrized by CFO

In the first simulation, the effect of CFO is analyzed. Hence, normalized CFO, ε, is calculated from (15). In this simulation, channel delay spread and phase noise are neglected. As can be seen from Fig. 10, as the frequency offset of the channel increases, BER performances decrease as well. This is because of the way how CFO increments the ICI without the CP-based channel synchronization. Both OFDM and SC-FDE systems experience the impact of severe frequency-selective fading channels even so there are certain contrasts between the performance of their decoders. For lower code rates such as $R = 1/3$ Turbo code, OFDM outperforms SC-FDE. For SC-FDE, the noise amendment loss increases with the average input SNR. When the channel is ineffective and the SNR is high, the equalizer tries to invert the nulls and, as a result, the noise in these ineffective locations is amplified. Conversely, OFDM combines the useful energy across all sub-carriers through turbo-coding and interleaving.

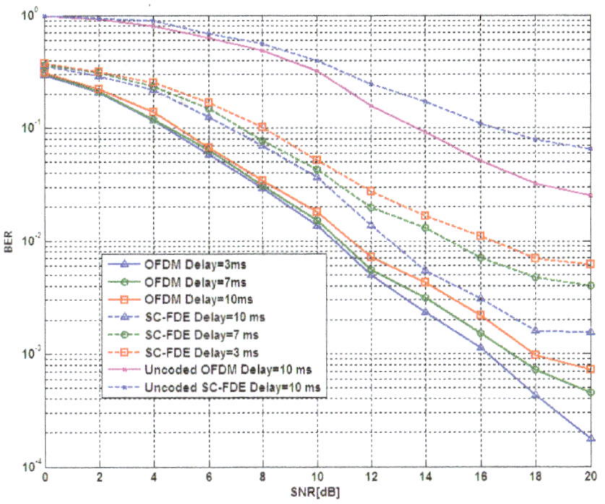

Fig. 11. BER performance versus SNR parametrized by STO.

In the second simulation, the effect of the channel delay spread, that is modeled with zero padding in each propagation path is analyzed. It is assumed that no CFO and phase noise exist. The simulation results are shown in the range of 3 ms to 10 ms channel delay spread. The $R = 1/3$ rate, 4 state (7,5) convolutional turbo encoder has d_{free}. Therefore, a coded OFDM system with this turbo code can achieve a diversity order of 5 without implementing any additional transmit/receive antennas, or using any other diversity techniques. Hence, especially when the channel order is larger, lower rate codes are required to achieve full diversity in OFDM systems. When this is the case, OFDM gives better performance than SC-FDE system because of the reduced effect of ISI.

For the third simulation, the effect of random fluctuations in the phase of a waveform due to the VCO at the -140 dBc/Hz, -100 dBc/Hz, and -70 dBc/Hz values is analyzed. For all simulations it can be seen that increasing the CFO, STO and phase noise effect clearly decreases the system performances especially for SC-FDE.

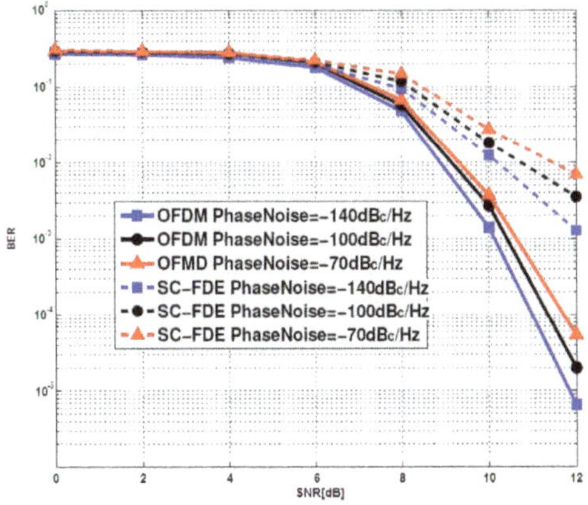

Fig. 12. BER performance versus SNR parametrized by phase noise.

6. Conclusion

In this paper, the performances of the turbo-coded SC and OFDM systems using FDE, MMSE channel estimation, CP-based synchronization over the Wideband Vogler-Hoffmeyer HF channel are simulated. The performance of the proposed systems were compared under only CFO, STO and phase noise effects. The simulation results confirm that turbo-coded OFDM performs significantly better than turbo-coded SC-FDE in HF channel model with the large diversity.

Acknowledgements

This work is partially supported by ASELSAN Inc. under Project Code: HBT-IA-2011-025.

References

[1] WEINSTEIN, S. B., et al. Data transmission by frequency-division multiplexing using the discrete Fourier transform. *IEEE Transactions on Communication Technology*, 1971, vol. 19, no. 5, p. 628 - 634.

[2] FALCONER, D., ARIYAVISTAKUL, S. L., BENYAMIN-SEEYAR, A., EDISON, B. Frequency domain equalization for single carrier broadband wireless systems. *IEEE Communications Magazine*, 2002, vol. 40, no. 4, p. 58 - 66.

[3] PANCALDI, F., et al. Single-carrier frequency domain equalization. *IEEE Signal Processing Magazine*, 2008, vol. 25, no. 5, p. 37 - 55.

[4] HASSA, E. S., ZHU, XU, EL-KHAMU, S. E., et al. Enhanced performance of OFDM and single-carrier systems using frequency domain equalization and phase modulation. In *National Radio Science Conference*. New Cairo (Egypt), 2009, p. 1-10.

[5] WATTERSON, C. C., JUROSHEK, J. R., BENSEMA, W. D. Experimental confirmation of an HF channel model. *IEEE Transactions on Communication Techniques*, 1970, vol. 18, no. 6, p. 792 - 803.

[6] HF Ionospheric Channel Simulators, CCIR Report 549-2. *Recommendations and Reports of the CCIR*, vol. 3, p. 59 - 67.

[7] HOFFMEYER, J. A., VOGLER, L. E. A new approach to HF channel modeling and simulation. In *Military Communications Conference (MILCOM)*. Monterey (CA, USA), 1990, vol. 3, p. 1199 - 1208.

[8] YANG GUAO, KE WANG A real-time software simulator of wideband HF propagation channel. In *International Conference on Communication Software and Networks*. Macau (China), 2009, p. 304 - 308.

[9] SHEN, Y., MARTINEZ, E. *Channel Estimation in OFDM Systems*, , rev. 1/2006. *Freescale Semiconductor*, 2008.

[10] COLERİ, S., ERGEN, M., PURI, A., BAHAI, A. Channel estimation techniques based on pilot arrangement in OFDM systems, *IEEE Transactions on Broadcasting*, 2002, vol. 48, no. 3.

[11] VAN DE BEEK, J.-J., EDFORS, O., SANDELL, M., WILSON, S. K., BÖRJESSON, P. O. On channel estimation in OFDM systems. In *Proceeding of 45th IEEE Vehicular Technology Conference*. Chicago (IL, USA), 1995, vol. 2, p. 815 - 819.

[12] SANDELL, M., VAN DE BEEK, J.-J., BÖRJESSON, P. O. Timing and frequency synchronization in OFDM systems using the cyclic prefix. In *International Symposium on Synchronization*, 1995, p. 16 - 19.

[13] POLAK, L., KRATOCHVIL, T. Exploring of the DVB-T/T2 performance in advanced mobile TV fading channels. In *36th International Conference on Telecommunications and Signal Processing (TSP2013)*. Rome (Italy), 2013, p. 768 - 772.

[14] MYUNG, H. G., LIM, J., GOODMAN, D. J. Peak-to-average power ratio of single carrier FDMA signals with pulse shaping. In The 17th Annual IEEE International Symposium on Personal, Indoor and Mobile Radio Communications (PIMRC). Helsinki (Finland), 2006, p. 1 - 5.

[15] LIU, Y., TAN, Z. Carrier frequency offset estimation for OFDM systems using repetitive patterns. *Radioengineering*, 2012, vol. 21, no. 3, p. 823 - 830.

[16] TAO, C., QIU, J., LIU, L. A novel OFDM channel estimation algorithm with ICI mitigation over fast fading channels. *Radioengineering*, 2010, vol. 19, no. 2, p. 347 - 355.

About Authors ...

Fatih GENÇ was born in Ankara, Turkey, in 1984. He received his B.S. degree in July 2007 and M.Sc. in September 2010, in Department of Electronic and Communication Engineering from Çankaya University, Turkey. His research interests include wireless communications, signal processing, HF channels, coding theory, FPGA and DSP. He is currently working for Ph.D. degree at the department of Electronic and Communication Engineering from Çankaya University, Turkey. He is also working at Telecommunications and Signal Processing Laboratory (TESLAB) in Gazi University as a project assistant.

Mustafa Anıl REŞAT was born in Ankara, Turkey, in 1988. He received his B.S. degree in January 2010, in the Department of Electrical and Electronics Engineering, TOBB University of Economics and Technology, Turkey. His research interests include SC-FDMA and OFDMA telecommunication systems, signal processing and high frequency communications. He is currently working for M.Sc. degree at the Department of Electrical and Electronics Engineering, Gazi University, Turkey. He is also working at Telecommunications and Signal Processing Laboratory (TESLAB) in Gazi University as a project assistant.

Asuman YAVANOĞLU has received her B.Sc. and M.Sc. degrees in Electrical Engineering from Erciyes University, Turkey, in 2005 and 2008 respectively. She is currently a Ph.D. student in Electrical and Electronics Engineering specializing in Telecommunications and Signal Processing at Gazi University, Ankara, Turkey, where she is working as a research assistant since 2006. Her research interests include wireless communication systems, MIMO communication and transmit/receive diversity of MIMO-OFDM

Özgür Ertuğ was born in Ankara, Turkey in 1975. He received his B.Sc. degree in 1997 from University of Southern California, USA, M.Sc. degree from Rice University in 1999 and Ph.D. degree from Midddle East Technical University in 2005. He is currently working as assistant professor in Electrical and Electronics Engineering Department of Gazi University. His main research interests lie in algorithm and architecture design as well as theoretical and simulation-based performance analysis of wireless communication systems especially in the physical layer.

7

Research on Channel Estimation and OFDM Signals Detection in Rapidly Time-Variant Channels

Min HUANG, Bingbing LI

The State Key Laboratory of Integrated Service Networks, Xidian University, XiˇŻan, 710071, P.R.China

mhuang@mail.xidian.edu.cn, bbli01@126.com

Abstract. *It is well known that iterative channel estimation and OFDM signals detection can significantly improve the performance of communication system. However, its performance is poor due to the modelling error of basis expansion model (BEM) being large enough and can not being ignored in rapidly time-variant channels. In this paper, channel estimation and OFDM signals detection are integrated into a real non-linear least squares (NLS) problem. Then the modified Broyden-Fletcher-Goldfarb-Shanno (MBFGS) algorithm is adopted to search the optimal solution. In addition, Cramer-Rao Bound (CRB) for our proposed approach is derived. Simulation results are presented to illustrate the superiority of the proposed approach.*

Keywords

Channel estimation, OFDM signals detection, a real NLS problem, MBFGS, CRB.

1. Introduction

Increasing demand for high spectral efficiency and high performance has led to the development of fourth-generation (4G) broadband wireless systems. A potential transmission technique for 4G is orthogonal frequency-division multiplexing (OFDM) which has recently become one of the most popular modulation techniques and has been adopted as the transmission technology in many wireless communication standards such as Wireless Fidelity (Wi-Fi), Worldwide Interoperability for Microwave Access (WiMAX), Long Term Evolution (LTE) standards, and the Digital Video Broadcasting (DVB) Project [1]. In the OFDM communication systems, Channel estimation and signal detection are all key techniques. In the last decade, they have been extensively studied, respectively [2]-[23].

It is well known that coherent detection schemes are superior to differentially coherent or noncoherent schemes in terms of power efficiency, if channel information can be established perfectly. In time-variant channels, for computational convenience, some papers [2]-[4] assume the channel is static within one or more consecutive OFDM symbols. In practice, this assumption will bring a certain amount of estimation error, especially in rapidly time-variant channels. However, the channel varies with time within a single OFDM symbol, leading to a big challenge. This is because the number of channel parameters which must be estimated is at least $N \times L$, where N and L denote the number of subcarriers and the number of channel taps, respectively, and it is larger than the number of the observation data in one OFDM symbol. Therefore, many existing works [5]-[11] resort to simplify channel model as a way of reducing the required number of channel parameters.

An alternative channel model is the Gauss-Markov model (GMM) [5], which models the time-variation of each tap by a Gauss-Markov process (usually only a first-order process is considered). However, it could only be appropriate for a slow-fading channel, not for a fast-fading channel [6]. Another popular channel model is the basis expansion model (BEM). In the BEM, the time-variation of each channel tap is expressed as a superposition of a few fixed basis functions, so that only $Q \times L$ BEM coefficients need to be estimated, where Q is the number of the basis functions. Several BEM variates are proposed in the literature, e.g., the complex-exponential BEM (CE-BEM) [7], the generalized CE-BEM (GCE-BEM) [8], the polynomial BEM (P-BEM) [9], the Karhunen-Loeve BEM (KL-BEM) [10], the discrete prolate spheroidal BEM (DPS-BEM) [11], and the others. Although the last two BEMs are closest to the true scenario, they require statistical channel knowledge, which has led to the model being usually unavailable in practice.

Based on the BEM models, the least squares estimator (LSE), the linear minimum mean square error estimator (LMMSEE) [12]-[14] and the best linear unbiased estimator (BLUE) are proposed in [6]. For the CE-BEM is constructed by a truncated Fourier series, it will cause the Gibbs phenomenon when it is used in these estimators [15]. In order to solve this problem, Hrycak proposes the orthogonal projection method [16] and the inverse reconstruction method [17]. Two methods based on P-BEM were proposed in [18] to estimate channel time-variations information in OFDM systems. The first one extracts these variations from the cyclic prefix (CP). The second one estimates that parameters by using the adjacent symbols. As the latter has more observation data, it is superior to the former. However, in these two schemes, the pilots have been smeared by the inter-carrier interference (ICI). Tao et al [19] proposed a channel estimation technique using the ICI self-cancellation to resolve this problem. Nevertheless, this scheme requires the number of pilots being twice as much, leading to low spectral efficiency.

In terms of OFDM signals detection, the nonlinear equalizer MMSE with Successive Detection proposed in [20] outperformed the linear equalizers. For the purpose of improving the performance of detection, Wang et al [21] proposed a detection method which whitened the residual ICI and the noise, while Sebesta et al [22] proposed another scheme based on the cyclic autocorrelation function of CP.

Recently, in order to further improve system performance, many works [24]-[27] resort to the iterative strategy. In slowly time-variant channels, due to the exchange of information between channel estimation and OFDM signals detection, the iterative strategy has achieved good performance. However, in the rapidly time-variant channels, i.e., in the high speed railway or the low altitude aircraft, since the modelling error of BEM is large enough, the subsequent iterative operation is not adequate to compensate for this error.

In this paper, we propose a new approach from the point of view of global optimization. First, channel estimation and OFDM signals detection are equivalent to a complex non-linear least squares (NLS) problem. For convenience, it is transformed to a real NLS problem on the premise of that its characteristic remains the same. Then, the modified Broyden-Fletcher-Goldfarb-Shanno (MBFGS) algorithm [28], which is proved that it could achieve the global solution even for nonconvex unconstrained optimization problems, is adopted to solve the real NLS problem. Lastly, in order to evaluate the quality of our proposed approach, Cramer-Rao Bound (CRB) is derived.

In a word, the contributions of our paper are given as follows:

- We construct a real NLS problem, which perfectly represents the channel estimation and OFDM signals detection in Rapidly Time-Variant Channels.
- We propose a method, which mainly employs the MBFGS algorithm to solve the real NLS problem.
- We derive the CRB for our proposed approach.

The rest of the paper is structured as follows: Section 2 briefly introduces the system model which includes the OFDM system model, the basis expansion model and the OFDM system based on BEM. In Section 3, we depict our proposed method. CRB for our estimation is given in Section 4. Simulation results are exhibited in Section 5. Conclusions are presented in Section 6.

The following notations are used throughout the paper. Boldface lowercase and uppercase letters are used for vectors and matrices, respectively. Superscripts T, H and $\dagger$ denote transpose, conjugate transpose and pseudo inverse, respectively. The notation $\hat{\boldsymbol{x}}$, diag($\boldsymbol{x}$) and $\boldsymbol{x}^{(k)}$ (or $\boldsymbol{X}^{(k)}$) are reserved for the estimated $\boldsymbol{x}$, the diagonal matrix whose main diagonal equals $\boldsymbol{x}$ and the vector $\boldsymbol{x}$ (or the matrix $\boldsymbol{X}$) in the k iteration, respectively. The matrix $\boldsymbol{F}$ denotes the fast Fourier transform (FFT) matrix and the matrix $\boldsymbol{F}^H$ denotes the inverse fast Fourier transform (FFT) matrix. Furthermore, we denote the $x \times x$ identity matrix as $\boldsymbol{I}_x$.

2. System Model

2.1 OFDM System Model

It is assumed that the synchronizations of frequency and time are perfect. The received signal after removing the cyclic prefix (CP) is given by

$$y(n) = \sum_{l=0}^{L-1} h_n^l d(n-l)_N + w(n), \quad 0 \le n \le N-1 \tag{1}$$

where h_n^l is the lth channel tap at the nth sample time, $d(n)$ is the nth transmitted sample, $(\cdot)_N$ represents a cyclic shift on the base of N, $w(n)$ is the additive white Gaussian noise (AWGN) with mean zero and variance σ_w^2, L is the total number of channel taps. In order to avoid the inter-symbol interference (ISI), it is assumed that the highest values of path delays are always less than or equal to the length of CP in this paper.

Collecting the samples of the received signal to form a vector $\boldsymbol{y} = [y(0), \cdots, y(N-1)]^T$ yields the following model

$$\boldsymbol{y} = \boldsymbol{H}_t \boldsymbol{d} + \boldsymbol{w} \tag{2}$$

where $\boldsymbol{d} = [d(0), \cdots, d(N-1)]^T$, $\boldsymbol{w} = [w(0), \cdots, w(N-1)]^T$, $\boldsymbol{H}_t$ is an $N \times N$ channel impulse response matrix in the time domain. Using $h_n^l = 0$ for $N > l \ge L$, $\boldsymbol{H}_t$ can be expressed as

$$\mathbf{H}_t = \begin{bmatrix} h_0^0 & 0 & \cdots & 0 & h_0^{L-1} & \cdots & h_0^1 \\ h_1^1 & h_1^0 & \ddots & \ddots & \ddots & \ddots & \vdots \\ \vdots & \ddots & \ddots & \ddots & \ddots & \ddots & h_{L-2}^{L-1} \\ h_{L-1}^{L-1} & \ddots & h_{L-1}^1 & h_{L-1}^0 & 0 & \ddots & 0 \\ 0 & h_L^{L-1} & \ddots & h_L^1 & h_L^0 & 0 & \vdots \\ \vdots & \ddots & \ddots & \ddots & \ddots & \ddots & 0 \\ 0 & \cdots & 0 & h_{N-1}^{L-1} & \cdots & h_{N-1}^1 & h_{N-1}^0 \end{bmatrix} = \sum_{l=0}^{L-1} \mathrm{diag}(\mathbf{h}_l^t)\mathbf{A}^l \tag{3}$$

where $\boldsymbol{h}_l^t = [h_0^l, \cdots, h_{N-1}^l]^T$ represents the lth channel tap within an OFDM symbol duration, and $\boldsymbol{A}$ is the $N \times N$ cyclic permutation matrix given by

$$\boldsymbol{A} = \begin{bmatrix} 0 & 0 & 0 & 1 \\ 1 & 0 & 0 & 0 \\ \ddots & \ddots & \ddots & \ddots \\ 0 & 0 & 1 & 0 \end{bmatrix}.$$

2.2 Basis Expansion Model

As can be seem from (3), it is very difficult to implement channel estimation in rapidly time-variant channels, since the number of parameters which need to be estimated is much larger than that of the observed data. A BEM [6] regarded as a simplified channel is employed, so that the number of estimated parameters is considerably reduced. Then, the lth channel tap $\boldsymbol{h}_l^t$ can be presented as

$$\begin{aligned}\boldsymbol{h}_l^t &= \sum_{q=0}^{Q-1} h_{q,l}\boldsymbol{b}_q + \boldsymbol{\varepsilon}_l \\ &= \boldsymbol{B}\boldsymbol{h}_l + \boldsymbol{\varepsilon}_l \end{aligned} \tag{4}$$

where $\boldsymbol{b}_q$ is the qth basis function, e.g., $\boldsymbol{b}_q = \begin{bmatrix} e^{(j2\pi 0/N)(q-Q/2)} & \cdots & e^{(j2\pi(N-1)/N)(q-Q/2)} \end{bmatrix}^T$ for CE-BEM [7] or $\mathbf{b}_q = \begin{bmatrix} 1^q & \cdots & N^q \end{bmatrix}^T$ for P-BEM [9], etc. $\boldsymbol{B} = [\boldsymbol{b}_0, \cdots, \boldsymbol{b}_{Q-1}]$ is an $N \times Q$ matrix that collects $Q(Q \ll N)$ orthonormal basis function $\boldsymbol{b}_q$ as columns, $\boldsymbol{h}_l = [h_{0,l}, \cdots, h_{Q,l}]^T$ represents the BEM coefficients for the lth tap and $\boldsymbol{\varepsilon}_l$ represents the corresponding modeling error. As shown in (4), due to the BEM, the lth tap channel $\boldsymbol{h}_l^t$ which needs to be estimated can be equivalent to $\boldsymbol{h}_l$. Therefore, the number of parameters for the lth tap decreases from N to Q.

2.3 OFDM System Model Based on BEM

Substituting (4) in (3), we can obtain

$$\begin{aligned}\boldsymbol{H}^t &= \sum_{l=0}^{L-1} \mathrm{diag}(\sum_{q=0}^{Q-1} h_{q,l}\boldsymbol{b}_q + \boldsymbol{\varepsilon}_l)\boldsymbol{A}^l \\ &= \sum_{q=0}^{Q-1} \mathrm{diag}(\boldsymbol{b}_q) \sum_{l=0}^{L-1} h_{q,l}\boldsymbol{A}^l + \sum_{l=0}^{L-1} \mathrm{diag}(\boldsymbol{\varepsilon}_l)\boldsymbol{A}^l \\ &= \sum_{q=0}^{Q-1} \mathrm{diag}(\boldsymbol{b}_q)\boldsymbol{F}^H\boldsymbol{\Delta}_q\boldsymbol{F} + \sum_{l=0}^{L-1} \mathrm{diag}(\boldsymbol{\varepsilon}_l)\boldsymbol{A}^l \end{aligned} \tag{5}$$

with

$$\boldsymbol{\Delta}_q = \mathrm{diag}\left(\boldsymbol{F}_L\left[h_{q,0}, \cdots, h_{q,L-1}\right]^T\right)$$

where $\boldsymbol{F}_L$ stands for the first L columns of $\sqrt{N}\boldsymbol{F}$.

In the light of (5), (2) can be written as

$$\boldsymbol{y} = \sum_{q=0}^{Q-1} \mathrm{diag}(\boldsymbol{b}_q)\boldsymbol{F}^H\boldsymbol{\Delta}_q\boldsymbol{F}\boldsymbol{d} + \sum_{l=0}^{L-1} \mathrm{diag}(\boldsymbol{\varepsilon}_l)\boldsymbol{A}^l\boldsymbol{d} + \boldsymbol{w}. \tag{6}$$

After carrying out an N-point FFT, (6) becomes

$$\begin{aligned}\boldsymbol{r} &= \boldsymbol{F} \sum_{q=0}^{Q-1} \mathrm{diag}(\boldsymbol{b}_q)\boldsymbol{F}^H\boldsymbol{\Delta}_q\boldsymbol{F}\boldsymbol{F}^H\boldsymbol{s} + \boldsymbol{F} \sum_{l=0}^{L-1} \mathrm{diag}(\boldsymbol{\varepsilon}_l)\boldsymbol{A}^l\boldsymbol{F}^H\boldsymbol{s} + \boldsymbol{F}\boldsymbol{w} \\ &= \sum_{q=0}^{Q-1} \boldsymbol{F}\mathrm{diag}(\boldsymbol{b}_q)\boldsymbol{F}^H\boldsymbol{\Delta}_q\boldsymbol{s} + \sum_{l=0}^{L-1} \boldsymbol{F}\mathrm{diag}(\boldsymbol{\varepsilon}_l)\boldsymbol{A}^l\boldsymbol{F}^H\boldsymbol{s} + \boldsymbol{\omega} \\ &= \boldsymbol{H}\boldsymbol{s} + \boldsymbol{\psi} + \boldsymbol{\omega} \end{aligned} \tag{7}$$

with

$$\boldsymbol{H} = \sum_{q=0}^{Q-1} \boldsymbol{F}\mathrm{diag}(\boldsymbol{b}_q)\boldsymbol{F}^H\boldsymbol{\Delta}_q, \boldsymbol{\psi} = \sum_{l=0}^{L-1} \boldsymbol{F}diag(\boldsymbol{\varepsilon}_l)\boldsymbol{A}^l\boldsymbol{F}^H\boldsymbol{s}$$

or

$$\begin{aligned}\boldsymbol{r} &= \sum_{q=0}^{Q-1} \boldsymbol{F}\mathrm{diag}(\boldsymbol{b}_q)\boldsymbol{F}^H\boldsymbol{\Delta}_q\boldsymbol{s} + \boldsymbol{\psi} + \boldsymbol{\omega} \\ &= \sum_{q=0}^{Q-1} \boldsymbol{F}\mathrm{diag}(\boldsymbol{b}_q)\boldsymbol{F}^H\mathrm{diag}(\boldsymbol{s})\boldsymbol{F}_L\boldsymbol{h} + \boldsymbol{\psi} + \boldsymbol{\omega} \\ &= \boldsymbol{P}\boldsymbol{h} + \boldsymbol{\psi} + \boldsymbol{\omega} \end{aligned} \tag{8}$$

with

$$\boldsymbol{P} = \left[\boldsymbol{F}\mathrm{diag}(\boldsymbol{b}_0)\boldsymbol{F}^H\mathrm{diag}(\boldsymbol{s})\boldsymbol{F}_L, \cdots, \boldsymbol{F}\mathrm{diag}(\boldsymbol{b}_{Q-1})\boldsymbol{F}^H\mathrm{diag}(\boldsymbol{s})\boldsymbol{F}_L\right]$$
$$\boldsymbol{h} = [h_{0,0}, \cdots h_{0,L-1}, \cdots, h_{Q-1,0}, \cdots h_{Q-1,L-1}]^T$$

where $\boldsymbol{s} = [s(0), \cdots, s(N-1)]^T$, which is the N-point FFT of $\boldsymbol{d}$, represents the transmit data in the frequency domain. $\boldsymbol{\omega} = [\omega(0), \cdots, \omega(N-1)]^T$, which is the N-point FFT of $\boldsymbol{w}$, represents the AWGN in the frequency domain.

Hijazi and Ros [24] proposed an iterative channel estimation and OFDM signals detection method, i.e., an alternating least square (ALS) algorithm. With (8), the detected information data which can be obtained by the LSE in the previous iterations, are used for refining the channel estimation by the least square (LS) equalizer [20] in the light of (7). In slowly time-variant channels, since the BEM modeling error in (4) is small enough, the interference item $\boldsymbol{\psi}$ in (7) and (8) can be negligible [6]. Consequently the LS method applied in (7) and (8) could be regarded as the optimal algorithm, which can be proved in Appendix A. In addition, according to the Lemma 1 proposed in [29], ALS algorithm could achieve a local optimum solution.

Lemma 1 *Supposing there are two resolved vectors $\boldsymbol{\alpha}$ and $\boldsymbol{\beta}$ in the given equations, alternating optimization algorithm (which solves for the optimal $\boldsymbol{\alpha}$ by fixing $\boldsymbol{\beta}$, and then solves for the optimal $\boldsymbol{\beta}$ by fixing $\boldsymbol{\alpha}$ and iterates in this manner until convergence) could obtain a local optimum solution.*

However, in rapidly time-variant channels, the interference item $\boldsymbol{\psi}$ in (7) and (8) can not be negligible for the BEM modeling error being no longer small [6]. Moreover, being different from $\boldsymbol{\omega}$, the interference item $\boldsymbol{\psi}$ associated with the BEM modeling error and the transmitted data is non-Gaussian distributed. According to Appendix A, since LS method can not be guaranteed to be the optimal algorithm in non-Gaussian, the ALS applied in [24] can not be guaranteed to achieve even a local optimum solution. For solving this problem, a new method is proposed in the next section.

3. Proposed Method

3.1 Construction of the Real NLS Problem

We assume there are K equally spaced pilots, s_{l_i}, at subcarriers $l_i = (i \times N)/K$, for $0 \le i < K$. All these pilots together form the pilot vector $\boldsymbol{s}_p = \left[s_{l_0}, \cdots, s_{l_{K-1}}\right]^T$. The remaining subcarriers in single OFDM symbol are reserved for the information data, which can be collected in the information vector $\boldsymbol{s}_d$. For convenience, we collect the unknown parameters in (7) or (8) to form a vector $\boldsymbol{x} = \left[\boldsymbol{s}_d{}^H, \boldsymbol{h}^H\right]^H \in \mathbb{C}^{M\times 1}$ with $M = N - K + Q \times L$. Consequently, (7) and (8) can be written as

$$\boldsymbol{r} = \boldsymbol{f}(\boldsymbol{x}) + \boldsymbol{\psi} + \boldsymbol{\omega} \tag{9}$$

where $\boldsymbol{f}$ is a nonlinear map from $\boldsymbol{x}$ to $\boldsymbol{Hs}$ or $\boldsymbol{Ph}$. For well estimating $\boldsymbol{x}$ in (9), the maximum likelihood estimation (MLE) regarded as optimal estimator is adopted and given by

$$\hat{\boldsymbol{x}} = \arg \min_{\boldsymbol{x}\in\mathbb{C}^{M\times 1}} \frac{1}{2} \|\boldsymbol{f}(\boldsymbol{x}) - \boldsymbol{r}\|_2^2. \tag{10}$$

Obviously, this is a complex NLS problem. In order to avoid the complex derivative in the later discussion, the problem should be transformed into a real NLS problem. Due to the unknown parameter vectors of the real and imaginary parts being independent of each other, we can define

$$\bar{\boldsymbol{x}} = \left[\mathrm{Re}[\boldsymbol{s}_d]^T, \mathrm{Im}[\boldsymbol{s}_d]^T, \mathrm{Re}[\boldsymbol{h}]^T, \mathrm{Im}[\boldsymbol{h}]^T\right]^T \in \mathfrak{R}^{2M\times 1}$$

$$\bar{\boldsymbol{r}} = \left[\mathrm{Re}[\boldsymbol{r}]^T, \mathrm{Im}[\boldsymbol{r}]^T\right]^T \in \mathfrak{R}^{2N\times 1}.$$

Then, (10) becomes

$$\bar{\boldsymbol{x}} = \arg \min_{\bar{\boldsymbol{x}}\in\mathfrak{R}^{2M\times 1}} \phi(\bar{\boldsymbol{x}}) \tag{11}$$

where $\phi(\bar{\boldsymbol{x}}) = \frac{1}{2}\|\boldsymbol{g}(\bar{\boldsymbol{x}}) - \bar{\boldsymbol{r}}\|_2^2$ with $\boldsymbol{g}$ is a nonlinear map from $\bar{\boldsymbol{x}}$ to $\left[\mathrm{Re}[\boldsymbol{Hs}]^T, \mathrm{Im}[\boldsymbol{Hs}]^T\right]^T$ or $\left[\mathrm{Re}[\boldsymbol{Ph}]^T, \mathrm{Im}[\boldsymbol{Ph}]^T\right]^T$.

3.2 Solution of the Real NLS Problem

It is known that BFGS algorithm [30] is a considerable popular Quasi-Newton method for solving the convex optimization problems. However, we could not directly adopt this algorithm to solve the real NLS problem for the problem being not necessarily a convex optimization problem. Fortunately, the MBFGS algorithm proposed in [28], has been proved that it could achieve the global solution even for nonconvex unconstrained optimization problems. Therefore, we adopt the MBFGS algorithm to solve the real NLS problem. The MBFGS uses several simple update formulas to sequentially update $\bar{\boldsymbol{x}}$, starting from a random vector $\bar{\boldsymbol{x}}^{(0)}$, until a stable solution is obtained. The update formulas are given at the top of next page.

Obviously, the estimated information vector and the BEM coefficients vector can be extracted from the estimated $\hat{\bar{\boldsymbol{x}}}$. However, we have assumed that the information data are continuous variables in (11) for convenience. In practical, they only take the discrete constellation points. To overcome this problem, we do a hard decision of $\hat{\boldsymbol{s}}_d$. Then, together with the pilots to update $\boldsymbol{P}$ in (8). Subsequently, we can achieve the more exact $\hat{\boldsymbol{h}}$ by the LS estimator [6]

$$\hat{\boldsymbol{h}} = \boldsymbol{P}^{\dagger}\boldsymbol{r}. \tag{14}$$

With the aid of $\hat{\boldsymbol{h}}$, renew the matrix $\boldsymbol{H}$ in (7). Then, after eliminating the effect of the pilots on the information data, (7) can be written as

$$\tilde{\boldsymbol{r}} = \boldsymbol{r} - \boldsymbol{H}_p\boldsymbol{s}_p = \boldsymbol{H}_d\boldsymbol{s}_d + \boldsymbol{\omega} \tag{15}$$

where $\boldsymbol{H}_p$ is a $N\times K$ matrix, which is carved out of $\boldsymbol{H}$ corresponding to the pilot subcarriers.

Lastly, the more exact information vector $\hat{\boldsymbol{s}}_d$ can be easily obtained by LS equalizer [20]

$$\hat{\boldsymbol{s}}_d = \boldsymbol{H}_d{}^{\dagger}\tilde{\boldsymbol{r}}. \tag{16}$$

3.3 Complexity Analysis

In this paper, the computational complexity of the proposed method is evaluated by the number of real multiplications. Assume that the operation of $N \times N$ complex matrix inversion needs N^3 complex multiplications and a complex multiplication is equivalent to three real multiplications. For our proposed method, the main complexity is in solving MBFGS, (14), (15) and (16). The number of real multiplication required to calculate each step of MBFGS is listed in Tab. 1. The computation of $\hat{\boldsymbol{h}}$ in (14) requires $3QLN + 6(QL)^2N + 3(QL)^3$ real multiplications. After updating the matrix $\boldsymbol{H}$ in (7) in the light of $\hat{\boldsymbol{h}}$, the computation of $\tilde{\boldsymbol{r}}$ in (15) requires $3NK$ real multiplications. Lastly, the computation of $\hat{\boldsymbol{s}}_d$ in (16) requires $9N^3 - 21KN^2 + 3N^2 + 15K^2N - 3KN - 3K^3$ real multiplications. Therefore, the whole algorithm requires $\mathrm{I}_m[9QN^3 + 15QN^2 + 3QLN^2 + 27N^2 + 6QL^2N + 3LN + 4MN + 16M^3 + 4M^2 + 2M\log(2M) + 6M] + 9N^3 - 21KN^2 + 3N^2 + 15K^2N + 6(QL)^2N + 3QLN - 3K^3 + 3(QL)^3$ real multiplications. Here, I_m represents the number of iterations in MBFGS. Note that, for large N, the computational complexity of the proposed method is $\sim O\left([9\mathrm{I}_mQ + 9]N^3\right)$.

On the other hand, in order to compare the complexity of different methods, the complexity of our proposed method, the LSE proposed in [6] and the ALS algorithm proposed in [24] are all presented in Tab. 2. Here, the symbol $\mathrm{I}_{[24]}$ denotes the number of iterations in [24]. Assuming $\mathrm{I}_{[24]} = \mathrm{I}_m$, it is concluded that these methods can be ordered in terms of the complexity in an ascending manner as the LSE, the ALS algorithm, and finally, our proposed method. However, the theory analysis above and the simulation results given in Section 5 confirm that these methods are ordered in terms of the performance in opposite direction.

Algorithm MBFGS

Step **1**: Initialize $k=0$, $\boldsymbol{\Omega}^{(0)}=\boldsymbol{I}_{2M}$, where $\boldsymbol{\Omega}$ denotes an approximation of the inverse Hessian of $\phi(\bar{\boldsymbol{x}}^{(k)})$.
Step **2**: In the light of (7) and (8), $\boldsymbol{P}^{(k)}$ and $\boldsymbol{H}^{(k)}$ can be constructed by $\bar{\boldsymbol{x}}^{(k)}$ and the pilot vector $\boldsymbol{s}_p$. Then, let $\nabla\phi(\bar{\boldsymbol{x}}^{(k)})$ denote the gradient of the $\phi(\bar{\boldsymbol{x}}^{(k)})$ with respect to $\bar{\boldsymbol{x}}^{(k)}$. And it can be easily obtained from

$$\nabla\phi(\bar{\boldsymbol{x}}^{(k)})=\begin{bmatrix}\mathrm{Re}[{\boldsymbol{H}_d}^{(k)}] & \mathrm{Re}[\boldsymbol{P}^{(k)}] & -\mathrm{Im}[{\boldsymbol{H}_d}^{(k)}] & -\mathrm{Im}[\boldsymbol{P}^{(k)}]\\ \mathrm{Im}[{\boldsymbol{H}_d}^{(k)}] & \mathrm{Im}[\boldsymbol{P}^{(k)}] & \mathrm{Re}[{\boldsymbol{H}_d}^{(k)}] & \mathrm{Re}[\boldsymbol{P}^{(k)}]\end{bmatrix}^T\begin{bmatrix}\mathrm{Re}[\boldsymbol{g}(\bar{\boldsymbol{x}}^{(k)})-\bar{\boldsymbol{r}}]\\ \mathrm{Im}[\boldsymbol{g}(\bar{\boldsymbol{x}}^{(k)})-\bar{\boldsymbol{r}}]\end{bmatrix} \tag{12}$$

where ${\boldsymbol{H}_d}^{(k)}$ is a $N\times(N-K)$ matrix, which is carved out of $\boldsymbol{H}^{(k)}$ corresponding to the information data subcarriers.
Step **3**: Compute the search direction $\boldsymbol{z}^{(k)}=-\boldsymbol{\Omega}^{(k)}\nabla\phi(\bar{\boldsymbol{x}}^{(k)})$.
Step **4**: Cast about the optimal step length $\lambda^{(k)}=\arg\min_{\lambda\geq 0}\phi(\bar{\boldsymbol{x}}^{(k)}+\lambda^{(k)}\boldsymbol{z}^{(k)})$.
Step **5**: Update solution $\bar{\boldsymbol{x}}^{(k+1)}=\bar{\boldsymbol{x}}^{(k)}-\lambda^{(k)}\boldsymbol{z}^{(k)}$.
Step **6**: Being similar to ***Step* 2**, $\boldsymbol{P}^{(k+1)}$ and $\boldsymbol{H}^{(k+1)}$ can be updated by $\bar{\boldsymbol{x}}^{(k+1)}$ and the pilot vector $\boldsymbol{s}_p$. Then, compute $\phi(\bar{\boldsymbol{x}}^{(k+1)})$ and check convergence. If $\left\|\phi(\bar{\boldsymbol{x}}^{(k+1)})\right\|<\varepsilon$ for sufficiently small values of ε, stop.
Step **7**: Compute $\nabla\phi(\bar{\boldsymbol{x}}^{(k+1)})$ by (12).
Step **8**: Set $\boldsymbol{p}^{(k+1)}=\lambda^{(k)}\boldsymbol{z}^{(k)}$ and $\boldsymbol{q}^{(k+1)}=\nabla\phi(\bar{\boldsymbol{x}}^{(k+1)})-\nabla\phi(\bar{\boldsymbol{x}}^{(k)})+t^{(k)}\left\|\nabla\phi(\bar{\boldsymbol{x}}^{(k)})\right\|\boldsymbol{p}^{(k)}$ with

$$t^{(k)}=1+\max\left\{0,-\frac{\left(\nabla\phi(\bar{\boldsymbol{x}}^{(k+1)})-\nabla\phi(\bar{\boldsymbol{x}}^{(k)})\right)^T\boldsymbol{p}^{(k)}}{\|\boldsymbol{p}^{(k)}\|^2}\right\},$$

then update the approximation of the inverse Hessian by

$$\boldsymbol{\Omega}^{(k+1)}=\left(\boldsymbol{I}_{2M}-\frac{\boldsymbol{p}^{(k)}\left[\boldsymbol{q}^{(k)}\right]^T}{\left[\boldsymbol{q}^{(k)}\right]^T\boldsymbol{p}^{(k)}}\right)\boldsymbol{\Omega}^{(k)}\left(\boldsymbol{I}_{2M}-\frac{\boldsymbol{q}^{(k)}\left[\boldsymbol{p}^{(k)}\right]^T}{\left[\boldsymbol{q}^{(k)}\right]^T\boldsymbol{p}^{(k)}}\right)+\frac{\boldsymbol{p}^{(k)}\left[\boldsymbol{p}^{(k)}\right]^T}{\left[\boldsymbol{q}^{(k)}\right]^T\boldsymbol{p}^{(k)}}. \tag{13}$$

Step **9**: Return to ***Step* 3** with $k=k+1$.

	The number of real multiplications
***Step* 1**	0
***Step* 2**	$6QN^3+9QN^2+3QL^2N+3QLN^2+3LN+4NM+N^2$
***Step* 3**	$4M^2$
***Step* 4**	$2M\log(2M)$
***Step* 5**	$2M$
***Step* 6**	$3QN^3+6QN^2+3QL^2N+4M$
***Step* 7**	$6QN^3+9QN^2+3QL^2N+3QLN^2+3LN+4NM+N^2$
***Step* 8**	$16M^3+26N^2$
***Step* 9**	0

Tab. 1. Computation complexity of each step in MBFGS.

	The number of real multiplications
Our proposed method	$\sim O\left([9\mathrm{I}_m Q+9]N^3\right)$
LSE	$\sim O(3QN^3)$
The ALS algorithm	$\sim O\left(6\mathrm{I}_{[24]}QN^3\right)$

Tab. 2. Computation complexity of each method.

4. Cramer-Rao Bound

In order to evaluate the quality of our proposed approach, the CRB for our estimation is derived in this section. Assuming the interference item $\boldsymbol{\psi}$ can be negligible, the received vector $\boldsymbol{r}$ in (7) or (8) is a complex circularly Gaussian process with mean $\boldsymbol{\mu}=\boldsymbol{Hs}$ or $\boldsymbol{\mu}=\boldsymbol{Ph}$ and variance $\boldsymbol{R}=\sigma_w^2\boldsymbol{I}_N$. According to the conclusion in [12], [31], its Fisher Matrix is given by

$$\begin{aligned}[\boldsymbol{J}]_{i,j}&=trace\left\{\boldsymbol{R}^{-1}\frac{\partial\boldsymbol{R}}{\partial\bar{\boldsymbol{x}}_i}\boldsymbol{R}^{-1}\frac{\partial\boldsymbol{R}}{\partial\bar{\boldsymbol{x}}_j}\right\}\\&\quad+2\mathrm{Re}\left[\frac{\partial\boldsymbol{\mu}^H}{\partial\bar{\boldsymbol{x}}_i}\boldsymbol{R}^{-1}\frac{\partial\boldsymbol{\mu}}{\partial\bar{\boldsymbol{x}}_j}\right]\\&=\frac{2}{\sigma_w^2}\mathrm{Re}\left[\frac{\partial\boldsymbol{\mu}^H}{\partial\bar{\boldsymbol{x}}_i}\frac{\partial\boldsymbol{\mu}}{\partial\bar{\boldsymbol{x}}_j}\right].\end{aligned} \tag{17}$$

For the sake of convenience, we can define

$$\bar{\boldsymbol{s}}_d=\left[\mathrm{Re}[\boldsymbol{s}_d]^T,\mathrm{Im}[\boldsymbol{s}_d]^T\right]^T,\bar{\boldsymbol{h}}=\left[\mathrm{Re}[\boldsymbol{h}]^T,\mathrm{Im}[\boldsymbol{h}]^T\right]^T.$$

As a result, (17) becomes

$$\boldsymbol{J}=\frac{2}{\sigma_w^2}\begin{bmatrix}\boldsymbol{J}_{\bar{\boldsymbol{s}}_d\bar{\boldsymbol{s}}_d} & \boldsymbol{J}_{\bar{\boldsymbol{s}}_d\bar{\boldsymbol{h}}}\\ \boldsymbol{J}_{\bar{\boldsymbol{h}}\bar{\boldsymbol{s}}_d} & \boldsymbol{J}_{\bar{\boldsymbol{h}}\bar{\boldsymbol{h}}}\end{bmatrix} \tag{18}$$

with

$$\boldsymbol{J}_{\bar{\boldsymbol{s}}_d\bar{\boldsymbol{s}}_d}=\mathrm{Re}\left[\frac{\partial\boldsymbol{\mu}^H}{\partial\bar{\boldsymbol{s}}_d}\frac{\partial\boldsymbol{\mu}}{\partial\bar{\boldsymbol{s}}_d}\right],\boldsymbol{J}_{\bar{\boldsymbol{s}}_d\bar{\boldsymbol{h}}}=\mathrm{Re}\left[\frac{\partial\boldsymbol{\mu}^H}{\partial\bar{\boldsymbol{s}}_d}\frac{\partial\boldsymbol{\mu}}{\partial\bar{\boldsymbol{h}}}\right],$$

$$\boldsymbol{J}_{\bar{\boldsymbol{h}}\bar{\boldsymbol{s}}_d}=\mathrm{Re}\left[\frac{\partial\boldsymbol{\mu}^H}{\partial\bar{\boldsymbol{h}}}\frac{\partial\boldsymbol{\mu}}{\partial\bar{\boldsymbol{s}}_d}\right],\boldsymbol{J}_{\bar{\boldsymbol{h}}\bar{\boldsymbol{h}}}=\mathrm{Re}\left[\frac{\partial\boldsymbol{\mu}^H}{\partial\bar{\boldsymbol{h}}}\frac{\partial\boldsymbol{\mu}}{\partial\bar{\boldsymbol{h}}}\right].$$

Since

$$\frac{\partial\boldsymbol{\mu}^H}{\partial\mathrm{Re}[\boldsymbol{s}_d]}=\frac{\partial\boldsymbol{s}^H\boldsymbol{H}^H}{\partial\mathrm{Re}[\boldsymbol{s}_d]}={\boldsymbol{H}_d}^H \tag{19}$$

and

$$\frac{\partial\boldsymbol{\mu}^H}{\partial\mathrm{Im}[\boldsymbol{s}_d]}=\frac{\partial\boldsymbol{s}^H\boldsymbol{H}^H}{\partial\mathrm{Im}[\boldsymbol{S}_d]}=-j{\boldsymbol{H}_d}^H, \tag{20}$$

then

$$\frac{\partial \mu^H}{\partial \bar{s}_d} = \frac{\partial s^H \boldsymbol{H}^H}{\partial \bar{s}_d} = \Xi_1 \boldsymbol{H}_d{}^H, \tag{21}$$

with

$$\Xi_1 = [\boldsymbol{I}_{N-K}, j\boldsymbol{I}_{N-K}]^H.$$

Similarly, we have

$$\frac{\partial \mu^H}{\partial \bar{\boldsymbol{h}}} = \frac{\partial \boldsymbol{h}^H \boldsymbol{P}^H}{\partial \bar{\boldsymbol{h}}} = \Xi_2 \boldsymbol{P}^H \tag{22}$$

with

$$\Xi_2 = [\boldsymbol{I}_{LQ}, j\boldsymbol{I}_{LQ}]^H.$$

Substituting (21) and (22) in (18), we obtain

$$\boldsymbol{J} = \frac{2}{\sigma_w^2}\begin{bmatrix} \mathrm{Re}\left[\Xi_1 \boldsymbol{H}_d{}^H \boldsymbol{H}_d \Xi_1{}^H\right] & \mathrm{Re}\left[\Xi_1 \boldsymbol{H}_d{}^H \boldsymbol{P} \Xi_2{}^H\right] \\ \mathrm{Re}\left[\Xi_2 \boldsymbol{P}^H \boldsymbol{H}_d \Xi_1{}^H\right] & \mathrm{Re}\left[\Xi_2 \boldsymbol{P}^H \boldsymbol{P} \Xi_2{}^H\right] \end{bmatrix}. \tag{23}$$

Then, the Cramer-Rao Bound for $\boldsymbol{h}$ can be expressed as

$$CRB_{\boldsymbol{h}} = \sum_{i=2(N-K)}^{2M-1} [\boldsymbol{J}^{-1}]_{i,i} \tag{24}$$

Note that the mean square error of the estimated $\boldsymbol{h}_l^t$ is given by

$$\begin{aligned} MSE_{\boldsymbol{h}_l^t} &= E\left[\left(\hat{\boldsymbol{h}}_l^t - \boldsymbol{h}_l^t\right)^H \left(\hat{\boldsymbol{h}}_l^t - \boldsymbol{h}_l^t\right)\right] \\ &= E\left[\left(\boldsymbol{B}\hat{\boldsymbol{h}}_l - \boldsymbol{B}\boldsymbol{h}_l\right)^H \left(\boldsymbol{B}\hat{\boldsymbol{h}}_l - \boldsymbol{B}\boldsymbol{h}_l\right)\right] \\ &= E\left[\left(\hat{\boldsymbol{h}}_l - \boldsymbol{h}_l\right)^H \boldsymbol{B}^H \boldsymbol{B} \left(\hat{\boldsymbol{h}}_l - \boldsymbol{h}_l\right)\right] \\ &= E\left[\left(\hat{\boldsymbol{h}}_l - \boldsymbol{h}_l\right)^H \left(\hat{\boldsymbol{h}}_l - \boldsymbol{h}_l\right)\right] \\ &= MSE_{\boldsymbol{h}_l} \end{aligned} \tag{25}$$

where $MSE_{\boldsymbol{h}_l}$ represents the mean square error of the estimated $\boldsymbol{h}_l$, for the columns of $\boldsymbol{B}$ being orthonormal.

Without loss of generality, we will consider the wide sense stationary uncorrelated scattering (WSSUS) model as the real channel mode, whose taps are independent with each other. Therefore, the mean square error of the estimated $\boldsymbol{h}$ can be written as

$$MSE_{\boldsymbol{h}} = L \times MSE_{\boldsymbol{h}_l}. \tag{26}$$

From (24), (25) and (26), we have

$$MSE_{\boldsymbol{h}_l^t} \geq \frac{1}{L} \sum_{2(N-K)}^{2M-1} [J^{-1}]_{i,i}. \tag{27}$$

5. Simulation Results

In this section, by using Matlab, we present simulation results to assess the QPSK-OFDM system performance based on the proposed algorithm. The parameters of the system selected are in concordance with the standard WiMAX IEEE 802.16e. The system operates with a 1.25 MHz bandwidth and is divided into 512 subcarriers. The carrier frequency is set to 3.5 GHz. The length of CP, as well as the number of pilots, is 64. The scheme given in [32], [33] will be used for generating the time-variant channels. We further assume the number of taps is 4. The mean power and the time delay of the lth path are $e^{-l/10}$ and $0.8 \cdot l\ \mu s$ for $l \in \{0, \cdots, L-1\}$, respectively. Moreover, all channel taps have a Jakes Doppler spectrum. In our proposed algorithm, the P-BEM will be adopted to simplify the time-variant channels, leading to estimated parameters reduction. As a rule of thumb, the Q depicted in (5) should satisfy $(Q-1)/2 = \lceil f_n \rceil + 1$ [34]. Here, f_n represent the normalized Doppler frequency shift. Therefore, in our simulation, $Q = 5$ is selected for $f_n = 0.2$ representing slowly time-variant cahnnels or $f_n = 0.8$ representing rapidly time-variant channels.

Note that, the receiver velocity v is related to the normalized Doppler frequency f_n by the formula [6], [16]

$$v = f_n \frac{cB}{f_c N}$$

where c, B and f_c represent the speed of light, the OFDM signal bandwidth and the carrier frequency, respectively. Hence, we can easily find that $f_n = 0.2$ and $f_n = 0.8$ are equivalent to a user moving at the speed of 151 km/h and 603 km/h with the WiMAX system parameters, respectively.

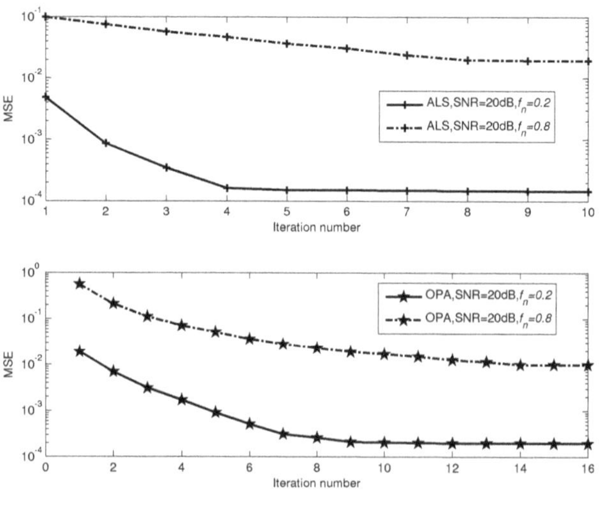

Fig. 1. Convergence characteristic of the ALS algorithm and OPA under various normalized Doppler shift. Solid curves: $f_n = 0.2$. Dashed curves: $f_n = 0.8$.

Fig. 1 shows the MSE of channel estimation for the ALS algorithm [24] and our proposed algorithm (OPA) under various numbers of iterations. Note, the solid lines and the dashed lines represent the simulations under the normalized Doppler frequency shift $f_n = 0.2$ and $f_n = 0.8$, respectively. From the graph, it can be seen that the ALS algorithm has to go through 4 iterations before convergence for

$f_n = 0.2$ while 8 iterations are needed for $f_n = 0.8$. OPA has to go through 9 iterations before convergence for $f_n = 0.2$ while 14 iterations are needed for $f_n = 0.8$. According to the complexity analysis in part 3 of Section 3, it can easily be found that the convergence speed of OPA is slower than that of the ALS algorithm, and consequently the complexity of OPA is about 3.5 times more than that of the ALS algorithm for $f_n = 0.2$ while about 2.7 times for $f_n = 0.8$. However, the simulation results given in Fig. 2 and Fig. 3 illustrate that OPA outperforms the ALS algorithm in rapidly time-variant channels.

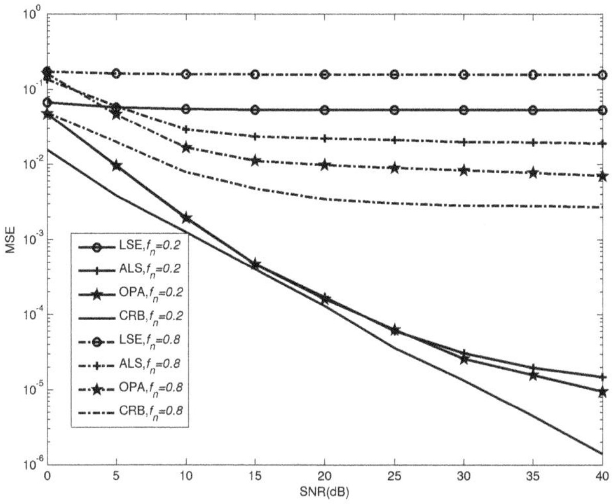

Fig. 2. MSE versus SNR for LSE, the ALS algorithm and OPA. Solid curves: $f_n = 0.2$. Dashed curves: $f_n = 0.8$.

In the condition of convergence, Fig. 2 shows the MSE of channel estimation for LSE, the ALS algorithm and OPA under various SNR. As a reference, we also plot the CRB for our proposed algorithm. The simulation results indicate that both the ALS algorithm and OPA are much better than LSE. This is due to LSE applying to the FDKD [35] pilot structure only and LSE not using the transmitted information data. Furthemore, when $f_n = 0.2$, the MSE of OPA is substantially the same as that of the ALS algorithm, which is relatively close to the CRB. When $f_n = 0.8$, the MSE of OPA is superior to that of the ALS algorithm. However, due to the BEM modeling error in (4) being relatively large under the rapidly time-variant channel, the MSE of OPA is still a little worse than that of CRB.

Fig. 3 shows the bit error rate (BER) performance with respect to the SNR for $f_n = 0.2$ and $f_n = 0.8$. It is obvious that the detection performance is almost in consistence with the corresponding channel estimation performance for the three methods. Especially, it is more apparent that the performance of our method is much better than that of the iterative method in the rapidly time-variant channel.

Furthermore, we operate our simulation in the Typical Urban (TU) channel [36] whose parameters are summarized in Tab. 3. The performance of convergence, estimation and detection for different algorithms are shown in Fig. 4, Fig. 5

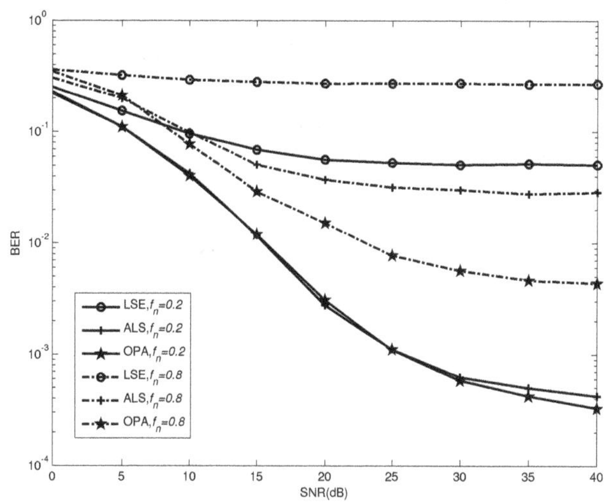

Fig. 3. BER versus SNR for LSE, the ALS algorithm and OPA. Solid curves: $f_n = 0.2$. Dashed curves: $f_n = 0.8$.

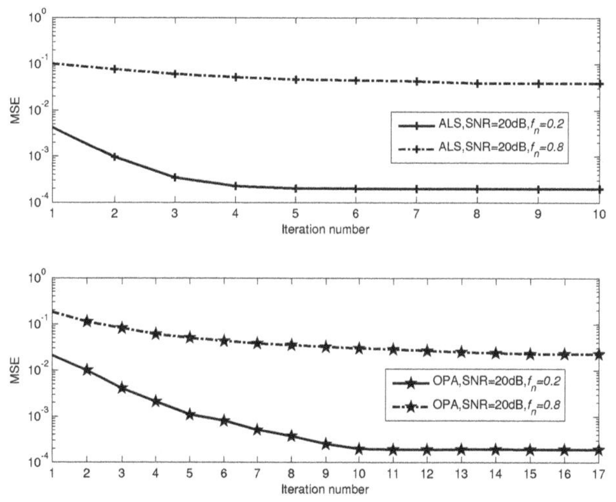

Fig. 4. Convergence characteristic of the ALS algorithm and OPA under various normalized Doppler shift values in the Typical Urban (TU) channel. Solid curves: $f_n = 0.2$. Dashed curves: $f_n = 0.8$.

Tap	Tap delay (μs)	Tap gain (dB)	Doppler spectrum
0	0	-3	Jakes
1	0.2	0	Jakes
2	0.5	-2	Jakes
3	1.6	-6	Gauss I
4	2.3	-8	Gauss II
5	5.0	-10	Gauss II

Tab. 3. Typical Urban (TU) channel parameters.

and Fig. 6, respectively. Fig. 4 indicates that the MSE of the ALS algorithm converges to the optimal point in 5 iterations for $f_n = 0.2$ and in 8 iterations for $f_n = 0.8$. Further, the MSE of OPA converges to the optimal point in 10 iterations for $f_n = 0.2$ and in 14 iterations for $f_n = 0.8$. As well as the description in Fig. 1, this shows that OPA keeps worse convergence performance than the ALS algorithm in

TU channel, leading to high computational complexity. Nevertheless, as shown in Fig. 2, Fig. 5 illustrates that the MSE performance of OPA in rapidly time-variant channels is still better than that of the ALS algorithm. In addition, comparing the results in Fig. 6 with those in Fig. 3, it is obvious that OPA also has lower BER than the ALS algorithm in the worse channel, i.e., the TU channel.

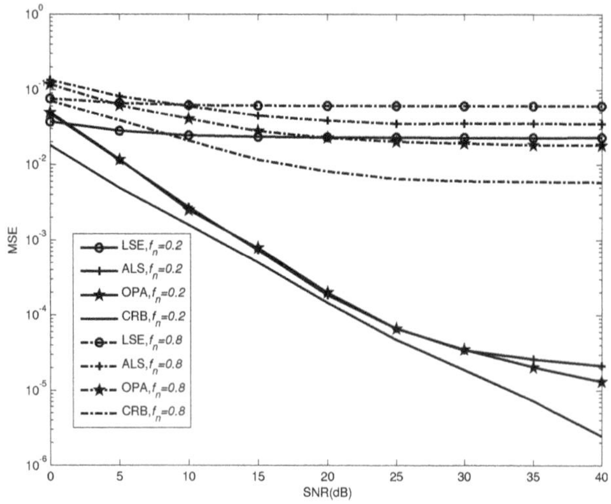

Fig. 5. MSE versus SNR for LSE, the ALS algorithm and OPA in the Typical Urban (TU) channel. Solid curves: f_n = 0.2. Dashed curves: f_n = 0.8.

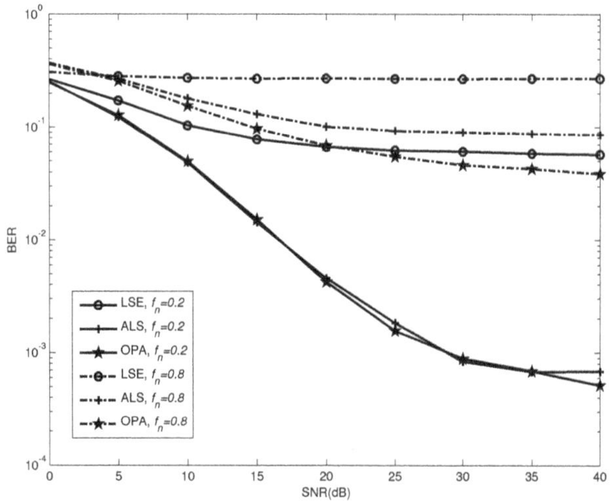

Fig. 6. BER versus SNR for LSE, the ALS algorithm and OPA in the Typical Urban (TU) channel. Solid curves: f_n = 0.2. Dashed curves: f_n = 0.8.

6. Conclusion

In this paper, we propose a novel algorithm for joint channel estimation and signal detection, which are considered a real NLS problem. Then, the MBFGS algorithm is adopted to solve the problem. Moreover, the CRB is derived for evaluating the quality of our proposed algorithm. Simulation results show that our proposed algorithm achieves better performance than the iterative method in rapidly time-variant channel.

Acknowledgements

This work was supported by the 111 Project (B08038), the National Natural Science Foundation of China under Grant No. 61271299 and the Shenzhen Kongqie talent program under Grate YFZZ20111013.

Appendix A

LS method can be regarded as the optimal algorithm in slowly time-variant channels.

First, we will prove that the LS method applied in (8) can be regarded as the optimal algorithm in slowly time-variant channels.

Since the BEM modeling error in (4) is very small in slowly time-variant channels, the interference item ψ in (8) can be negligible. Therefore, (8) can be rewritten as

$$\boldsymbol{r} = \boldsymbol{H}\boldsymbol{s} + \omega \tag{28}$$

With $\boldsymbol{H}$, the transmitted data $\boldsymbol{s}$ in (28) can be estimated by maximum likelihood estimator (MLE) which is considered the optimal estimator [12]. The MLE estimate is obtained by maximizing the likelihood function of the received data $\boldsymbol{r}$ given transmitted data $\boldsymbol{s}$, which is given by

$$p(\boldsymbol{r}|\boldsymbol{s}) = \frac{1}{(2\pi)^{N/2}|\boldsymbol{R}|^{1/2}} \exp\left[-\frac{1}{2}(\boldsymbol{r}-\boldsymbol{H}\boldsymbol{s})^H \boldsymbol{R}^{-1}(\boldsymbol{r}-\boldsymbol{H}\boldsymbol{s})\right]. \tag{29}$$

For computational convenience, the MLE estimate can be obtained by the log-likelihood function, which can be expressed as

$$\ln p(\boldsymbol{r}|\boldsymbol{s}) = \ln \frac{1}{(2\pi)^{N/2}|\boldsymbol{R}|^{1/2}} - \frac{1}{2}(\boldsymbol{r}-\boldsymbol{H}\boldsymbol{s})^H \boldsymbol{R}^{-1}(\boldsymbol{r}-\boldsymbol{H}\boldsymbol{s}). \tag{30}$$

By setting the derivative of the log-likelihood function with respect to $\boldsymbol{s}$ to zero, we have

$$\begin{aligned}\frac{\partial \ln p(\boldsymbol{r}|\boldsymbol{s})}{\partial \boldsymbol{s}} &= -\frac{1}{2}\frac{\partial(\boldsymbol{r}-\boldsymbol{H}\boldsymbol{s})^H \boldsymbol{R}^{-1}(\boldsymbol{r}-\boldsymbol{H}\boldsymbol{s})}{\partial \boldsymbol{s}} \\ &= -\frac{1}{2}\boldsymbol{H}^H \boldsymbol{R}^{-1}(\boldsymbol{r}-\boldsymbol{H}\boldsymbol{s}) \\ &= 0\end{aligned} \tag{31}$$

which leads to the MLE solution for $\boldsymbol{s}$ given by

$$\begin{aligned}\boldsymbol{s} &= (\boldsymbol{H}^H \boldsymbol{R}^{-1}\boldsymbol{H})^{-1}\boldsymbol{H}^H \boldsymbol{R}^{-1}\boldsymbol{r} \\ &= (\boldsymbol{H}^H \boldsymbol{H})^{-1}\boldsymbol{H}^H \boldsymbol{r} \\ &= \boldsymbol{H}^{\dagger}\boldsymbol{r}.\end{aligned} \tag{32}$$

As can be seen in the equation above, the solution for MLE is the same as that for LS estimator. Hence, the LS method applied in (8) can be regarded as the optimal algorithm in slowly time-variant channels.

Similarly, the LS method applied in (7) also can be regarded as the optimal algorithm in slowly time-variant channels.

References

[1] POLAK, L., KRATOCHVIL, T. DVB-H and DVB-SH-A performance and evaluation of transmission in fading channels. In *Proceedings of the 34th International Conference on Telecommunication and Signal Processing (TSP2011)*. Budapest (Hungary), 2011, p. 549 - 553.

[2] SANZI, F., JELTING, S., SPEIDEL, J. A comparative study of iterative channel estimators for mobile OFDM systems. *IEEE Transactions on Wireless Communications*, 2003, vol. 2, no. 5, p. 849 - 859.

[3] NEVAT, I., YUAN, J. Error propagation mitigation for iterative channel tracking, detection and decoding of BICM-OFDM systems. In *4th International Symposium on Wireless Communication Systems (ISWCS)*. Trondheim (Norway), 2007, p. 75 - 80.

[4] AL-NAFFOURI, T. An EM-based forward-backward Kalman filter for the estimation of time-variant channels in OFDM. *IEEE Transactions on Signal Processing*, 2007, vol. 55, no. 7, p. 3924 - 3930.

[5] DONG, M., TONG, L., SADLER, B. Optimal insertion of pilot symbols for transmissions over time-varying flat fading channels. *IEEE Transactions on Signal Processing*, 2004, vol. 52, no. 5, p. 1403 - 1418.

[6] TANG, Z., CANNIZARO, R. C., LEUS, G., BANELLI, P. Pilot-assisted time-varying channel estimation for OFDM systems. *IEEE Transactions on Signal Processing*, 2007, vol. 55, no. 5, p. 2226 - 2238.

[7] TSATSANIS, M. K., GIANNAKIS, G. B. Modeling and equalization of rapidly fading channels. *International Journal of Adaptive Control and Signal Processing*, 1996, vol. 10, no. 2 - 3, p. 159 - 176.

[8] LEUS, G. On the estimation of rapidly time-varying channels. In *Proceedings of 12th European Signal Processing Conference (EUSIPCO)*. Vienna (Austria), 2004, p. 2227 - 2230.

[9] TOMASIN, S., GOROKHOV, A., YANG, H., LINNARTZ, J. P. Iterative interference cancellation and channel estimation for mobile OFDM. *IEEE Transactions on Wireless Communications*, 2005, vol. 4, no. 1, p. 238 - 245.

[10] VISINTIN, M. Karhunen-Loeve expansion of a fast Rayleigh fading process. *IEEE Electronics Letters*, 1996, vol. 32, no. 8, p. 1712 - 1713.

[11] ZEMEN, T., MECKLEMBRÄUKER, C. F. Time-variant channel estimation using discrete prolate spheroidal sequences. *IEEE Transactions on Signal Processing*, 2005, vol. 53, no. 9, p. 3597 - 3607.

[12] KEY, S. M. *Fundamentals of Statistical Signal Processing: Estimation Theory*. Englewood Cliffs (NJ, USA): Prentice-Hall, 1993.

[13] HUANG, L., LONG, T., MAO, E., SO, H. C. MMSE-based MDL method for robust estimation of number of sources without eigendecomposition. *IEEE Transactions on Signal Processing*, 2009, vol. 57, no. 10, p. 4135 - 4142.

[14] HUANG, L., LONG, T., MAO, E., SO H. C. MMSE-based MDL method for accurate source number estimation. *IEEE Signal Processing Letters*, 2009, vol. 16, no. 9, p. 798 - 801.

[15] TADMOR, E. Filters, mollifiers and the computation of the Gibbs phenomenon. *Acta Numerica*, 2007, vol. 16, p. 305 - 378.

[16] HRYCAK, T., DAS, S., MATZ, G., FEICHTINGER, H. Practical estimation of rapidly varying channels for OFDM systems. *IEEE Transactions on Communications*, 2011, vol. 59, no. 11, p. 3040 - 3048.

[17] HRYCAK, T., DAS, S., MATZ, G. Inverse methods for reconstruction of channel taps in OFDM systems. *IEEE Transactions on Signal Processing*, 2012, vol. 60, no. 5, p. 2666 - 2671.

[18] TAO, C., QIU, J., LIU, L. A novel OFDM channel estimation algorithm with ICI mitigation over fast fading channels. *Radioengineering*, 2010, vol. 19, no. 2, p. 347 - 355.

[19] MOSTOFI, Y., COX, D. ICI mitigation for pilot-aided OFDM mobile systems. *IEEE Transactions on Wireless Communications*, 2005, vol. 4, no. 2, p. 765 - 774.

[20] CHOI, Y. S., VOLTZ, P. J., CASSARA, F. A. On channel estimation and detection for multicarrier signals in fast and selective Rayleigh fading channels. *IEEE Transactions on Communications*, 2001, vol. 49, no. 8, p. 1375 - 1387.

[21] WANG, H. W., LIN, D. W., SANG, T. H. OFDM signal detection in doubly selective channels with whitening of residual intercarrier interference and noise. *IEEE Journal on Selected Areas in Communications*, 2012, vol. 30, no. 4, p. 684 - 694.

[22] SEBESTA, V., MARSALEK, R., FEDRA, Z. OFDM signal detector based on cyclic autocorrelation function and its properties. *Radioengineering*, 2011, vol. 20, no. 4, p. 926 - 931.

[23] LI, R., HUANG, L., SHI, Y., SO, H. C. Gerschgorin disk-based robust spectrum sensing for cognitive radio. In *Proceedings of the International Conference on Acoustics, Speech, and Signal Processing (ICASSP)*. Florence (Italy), 2014, p. 7328 - 7332.

[24] HIJAZI, H., ROS, L. Polynomial estimation of time-varying multipath gains with intercarrier interference mitigation in OFDM systems. *IEEE Transactions on Vehicular Technology*, 2009, vol. 58, no. 1, p. 140 - 151.

[25] PANAYIRCI, E., SENOL, H., POOR, H. V. Joint channel estimation, equalization, and data detection for OFDM systems in the presence of very high mobility. *IEEE Transactions on Signal Processing*, 2010, vol. 58, no. 8, p. 4225 - 4238.

[26] ABOUTORAB, N., HARDJAWANA, W., VUCETIC, B. A new iterative Doppler-assisted channel estimation joint with parallel ICI cancellation for high-mobility MIMO-OFDM systems. *IEEE Transactions on Vehicular Technology*, 2012, vol. 61, no. 4, p. 1577 - 1589.

[27] HARDJAWANA, W., LI, R., VUCETIC, B., LI, Y. A new iterative channel estimation for high mobility MIMO-OFDM systems. In *Proceedings of IEEE 71st Vehicular Technology Conference (VTC 2010 - Spring)*. Taipei (Taiwan), 2010, p. 1 - 5.

[28] LI, D. H., FUKUSHIMA, M. A modified BFGS method and its global convergence in nonconvex minimization. *Journal of Computational and Applied Mathematics*, 2001, vol. 129, p. 15 - 35.

[29] AGRWAL, R., CIOFFI, J. M. Beamforming design for the MIMO downlink for maximizing weighted sum-rate. In *Proceedings of International Symposium on Information Theory and Its Applications (IISITA)*. Auckland (New Zealand), 2008, p. 1 - 6.

[30] NOCEDAL, J., WRIGHT, S. J. *Numerical Optimization*. New York: Springer-Verlag, 2006.

[31] HUANG, J. Y., WANG, P., SHEN, X. F. CRLBs for pilot-aided channel estimation in OFDM system under Gaussian and non-Gaussian mixed noise. *Radioengineering*, 2012, vol. 21, no. 4, p. 1117 - 1124.

[32] ZHENG, Y. R., XIAO, C. Simulation models with correct statistical properties for Rayleigh fading channels. *IEEE Transactions on Communications*, 2003, vol. 51, no. 6, p. 920 - 928.

[33] HUANG, L., SO, H. C., QIAN, C. Volume-based method for spectrum sensing. *Digital Signal Processing*, 2014, vol. 28, p. 48 - 56.

[34] SCHNITER, P. Low-complexity equalization of OFDM in doubly selective channels. *IEEE Transactions on Signal Processing*, 2004, vol. 52, p. 1002 - 1011.

[35] KANNU, A. P., SCHNITER, P. MSE-optimal training for linear time-varying channels. *Proceedings of IEEE International Conference on Acoustics, Speech, and Signal Processing (ICASSP)*. 2005, p. 789 - 792.

[36] PÄTZOLD, M. *Mobile Fading Channels*. New York: Wiley-IEEE Press, 2002.

About Authors ...

Min HUANG was born in 1985. He received the B.S. and M.E. degrees in Electrical Communication from Xidian University, China in 2008 and 2010, respectively. Now, he is working toward the doctor degree in the State Key Laboratory of Integrated Service Networks, Xidian University. His research interest includes wireless communication, digital signal processing.

Bingbing LI was born in 1955. He received the Ph.D. degree in Communication and Information Systems from Xidian University in 1995. He is now a professor of State Key Laboratory of ISN at the Department of Communication Engineering in Xidian University. He has worked at the DTV standardization as a DTV technical expert. His research interests include digital video broadcasting systems and mobile communication techniques.

8

Low Complexity Noncoherent Iterative Detector for Continuous Phase Modulation Systems

Bing LI[1], *Baoming BAI*[1,2]

[1]State Key Lab. of ISN, Xidian University, Xi'an, 710071, P. R. China
[2]Science and Technology on Information Transmission and Dissemination in Communication Networks Laboratory, Shijiazhuang, 050002, P. R. China

libingprc@gmail.com, bmbai@mail.xidian.edu.cn

Abstract. *This paper focuses on the noncoherent iterative detection of continuous phase modulation. A class of simplified receivers based on Principal-Component-Analysis (PCA) and Exponential-Window (EW) is developed. The proposed receiver is evaluated in terms of minimum achievable Euclidean distance, simulated bit error rate and achievable capacity. The performance of the proposed receiver is discussed in the context of mismatched receiver and the equivalent Euclidean distance is derived. Analysis and numerical results reveal that the proposed algorithm can approach the coherent performance and outperforms existing algorithm in terms of complexity and performance. It is shown that the proposed receiver can significantly reduce the detection complexity while the performance is comparable with existing algorithms.*

Keywords

Continuous phase modulation, noncoherent iterative detection, time-varying phase noise, serially concatenated systems, minimum achievable distance, principal component analysis, mismatched receiver, wireless communication.

1. Introduction

Continuous-Phase-Modulation (CPM) is a class of coded modulations with good power and bandwidth efficiency [1], which has been widely used in wireless communication systems. As shown in some previous works [2–5] that CPM can offer near capacity performance in various scenarios such as satellite communication, deep space communication, optical communication, digital video broadcasting (DVB) [6–9] and the discussion has generalized to CPM-based multiuser systems very recently [10–14], which show that CPM-based system can significantly improve the spectral-efficiency.

However, previous discussion presumed coherent detection which requires perfect acquisition of channel state information at the receiver side. This is usually unavailable in practice. On the other hand, noncoherent detection makes itself an attractive strategy due to the fact that no explicit phase estimation is required. It is firstly shown in [15] that noncoherent receiver using multiple-symbol differential detection can perform close to the coherent receiver with exponential complexity. As a matter of fact, it was analytically explained in [1] that the performance of the maximum-likelihood noncoherent detection is actually identical to coherent receiver in terms of Euclidean distance. Inspired by this work, some simplified noncoherent detectors were later developed in [16–20]. It is shown that near coherent performance is obtained with significantly reduced complexity.

Though achieving near coherent performance, existing algorithms aforementioned exhibit considerable complexity for CPMs of large alphabet size, i.e. $M \geq 4$. Therefore, in this paper we develop a reduced complexity noncoherent iterative receiver for uncoded and coded CPM system. The proposed receiver is built upon two modules: a low-dimensional front-end and a trellis-based detector. The front-end is based upon Principal-Component-Analysis (PCA) [21] which is in particular suited for partial response system based on the technique reported in [22, 23]. The trellis-based detector followed employs Exponential-Window (EW) [24] to further reduce the complexity of detection complexity.

Another concern is to evaluate the performance of the proposed receiver subject to time-varying phase noise, which could be introduced by the channel or inaccurate estimated carrier frequency. To achieve a better performance, we employ a factor to weight the soft metric delivered from the outer convolutional decoder to the inner CPM detector. The optimum value of this factor is obtained through exhaustive simulations. It turns out that the performance of the proposed receiver offers better performance tackling the time-varying phase noise if the factor is properly chosen.

This paper is organized as follows. The system model is presented in Section 2, Section 3 presents the simplified receivers and derives the equivalent Euclidean distance. The proposed receiver is compared to some existing ones in terms of complexity. Section 4 gives the numerical and simulated results and Section 5 concludes the paper.

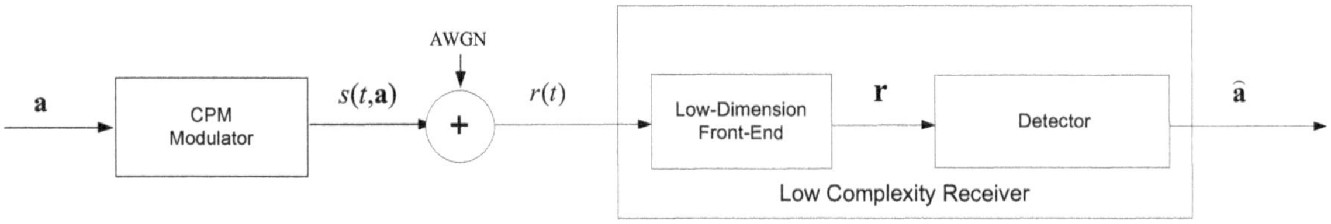

Fig. 1. System model.

2. System Model

The system model given in Fig. 1. The modulator generates the transmitted signal $s(t,\mathbf{a})$ given the M-ary information sequence $\mathbf{a}$=$\{c_0,c_1,\ \ldots\ ,c_{N-1}\}$, and the complex baseband signal of CPM in the time interval $nT < t < (n+1)T$ is defined as [1]

$$s(t,\mathbf{a}) = \sqrt{\frac{2E_s}{T}}\exp\{j2\pi h\sum_{i=0}^{n} a_i q(t-iT)\} \tag{1}$$

where E_s is the energy per symbol, T is the symbol interval, $h = k/p$ is the modulation index (k and p are relatively prime integers), and the symbols a_i are assumed independent and takes on values from the M-ary alphabet $\{\pm1,\pm3,\ \ldots\ ,\pm(M-1)\}$. The function $q(t)$ is the phase response and its derivative is the frequency pulse, assumed of duration L. The information bearing phase $\theta(t,\mathbf{a})$ is defined accordingly as [1]

$$\begin{aligned}\theta(t,\mathbf{a}) &= 2\pi h\sum_{n=0}^{k} a_n q(t-nT)\\ &= \pi h\sum_{n=0}^{k-L} a_n + 2\pi h\sum_{k=N-L+1}^{k} a_n q(t-nT)\\ &= \theta_n + \theta(t),\end{aligned} \tag{2}$$

$$\theta_n \triangleq [\pi h\sum_{i=0}^{n-L} a_i]_{mod\ 2\pi}, \tag{3}$$

$$\theta(t) \triangleq 2\pi h\sum_{i=n-L+1}^{n} a_i q(t-iT) \tag{4}$$

where θ_n is the *accumulated phase*, and $\theta(t)$ is the *incremental phase* within one interval. At the receiver side, coherent trellis defined accordingly as shown in [1].

In this paper, we focus on the Additive-White-Gaussian-Noise (AWGN) channel wherein the received signal $r(t)$ reads

$$r(t) = s(t,\mathbf{a})e^{j\varphi_n} + n(t) \tag{5}$$

where $\varphi_{n\ \mathrm{mod}\ 2\pi} \in [0,2\pi]$ is the phase noise assumed random and $n(t)$ is a zero-mean circularly symmetric white Gaussian noise process of two-sided power spectral density $N_0/2$. The phase noise is modeled as a discrete time random walk (Wiener) process defined as [24]:

$$\varphi_n = \varphi_{n-1} + \sigma_n \tag{6}$$

where φ_n are assumed to be i.i.d Gaussian random variable. The normalized Euclidean distance is defined as

$$d^2(\mathbf{a},\mathbf{b}) = \frac{1}{2E_b}\int_0^{NT} |s(t,\mathbf{a}) - s(t,\mathbf{b})|^2 dt, \tag{7a}$$

$$d^2_{min} = \min_{all\ \mathbf{a}\neq\mathbf{b}} d^2(\mathbf{a},\mathbf{b}) \tag{7b}$$

where E_b is the average energy per information bit.

3. Low Complexity Iterative Receiver

The proposed receiver consists of two modules: front-end and detector. The front-end adopted is in fact a mismatched filter. The main idea is using an alternative signal space $s_R(t,\mathbf{a})$, whose size is much smaller than the original signal space $s(t,\mathbf{a})$. These filters should be optimized first such that the minimum achievable distance is maximized. The detector followed is also defined over $s_R(t,\mathbf{a})$ which reduces the search effort for optimum detection. As we shall see latter, the proposed receiver can reduce the complexity significantly while results in little performance loss.

3.1 Generalized PCA

Similar to [14, 21], the method presented here is based on eigenvalue analysis. The main idea is utilizing an alternative low-dimensional signal space $s_R(t,\mathbf{a})$ by shortening L to L_R at the receiver side. Comparing to the conventional method, the dimensionality is reduced from $D = M^L$ to $K(<< D)$. The details are presented below.

1. Calculate the correlation matrix $R_{D\times D}$ which is defined as
$R_{D\times D} = \langle s_R(t,\mathbf{a}), s_R(t,\mathbf{a}))\rangle$,
$(s_R(t,,\mathbf{a}) = [s_{R_0}(t,\mathbf{a}),\ldots,s_{R_{D-1}}(t,\mathbf{a})]);$
2. Using eigenvalue decomposition we have
$R_{D\times D} = Q_1 diag[\lambda_0,\ldots,\lambda_{D-1}]\,Q_1^H;$

3. The orthogonal basis are obtained as $\beta(t) = diag\,[\lambda_0, \ldots, \lambda_{D-1}]\, Q_1 s(t)$;
4. The constellations are $\mathbf{s} = \langle s(t,\mathbf{a}), \varphi(t)\rangle$.

Only those basis having positive eigenvalue are considered effective (of which the number is D) to construct the low-dimensional front-end for CPMs.

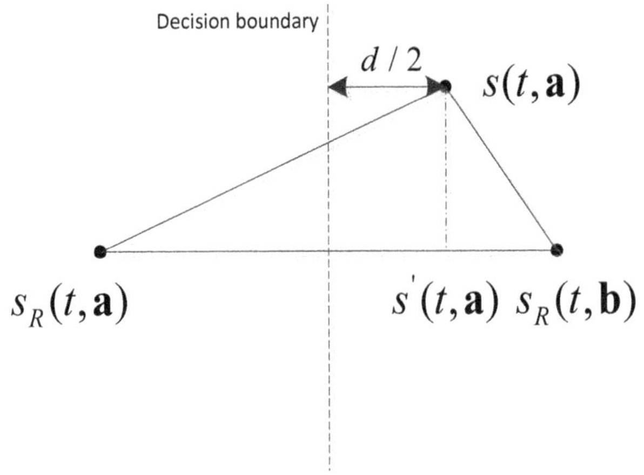

Fig. 2. Geometric interpretation of d.

Another module of the receiver is the mismatched detector. Now the performance of the proposed technique is evaluated in terms of equivalent minimum achievable Euclidean distance d^2. Define $\boldsymbol{r}$, $\boldsymbol{s_T}$, $\boldsymbol{n}$, $\boldsymbol{s}$ and $\boldsymbol{s'}$ the vector representation of $r(t)$, $n(t)$, $s(t,\mathbf{a})$, $s_R(t,\mathbf{a})$ and $s_R(t,\mathbf{b})$ given a set of orthogonal bases $\beta(t) = \{\beta_1(t), \ldots, \beta_D(t)\}$, respectively. The detection problem is now formulated as [14]

$$|\boldsymbol{r}-\boldsymbol{s}|^2 \underset{\mathbf{a}}{\overset{\mathbf{b}}{\gtrless}} |\boldsymbol{r}-\boldsymbol{s'}|^2 \tag{8}$$

which is equivalent to

$$|\boldsymbol{s}|^2 - |\boldsymbol{s'}|^2 - 2\Re < \boldsymbol{r},\boldsymbol{s} > +2\Re < \boldsymbol{r},\boldsymbol{s'} > \underset{\mathbf{a}}{\overset{\mathbf{b}}{\gtrless}} 0. \tag{9}$$

Let $y = |\boldsymbol{s}|^2 - |\boldsymbol{s'}|^2 - 2\Re < \boldsymbol{r},\boldsymbol{s} > +2\Re < \boldsymbol{r},\boldsymbol{s'} >$, which is a Gaussian random variable by definition. The mean m and variance σ^2 of y are, respectively

$$m = |\boldsymbol{s_T}-\boldsymbol{s}|^2 - |\boldsymbol{s_T}-\boldsymbol{s'}|^2 \tag{10}$$

and

$$\sigma^2 = 2N_0\,|\boldsymbol{s}-\boldsymbol{s'}|^2. \tag{11}$$

Therefore, the probability that the transmitted sequence $\mathbf{a}$ is wrongly detected as $\mathbf{b}$ is

$$\begin{aligned}\mathbf{P}(y>0) &= Q\left(\sqrt{\frac{m^2}{\sigma^2}}\right)\\ &= Q\left(\sqrt{\frac{\left[|\boldsymbol{s_T}-\boldsymbol{s}|^2 - |\boldsymbol{s_T}-\boldsymbol{s'}|^2\right]^2}{2E_b\,|\boldsymbol{s}-\boldsymbol{s'}|^2}\cdot\frac{E_b}{N_0}}\right)\end{aligned} \tag{12}$$

where E_b is the average transmitted energy per information bit. The equivalent Euclidean distance is readily recognized as

$$\begin{aligned}d^2 &= \frac{1}{2E_b}\frac{\left[|\boldsymbol{s_T}-\boldsymbol{s}|^2 - |\boldsymbol{s_T}-\boldsymbol{s'}|^2\right]^2}{|\boldsymbol{s}-\boldsymbol{s'}|^2}\\ &= \frac{1}{2E_b}\left[\frac{|s(t,\mathbf{a})-s_R(t,\mathbf{b})|^2 - |s(t,\mathbf{a})-s_R(t,\mathbf{a})|^2}{|s_R(t,\mathbf{a})-s_R(t,\mathbf{b})|}\right]^2.\end{aligned} \tag{13}$$

As expected, (13) coincides with the result in [21, 25]. It is noticed that d^2 is positive by definition but is not additive. Therefore, no efficient method but an exhaustive search is employed to find this quantity in most cases. A geometric interpretation of d is shown in Fig. 2. Except minimum achievable distance d^2, the performance of the PCA-based receiver is measured in terms of the average distance loss which defined as

$$\delta = \frac{1}{M^{2L}}\sum_{i=0}^{M^L}\sum_{j=0}^{M^L} d_{ij} - d'_{ij} \tag{14}$$

where d_{ij} and d'_{ij} denotes the Euclidean distance over one symbol interval between two signals, for the optimal (full-rank) receiver, and for the D rank approximation, respectively. Another parameter indicating the performance loss is the average energy loss which is defined as

$$\varepsilon = \sum_{i=0}^{M^L}\lambda_i. \tag{15}$$

3.2 Achievable Capacity

The capacity of a communication system is defined as the maximum mutual information between the channel input and the channel output over all possible input distributions. Unfortunately, unlike the conventional memoryless modulation, it is impossible to obtain a closed-form expression for CPM based system. However, some recent results [14, 26] reveal that the capacity of such a system can be calculated numerically. Therefore, in this paper we generalize the discussion to noncoherent detection.

At the ith epoch, designate the transmitted symbol, assuming the input a_i is uniformly distributed, the average information rate C is calculated as [14]

$$\begin{aligned}C &= \lim_{N\to\infty}\frac{1}{N}I\left(a_1^N, r_1^N\right)\\ &= \lim_{N\to\infty}\frac{1}{N}H\left(a_1^N\right) - H\left(a_1^N|r_1^N\right)\\ &= \lim_{N\to\infty}\frac{1}{N}\sum_{i=1}^{N}H(a_i) - \sum_{i=1}^{N}H(a_{i-1}|a_1^{i-1}, r_i^N)\\ &= \lim_{N\to\infty}\frac{1}{N}\sum_{i=1}^{N}\log_2 M - \sum_{i=1}^{N}H(a_i|a_1^{i-1}, r_i^N)\\ &\le \log_2 M \quad (SNR\to\infty)\end{aligned} \tag{16}$$

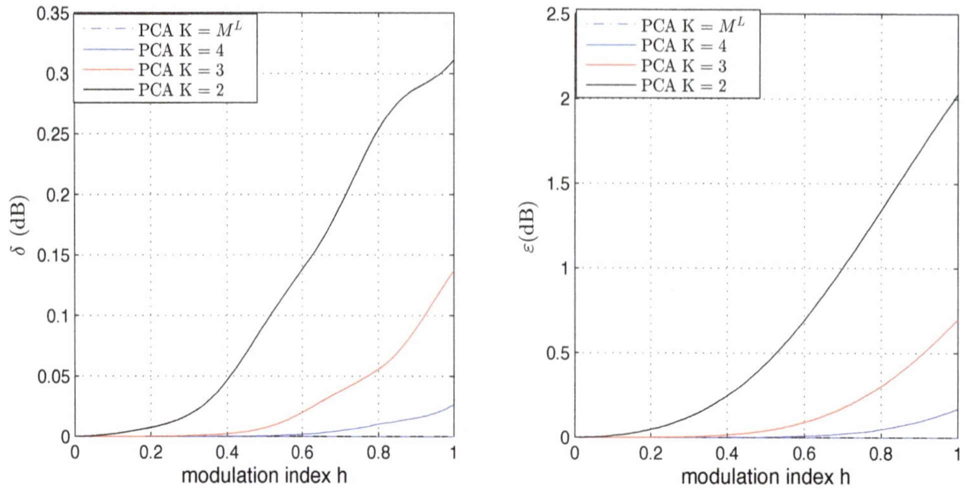

Fig. 3. The performance degradation by reducing the rank from $M^L = 16$ to $K = 2,3,4$.

where M is the modulation level. The chain rule [26] and $H(a_i|a_1^N) = H(a_i) = \log_2 M$ are used. The upper bound $\log_2 M$ is achievable as the Signal-to-Noise-Ratio (SNR) is sufficiently large. The problem is now to calculate $H(a_i|a_1^N, r_i^N)$ which is rewritten as

$$H(a_i|a_1^{i-1}, r_i^N) = -E\left[\log_2 p\left(a_i|a_1^{i-1}, r_1^N\right)\right]. \quad (17)$$

The quantity $H(a_i|a_1^{i-1}, r_i^N)$ is usually obtained through Monte-Carlo calculation [26] and the APP $p\left(a_i|a_1^{i-1}, r_1^N\right)$ is obtained employing the N-SISO algorithm proposed. Therefore, the performance of N-SISO can also be evaluated by the achievable capacity which will be shown later.

3.3 Serially Concatenated CPM

It is shown in [2–4] that serially concatenated CPM offers near capacity performance. A detailed discussion in [2–4] reveals that the performance is significantly improved due to the interleaver-gain. In this paper, both uncoded and coded CPM are considered. We show that the proposed noncoherent CPM receiver can successfully be applied to the concatenated system.

The details of the coded system are demonstrated in Fig. 4. The information sequence **u** is first coded by the outer encoder. The interleaved code **a** is modulated by CPM and transmitted to the AWGN channel. At the receiver side, the CPM receiver based on N-SISO generates the *a posterior probability* (APP) $p(a_i|\mathbf{r})$ of the code symbol a_i. The APPs are then deterleaved and passed to the outer decoder which manages to generate the prior information of the code symbol $p(a_i)$. The a priori information $p(a_i)$ is weighted by a factor before feeding it to the CPM receiver. This process repeats several times and finally obtains a decision of **u** designated as $\hat{\mathbf{u}}$. This process is visualized in Fig. 4. It should be pointed out that the performance of such a system relies heavily on the CPM receiver when noncoherent detection is considered. Therefore, our goal is to design a CPM receiver

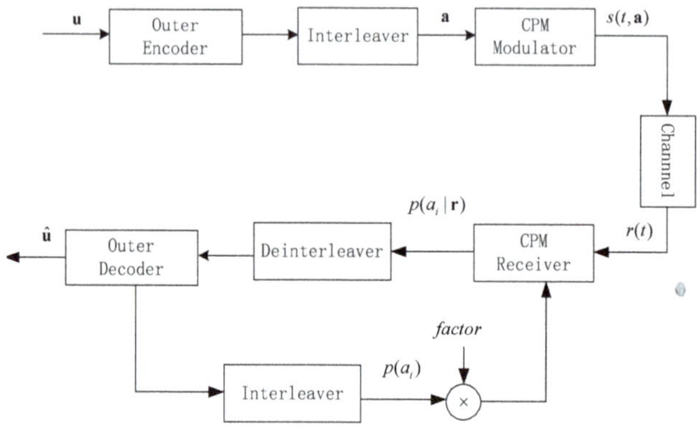

Fig. 4. The system model of the serially concatenated CPM.

which is able to successfully abstract the APPs noncoherently when subject to a time-varying phase noise.

3.4 Noncoherent Iterative Detector

The noncoherent soft-input-soft-output (N-SISO) detector presented in this paper is actually a bi-directional Viterbi algorithm employing forward and backward search to calculate the APP of each transmitted symbol. This quantity is later used to make hard decision (i.e., uncoded system) or fed to decoder for joint iterative detection (i.e., serially concatenated CPM). The details of this algorithm are presented below.

Let $s \triangleq (a_{n-Q}, a_{n-Q+1}, \dots, a_{n-1})$ be the state of the noncoherent trellis at k-th epoch, wherein Q is an integer incorporating the phase memory. The corresponding sufficient statistics from epoch 0 to epoch n are denoted by $\mathbf{r}_0^n = (\mathbf{r}_0, \mathbf{r}_1, \dots, \mathbf{r}_n)$. With s' and s being the start and end state, respectively. We have the following definitions of the branch metric [1,24]

$$\Gamma_n(s',s) \propto \ln \frac{I_0\left(\left|\frac{2}{N_0}\sum_{i=0}^{n}\mathbf{r}_i\mathbf{s}_i^H\right|\right)}{I_0\left(\left|\frac{2}{N_0}\sum_{i=0}^{n-1}\mathbf{r}_i\mathbf{s}_i^H\right|\right)} \quad (18)$$

where $\mathbf{s}_n^H$ is the conjugate transpose of $\mathbf{s}_n$, and $I_0(|x|)$ is the zeroth-order modified Bessel function of the first kind, which can be calculated approximately by $I_0(|x|) \cong e^{|x|}$.

Therefore, the accumulated metric $A_{n+1}(s)$ is updated recursively as

$$A_{n+1}(s) = \max_{s'}(A_n(s') + \Gamma_n(s',s)). \quad (19)$$

Since $I_0(|x|)$ is nonlinear, it is difficult to calculate effectively over the entire received sequence. However, by introducing EW [24] into the branch metric calculation, define *beginning phase* for state s at epoch k as

$$\phi_n(s) \triangleq 2\pi h \sum_{i=0}^{n-L} \hat{a}_i \mod 2\pi \quad (20)$$

where $\{\hat{a}_i\}$ is the sequence associated to the survival state s, which is updated in a per-survivor processing (PSP) [25] approach. Then applying the EW into calculating branch metric, we obtain

$$\Gamma_n(s',s) = \frac{2}{N_0}(|q_{n-1}(s') + e^{j\phi_n(s)}\mathbf{r}_k\mathbf{s}_k^H| - |q_{n-1}(s')|) \quad (21)$$

where $I_0(|x|) \cong e^{|x|}$ is used, and $q_{n-1}(s')$ is designated as *phase reference symbol* which corresponds to an unlimited phase memory increasing with time. This quantity can be recursively updated as

$$q_n(s) \triangleq \eta \cdot q_{n-1}(s') + e^{j\phi_n(s)}\mathbf{r}_n\mathbf{s}_n^H \quad (22)$$

where $\eta \in (0,1)$ is the so-called *forgetting factor*. By adjusting the *forgetting factor*, the branch metric is actually flexible against time-varying phase noise. Generally speaking, the smaller η is the more robust the algorithms is under time-varying channels. Moreover, we use the technique of reduced state sequence detection (RSSD) [27] to further reduce the complexity of N-SISO. The proposed was previously successfully used in high mobility systems, where the phase noise is time varying. The APP of a_i could be evaluated recursively based on the N-SISO proposed below.

4. Numerical and Simulated Results

The performance of the proposed receiver is evaluated in this section in terms of complexity and simulated Bit-Error-Rate (BER). Firstly, the PCA-based receiver is evaluated using the measurements δ and ε illustrated by quaternary CPM2RC in Fig. 3, which shows $K = 4$ is good enough to obtain near optimum performance, while the rank of signals is reduced significantly. The maximum loss no matter in terms of δ or ε is marginal ($\ll 0.1$).

Algorithm of Noncoherent Soft-Input Soft-Output (N-SISO)

Define
$A_n(s) = \log(\alpha_n(s))$,
$B_n(s) = \log(\beta_n(s))$,
$\Gamma_n(s',s) = \log(\gamma_n(s',s))$.
Initialization
$A_n(0) = Constant \quad (Constant \gg 1)$,
$B_n(0) = Constant \quad (Constant \gg 1)$,
$\varphi_n(s) = 0 \quad (any \quad s \in S)$
in which $\varphi_n(s)$ is the *beginning phase* of state s at time kT.
Forward Recursion
$A_n(s) = \max_{s'}(A_{n-1}(s') + \Gamma_{n-1}(s',s))$,
$\varphi_n(s) = \varphi_{n-1}(s') + 2\pi h \sum_{i=n-L+1}^{n} \hat{a}_i q(t-iT)$,
$\hat{a}_i$ is the tentative decision, *beginning phase* is updated according to the corresponding survival path. Note, $\Gamma_{n-1}(s',s)$ could not be obtained until forward recursion is completed. After forward recursion is finished, all $\Gamma_{n-1}(s',s)$ are stored for backward recursion.
Backward Recursion
$B_n(s') = \max_{s}(B_n(s)) + \Gamma_{n-1}(s',s))$,
Output Soft Metric
$\ln P(a_i = a|r_0^{N-1}) = \max_{(s',s)}(A_n(s') + \Gamma_n(s',s) + B_{n+1}(s))$.
This value is for hard-decision, if fed to outer decoder, it has to subtract $\ln P(a_i)$

We then evaluate the performance of the proposed receiver in terms of minimum achievable distance, which is shown in Fig. 5. It is readily seen that the performance loss using PCA-based receiver is negligible by shortening $L = 2$ to $L_R = 1$. The maximum loss is usually no more than 0.2 dB. When modulation index h is approaching 1, both optimum and PCA receiver experience significant loss due to the fact that $h = 1$ is the so called *weak index* [1]. Therefore, we shall avoid this quantity to make sure the minimum achievable distance is maximized.

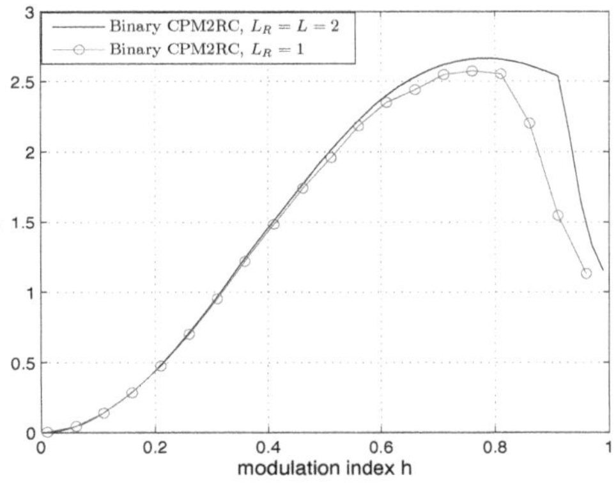

Fig. 5. The minimum achievable distance of optimum receiver and PCA receiver.

Then we proceed to the discussion of the noncoherent detection of both uncoded and Serially-Concatenated-CPM

(SCCPM) systems [2–4]. The outer coding scheme is $(7,5)_8$ convolutional code. The slow time-varying phase noise is considered in this paper. The optimal η is 0.95, Q is set to be 2 and RSSD is adopted.

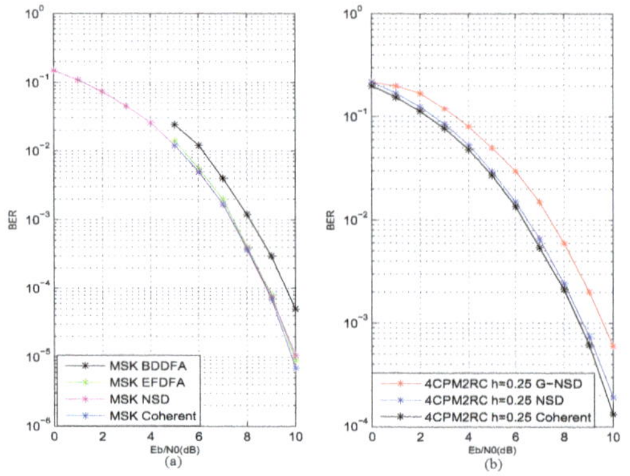

Fig. 6. BER comparison of the proposed N-SISO and some existing noncoherent detection algorithms for binary CPM1REC with $h = 0.5$ and quaternary CPM2RC with $h = 0.25$ [14]. The algorithms for comparison are from [16](BDDFA, EFDFA NSD) and [28] (G-NSD).

The noncoherent detection of uncoded CPMs such as binary CPM of 1REC (i.e., MSK) and quaternary of 2RC (i.e., 4CPM2RC) is shown in Fig. 6. It is observed that all algorithms obtain the near coherent performance eventually. However, the proposed algorithm performs slightly better than other existing algorithms [16, 17, 28]. Though N-SISO and G-NSD receiver require 16 and 64 states, respectively, it can be seen that N-SISO performs about 0.5 dB better than G-NSD. Based on the results in Fig. 6, it can be concluded that N-SISO not only reduces the complexity of branch metric but has a better performance even with fewer state number, due to the fact that N-SISO can fully exploit the Markovity/memory of CPM by using EW.

The discussion is now generalized to coded CPMs. The APPs obtained by detector are first weighted by a factor and then passed to the decoder. For serially concatenated MSK (SCMSK), an improvement of about 5 dB is obtained by increasing the number of iterations to 5. For 10 iterations, performance loss compared to coherent detection is narrowed down to 1 dB. It is also seen that the improvement due to the refined factor is quite obvious. However, in most cases this quantity can not be predetermined but obtained through exhaustive simulations.

In Fig. 8, a serially concatenated quaternary CPM (SCQCPM) is evaluated. First, a 8-state N-SISO is tested, which is not good enough for N-SISO even with more iterations. As a contrast, a 16-state N-SISO with 5 iterations is a good trade-off between the performance and complexity. This result reveals that RSSD affects more on the '*quality*' of the reliability.

An interesting phenomenon is that for full response CPM (SCMSK), doubling the number of iterations could improve the performance by about 0.5 dB, but for partial response system (SCQCPM) only 0.1 dB is obtained. Thus, if extra improvement of SCQCPM is expected, state number has to be further increased. But in practice, 5 iterations may meet the requirement, more iterations gain a little but bring more time delay and complexity.

Finally, we compare the proposed detector with some previously developed detectors in terms of complexity and simulated bit error rate. In this case, the time-vary phase noise is considered. The stand derivation of the phase noise ϕ_n is 5° or 15°, corresponding to moderate and strong phase noise, respectively. The optimal *forgetting factor* is obtained by simulations. It is seen from Fig. 9 that the proposed receiver actually is more robust than existing detectors such as Tikhonov [17], dp-BCJR [17], GA [20] and A-SISO [18] proposed before.

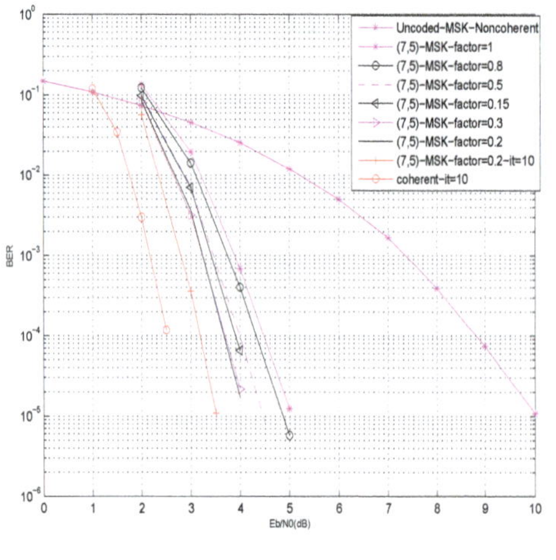

Fig. 7. Noncoherent iterative detection of SCMSK, $N = 1024$, S-Random interleaver. In this figure, 'it' stands for the number of iterations.

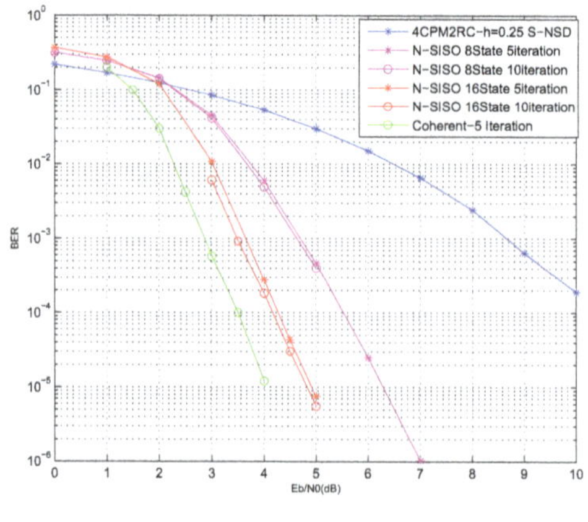

Fig. 8. Noncoherent iterative detection of SCQCPM with symbol interleaver. $N = 1024$, S-Random interleaver.

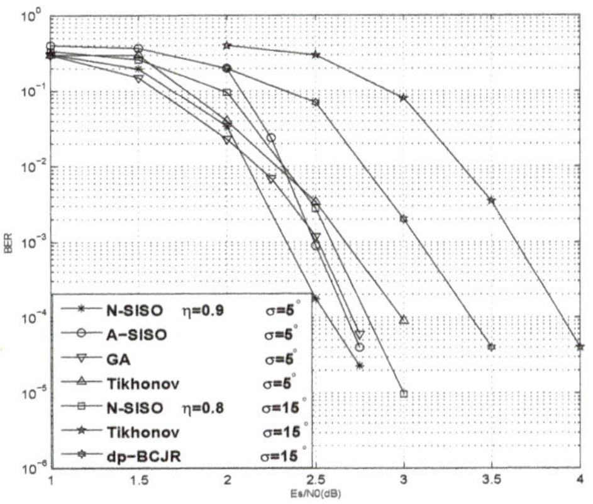

Fig. 9. Noncoherent iterative detection of SCQCPM under strong phase noise.

Proposed	Tikhonov [17]	dp-BCJR [17]	GA [20]
$M\|S\|(6M+10)$	$p(16pM+6M+14)$	$61MD$	$7D+56$

Tab. 1. Computational complexity of proposed and existing algorithms.

The complexity of the proposed algorithm and the algorithms developed in [17, 18, 20] is compared in Tab. 1, wherein D denotes the discretization levels [17] and $|S|$ is the number of states required by the detector. The numerical results were previously partially reported in [17]. The computational complexity is evaluated in terms of number of operations per symbol including additions and multiplications between two real arguments. For dp-BCJR, it requires at least $D = 16$ discretization levels to obtain a reliable phase estimation and thus has a higher complexity than Tikhonov. It can be seen that the computational complexity of N-SISO and other algorithms varies with the parameters of CPM. However, for most binary CPMs, N-SISO has lower complexity. As the M increases, the proposed algorithm would have a comparable complexity with Tikhonov, which is still lower than dp-BCJR.

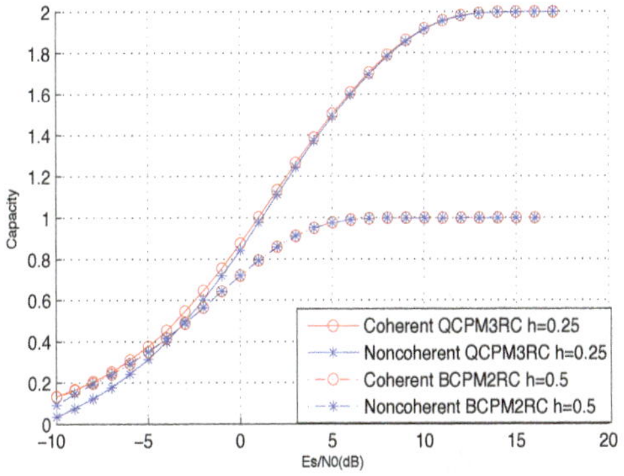

Fig. 10. The capacity of noncoherent detection vs. coherent detection for binary and quaternary CPMs [14].

The aforementioned results is explained in Fig. 10 where the capacities (bits/channel use) employing coherent and noncoherent detection are demonstrated. It is observed that the coherent detection and noncoherent detection (i.e., N-SISO) actually have the same capacity as $SNR \to \infty$. When SNR is low, the noncoherent detection is worse than coherent detection, but this gap is marginal. This reflects the fact that the minimum achievable distance of coherent and noncherent detection is identical [1].

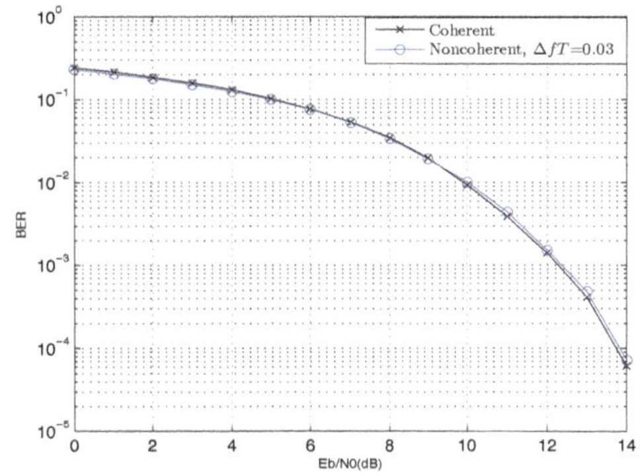

Fig. 11. The performance of an octal CPM2RC employing N-SISO when the carrier frequency is mi-estimated.

The performance of N-SISO subject to mis-estimated carrier frequency is also considered. Here, $\Delta fT = (f_c - \hat{f}_c)T$, wherein f_c and $\hat{f}_c$ are the true carrier and the estimated carrier, respectively. Due to the existence of ΔfT, there is always a time-varying and unknown phase noise $2\pi\Delta ft$ each interval. It is observed in Fig. 11 that the proposed algorithm successfully suppresses the phase noise and obtains a performance approaching coherent detection.

5. Conclusions

In this paper, we proposed a simplified noncoherent iterative receiver for CPM systems. The techniques of exponential window, principal components analysis, singular value decomposition and reduced-state sequence detection are generalized from coherent detection to noncoherent detection. Numerical and analytical results reveal that the proposed receiver can approach coherent performance. In the case of tackling time-varying phase noise, the proposed receiver has a better performance than some existing detectors. The proposed receiver can be generalized to other CPM-based systems such as satellite communication system, deep space communication system, optical communication system, digital video broadcasting (DVB), and multiuser system [10–13] to build simplified receivers.

Acknowledgements

The authors want to thank the anonymous reviewers for their contructive comments. This work was supported in part by the 973 Program of China under Grant 2012CB316100, NSFC under Grant 61372074 and the Open Research Fund of the Science and Technology on Information Transmission and Dissemination in Communication Networks Laboratory (ITD-U12006).

References

[1] ANDERSON, J. B., AULIN, T., SUNDBERG, C.-E. *Digital Phase Modulation.* New York: Plenum Press, 1986.

[2] MOQVIST, P., AULIN, T. M. Serially concatenated continuous phase modulation with iterative decoding. *IEEE Transactions on Communications*, 2001, vol. 49, no. 11, p. 1901 - 1915.

[3] XIAO, M., AULIN, T. M. Serially concatenated continuous phase modulation with convolutional codes over rings. *IEEE Transactions on Communications*, 2006, vol. 54, no. 8, p. 1385 - 1396.

[4] ALEXANDER, G. I. A, NOUR, C. A., DOUILLARD. C. Serially concatenated continuous phase modulation for satellite communications. *IEEE Transactions on Wireless Communications*, 2009, vol. 8, no. 6, p. 3260 - 3269.

[5] RIMOLDI, B. E. A decomposition approach to CPM. *IEEE Transactions on Information Theory*, 1988, vol. 34, p. 260 - 270.

[6] SIMON, M. K. *Bandwidth-Efficient Digital Modulation With Application to Deep-Space Communication.* New York: Wiley, 2003.

[7] LI, L., SIMON, M. Performance of coded OQPSK and MIL-STD. SOQPSK with iterative decoding. *IEEE Transaction on Communications*, 2004, vol. 52, no. 11, p. 1890 - 1900.

[8] DETWILER, T. F., SEARCY, S. M., RALPH, S. E., BASCH, B. Continuous phase modulation for fiber-optic links. *Journal of Lightwave Technology*, 2011, vol. 29, no. 24, p. 3659 - 3671.

[9] *Digital Video Broadcasting (DVB), Second Generation DVB Interactive Satellite System; Part 2: Lower Layers for Satellite Standard.* DVB document A155-2, 2011.

[10] LI, B., AULIN, T., BAI, B. Efficient algorithms for calculating Euclidean distance spectra of multi-user continuous phase modulation systems. in *Proceedings of IEEE International Symposium on Information Theory (ISIT).* Cambridge (MA, USA), 2012, p. 2391 - 2395.

[11] PIEMONTESE, A., GRAELL I AMAT, A., COLAVOLPE, G. Frequency packing and multiuser detection for CPMs: How to improve the spectral efficiency of DVB-RCS2 systems. *IEEE Wireless Communications Letters*, 2013, vol. 2, no. 1, p. 74 - 77.

[12] LI, B., AULIN, T., BAI, B. Simplified detection of spectrally-efficient multiuser CPM systems. In *Proceedings of IEEE Wireless Communications and Networking Conference (WCNC).* Istanbul (Turkey), 2014, to be presented.

[13] MOQVIST, P. *Multiuser Serially Concatenated Continuous Phase Modulation*, Ph.D. thesis. Gothenburg (Sweden): Chalmers University of Technology, 2002. [Online] Available at: http://www.chalmers.se/cse/EN/research/research-groups/publications-from/phd-theses

[14] LI, B., BAI, B., AULIN, T., LI, Q. Advanced continuous phase modulation for high mobility communications. *Chinese Science Bulletin*, 2014, in press.

[15] SIMON, M. K., DIVSALAR, D. Maximum-likelihood block detection of noncoherent continuous phase modulation. *IEEE Transactions on Communications*, 1993, vol. 41, no. 1, p. 90 - 98.

[16] RAPHAELI, D., DIVSALAR, D. Multiple-symbol noncoherent decoding of uncoded and convolutionally coded continuous phase modulation. *Journal of Communications and Networks*, 1999, vol. 1, no. 4, p. 238 - 248.

[17] BARBIERI, A., COLAVOLPE, G. Simplified soft-output detection of CPM signals over coherent and phase noise channels. *IEEE Transactions on Wireless Communications*, 2007, vol. 6, no. 7, p. 2486 - 2496.

[18] BOKOLAMULLA, D., LIM, T. J., AULIN, T. Iterative decoding of serially concatenated CPM in fading channels with noisy channel state information. *IEEE Transactions on Communications*, 2009, vol. 57, no. 4, p. 1079 - 1086.

[19] LI, B., BAI, B., HUANG, M. Y. A robust noncoherent iterative detection algorithm for serially concatenated CPM. In *Proceedings of IEEE International Symposium on Turbo Codes & Iterative Information Processing (ISTC).* Brest (France), 2010, p. 1 - 5.

[20] ZHAO, Q., KIM, H., STÜBER, G. L. Innovation-based MAP estimation with application to phase synchronization for serially concatenated CPM. *IEEE Transactions on Wireless Communications*, 2006, vol. 5, no. 5, p. 1033 - 1043.

[21] MOQVIST, P., AULIN, T. Orthogonalization by principal components applied to CPM. *IEEE Transactions on Communications*, 2003, vol. 51, no. 11, p. 1838 - 1845.

[22] AULIN, T., SUNDBERG, C.-E., SVENSSON, A. Simple Viterbi detectors for partial response continuous phase modulated signals. In *National Telecommunication Conference Record.* New Orleans (LA, USA), 1981, p. A7.6.1 - A7.6.7.

[23] SVENSSON, A., SUNDBERG, C.-E., AULIN, T. A class of reduced-complexity Viterbi detectors for partial response continuous phase modulation. *IEEE Transactions on Communications*, 1984, vol. 32, no. 10, p. 1079 - 1087.

[24] SCHOBER, R., GERSTACKER, W. H. Metric for noncoherent sequence estimation. *Electronic Letters*, 1999, vol. 35, no. 25, p. 2178 - 2179.

[25] RAHELI, R., POLYDOROS, A., TZOU, C. K. Per-survivor processing: A general approach to MLSE in uncertain environments. *IEEE Transaction on Communications*, 1995, vol. 43, no. 2/3/4, p. 354 - 364.

[26] ARNOLD, D. M., LOELIGER, H.-A., VONTOBEL, P. O., KAVČIĆ, A., ZENG. W. Simulation-based computation of information rates for channels with memory. *IEEE Transactions on Information Theory*, 2006. vol. 52, no. 8, p. 3498 - 3508.

[27] SVENSSON, A. Reduced state sequence detection of partial response continuous phase modulation. *Proceedings of IEE*, 1991, vol. 138, p. 256 - 268.

[28] COLAVOLPE, G., RAHELI, R. Noncoherent sequence detection of continuous phase modulations. *IEEE Transactions on Commuunications*, 1999, vol. 47, no. 9, p. 1303 - 1307.

About Authors . . .

Bing LI was born in 1983. He received the M.Sc. from Xidian University, China in 2009. He is now pursing the Ph.D degree at Xidian University. He was a guest-doctor at Chalmers University of Technology from 2010 to 2012. His research interests include coded modulation, turbo coding, iterative detection and multiuser signaling.

Baoming BAI was born in 1964 and received his M.Sc and Ph.D from Xidian University in 1990 and 2000, respectively. His research interests include channel coding, network coding and deep space communications.

9

The Impact of Interference on GNSS Receiver Observables – A Running Digital Sum Based Simple Jammer Detector

Mohammad Zahidul H. BHUIYAN[1], *Heidi KUUSNIEMI*[1], *Stefan SÖDERHOLM*[1], *Esa AIROS*[2]

[1]Dept. of Navigation and Positioning, Finnish Geodetic Institute, Kirkkonummi, Finland
[2]Defence Forces Technical Research Centre, Riihimäki, Finland

zahidul.bhuiyan@fgi.fi, heidi.kuusniemi@fgi.fi, stefan.soderholm@fgi.fi, esa.airos@mil.fi

Abstract. *A GNSS-based navigation system relies on externally received information via a space-based Radio Frequency (RF) link. This poses susceptibility to RF Interference (RFI) and may initiate failure states ranging from degraded navigation accuracy to a complete signal loss condition. To guarantee the integrity of the received GNSS signal, the receiver should either be able to function in the presence of RFI without generating misleading information (i.e., offering a navigation solution within an accuracy limit), or the receiver must detect RFI so that some other means could be used as a countermeasure in order to ensure robust and accurate navigation. Therefore, it is of utmost importance to identify an interference occurrence and not to confuse it with other signal conditions, for example, indoor or deep urban canyon, both of which have somewhat similar impact on the navigation performance. Hence, in this paper, the objective is to investigate the effect of interference on different GNSS receiver observables in two different environments: i. an interference scenario with an inexpensive car jammer, and ii. an outdoor-indoor scenario without any intentional interference. The investigated observables include the Automatic Gain Control (AGC) measurements, the digitized IF (Intermediate Frequency) signal levels, the Delay Locked Loop and the Phase Locked Loop discriminator variances, and the Carrier-to-noise density ratio (C/N_0) measurements. The behavioral pattern of these receiver observables is perceived in these two different scenarios in order to comprehend which of those observables would be able to separate an interference situation from an indoor scenario, since in both the cases, the resulting positioning accuracy and/or availability are affected somewhat similarly. A new Running Digital Sum (RDS) -based interference detection method is also proposed herein that can be used as an alternate to AGC-based interference detection. It is shown in this paper that it is not at all wise to consider certain receiver observables for interference detection (i.e., C/N_0); rather it is beneficial to utilize certain specific observables, such as the RDS of raw digitized signal levels or the AGC-based observables that can uniquely identify a critical malicious interference occurrence.*

Keywords

GNSS, jamming, in-car jammer, interference detection, receiver observables.

1. Introduction

Global Navigation Satellite System (GNSS) based positioning has an immense role in modern society. Reliable navigation is imperative in more and more applications nowadays on land, sea, and air. A major dependency on reliable localization has been emerging, especially within safety-critical applications [1]. GNSS signals, as well as many other Radio Frequency (RF) signals, are however extremely susceptible to unintentional and intentional malicious interference. Since GNSS signals are also very weak, after travelling the distance of about 20000 km from the satellite to the Earth, they can be very difficult to recover when subject to interference.

Applications using GNSS based positioning for road tolling, insurance billing, or logistics have increased recently in quantity. Simultaneously, despite being illegal, intentional jamming of the related satellite navigation receivers has become temptingly easy. Though illegal, affordable jammer devices can easily be purchased online or built according to widely attainable online recipes. The increase in the amount of satellite navigation jammers is alarming, especially due to the serious damage they may cause. Because satellite positioning is in such a vital role in many applications, jammers may cause great damage if not detected and their effects mitigated. The typical usage environment of jammers is in cars, where they transmit a jamming signal usually on the civilian L1/E1-band where the accessible GPS C/A and the upcoming civilian Galileo codes are located. Civilian in-car jammers pose a severe threat to the trustworthiness of GNSS receivers [2]-[4]. High-power jammers may not only hinder the usage of GNSS in the vicinity of the jammer but also paralyze GNSS usage over a larger area.

The jamming signal may deteriorate the position solution or totally induce loss of lock of GNSS signals depending on the perceived Jamming-to-Signal (J/S) power ratio at the receiver [5], [6]. Different receivers react differently to jamming depending on the properties of the jamming signal. The basic functionalities in most GNSS receivers are fairly similar, but the internal architecture and algorithms vary. Different kinds of filtering may for example mitigate the effect of the jammer on the positioning accuracy and availability. Intentional GNSS jamming raises the noise floor and thus reduces the Carrier-to-Noise density ratio (C/N_0) of the received signals. This effect is similar to the phenomenon perceived in the context of multipath propagation or general signal attenuation due to for example foliage: in such a case the received signals become weaker, also resulting in lower C/N_0 measurements. When the C/N_0 is low enough, the receiver cannot anymore generate ranging measurements and the position solution cannot be computed.

In order to mitigate the effects of interference from intentional or unintentional sources, reliable interference detection must be conducted first. In [7] and [8], interference detection is performed based on a combination of several receiver observables, i.e., correlator output power, variance of correlator output power, carrier phase vacillation, and AGC (Automatic Gain Control) values. The studies reported in [8] are mainly based on simulated signals for different kinds of interference, for example, AWGN, Continuous Wave (CW), pulsed broadband, pulsed CW, etc. It is concluded in [8] that the correlator output power showed the best consistent performance under varying level as well as varying sources of interference. The impact of various types of interference on AGC circuit is also studied in [9], [10], and it is concluded that the AGC can be used as an interference assessment tool for GNSS receivers.

The objective of this research is to investigate the effect of inexpensive in-car jammers on different GNSS receiver observables. The investigated observables include the Automatic Gain Control (AGC) measurements, the digitized IF (Intermediate Frequency) signal levels, the Delay Locked Loop, the Phase Locked Loop discriminator variances, and the Carrier-to-noise density ratio (C/N_0) measurements. The reason for choosing these observables is motivated by the fact that all observables except AGC are instantly available at the signal processing stage, and can be utilized solely in a software-defined GNSS receiver. The software-defined receivers are useful tools for research purposes and are beneficial for testing and implementing various algorithms before hardware implementation. However, it is little bit tricky to obtain access to AGC observables, as the front-end manufacturers do not usually offer access to an AGC circuit. In this study, the AGC voltage test pin was used to measure the level of the gain control in the various scenarios. The selection of the observables assessed in this paper for interference detection is also in line with other previous studies on the topic mentioned in [7]-[9]. The behavioral pattern of these receiver observables is perceived in the presence of interference and in indoor signal condition without any intentional jamming. The authors also introduce a unique Running Digital Sum (RDS) based interference detection method for GNSS signals. In this method, the interference detection is performed via a Running Digital Sum (RDS) [11], [12] check of the digitized signal after the Analog-to-Digital Conversion (ADC) at the intermediate frequency. As shown later, the introduced RDS based interference detection method can successfully distinguish the intentional interference occurrence from that of a weak indoor signal condition.

The remaining of this paper is organized as follows. Section 2 briefly characterizes the cheap car-jammer used in the experiment. In Section 3, all the considered receiver observables used for jamming detection are discussed. Experimental setup and result analysis are presented in Sections 4 and 5, respectively. Finally, some concluding remarks are made in Section 6 based on the experimental results.

2. Jammer Characterization

The increase in the amount of satellite navigation jammers is alarming due to the serious harm they may cause. The typical usage environment of jammers is in cars, where they act as so called personal privacy devices and transmit a jamming signal usually on the civilian L1/E1-band. A car jammer, the Covert GPS L1 jammer is used in the experiment. The Covert L1 jammer is shown in Fig. 1.

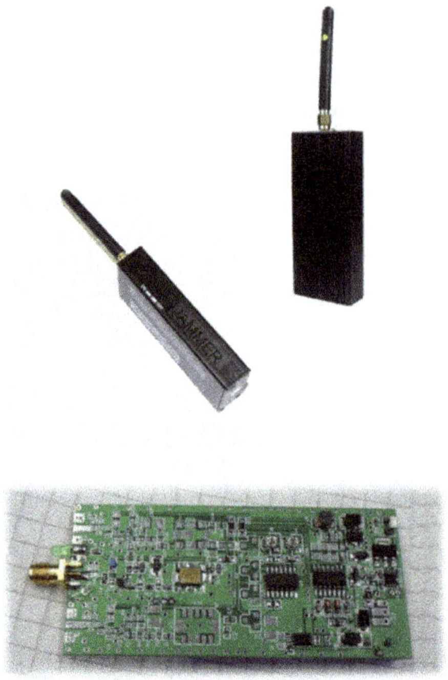

Fig. 1. Covert GPS L1 jammer.

A constrained usage permission of the jammer was obtained from the Finnish Communications Regulatory

Authorities. According to the permission, the jammer has to be used within the laboratory of the Finnish Geodetic Institute (FGI) only for the purpose of research. The output power of the jammer was measured to be around +18 dBm, and it transmits chirp-like signals. A chirp signal (also known as sweep signal) is a signal in which the frequency increases ('up-chirp') or decreases ('down-chirp') with time. The Covert L1 jammer transmits a chirp signal with multi saw-tooth functions having a center frequency at approximately 1.577 GHz with a bandwidth of about 16.3 MHz. The power spectrum and the instantaneous frequency of the jammer are shown in Figs. 2 and 3, respectively. The characteristics of Covert L1 jammer coincide very well with the findings reported in [2], [3].

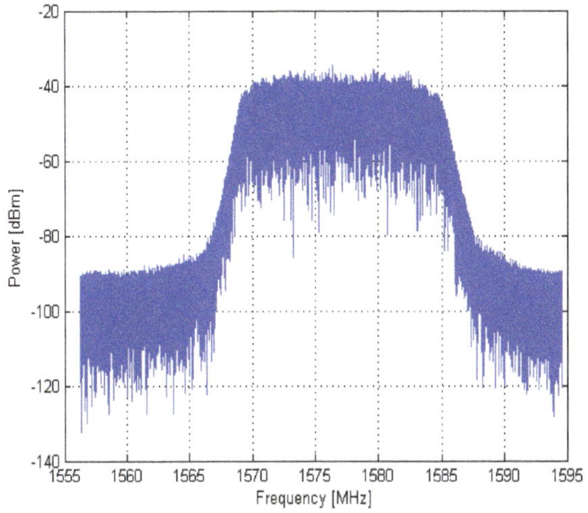

Fig. 2. Power spectrum of the Covert GPS L1 jammer.

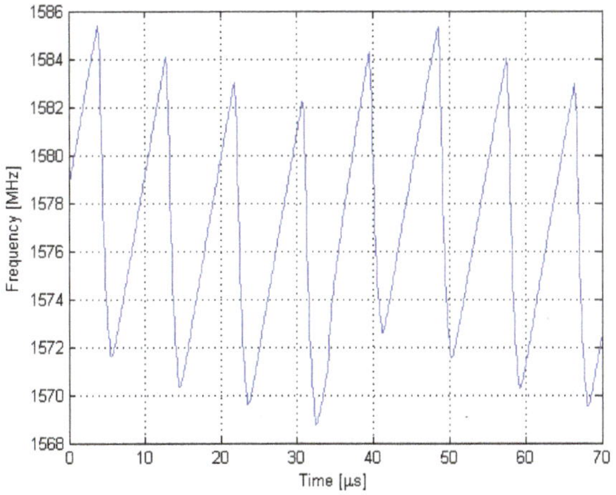

Fig. 3. Instantaneous frequency of the Covert GPS L1 jammer.

3. Observables Used for Jammer Detection

The following receiver observables are considered throughout the experiment for jamming detection:

i) Carrier-to-Noise density ratio,

ii) Running DLL variance,

iii) Running PLL variance,

iv) Automatic Gain Control level changing rate, and

v) Running digital sum of the IF data samples.

As mentioned earlier, the reason for choosing these observables is motivated by the fact that all observables except AGC are instantly available at the signal processing stage, and can be utilized solely in a software-defined GNSS receiver. A brief overview on each of these receiver observables is presented in what follows.

3.1 Carrier-to-Noise Density Ratio

Carrier-to-Noise density ratio (C/N_0) is used to measure the strength of a received GNSS signal. The Carrier-to-Noise density ratio (C/N_0) estimation is performed based on the ratio of the signal's wideband power to its narrow-band power as mentioned in [5]:

$$C/N_0 = 10\log 10\left(\frac{1}{T}\frac{\hat{\mu}_{NP}-1}{M-\hat{\mu}_{NP}}\right) \tag{1}$$

where T is the code integration time in seconds (i.e., 0.001 s for GPS L1 C/A signal); M is the total number of T blocks used for coherent integration (usually a fair choice of $M = 20$ is used as the data bit duration for GPS L1 C/A signal is 20 ms), and $\hat{\mu}_{NP}$ is the mean normalized power, as expressed in the following equation:

$$\hat{\mu}_{NP} = \frac{1}{K}\sum_{k=1}^{K} NP_k \tag{2}$$

where NP is the normalized power between narrow-band power and wide-band power:

$$NP_k = \frac{NBP_k}{WBP_k} \tag{3}$$

where NBP and WBP can be expressed as follows:

$$NBP_k = \left(\sum_{i=1}^{M} I_{P_i}\right)_k^2 + \left(\sum_{i=1}^{M} Q_{P_i}\right)_k^2, \tag{4}$$

$$WBP_k = \left(\sum_{i=1}^{M} (I_{P_i}^2 + Q_{P_i}^2)\right)_k \tag{5}$$

where I_{P_i} and Q_{P_i} are the prompt correlation outputs at the tracking stage from the in-phase and quadrature arms, respectively. The values used in the experiments for M and K are 20 and 50, respectively.

3.2 Running DLL Variance

A non-coherent Early-Minus-Late (EML) discriminator is used in this particular experiment as the Delay

Lock Loop. According to [13], the EML discriminator can be written as:

$$DLL_i = \frac{\sqrt{I_{E_i}^2 + Q_{E_i}^2} - \sqrt{I_{L_i}^2 + Q_{L_i}^2}}{\sqrt{I_{E_i}^2 + Q_{E_i}^2} + \sqrt{I_{L_i}^2 + Q_{L_i}^2}} \tag{6}$$

where I_{E_i} and Q_{E_i} are the in-phase and quadrature correlation outputs of the early correlators, respectively (i.e., 0.25 chips early from the prompt correlation), I_{L_i} and Q_{L_i} are the in-phase and quadrature correlation outputs of the late correlators, respectively (i.e., 0.25 chips late from the prompt correlation). The DLL discriminator variance is then calculated from (6) for a running window of $N = 1000$ points.

3.3 Running PLL Variance

A two-quadrant 'ATAN' Costas discriminator is used in this experiment as a Phase Locked Loop. According to [13], the ATAN Costas discriminator can be written as:

$$PLL_i = ATAN\left(\frac{Q_{P_i}}{I_{P_i}}\right) \tag{7}$$

where I_{P_i} and Q_{P_i} are the prompt correlation outputs from the in-phase and quadrature arms, respectively. The Costas PLL discriminator variance is then computed from (7) for a running window of $N = 1000$ points.

3.4 Automatic Gain Control

Automatic Gain Control (AGC) is a key element in a GNSS receiver. The main responsibility of an AGC is to adjust the incoming signal power such that the quantization losses are kept as minimum as possible. Therefore, the AGC operation is usually directly coupled with the ADC. In case of a GNSS receiver, where the signal power remains below that of the thermal noise floor, the AGC is mostly driven by the ambient noise environment rather than the signal power. As a result, AGC can be utilized as an important tool for assessing the operating environment of any GNSS receiver.

Due to restrictions on emissions in and near the GNSS bands, it is quite likely that the AGC gain exclusively depends on the ambient noise environment rather than the GNSS signal power, as is expected in a typical interference-free situation. However, in case of an unlikely presence of interference, the AGC gain drops sharply in response to increased power in the GNSS band. This sharp immediate change in the AGC gain pattern can be utilized to indicate an intentional interference occurrence, as shown in Fig. 7. A metric, termed as AGC level changing rate, is used in this experiment for jamming detection. The AGC level changing rate can be calculated as follows:

$$\tau_i = \frac{x_i - x_{i-1}}{t_i - t_{i-1}}; i \ge 1 \tag{8}$$

where x_i is the measured AGC level at time t_i.

3.5 Running Digital Sum

The GNSS signal Interference Detection is performed via a Digital Sum (DS) check of the digitized signal after analog-to-digital conversion at the intermediate frequency. The DS is a function that sums the digital levels of the received digitized signal after ADC. The Digital Sum (DS) can be written as follows [11], [12]:

$$DS(k) = \frac{\sum_{i=1}^{k} a_i}{k} \tag{9}$$

where a_i is the digitized signal samples after ADC. For a 2-bit real quantization, a_i can take any values from the set [±1; ±3]. Before calculating the digital sum, the bin distribution of a_i is converted from [±1; ±3] to [±1] as follows:

$$a_i = +1; a_i \ge 1, \tag{10}$$

$$a_i = -1; a_i \le -1. \tag{11}$$

This is done in order to make sure that all digital levels have similar contributions to the final digital sum count. An example DS count is shown in Fig. 4 in a normal jamming-free scenario and a jamming scenario, where the DS counts are 1.3% in a normal jamming-free scenario and 10.8% in a jamming scenario with a maximum Jamming-to-Signal (J/S_{max}) ratio of 25 dB. The example DS counts are obtained with a front-end module from Sparkfun Electronics, named as SiGe GN3S sampler v3 [14]. The DS counts of the digitized signal levels after ADC in a nominal jamming-free environment should always be as close as possible to zero. In other words, the quantized bin distribution after ADC should be balanced such that there are

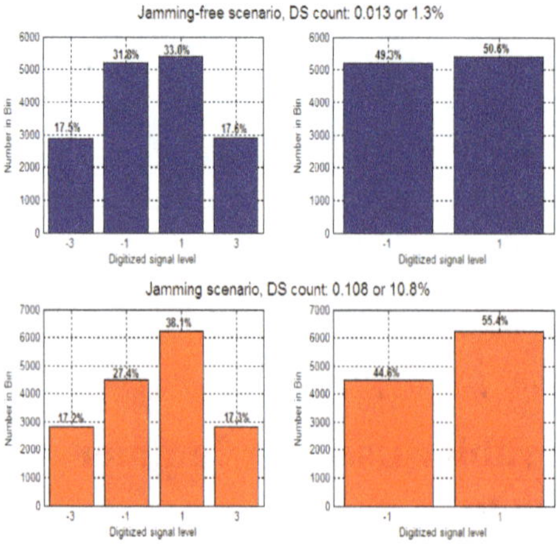

Fig. 4. Bin distribution of the digitized GNSS signal samples for 1 millisecond long data.

almost equal numbers of '+' and '–' levels in the digitized signal to avoid the presence of any DC bias in the signal. However, a small DC bias can always be present in a front-end module, but in any case it should always be fixed to a certain number. For example, the DC bias for the used SiGe GN3S sampler v3 front-end is less than 2%.

Finally, a Running DS (RDS) is calculated for smoothing the observations by the following equation:

$$RDS(j) = \frac{\sum_{k=j}^{j+N-1} |DS(k)|}{N}; \quad j = 1,2,\cdots \qquad (12)$$

where a running window of $N = 1000$ is used in the experiment.

4. Experimental Setup

A software defined GNSS receiver platform, the FGI-GSRx, is used to process the raw IF data samples in post-mission. The FGI-GSRx has been developed for the analysis and validation of novel algorithms for an optimized GNSS navigation performance. The FGI-GSRx development was started from an open-source software radio platform introduced in [15]. In this particular experiment, besides the DLL and the PLL implementation, the C/N_0 estimation technique and the proposed RDS based interference detection method are implemented. A USB front-end module from Sparkfun Electronics, named as SiGe GN3S sampler v3 [14], is used to capture the raw GPS L1 C/A signal. The configuration of the SiGe radio front-end used in the experiment is mentioned in Tab. 1.

Intermediate Frequency	4.092 MHz
Front-end Bandwidth	2 MHz
Sampling Frequency	16.368 MHz
Number of Quantization bits	2 bits

Tab. 1. Front-end configuration for SiGe GN3S sampler v3.

Two different test scenarios are considered in the experiment: *i.* an intentional interference scenario with a maximum Jamming-to-Signal ratio (J/S_{max}) of 25 dB, and *ii.* an outdoor-indoor scenario without any intentional interference. The jamming-to-signal ratio, usually expressed in dB, is the ratio of the power of a jamming signal to that of a desired GNSS signal at a given point of a positioning receiver. The jammer signal was also constantly monitored with a spectrum analyzer during the test.

Many RF front ends today have implemented an integrated analog ADC and therefore it is almost impossible to read the settings and values from the AGC in the radio. Fortunately, for the purpose of calibrating the RF chain in a GNSS receiver, the RF often has a test pin for measuring the level of the AGC. This is intended to be used either with no RF input (or no signal) or with some predefined calibration signal. The SiGe RF front end that is used in the GN3S sampler also has such a calibration pin [16]. The datasheet describes the above mentioned two test procedures, where the calibration signal level was set to -88 dBm. According to the datasheet, the test point should provide an output voltage of 1.2 V when no signal is present (maximum AGC gain) and 0.8 V when an L1 signal of -88 dBm is present at the RF input (minimum AGC gain). Unfortunately not much more information is available, but based on this and our measurements we have determined that when the noise level is increased the AGC voltage level is decreased.

5. Result Analysis

The impact of interference from the analyzed jammer on different GNSS receiver observables is shown in Figs. 4 to 7. The Covert GPS L1 jammer was turned on after about 56 seconds from the start of GNSS data capture with the SiGe front-end. The sudden drop in C/N_0 values for all the tracked satellites at the 56th second is clearly visible in Fig. 5. The approximate J/S_{max} is measured by monitoring the spectrum analyzer. This is also evident from Fig. 5 that the loss of C/N_0 due to jamming is at least 15 dB or more for all the tracked satellites.

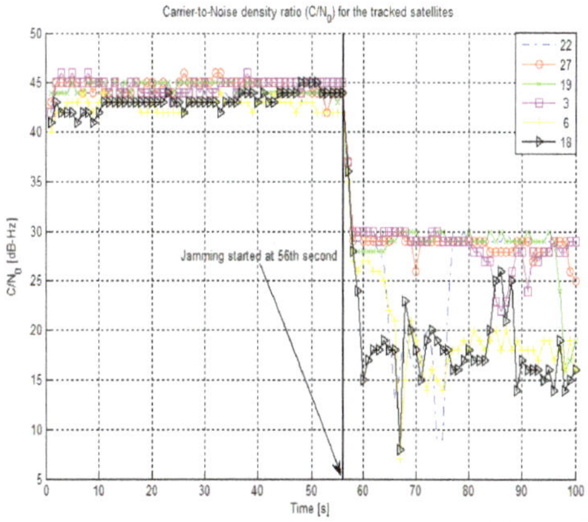

Fig. 5. C/N_0 for the tracked satellites in a jamming scenario with $J/S_{max} = 25$ dB.

The running DLL variance and the running PLL variance for one of the tracked satellites (i.e., PRN 27) in the presence of interference are shown in Figs. 6 and 7, respectively. The RDS of the digitized IF samples of the GNSS signal after ADC is also shown in Fig. 8. As shown in the figures, all these observables can successfully detect the interference occurrence almost immediately. The detection thresholds in all the above cases are computed against a false alarm probability of 0.01. However, it takes about 1 second for each of these techniques to stabilize due to the coarse code and frequency estimation at the acquisition stage, and therefore, they can only offer a detection decision after that time period. The output from the AGC block of the SiGe front-end is measured at 1 Hz rate via a PC-based PicoScope oscilloscope [17] connected through an USB port. The sudden drop of AGC levels during 56 to

59 second is evident from Fig. 9. Due to the presence of jamming signal on the GPS L1 band, the AGC block reacts almost immediately by lowering down the AGC gain values to a minimal level. The AGC level changing rate is shown Fig. 10. In a nominal environment where the temperature changes are steady, the only reason for such a huge drop in AGC is due to the presence of other unwanted RF signal in the GNSS spectrum. Therefore, the AGC level changing rate can be utilized to trigger any malicious interference occurrence by monitoring the AGC gain variations with respect to time.

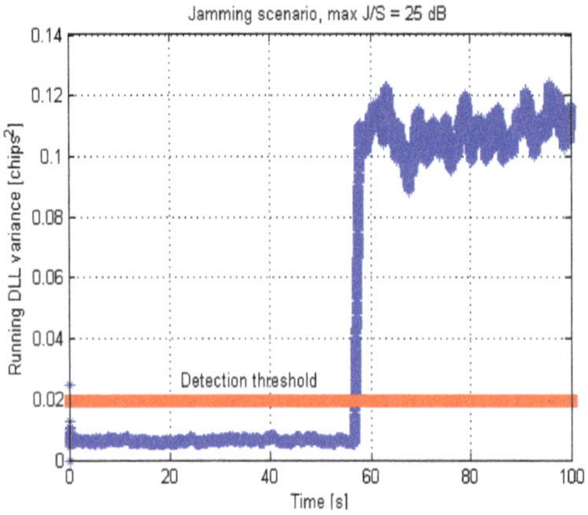

Fig. 6. Running DLL variance in a jamming scenario with $J/S_{max} = 25$ dB.

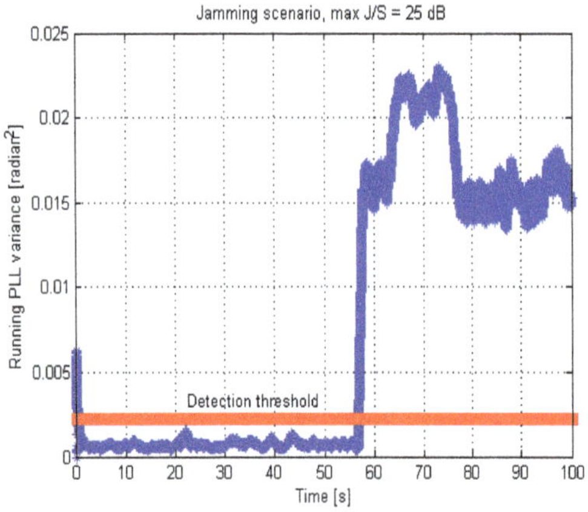

Fig. 7. Running PLL variance in a jamming scenario with $J/S_{max} = 25$ dB.

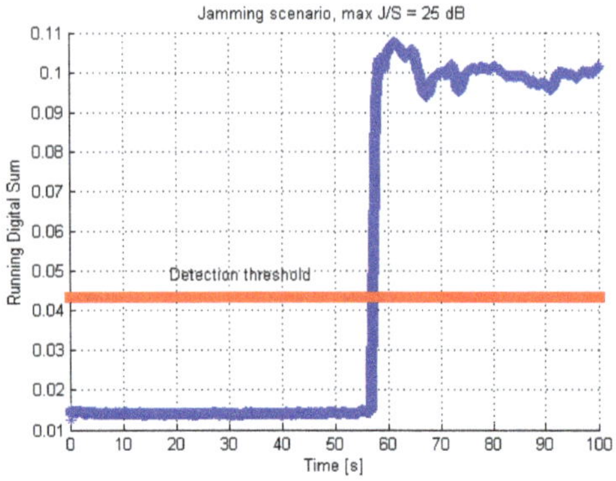

Fig. 8. RDS of the digitized IF samples in a jamming scenario with $J/S_{max} = 25$ dB.

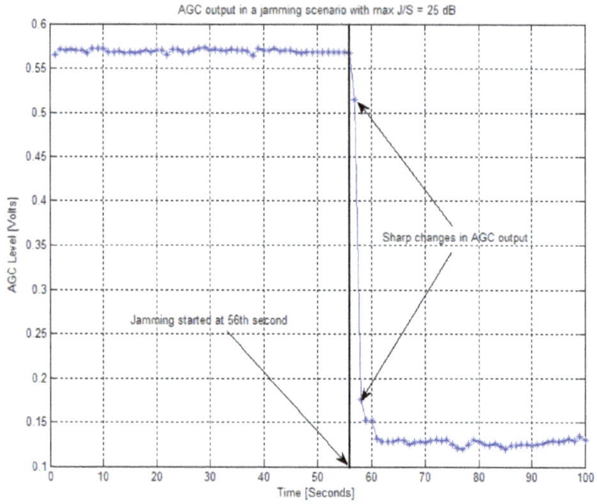

Fig. 9. AGC level in a jamming scenario with $J/S_{max} = 25$ dB.

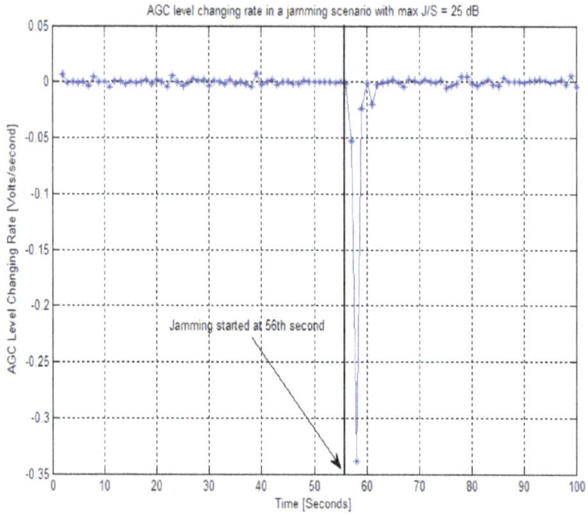

Fig. 10. AGC level changing rate in a jamming scenario with $J/S_{max} = 25$ dB.

The behavioral pattern of the analyzed receiver observables in a typical outdoor-indoor scenario are shown in Figs. 11 to 15. The tester was standing with the test equip-

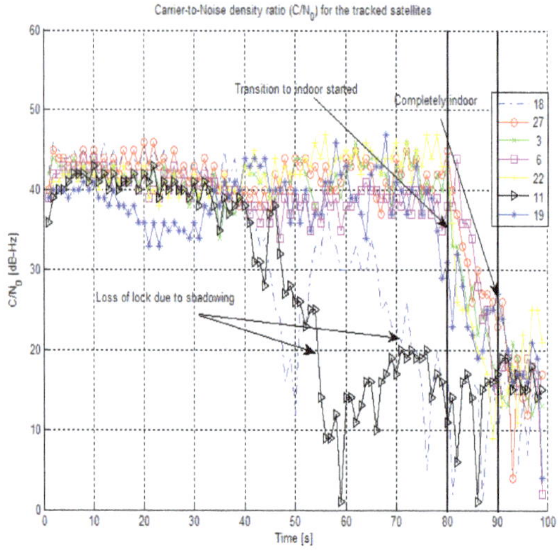

Fig. 11. C/N_0 for the tracked satellites in an outdoor-indoor scenario.

ment outside a typical three-story office building for about 40 seconds, after which he started walking into the office entrance for another 40 seconds. During the following 10 seconds (in between 80th and 90th seconds in the test), the tester was in the transition phase to walk from the outdoor to deep inside building made of concrete and steel, as can be seen from the sudden drop of C/N0 in Fig. 11.

The running DLL variance and the running PLL variance of one of the tracked satellites (i.e., PRN 27) are shown in Figs. 12 and 13, respectively. It can be seen from these figures that both the variances increase due to the raise in the noise level, as the space-based GNSS signal faded away while moving towards indoor. The computed RDS of the digitized IF samples and the logged AGC level changing rate in this outdoor-indoor scenario are shown in Figs. 14 and 15, respectively. It can be seen from the figures that the computed RDS and the AGC gain is not at all affected by the increase in the noise level due to the signal fading while moving towards indoor.

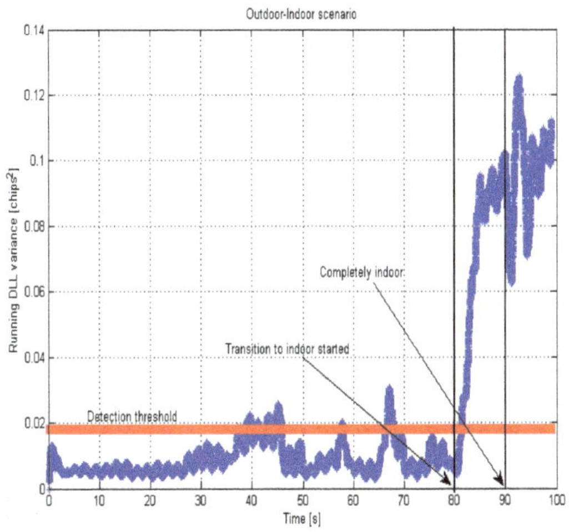

Fig. 12. Running DLL variance in an outdoor-indoor scenario.

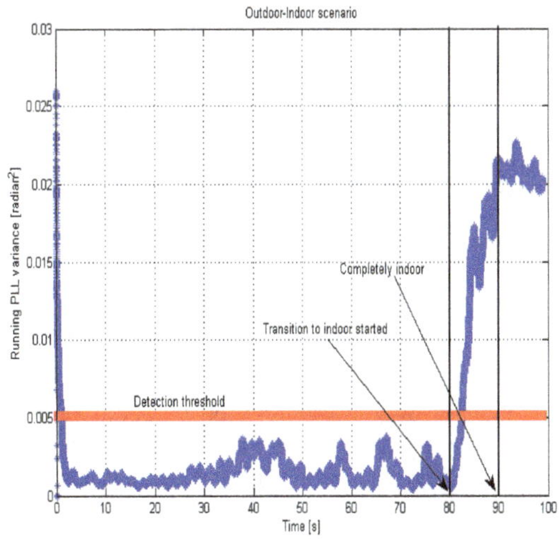

Fig. 13. Running PLL variance in an outdoor-indoor scenario.

The above results demonstrate the fact that the interference detection should only be based on such receiver observables which get affected only by the presence of interference, not by any other error source like shadowing or weak signal condition. Hence, the most suitable receiver observables for interference detection are either the AGC output levels or the RDS count of the digitized signal levels. However, it is not always trivial to get access to AGC output, as it resides in the RF chain within the ADC block. Fortunately, the digitized signal levels are the output from the ADC block, which are then utilized by the acquisition and tracking blocks for further receiver-specific processing. While doing the receiver-specific processing, a simple RDS count on the digitized signal levels can be done to identify the presence of any malicious interference occurrence.

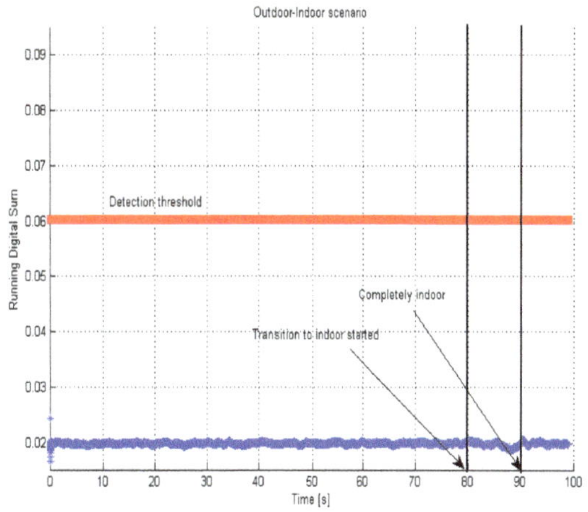

Fig. 14. RDS of the digitized IF samples in an outdoor-indoor scenario.

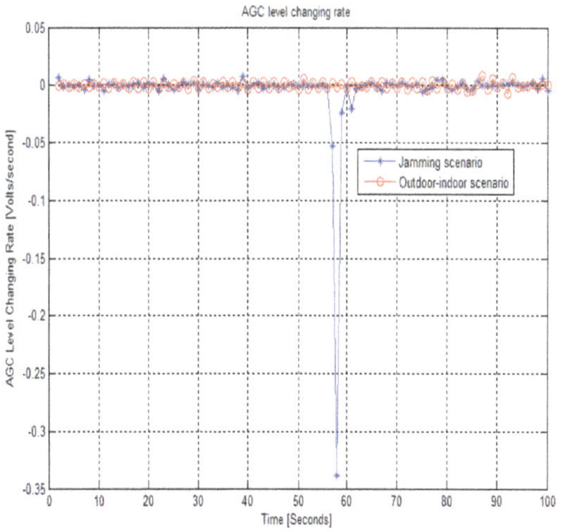

Fig. 15. AGC level changing rate in an outdoor-indoor scenario plotted on top of that of the jamming scenario presented in Fig. 10.

6. Conclusions

Five different receiver observables are investigated as candidate decision statistic for interference detection in two different environments: in the presence of intentional interference from an inexpensive car-jammer and in an outdoor-indoor environment without any intentional interference. The behavior of these observables in the above conditions and their ability to uniquely identify an interference occurrence were addressed. It was shown that three of the receiver observables, i.e., C/N_0, running DLL variance, and running PLL variance, cannot really distinguish the intentional interference occurrence from that of a weak signal condition as they react similarly in both the cases. It was also concluded that the proposed running digital sum of the digitized IF samples and the AGC output levels, can uniquely identify an intentional interference occurrence of a jammer as these observables do not get affected by the additive white Gaussian noise. The future work includes investigation of the proposed RDS based interference detection method in the presence of different interference sources (i.e., continuous wave interference, pulsed interference, etc.) with a variety of receiver front-ends from different manufacturers.

Acknowledgements

This research has been conducted within the project DETERJAM (Detection, analysis, and risk management of satellite navigation jamming) funded by the Scientific Advisory Board for Defense of the Finnish Ministry of Defense and the Finnish Geodetic Institute, Finland.

References

[1] PULLEN, S., GAO, G. GNSS jamming in the name of privacy, potential threat to GPS aviation. *Inside GNSS*, USA, 2012, p. 34–43.

[2] KRAUS, T., BAUERNFEIND, R., EISSFELLER, B. Survey of in-car jammers – analysis and modeling of the RF signals and IF samples (suitable for active signal cancellation). In *Proceedings of ION GNSS*. USA, 2011, p. 430–435.

[3] MITCH, R.H., DOUGHERTY, R.C., PSIAKI, M.L., POWELL, S.P., O'HANLON, B.W., BHATTI, J.A., HUMPHREYS, T.E. Signal characteristics of civil GPS jammers. In *Proceedings of ION GNSS*. USA, 2011, p. 1907–1919.

[4] KUUSNIEMI, H., AIROS, E., BHUIYAN, M.Z.H., KRÖGER, T. GNSS jammers: how vulnerable are consumer grade satellite navigation receivers? *European Journal of Navigation*, 2012, vol. 10, no. 2, p. 14–21.

[5] PARKINSON, B.W., SPILKER, J.J. Jr. *Global Positioning System: Theory and Applications*. American Institute of Aeronautics, Vol. 1, 370 L'Enfant Promenade, SW, Washington, DC, USA, 1996, p. 390–392.

[6] MOTELLA, B., SAVASTA, S., MARGARIA, D., DOVIS, F. A Method to assess robustness of GPS C/A code in presence of CW interferences. *International Journal of Navigation and Observation*, 2010, vol. 2010, Article ID 294525, 8 pages.

[7] NDILI, A. Robust GPS autonomous signal quality monitoring. *PhD thesis*. USA, 1998, Department of Mechanical Engineering, Stanford University.

[8] NDILI, A., ENGE, P. GPS receiver autonomous interference detection. In *Proceedings of IEEE PLANS*. USA, 1998.

[9] BASTIDE, F., AKOS, D. M., MACABIAU, C., ROTURIER, B. Automatic Gain Control (AGC) as an interference assessment tool. In *Proceedings of the 16th International Technical Meeting of the Satellite Division of the Institute of Navigation*. USA, 2003, p. 2042–2053.

[10] ISOZ, O., BALAEI, A.T., AKOS, D.M. Interference detection and localization in the GPS L1 band. In *Proceedings of the International Technical Meeting of the Institute of Navigation*. USA, 2010, p. 925–929.

[11] BISSELL, C., CHAPMAN, D. *Digital Signal Transmission*. Cambridge University Press, The Edinburgh building, Cambridge CB2 2RU, UK, 1992.

[12] SMITH, D. R. *Digital Transmission Systems*. 3rd ed. Kluwer Academic Publishers, 3300 AH Dordrecht, The Netherlands, 2004, p. 260–261.

[13] KAPLAN, E.D., HEGARTY, C. *Understanding GPS - Principles and Applications*. 2nd ed. Boston (USA): Artech House Publishers, 2006. Chapter 5.

[14] SPARKFUN E. SiGe GN3S sampler v3. [Online] Cited 2013-10-29. Available at: https://www.sparkfun.com/products/10981.

[15] BORRE, K., AKOS, D. M., BERTELSEN, N., RINDER, P., JENSEN, S.H. *A software-Defined GPS and Galileo Receiver: A Single-Frequency Approach*. 1st ed. Applied and Numerical Harmonic Analysis. Boston (USA): Birkhäuser Verlag GmbH, 2006.

[16] SPARKFUN E. SiGe SE4120 Datasheet. [Online] Cited 2013-10-21. Available at: http://media.digikey.com/pdf/Data%20Sheets/Skyworks%20PDFs/SE4120L.pdf

[17] PICOTECH PicoScope 4000 Series High-Resolution USB Oscilloscopes. [Online] Cited 2013-11-01. Available at: http://www.picotech.com/picoscope4000.html.

About Authors ...

Mohammad Zahidul H. BHUIYAN received his M.Sc. degree in 2006 and Ph.D. degree in 2011 from the Department of Communications Engineering, Tampere University of Technology, Finland. Dr. Bhuiyan is now working in the Department of Navigation and Positioning at the Finnish Geodetic Institute as a Specialist Research Scientist with research interests covering various aspects of GNSS receiver design and sensor fusion algorithms for seamless outdoor/indoor positioning.

Dr. Heidi KUUSNIEMI is a Research Manager at the Department of Navigation and Positioning at the Finnish Geodetic Institute, where she leads the research group on satellite and radio navigation. She is also a Lecturer at the Department of Surveying Sciences at Aalto University, Finland. She received her M.Sc. degree in 2002 and D.Sc. (Tech.) degree in 2005 from Tampere University of Technology, Finland. Her doctoral studies on personal satellite navigation were partly conducted at the Department of Geomatics Engineering at the University of Calgary, Canada. Her research interests cover various aspects of GNSS navigation, quality control, software defined receivers, multi-sensor fusion algorithms for seamless outdoor/indoor positioning, and GNSS interference mitigation methods. She is the President of the Nordic Institute of Navigation since 2011, and a member of the IEEE and the ION.

Mr. Stefan SÖDERHOLM is currently working as a Researcher at the Finnish Geodetic Institute, Department of Navigation and Positioning. He is working in many different areas focusing on improving the performance in consumer grade GNSS receivers. Before joining the Institute in September 2013 he was working at Fastrax Ltd since the end of 2000. The first 4 years at Fastrax he was developing GPS signal processing algorithms, navigation filters and GPS receiver integrity monitoring algorithms. From 2004 he headed the software development team and 2008 he became Vice President of R&D in the company. In October 2012 Fastrax was acquired by u-Blox AG and before transferring to his current employer Stefan worked as a project manager and software developer in the Algorithm and Signal Processing team at uBlox. Stefan received his M.Sc. degree from Åbo Akademi University, Department of Experimental Physics, in 1991 and his Licentiate degree from University of Turku, Dept. of Applied Physics, in 1996. Before joining Fastrax in 2000, Stefan was involved in various research projects in the fields of FT-IR and FT-Raman spectroscopy as well as molecular modeling and laser physics. In the field of GPS, Stefan has been the co-author of 9 conference papers and given numerous presentations on various topics. In other fields like FT-spectroscopy Stefan was the main author in two journal papers and numerous conference papers.

Mr. Esa AIROS received his M.Sc. degree in Electrical Engineering from the University of Lappeenranta. He is currently working as a research scientist at the Defense Forces Technical Research Centre. His research interests cover various electronic warfare applications including also GNSS.

10

Digital Offset Calibration of an OPAMP Towards Improving Static Parameters of 90 nm CMOS DAC

Daniel ARBET, Gabriel NAGY, Viera STOPJAKOVÁ, Martin KOVÁČ

Institute of Electronics and Photonics, Slovak University of Technology,
Ilkovičova 3, 812 19 Bratislava, Slovakia

daniel.arbet@stuba.sk

Abstract. *In this article, an on-chip self-calibrated 8-bit R-2R digital-to-analog converter (DAC) based on digitally compensated input offset of the operational amplifier (OPAMP) is presented. To improve the overall DAC performance, a digital offset cancellation method was used to compensate deviations in the input offset voltage of the OPAMP caused by process variations. The whole DAC as well as offset compensation circuitry were designed in a standard 90 nm CMOS process. The achieved results show that after the self-calibration process, the improvement of 48% in the value of DAC offset error is achieved.*

Keywords

Self-calibration, digital to analog converter, input offset voltage trimming.

1. Introduction

Design of analog and mixed-signal integrated circuits (ICs) in nanotechnologies becomes quite difficult due to presence of imperfections in the manufacture process. Unfortunately, ICs designed in advanced nanoscale technologies exhibit a high sensitivity to significant process parameter variations. Therefore, in these technologies, it is rather difficult to design high performance integrated circuits using standard design techniques and approaches [1], [2].

Analog-to-digital converters (ADC) and their digital-to-analog counterparts are probably the most commonly used mixed-signal circuits. Usually, these circuits represent only subcircuits of complex integrated systems, containing both digital and analog signal domains. However, undesired deviations in the value of devices and parameters of circuits used in the converters (ADC and DAC) might seriously influence their static parameters. In some PVT (power, voltage, temperature) corners, even the overall functionality of a DAC might be disrupted. Additionally, standard calibration methods usually require usage of Automatic Test Equipment (ATE) that can be rather expensive. Therefore, development of new alternative calibration/trimming approaches, which help to improve parameters of a designed integrated system (a DAC in our case), is a rather important task and challenging issue.

The important DC parameter of operational amplifiers (which are commonly used in binary-weighted R-2R ladder DACs) is the input offset voltage, and its deviation may be greater than $\pm$20 mV in a standard CMOS nanometer technology. The input offset voltage of the OPAMP presented in [3] exhibits a value in the range from -50 mV to +50 mV that is not acceptable for high-performance DACs. To solve this problem and to make the design robust and resistant to the process variations, a programmable biasing network was introduced and used in [4]. The other solution is a digital calibration of the OPAMP essential parameters [5]. In [6], an adaptive self calibration test was used in order to improve the linearity of a DAC.

In this paper, a digital offset trimming method has been used to compensate the input offset voltage of an OPAMP in order to improve the overall static parameters of an 8-bit binary-weighted R-2R ladder DAC designed in standard 90 nm CMOS technology. Another method for additional correction of the input offset voltage was presented in [7], where the offset mean value was reduced. Using this method, the mean value of the input voltage offset is kept near zero volts, which is a typical result of analog offset compensation methods.

2. Preliminary Work

Calibration techniques are frequently used to compensate the undesired deviation in the value of circuit's parameters. Currently, many techniques such as special layout techniques, dynamic element matching, error correction as well as analog or digital calibrations have been used to improve the accuracy of the data converters. However, each of them has certain limitations. Dedicated layout techniques are easy to implement because they do not require additional circuitry but they increase the area overhead and interconnection complexity, and might influence dynamic linearity of the converter. Dynamic element matching techniques are usually used in low-bit DACs because the complexity of the encoding circuits becomes significant for higher number of bits [8]. Self-trimming error correction techniques presented in [9], [10] require several additional circuits (ADC, digital correction circuit, memory storage, etc.), and, therefore, are not the suitable solution in terms of the chip area. Digital calibration techniques for improving the accuracy of ADCs and DACs are presented in [11]. In [12], digital calibration techniques for sigma-delta (Σ-Δ) data converters were introduced.

Calibration techniques for the input offset voltage of an OPAMP, which is a basic building block of the DAC, are presented in [13]. Chopper stabilization technique used in [14], [15] is based on the transposition of the signal to a higher frequency, where the effect of the input voltage offset is negligible. After the transposition to higher frequencies, the signal is amplified and demodulated back to the baseband. Basic principle of so-called *auto-zero technique* is to sample the undesired effect and then to subtract it during the second phase, when the input signal is processed by an imperfect amplifier [15]. A similar method is also *correlated double sampling* (CDS) used in [16]. However, both the autozero and CDS techniques are nor suitable for applications where a continuous-time output is required because during the sampling process the system must be disconnected from the signal path. This disadvantage can be eliminated by *ping-pong* technique [17], [18], which is used in [18] to calibrate a rail-to-rail amplifier. The amplifier is duplicated and one of them is calibrated while the other is amplifying the input signal. After calibration the role of both amplifiers is exchanged. An improved ping-pong technique that reduces the undesired glitches during the transition from one amplifier to another is presented in [19]. However, digital techniques represent the best alternative for calibration of different analog and mixed-signal systems, because of their versatility. Therefore, these techniques are usually used for calibration of the input offset voltage of the OPAMP.

Digital offset trimming techniques usually use the binary-weighted current network, described in [5], [20], [21]. The output current of such a converting network is used to compensate the undesired influence (e.g. offset voltage) of process parameter variations. Methods based on digital trimming principles have compensation limitations determined by the converting network resolution [7]. Therefore, also the value of the OPAMP input offset voltage achieved using the compensation is a function of the converting network resolution and the offset level to be compensated. The resolution can be expressed as follows:

$$\text{Resolution} = \frac{\text{Full Scale}}{1.\text{LSB}} \leq \frac{V_{off\_uncomp}}{V_{off\_comp}} \tag{1}$$

where V_{off_uncomp} and V_{off_comp} is the input offset voltage of the OPAMP before and affter the calibration process, respectivelly.

Depending on the used converting network, the different network resolution can be achieved, since it is given by the number of bits and weight type of the converting network. Using the redundance R, the resolution can be expressed by (2).

$$\text{Resolution} = \frac{b_1 + \sum_{i=1}^{n} b_i}{b_1} = \frac{R^n}{R-1}. \tag{2}$$

Redundancy parameter R of the binary weighted M-2M converting network is $R = 2$. In the case of a 6-bit converter, it corresponds to resolution of 64 steps. This is, at the same time, the maximum resolution for a 6-bit converter. Linearity and sensitivity, influenced mainly by components mismatch, are the most critical characteristics of the binary weighted converting network. Therefore, redundant converting networks described in detail in [5], [20], [21] are often used. Redundancies $R = 1.77$ and $R = 1.86$ are achieved for the converting networks of M-3M and M-2.5M types, respectively. Resolution of M-2.5M converting network type is 48 steps.

For the digital trimming, one thing is always common independently of the used converting network type. Maximum to minimum offset voltage ratio is always given by the network resolution, and the offset mean value is near half of that ratio. The mean value of the offset voltage is obviously different from zero volts. Therefore, a correction circuit was developed and used to reduce this feature [7]. After its intervention, the mean offset value can be kept close to zero volts. Similar result can be achieved using analog compensation methods [7].

3. Self-Calibrated DAC

Basic block diagram of the proposed self-calibrated DAC, designed in 90 nm CMOS technology, is shown in Fig. 1. The whole system consists of the converter itself (yellow parts) and on-chip additional hardware necessary for the self-calibration process (blue blocks), including control logic and the trimming block. Compensation of the OPAMP input offset voltage as well as additional circuitry needed for this purpose are described in more details in the next Section.

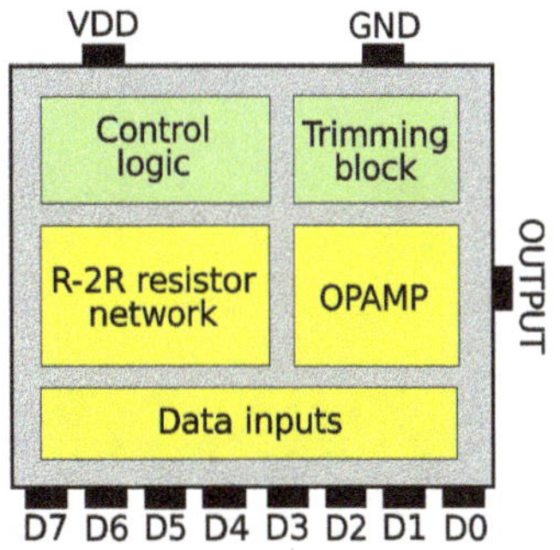

Fig. 1. Block diagram of the self-calibrated R-2R DAC.

3.1 OPAMP Offset Compensation

The circuit diagram of the OPAMP (representing the basic building block of the DAC) is shown in Fig. 2. Input stage together with the folded cascode represent the first

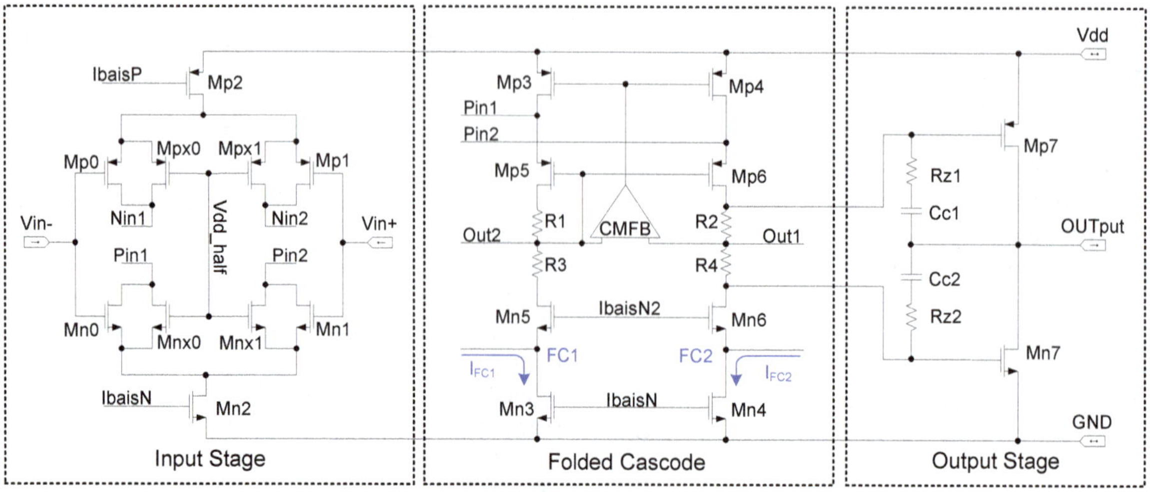

Fig. 2. Schematic diagram of the OPAMP used in DAC.

stage, while the push-pull inverter forms the output stage of the OPAMP. Rail-to-rail input and output stages allow the OPAMP to operate within the input and output voltages near to the voltage supply rails. Folded cascode stage was used to achieve high gain. To stabilize the operational point of the folded cascode and the push-pull inverter as well as to achieve a high gain of the output stage, the Common-Mode Feedback (CMFB) network was used. Detailed description of individual parts of the OPAMP can be found in [22].

Since device mismatch may influence the operational point of all transistors forming the OPAMP, it has to be taken into account that the input offset voltage represents a critical parameter of the OPAMP and thus, also the whole DAC. To improve this parameter, the digital compensation of the OPAMP's input offset was employed.

The offset compensation process is based on generation of compensation currents that are pumped into the respective branch of the folded cascode structure of the OPAMP. Such compensation currents can be generated by a converting network. Two different topologies for digital offset trimming based on the converting network, one using successive approximation register (SAR) and the other employing a simple counter have been used in [7]. Results presented in [7] indicate that M-2M converting network (depicted in Fig. 3) is more appropriate to digital trimming. However, the main drawback of the M-2M structure is the imbalance in the current division [5]. Since the compensation currents (I_{FC1} and I_{FC2}) in the M-2M compensating network (Fig. 3) will never be identical, an extra correction circuit (to be described later in Section 3.3) was employed to make them equal.

3.2 DAC Circuit Diagram

The circuit diagram of the whole 8-bit self-calibrated R-2R DAC with additional circuitry necessary for the OPAMP input offset compensation (within the DAC design) is shown in Fig. 4. Transmission gates or so-called T-gates

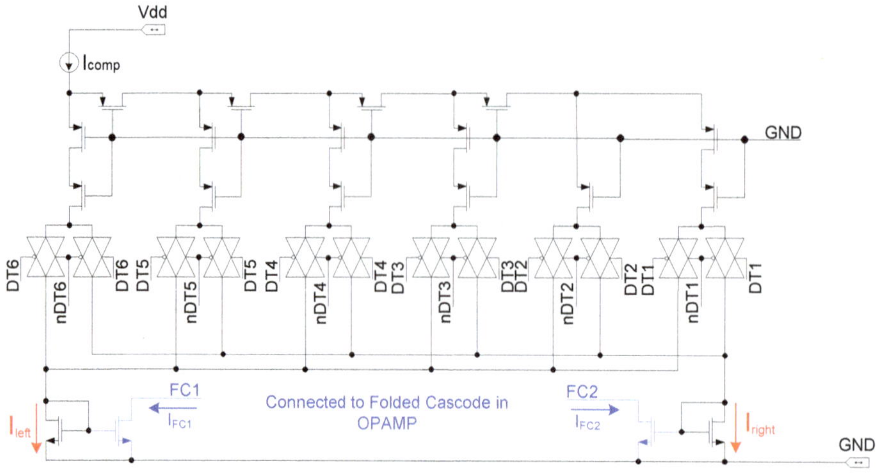

Fig. 3. M-2M converting network.

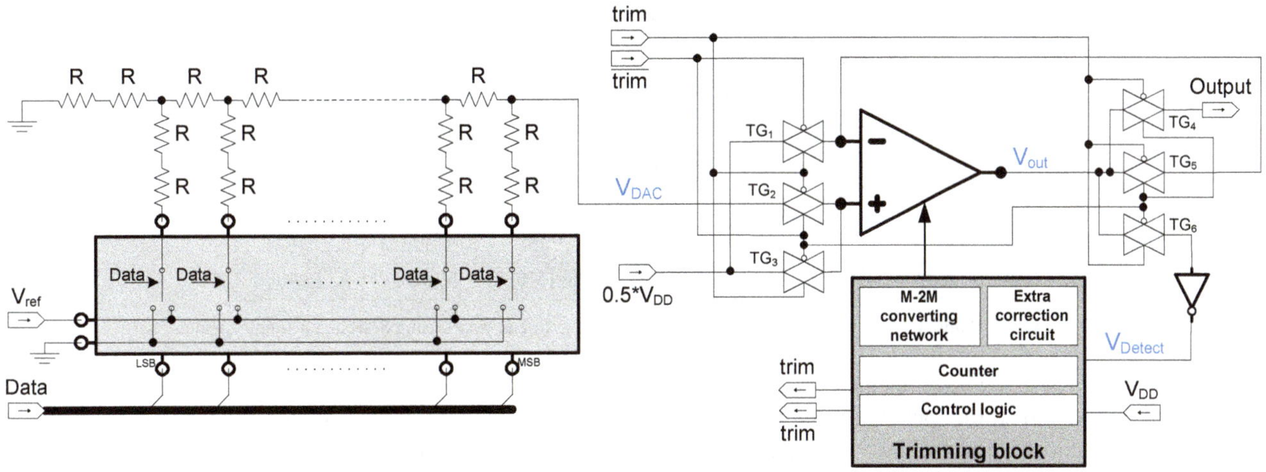

Fig. 4. Circuit diagram of the self-calibrated 8-bit R-2R DAC.

(TG) are used to switch the OPAMP from functional mode to trimming mode and to disconnect the OPAMP from the R-2R resistor network (a part of the DAC). Another extra hardware is the trimming block, which includes also control logic that is realized using a serial counter and converting network that generates the compensating current for the respective side of the folded cascode stage. Control logic together with the counter generates control signals for T-gates used in both the M-2M converting network and extra correction circuits. A voltage comparator is used to compare the OPAMP output voltage to the reference voltage and to indicate the completion of the trimming process [7]. In this case, a simple invertor plays the role of the voltage comparator.

When the supply voltage (V_{DD}) is turned on, control logic generates control signals ($trim$ and $\overline{trim}$) for trimming circuitry and starts the process of the digital compensation of the OPAMP input offset voltage. During the calibration process, the OPAMP is disconnected from the rest of the system. When the OPAMP output voltage (V_{detect}) is equal to $V_{DD}/2$, control logic terminates the calibration process. As soon as the calibration process is completed, control logic turns off the trimming block and the DAC is ready for use.

3.3 Extra Correction Circuit

The whole philosophy of the proposed extra correction circuit is based on the manual equalization of the compensation currents that are then injected into the respective side of the OPAMP's folded cascode. In the situation that the OPAMP offset voltage is zero, offset compensation current equivalent to 1 LSB would be generated. Since according to (3), the standard offset compensation currents in the two branches of the compensating network will never be identical, as already mentioned above. Therefore, a correction circuit was proposed and applied (4), to shift the mean offset value close to zero (Fig. 5).

No correction:

$$I_{left} = 32.\text{LSB}, \quad I_{right} = 31.\text{LSB},$$

$$I_{left} \neq I_{right}. \tag{3}$$

With correction:

$$I_{left} = 32.\text{LSB}, \quad I_{right} = 31.\text{LSB} + I_{correction},$$
$$I_{left} = I_{right}. \tag{4}$$

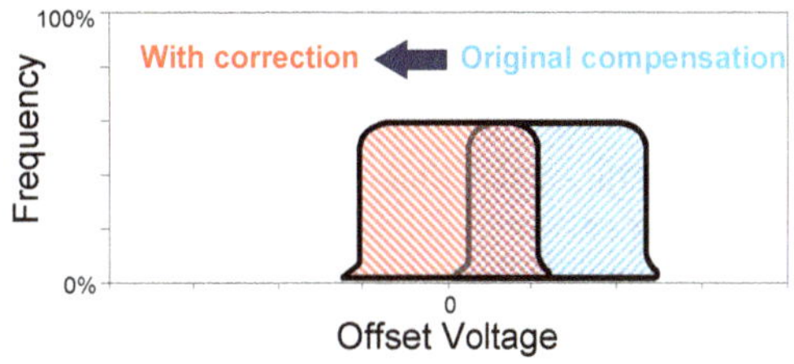

Fig. 5. Principle of the offset mean value shift by a correction circuit.

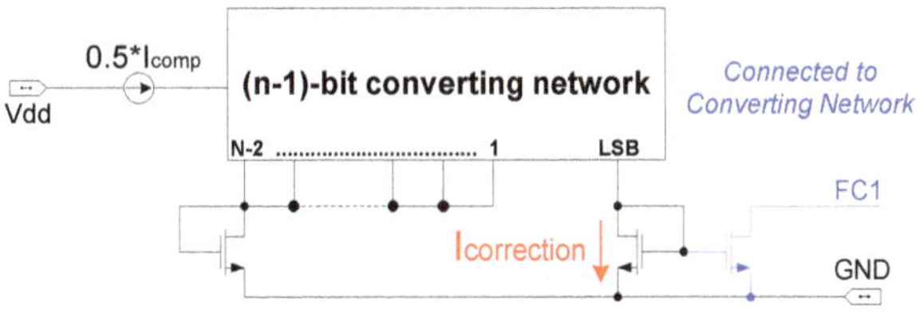

Fig. 6. Correction circuit generating the correction current.

With the extra correction circuit, standard trimming procedure is used. Differential current, generated by the converting network, is stepwise injected in the left and right sides of the folded cascode, till the output voltage is dissimilar to the reference voltage. Standard trimming procedure is finished when the output voltage value is equal to the reference voltage. When the trimming process ends, 1 LSB correction current $I_{correction}$ will be kept injecting in one side the folded cascode. To provide this, it was necessary to employ one additional converting network (Fig. 6). Taking into account the energy saving aspect, it was sufficient to use one order lower converter with half the maximum compensation current.

	Not trimmed circuit	Counter		SAR	
		No correction	correction circuit	No correction	correction circuit
sd [μV]	5517	343.15	343.55	333.97	331.07
mu [μV]	2	688.95	78.82	480.92	-113.16
Min [mV]	-17.357	0.055	-0.695	-0.012	-0.668
Max [mv]	18.216	1.547	0.945	1.396	0.754
Improvement	—	**95.81%**	**95.84%**	**96.04%**	**96%**

Tab. 1. Comparison of the OPAMP digital offset trimming based on the M-2M converting network using SAR and a simple counter.

The binary weighted M-2M converting network is very suitable for the presented correction approach since in this type of networks, the compensation current values for two succeeding states are approximately identical. On the other hand, redundant converting networks exhibit a difference between two succeeding states, and therefore, such topologies are not optimal for the offset compensation according to Fig. 5.

Generally, control signals that switch T-gates used in the M-2M converting network can be generated by a binary counter or by the successive approximation register (SAR). Comparison of results achieved for the OPAMP digital offset trimming based on the M-2M converting network using SAR and a simple counter is summarized in Tab. 1, and it was presented in [7]. It can be observed that the application of the correction circuit has no negative effect on the standard deviation of the offset voltage. On the other hand, the offset mean value is near the value of untrimmed circuit. The combination of M-2M type converting network and the control logic realized with SAR can achieve the best results, however, the version with control logic realized by a counter provides the lowest mean value of the offset.

4. Achieved Results

The proposed self-calibrated DAC was simulated and evaluated in CADENCE design environment for the power supply voltage of 2.5 V and temperature of 25°C, while performing 200 runs of Monte Carlo (MC) analysis. The value of main static parameters such as integral nonlinearity (INL), differential nonlinearity (DNL), offset error (OE), gain error (GE) and full scale error (FSE) of the designed DAC have been analyzed before and after applying trimming process. The obtained results are presented in this section.

Fig. 7 shows the deviation of INL parameter as a function of the DAC input data code. One can observe that in the worst case, deviation of INL (caused by process fluctuation) before the trimming process is in the range from -2.292 LSB to 2.286 LSB. After running the trimming, the deviation of INL is kept within the range from -1.298 LSB to 1.341 LSB, which represents about 42% improvement in comparison to the original circuit parameter. In the best case, six times better value of the INL parameter can be achieved that is an excellent result with respect to significant fluctuation of the process parameters expected in 90 nm technology.

Dependence of DNL parameter on the DAC input data code is shown in Fig. 8. The best improvement was achieved for low values of the input data code. The obtained results also show that with the OPAMP offset trimming process, the improvement of only 18% is achieved for DNL. This is due to the fact that the DNL parameter does not directly depend on the input offset voltage of the OPAMP.

Fig. 9 shows the OE parameter of the DAC before and after calibration process obtained by MC analysis. With no calibration process being applied, the OE deviation was in the range from -1.67 LSB to 2.29 LSB. After the calibration, the OE value falls into the range from -1.02 LSB to 1.04 LSB, which represents improvement of 48% of the

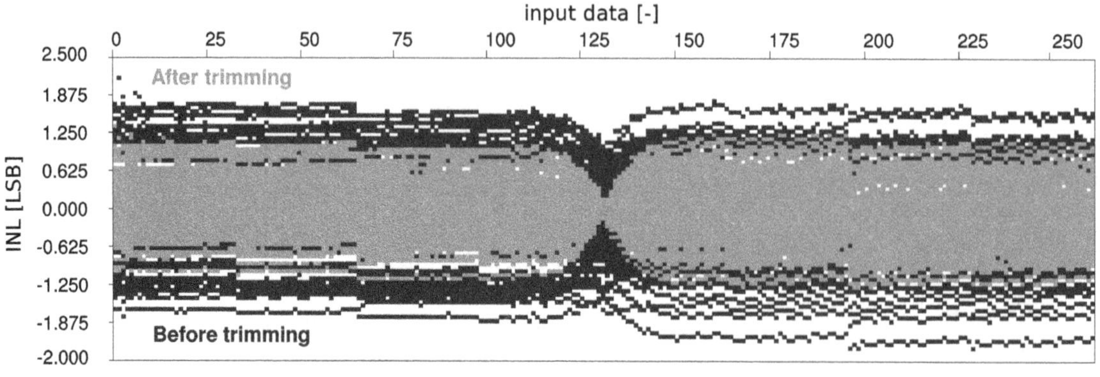

Fig. 7. Deviation of INL versus the DAC input data code.

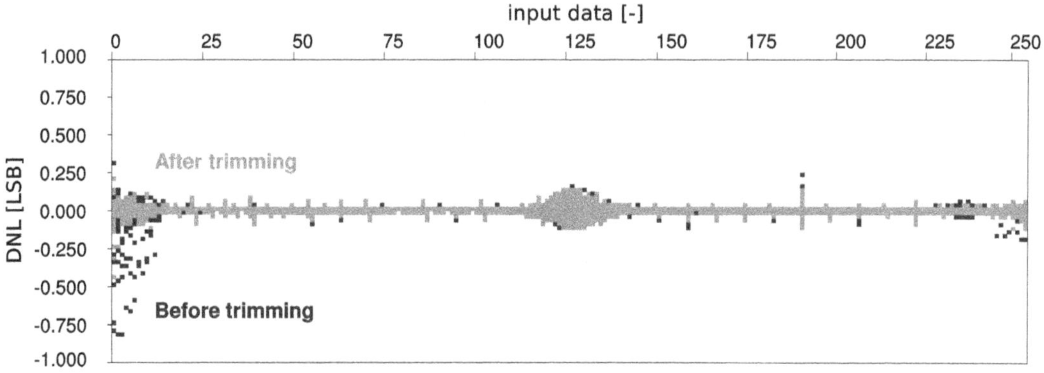

Fig. 8. Deviation of DNL versus the DAC input data code.

DAC offset error. We would like to note that for the offset error, the best improvement can be reached since this parameter of the proposed DAC directly depends on the OPAMP's input offset voltage.

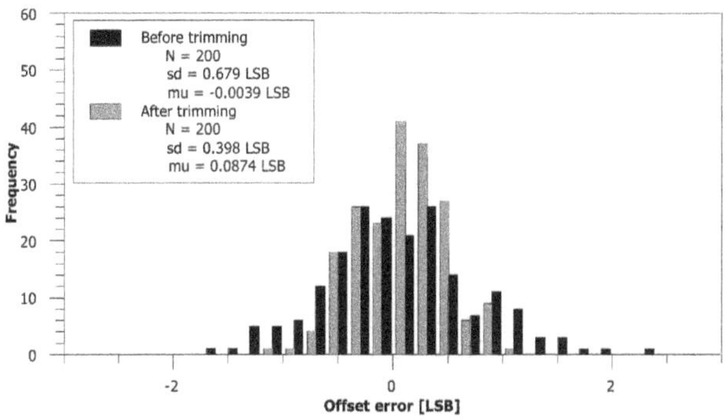

Fig. 9. Deviation of the offset error.

Similarly, the improvement of 38% and 31% can be achieved for GE and FSE, respectively. Deviation of gain error before and after calibration process obtained by MC analysis is depicted in Fig. 10. Before calibration, the deviation of GE was in the range from -5.48 LSB to 1.5 LSB, while the calibration process brings improvement of 38% of the DAC gain error. Similarly, the improvement of 31% can be achieved for FSE after calibration process. Deviation of FSE obtained by MC analysis is shown in Fig. 11.

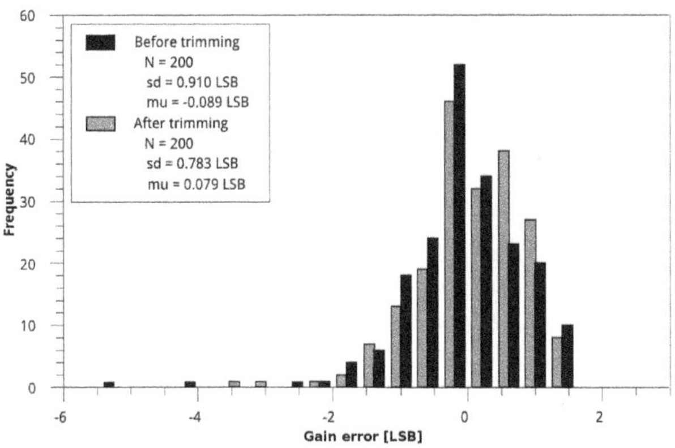

Fig. 10. Deviation of the gain error.

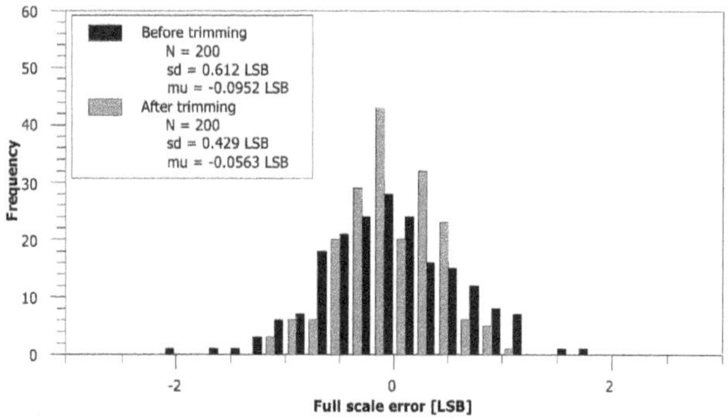

Fig. 11. Deviation of the full scale error.

The main static parameters of the designed DAC are summarized in Tab. 2, which also presents the worst case improvement of these parameters achieved by performing on-chip self-calibration of the converter. The best improvement of 48% was obtained for the offset error. Good results were also obtained for INL (42.4%), gain error (38.1%) and full scale error (31.1%) parameters. The less improved parameter is the differential nonlinearity (DNL), where the improvement of 18.6% was achieved.

Parameter	Before trimming	After trimming	Improvement (worst case)
INL [LSB]	-2.29 ÷ 2.29	-1.30 ÷ 1.34	**42.4%**
DNL [LSB]	-0.81 ÷ 0.67	-0.56 ÷ 0.64	**18.6%**
OE [LSB]	-1.67 ÷ 2.29	-1.02 ÷ 1.04	**48.0%**
GE [LSB]	-5.48 ÷ 1.50	-3.26 ÷ 1.55	**38.1%**
FSE [LSB]	-2.08 ÷ 1.60	-1.18 ÷ 1.10	**31.1%**

Tab. 2. Main static parameters of the DAC.

5. Discussion and Conclusion

The comparison of the achieved results to other works in terms of the INL parameter improvement is presented in Tab. 3. Nevertheless, this comparison might not be quite relevant because approaches presented in the other publications are based on calibration of the whole DAC, while this work utilizes the OPAMP offset voltage trimming only. As one can observe, only by 7.6% lower improvement of the INL parameter was achieved with respect to [23]. On the other hand, in [24], up to 30% higher improvement was reported. However, it is important to note that a DAC published in [24] has a higher resolution (12-bit) and has been calibrated completely as a whole system.

	Before trimming	After trimming	Improvement
This work	-2.29 ÷ 2.29	-1.30 ÷ 1.34	**42.4%**
[23] *8-bit*	-0.4 ÷ 0.4	-0.2 ÷ 0.2	**50.0%**
[24] *12-bit*	-1.4 ÷ 1.5	-0.4 ÷ 0.4	**74.4%**

Tab. 3. Comparison of INL parameter to other works.

Tab. 4 presents the comparison of the proposed DAC to other published works in terms of area overhead. Although DACs presented in [9], [24], [25] have a higher resolution, additional calibration hardware brings a larger area overhead, while in our case, the area overhead is only 8%. Moreover, the proposed DAC was designed in 90 nm CMOS technology were the chip area is rather expensive, and therefore, the area overhead of only 8% represents a significant result.

Parameter	**This work**	**[9]**	**[25]**	**[24]**
Technology	90 nm	0.13 μm	0.18 μm	0.25 μm
Area [mm^2]	0.01	0.1*	1.00	1.14
Overhead	8%	-	50%	25%

* only active area reported

Tab. 4. Comparison in terms of area overhead .

From the results presented in Tab. 3 and Tab. 4, it can be observed that only the DAC published in [24] is comparable to the DAC presented in this work. However, the DAC presented in [24] has a higher resolution and was designed in 0.25 μm technology, where the process parameter fluctuation is significantly smaller than in 90 nm technology. Therefore, the proposed approach based on calibration of the OPAMP input offset voltage represent a proper solution for calibration of DAC in nanoscale technologies.

The main advantages of the proposed self-calibrated DAC design include mainly easy implementation and possible fully on-chip realization of the self-calibration. Since the calibration process runs automatically with each turn-on of the DAC, deviation in the value of the OPAMPs input offset voltage caused by aging effects might be compensated as well. The proposed self-calibration offers the improvement of all static parameters of the DAC up to at least 30%, except for the DNL parameter. Drawback of the proposed approach is that dynamic parameters of the DAC are not improved since they do not directly depend on the input offset voltage of the OPAMP. Therefore, it is important to underline that the achieved results can be improved further if calibration of the R-2R ladder resistor network is carried out as well.

To summarize the performed work and achieved results, it can be stated that the on-chip digital offset compensation of the OPAMP has been used to improve the static parameters of the 8-bit binary weighted R-2R ladder DAC designed in a standard 90 nm CMOS technology. For this purpose, additional hardware including control logic has been inserted into the DAC, with no extra pins needed. Add-on hardware requires the area of 330 μm^2, which represents the overhead of 8% (with respect to the whole DAC). Nevertheless, if the DAC is used in a complex mixed-signal system this area overhead will become negligible.

The future work will be focused on the minimization of the area requirements for additional hardware as well as on the investigation of the undesired influence of the additional hardware on the DAC parameters in its functional mode.

Acknowledgements

This work was supported in part by the Ministry of Education, Science, Research and Sport of the Slovak Republic under grant VEGA 1/0823/13, by the ENIAC JU under project E2SG (Agr. No. 296131), and by the EC under FP7 ICT Project SMAC (Agr. No. 288827).

References

[1] LEWYN, L., YTTERDAL, T., WULFF, C., MARTIN, K. Analog circuit design in nanoscale CMOS technologies. *Proceedings of the IEEE*, 2009, vol. 97, no. 10, p. 1687 - 1714.

[2] KARMANI, M., KHEDHIRI, C., HAMDI, B. Design and test challenges in nano-scale analog and mixed CMOS technology. *International Journal of VLSI design and Communication Systems (VLSICS)*, 2011, vol. 2, no. 2, p. 33 - 43.

[3] MACHOWSKI, W., JASIELSKI, J. Offset compensation for voltage- and current amplifiers with CMOS inverters. In *Proceedings of the 19th International Conference on Mixed Design of Integrated Circuits and Systems (MIXDES)*. 2012, p. 382 - 385.

[4] DELA CRUZ, S., DELOS REYES, M., GAFFUD, T., ABAYA, T., GUSAD, M., ROSALES, M. Design and implementation of operational amplifiers with programmable characteristics in a 90 nm CMOS process. In *European Conference on Circuit Theory and Design (ECCTD)*. 2009, p. 209 - 212.

[5] PASTRE, M., KAYAL, M. Methodology for the digital calibration of analog circuits and systems using sub-binary radix DACs. In *16th International Conference on Mixed Design of Integrated Circuits Systems (MIXDES)*. 2009, p. 456 - 461.

[6] JIANG, W., AGRAWAL, V. D. Built-in adaptive test and calibration of DAC. In *Proceedings of 18th IEEE North Atlantic Test Workshop*. 2009, p. 124 - 127.

[7] NAGY, G., ARBET, D., STOPJAKOVA, V. Digital methods of offset compensation in 90 nm CMOS operational amplifiers. In *IEEE 16th International Symposium on Design and Diagnostics of Electronic Circuits Systems (DDECS)*. 2013, p. 124 - 127.

[8] SAEEDI, S., MEHRMANESH, S., ATARODI, M. A low voltage 14-bit self-calibrated CMOS DAC with enhanced dynamic linearity. *Analog Integrated Circuits and Signal Processing*, 2005, vol. 43, no. 2, p. 137 - 145.

[9] CONG, Y., GEIGER, R. A 1.5-V 14-bit 100-MS/s self-calibrated DAC. *IEEE Journal of Solid-State Circuits*, 2003, vol. 38, no. 12, p. 2051 - 2060.

[10] BUGEJA, A., SONG, B.-S. A self-trimming 14-b 100-MS/s CMOS DAC. *IEEE Journal of Solid-State Circuits,* 2000, vol. 35, no. 12, p. 1841 - 1852.

[11] MOON, U.-K., TEMES, G., STEENSGAARD, J. Digital techniques for improving the accuracy of data converters. *IEEE Communications Magazine*, 1999, vol. 37, no. 10, p. 136 - 143.

[12] SILVA, J., WANG, X., KISS, P., MOON, U., TEMES, G. Digital techniques for improved $\Delta\Sigma$ data conversion. In *Proceedings of the IEEE Custom Integrated Circuits Conference.* 2002, p. 183 - 190.

[13] PASTRE, M., KAYAL, M. *Methodology for the Digital Calibration of Analog Circuits and Systems: With Case Studies*, 1st ed. Springer, 2009.

[14] ENZ, C., VITTOZ, E., KRUMMENACHER, F. A CMOS chopper amplifier. *IEEE Journal of Solid-State Circuits*, 1987, vol. 22, no. 3, p. 335 - 342.

[15] ENZ, C., TEMES, G. Circuit techniques for reducing the effects of op-amp imperfections: Autozeroing, correlated double sampling, and chopper stabilization. *Proceedings of the IEEE*, 1996, vol. 84, no. 11, p. 1584 - 1614.

[16] WHITE, M., LAMPE, D., BLAHA, F., MACK, I. Characterization of surface channel CCD image arrays at low light levels. *Solid-State Circuits, IEEE Journal*, vol. 9, no. 1, Feb 1974, p. 1–12.

[17] YU, C.-G., AND GEIGER, R. An automatic offset compensation scheme with ping-pong control for CMOS operational amplifiers. *IEEE Journal of Solid-State Circuits*, 1994, vol. 29, no. 5, p. 601 - 610.

[18] OPRIS, I., KOVACS, G. T. A. A rail-to-rail ping-pong op-amp. *IEEE Journal of Solid-State Circuits*, 1996, vol. 31, no. 9, p. 1320 - 1324.

[19] KAYAL, M., SAEZ, R., DECLERCQ, M. An automatic offset compensation technique applicable to existing operational amplifier core cell. In *Proceedings of the IEEE Custom Integrated Circuits Conference.* 1998, p. 419 - 422.

[20] SCANDURRA, G., CIOFI, C., CAMPOBELLO, G., CANNATA, G. On the Calibration of DA Converters Based on R/βR Ladder Networks. *IEEE Transactions on Instrumentation and Measurement*, 2009, vol. 58, no. 11, p. 3901 - 3906.

[21] CIOFI, C., SCANDURRA, G., CAMPOBELLO, G., CANNATA, G. On the calibration of AD and DA converters based on R/βR ladder networks. In *13th IEEE International Conference on Electronics, Circuits and Systems (ICECS)*. 2006, p. 958 - 961.

[22] ARBET, D., NAGY, G., GYEPES, G., STOPJAKOVA, V. Design of rail-to-rail operational amplifier with offset cancelation in 90 nm technology. In *International Conference on Applied Electronics (AE)*. 2012, p. 17 - 20.

[23] VARGHA, B., ZOLTAN, I. Calibration algorithm for current-output R-2R ladders. *IEEE Transactions on Instrumentation and Measurement*, 2001, vol. 50, no. 5, p. 1216 - 1220.

[24] RADULOV, G., QUINN, P., HEGT, H., VAN ROERMUND, A. An on-chip self-calibration method for current mismatch in D/A converters. In *Proceedings of the 31st European Solid-State Circuits Conference (ESSCIRC)*. 2005, p. 169 - 172.

[25] TIILIKAINEN, M. A 14-bit 1.8-V 20-mW 1-mm2 CMOS DAC. *IEEE Journal of Solid-State Circuits*, 2001, vol. 36, no. 7, p. 1144 - 1147.

About Authors . . .

Daniel ARBET received the M.S. and the Ph.D. degrees in Microelectronics from Slovak University of Technology in Bratislava, Slovakia in 2009 and 2013, respectively. Since October 2013 he has been a researcher at the Institute of Electronics and Photonics of Slovak University of Technology. He has published more than 20 papers. His main research interests are low-voltage low-power analog design, ASIC design, on-chip parametric testing, analog BIST and test implementation.

Gabriel NAGY received the M.S. degree in Electronics from Slovak University of Technology in Bratislava, Slovakia in 2012. Since September 2012 he has been a Ph.D. student at the Institute of Electronics and Photonics of the same university. His main research interests are low-voltage low-power analog IC design and energy-efficient integrated systems.

Viera STOPJAKOVÁ received the M.S. degree and the Ph.D. degrees in Electronics from Slovak University of Technology in Bratislava, Slovakia, in 1992, and 1997, respectively. Currently, she is a full professor at the Institute of Electronics and Photonics of the same institution. She has been involved in several EU funded research projects under different funding schemes such as TEMPUS, ESPRIT, Copernicus, FP, ENIAC-JU, etc. She has published over 100 papers in scientific journals and in proceedings of international conferences. She is a co-inventor of two US patents in the field of on-chip supply current testing. Her main research interests include ASIC design, on-chip testing, design and test of mixed-signal circuits, energy harvesting, smart sensors and biomedical monitoring.

Martin KOVÁČ received his M.S. degree in Electronics from Slovak University of Technology in Bratislava, Slovakia in 2013. He is currently a Ph. D. student under the supervision of Prof. Viera Stopjaková. His research is centered on development of low-voltage low-power analog IC design and on-chip parametric testing. Current research work includes design of energy-efficient RF integrated systems for biomedical applications.

11

All-Pole Recursive Digital Filters Design Based on Ultraspherical Polynomials

Nikola STOJANOVIĆ[1], *Negovan STAMENKOVIĆ*[2], *Vidosav STOJANOVIĆ*[1]

[1] University of Niš, Faculty of Electronics, A. Medvedeva 14, 18000 Niš, Serbia
[2]University of Prishtina, Faculty of Natural Science and Mathematics, 28220 K. Mitrovica, Serbia

nikola.stojanovic@elfak.ni.ac.rs, negovan.stamenkovic@pr.ac.rs

Abstract. *A simple method for approximation of all-pole recursive digital filters, directly in digital domain, is described. Transfer function of these filters, referred to as Ultraspherical filters, is controlled by order of the Ultraspherical polynomial, ν. Parameter ν, restricted to be a non-negative real number ($\nu \geq 0$), controls ripple peaks in the passband of the magnitude response and enables a trade-off between the passband loss and the group delay response of the resulting filter. Chebyshev filters of the first and of the second kind, and also Legendre and Butterworth filters are shown to be special cases of these all-pole recursive digital filters. Closed form equations for the computation of the filter coefficients are provided. The design technique is illustrated with examples.*

Keywords

All-pole IIR filter, lowpass filters, highpass filters, ultraspherical filter, approximation theory.

1. Introduction

The ultraspherical (or Gegenbauer) orthogonal polynomials have already been used in low-pass FIR filter design in time domain [1], [2] and as wavelet functions [3]. However, recursive digital filters can be designed either through application of bilinear transformation on continuous-time filter [4], or directly in the z-domain [5].

In the first approach, the starting point is designing of recursive filters in the continuous-time domain (analog prototype), in addition to designing continuous-time filters based on ultraspherical polynomials [6]. Lastly, transfer function of recursive filter is obtained by using the bilinear transformation. This method requires that all zeros lie at $z = -1$ or on the unit circle.

The second approach is desirable especially for or all-pole (autoregressive) digital filters which have no counterpart in the continuous-time domain. All-pole transfer function class is an important filter category in which low-pass transfer function contains all its zeros at the origin in the z-plane. Those transfer functions are easier to implement than transfer functions that contain only finite zeros on the unit circle, such as elliptic filters.

Discussion in this paper has been restricted to direct design of all-pole digital filters based on ultraspherical polynomials.

Direct design of the recursive digital filters has first been proposed by Rader and Gold [7]. They have shown that characteristic function of these filters is trigonometric polynomial of $\omega/2$, where ω is the digital frequency in radians. They have also concluded that the square of the amplitude characteristic must be rational function of z, where denominator is an image mirror polynomial. Choosing different trigonometric functions for frequency variable, different types of IIR filters can be obtained. Based on these results, direct synthesis of the transitional Butterworth-Chebyshev (TBC) and Butterworth-Legendre filters has been proposed in [8], [9]. These TBC filters are the generalization of the results of previously given continuous-time TBC filters [10], obtained by a mixture of the Butterworth and the Chebyshev components.

Later, other types of orthogonal polynomial approximations for designing continuous-time and IIR digital filters have been used, such as Bessel [11], Jacobi [12], ultraspherical [6] and Pascal polynomials [13]. These approximations are also referred to as polynomial approximations due to the fact that characteristic functions are polynomials. Only Butterworth [7], Chebyshev [14] and transitional Butterworth-Chebyshev [10] continuous-time filters have counterparts in the discrete-time domain.

In this paper a direct method for designing the all-pole recursive digital filters using ultraspherical polynomials, is presented. The frequency responses of ultraspherical filters span between Butterworth to Chebyshev, as the order ν of ultraspherical polynomials goes from infinity to zero. Transition between Butterworth to Chebyshev transfer function is continuous, in contrast with the classical TBC filter where transition is gradual. The order of ultraspherical polynomials, ν, restricted to be a non-negative number, enables a trade-off between the stopband attenuation, passband ripples and group delay deviation of the resulting filter.

The rest of this paper is organized as follows. In Section 2, we derive filter coefficients in closed form and cutoff slope for the proposed design of the all-pole digital filters. Section 3 presents design examples to illustrate the effectiveness of the proposed approach, and finally the conclusions of this paper are presented in Section 4.

2. Approximation

The squared amplitude characteristic of the ultraspherical filters can be expressed as a real function of frequency variable x by using the Feldtkeller's equation [15, Chap. 2]:

$$|H_n(x)|^2 = \frac{1}{1+\varepsilon^2\left[\frac{C_n^\nu(x)}{C_n^\nu(1)}\right]^2} \tag{1}$$

where $C_n^\nu(x)$ is an ultraspherical, also known as Gegenbauer, polynomial (entire even or odd) of order ν (ν is a real number) and degree n. Usually, ε is a design parameter related to the maximum passband attenuation a_{max} (in dB) as $\varepsilon = \sqrt{10^{0.1a_{max}}-1}$.

Formally, ultraspherical polynomials of degree n, $C_n^\nu(x)$, can be defined by the explicit expression [16]:

$$C_n^\nu(x) = \frac{1}{\Gamma(\nu)}\sum_{k=0}^{\lfloor n/2\rfloor}\frac{(-1)^k\Gamma(\nu+n-k)}{k!(n-2k)!}(2x)^{n-2k} \tag{2}$$

or by the recurrence formula:

$$C_n^\nu(x) = \frac{1}{n}[2x(n+\nu-1)C_{n-1}^\nu(x)-(n+2\nu-2)C_{n-2}^\nu(x)] \tag{3}$$

where $C_0^\nu(x)=1$, $C_1^\nu(x)=2\nu x$ and ν acts as a free parameter. Furthermore, $C_n^\nu(x)$ is an even function of x for n even, and odd function of x for n odd. It also has n single zero locations in interval $x\in(-1,1)$.

The ultraspherical polynomials are related to the Chebyshev polynomials of the first kind, $T_n(x)$, to the Legendre polynomials, $P_n(x)$, to the Chebyshev polynomials of the second kind, $U_n(x)$, and to the characteristic polynomial of the Butterworth filter, $B_n(x)$, by following relations [16]:

$$\begin{aligned} T_n(x) &= \frac{n}{2}\lim_{\nu\to 0}\frac{C_n^\nu(x)}{\nu},\\ P_n(x) &= C_n^{0.5}(x),\\ U_n(x) &= C_n^1(x),\\ B_n(x) &= \lim_{\nu\to\infty}\frac{C_n^\nu(x)}{C_n^\nu(1)} = x^n. \end{aligned} \tag{4}$$

Thus, the ultraspherical responses span between Butterworth to Chebyshev response, as the order ν goes from infinity to zero.

Since, in approximation of all-pole transfer function in (1), term $C_n^\nu(x)=\sum_{i=0}^{n} g_{n-i}x^{n-i}$ is polynomial, the corresponding squared amplitude characteristic of all pole transfer function takes the following form:

$$|H_n(x)|^2 = \frac{1}{c_{2n}x^{2n}+c_{2n-2}x^{2n-2}+\cdots+c_2x^2+c_0} \tag{5}$$

where

$$c_i = \frac{\varepsilon^2}{(C_n^\nu(1))^2}\sum_{j=0}^{i} g_j g_{i-j}$$

for $i=1,\ldots,2n$ and for $i=0$ holds $c_0=g_0^2+1$ for n even, but $c_0=1$ if n is odd. By convention, $g_{n+1}=\ldots=g_{2n}=0$. Therefore, the magnitude response of all-pole transfer functions (5) is a complete even polynomial.

If x is continuous-time angular frequency $x^2=-s^2$, then function (5) is magnitude characteristic of the continuous time lowpass transfer function. On the other hand, for obtaining the lowpass all-pole discrete-time transfer function a suitable rational function for the frequency variable x is [17]:

$$x^2 = \frac{(z-1)^2}{-4\alpha^2 z} \tag{6}$$

where $\alpha=\sin(\omega_c/2)$ and ω_c is the normalized lowpass cutoff digital frequency in π units. If we want high-pass all-pole filter design, for frequency variable x should be used:

$$x^2 = \frac{(z+1)^2}{4\beta^2 z} \tag{7}$$

where $\beta=\cos(\varpi_c/2)$ and ϖ_c is the normalized highpass cutoff digital frequency in π units. This high-pass approximation is performed by using transformation $z\to -z$ on the lowpass transfer function. The poles of resulting highpass filter are obtained by changing angle by $\pi-\varphi$ where φ is the angle of the lowpass filter pole. This implies that $\omega_c+\varpi_c=\pi$.

Substituting (6) into (5), function $G(z)=H(z)H(1/z)$ is obtained, which is equal to $|H(e^{j\omega})|^2$ when it is evaluated along the unit circle:

$$G(z) = \frac{1}{c_{2n}\frac{(z-1)^{2n}}{(-4\alpha^2 z)^n}+\cdots+c_4\frac{(z-1)^4}{(-4\alpha^2 z)^2}+c_2\frac{(z-1)^2}{-4\alpha^2 z}+c_0}. \tag{8}$$

As can be seen, $G(z)$ is a rational function of z with zero of order n at the origin. Equation (8) can be rewritten in the following form:

$$G(z) = \frac{z^n}{c_{2n}\frac{(z-1)^{2n}}{(-4\alpha^2)^n}+\cdots+c_2\frac{(z-1)^2}{-4\alpha^2}z^{n-1}+c_0z^n}. \tag{9}$$

Note that the component $(z-1)^m$ is a mirror-image polynomial, and that the sum of the mirror-image polynomial of degree m and the mirror-image polynomial of degree $(m-2r)$, multiplied by z^r, is a mirror-image polynomial of degree m. Applying this property, it follows that denominator of $G(z)$ is the mirror-image polynomial of degree $2n$:

$$G(z) = \frac{z^n}{d_0z^{2n}+d_1z^{2n-1}+\cdots+d_nz^n+\cdots+d_1z+d_0}. \tag{10}$$

Relation between coefficients d_i and coefficients c_{2i} is given in closed form by:

$$d_{2n-i} = \sum_{j=0}^{2n-i} \frac{(-1)^j c_{2(i+j-n)}}{(-4\alpha^2)^{i+j-n}} \binom{2(i+j-n)}{j} \tag{11}$$

for $i = n, n+1, \ldots, 2n$.

Poles of the transfer function $H_n(z)$ are obtained by equating the denominator of (10) with zero, and solving it by numerical technique. Since the roots occur in reciprocal pairs, the poles of all-pole ultraspherical filter, $H(z)$, are the roots z_i that lie inside the unit circle:

$$H_n(z) = \frac{h_0 z^n}{\prod_{i=1}^{n}(z - z_i)} = \frac{h_0 z^n}{\sum_{i=0}^{n} a_{n-i}\, z^{n-i}} \tag{12}$$

where $h_0 = \sum_{i=0}^{n} a_i / \sqrt{\sum_{i=0}^{2n} c_i}$ is constant which ensures that amplitude $|H_n(e^{j\omega})|$ is bounded above by unity.

These types of filters can not be obtained from analogue domain by applying the bilinear transformation.

2.1 Cut-off Slope

For filters considered here, a comparison of steepness of their slopes at the cutoff frequency (cutoff slope), can be made by calculating the slopes:

$$S = \frac{\mathrm{d}}{\mathrm{d}\omega} \frac{1}{\sqrt{1+\varepsilon^2 \left[\frac{C_n^\nu(x)}{C_n^\nu(1)}\right]^2}} \Bigg|_{\omega=\omega_c} \tag{13}$$

at the cutoff frequency $\omega = \omega_c$ for equal attenuation in the pass-band, a_{max} [6]. Since on the unit circle, $z = \exp(j\omega)$, the frequency variable (6) on the real frequency is:

$$x = \frac{1}{\alpha} \sin \frac{\omega}{2}.$$

By implying the relation [16]:

$$\frac{\mathrm{d}}{\mathrm{d}x} C_n^\nu(x) = 2\nu C_{n-1}^{\nu+1}(x)$$

and after simple mathematical manipulation follows:

$$S = -\frac{\varepsilon^2 \nu}{(1+\varepsilon^2)^{3/2}} \frac{C_{n-1}^{\nu+1}(1)}{C_n^\nu(1)} \cot \frac{\omega_c}{2}. \tag{14}$$

The cutoff slope depends on the width of the passband, ω_c, and it is steeper if the passband is narrower. When the normalized passband is π, the cutoff slope is equal to zero. In comparison to standard approximation, which uses bilinear transformation [8], this all-pole approximation is suitable for the design of narrow-band lowpass recursive digital filters because it uses $n+1$ multipliers less than for their implementation [18]. For example, if the pass-band edge is less than 0.2π, then both filters have approximately the same slope.

The cutoff slope of highpass all-pole filters depends also on the cutoff frequency:

$$S = \frac{\varepsilon^2 \nu}{(1+\varepsilon^2)^{3/2}} \frac{C_{n-1}^{\nu+1}(1)}{C_n^\nu(1)} \tan \frac{\varpi_c}{2}. \tag{15}$$

If cutoff frequency, ϖ_c, increases then cutoff slope also increases. Since the highpass filter has passband above the cutoff frequency, then passband decreases if cutoff frequency increases. In comparison with standard approximation, which uses bilinear transformation, this approximation is suitable also for design narrow-band high pass all-pole digital filter because it saves $(n+1)$ multipliers.

Based on the above-mentioned cutoff slope, cascading low pass filter with high pass filter for the bandpass filter producing is not suitable.

3. Design Examples

Derived equations have been used for calculation of the magnitude and the group delay responses of the Chebyshev of the first and of the second kind, Legendre and Butterworth filters, for degree $n = 8$ and for different values of the parameter ν.

The coefficients of the eight degree transfer functions are given in Tab. 1 ($\nu = 0, 0.5, 1$ and ∞) and corresponding digital frequency responses are displayed in Fig. 1. The frequency is normalized so that the passband edge is $\omega_c = 0.3\pi$ and the maximum passband attenuation is $a_{max} = 2$ dB ($\varepsilon = 0.7647831$).

Coeff.	$A(z) = a_8 z^8 + a_7 z^7 + \cdots + a_1 z + a_0$			
	$\nu = 0$	$\nu = 0.5$	$\nu = 1$	$\nu \to \infty$
a_8	1.000000	1.000000	1.000000	1.000000
a_7	−5.789367	−5.353353	−5.059713	-3.381678
a_6	15.871965	13.635670	12.229774	5.649514
a_5	-26.6694584	-21.321581	-18.172022	-5.830866
a_4	29.891276	22.232672	18.004784	3.988422
a_3	-22.827204	-15.767002	-12.118705	-1.829804
a_2	11.593282	7.411023	5.394609	0.545467
a_1	-3.584770	-2.109682	-1.449659	-0.096038
a_0	0.518558	0.278735	0.179975	0.007612
h_o	0.003399	0.006344	0.009009	0.052630

Tab. 1. Polynomial coefficients for the eight degree ultraspherical filters for different order ν.

When ν is gradually changing from zero to infinity we have a continuous transitional Butterworth-Chebyshev all-pole approximation of recursive digital filters which covers Chebyshev second kind and Legendre approximation. If the degree of the filter is given, transitional region can be continually adjusted with order (ν) of ultraspherical polynomial.

Figure 1 shows the attenuation characteristics of eight degree ultraspherical all-pole filter with $\omega_c = 0.3\pi$ for various values of ν. In Fig. 1 it can be shown that the proposed ultraspherical filter with $\nu > 1$, has very small ripple in the passband and lower group delay variation in comparison to

Chebyshev filter. As might be expected, the Chebyshev filter ($\nu = 0$) has best performance in the stopband. It can be concluded that case $\nu = 1$ is better from the standpoint of amplitude response, but it has a poorer group delay response than the Butterworth filter ($\nu \to \infty$). Order of ultraspherical polynomial, ν, enables trade-off between ripple peaks in passband and delay response of filter.

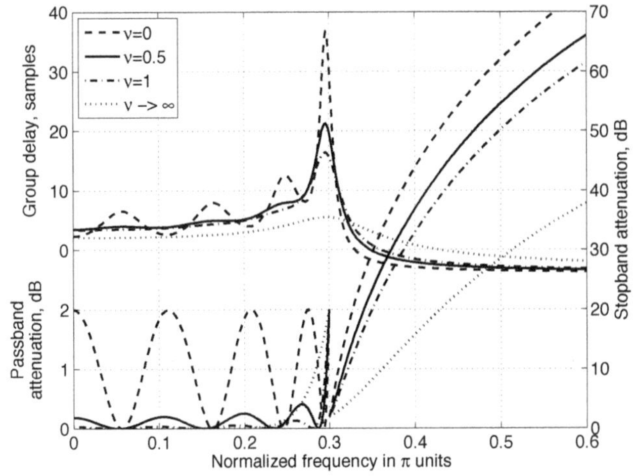

Fig. 1. Attenuation responses and group delay characteristic of the eight-degree ultraspherical all-pole digital filters.

If group distortion is too great, then group delay corrector is available [19].

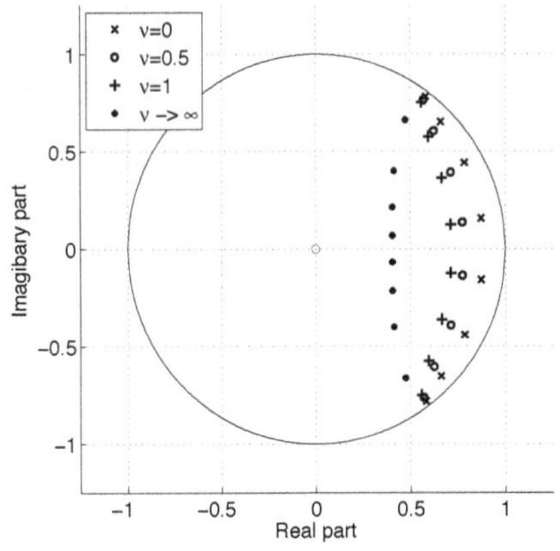

Fig. 2. The pole plot of the eight-degree ultraspherical all-pole digital filters with passband edge $\omega_c = 0.3\pi$.

Figure 2 gives the pole-zero diagram of the eight order ultraspherical recursive digital filters. It is shown that dominant poles of ultraspherical filters for $\nu \leq 1$ are positioned very close to each other, but their dominant pole quality factors (Q-factors) are significantly different.

For example, modulus of dominant poles for thirteenth-degree ultraspherical filters for $\nu = 0$, 0.5 and 1 are 0.98553, 0.97255 and 0.96206, respectively, but their Q-factors[1] are 42.8962708, 22.4701481 and 16.1788120, respectively. As it is known [20, Chapter 5], the sensitivity in the passband increases with pole Q-factor. Thus, the sensitivity in the passband decreases as the order of the ultraspherical polynomial increases.

4. Conclusion

Polynomial approximations, such as Butterworth and Chebyshev, leading to all-pole transfer functions, are extensively used in analog and IIR digital filter design. The Ultraspherical polynomials, $C_n^\nu(x)$, are used to present new all-pole IIR discrete-time filter approximation. These filters, which can be referred to as Gegenbauer filters, include as special cases Butterworth ($\nu \to \infty$), Chebyshev second kind ($\nu = 1$), Legendre ($\nu = 0.5$) and Chebyshev first kind ($\nu = 0$) discrete time all-pole filters, amongst others. The order of ultraspherical polynomials, ν, enables a trade-off between the stopband attenuation, the group delay behavior and the passband ripples of the resulting filter. As expected, the group delay becomes more constant as ν deviates from zero (Chebyshev of the first kind) to infinity (Butterworth). The coefficients of the eight order transfer function are tabulated for $\nu = 0, 0.5, 1$ and ∞.

It should be noted that other combinations of ultraspherical polynomials can be used in (1). For example, a product of lower degree ultraspherical polynomials yielding a new one of the same order. Thus, another transfer function is given by

$$|H_n(x)|^2 = \frac{1}{1 + \varepsilon^2 \left[\frac{C_k^\nu(x) C_{n-k}^\nu(x)}{C_k^\nu(1) C_{n-k}^\nu(1)}\right]^2} \quad (16)$$

for $k = 0, 1, \ldots, n/2$.

Acknowledgements

This work was supported and funded by the Serbian Ministry of Science and Technological Development under the project No. 32009TR.

References

[1] DECZKY, A. G. Unispherical windows. In *Proceedings of the IEEE International Symposium on Circuits and Systems (ISCAS)*. Caracas (Venezuela), 2001, p. 85 - 88.

[2] ROWINSKA-SCHWARZWELLER, A., WINTERMANTEL, M. On designing FIR filters using windows based on Gegenbauer polynomials. In *Proceedings of IEEE International Symposium on Cir-*

[1] Z-plane may refer to $z = e^{sT}$ or $re^{j\theta} = e^{\sigma T} e^{j\omega}$. Further, $\sigma T = \ln r$ and $\omega = \theta$. Finally,

$$Q = -\frac{\sqrt{\sigma^2 + \omega^2}}{2\sigma} = -\frac{\sqrt{\ln^2 r + \theta^2}}{2 \ln r}.$$

cuits and Systems (ISCAS). Scottsdale (AZ, USA), 2002, vol. 1, p. I-413 - I-416.

[3] SOARES, L. R., DE OLIVEIRA, H. M., SOBRAL CINTRA, R. J. D. Applications of non-orthogonal filter banks to signal and image analysis. In *Proceedings of 2006 IEEE PES Transmission and Distribution Conference and Exposition (TDC)*. Caracas (Venezuela), 2006.

[4] ANTONIOU, A. *Digital Signal Processing: Signals, Systems, and Filters*. New York: McGraw-Hill, 2006.

[5] THIRAN, J.-P. Recursive digital filters with maximally flat group delay. *IEEE Transactions on Circuit Theory*, 1971, vol. 18, no. 6, p. 659 - 664.

[6] JOHNSON, D., JOHNSON, J. Low-pass filters using ultraspherical polynomials. *IEEE Transactions on Circuits Theory*, 1966, vol. 13, no. 4, p. 364 - 369.

[7] GOLD, B., RADER, C. Digital filter design techniques in the frequency domain. *Proceedings of the IEEE*, 1967, vol. 55, no. 2, p. 149 - 171.

[8] NIKOLIĆ, S., STOJANOVIĆ, V. Transitional Butterworth-Chebyshev recursive digital filters. *International Journal of Electronics*, 1996, vol. 80, no. 1, p. 13 - 20.

[9] STAMENKOVIĆ, N., STOJANOVIĆ, V. On the design transitional Legendre–Butterworth filters. *International Journal of Electronics Letters*, 2014, vol. 2, no. 3.

[10] BUDAK, A., ARONHIME, P. Transitional Butterworth-Chebyshev filters. *IEEE Transactions on Circuits Theory*, 1971, vol. 18, no. 5, p. 413 - 415.

[11] THOMSON, W. Delay networks having maximally flat frequency characteristics. *Proceedings of the IEEE*, 1949, vol. 96, no. 44, p. 487 - 490.

[12] PAVLOVIĆ, V., ILIĆ, A. New class of filter functions generated most directly by the Cristoffel-Darboux formula for classical orthonormal Jacobi polynomials. *International Journal of Electronics*, 2011, vol. 98, no. 12, p. 1603 - 1624.

[13] DIMOPOULOS, H. G., SARRI, E. The modified Pascal polynomial approximation and filter design method. *International Journal of Circuit Theory and Applications*, 2012, vol. 40, no. 2, p. 145 - 163.

[14] SOLTIS, J. J., SID-AHMED, M. A. Direct design of Chebyshev-type recursive digital filters. *International Journal of Electronics*, 1992, vol. 70, no. 2, p. 423 - 419.

[15] CHEN, W.-K. *The Circuits and Filters Handbook*, 3rd ed. Boka Raton: CRC Press, 2009.

[16] ABRAMOWITZ, M., STEGUN, I. *Handbook of Mathematical Functions with Formulas, Graphs, and Mathematical Tables*, 9 ed. New York, Dover: National Bureau of Standards Applied Mathematics Series 55, 1972.

[17] STOJANOVIĆ, V., NIKOLIĆ, S. Direct design of of sharp cutoff low-pass recursive digital filters. *International Journal of Electronics*, 1998, vol. 85, no. 5, p. 589 - 596.

[18] HARRIS, F., LOWDERMILK, W. Implementing recursive filters with large ratio of sample rate to bandwidth. In *Proceedings of the Forty-First Asilomar Conference on Signals, Systems and Computers (ACSSC)*. Pacific Grove (CA, USA), 2007, p. 1149 - 1153.

[19] ZAPLATÍLEK, K., ŽIŠKA, P., HÁJEK, K. Practice utilization of algorithms for analog filter group delay optimization. *Radioengineering*, 2007, vol. 16, no. 1, p. 7 - 15.

[20] DARYANANI, G. *Principles of Network Synthesis and Design*. New York: John Wiley and Sons, 1976.

About Authors ...

Nikola STOJANOVIĆ was born in 1973. He received his M.Sc. degree in Electronics and Telecommunication at the Faculty of Technical Sciences, University of Priština, Kosovska Mitrovica in 1997, and M.Sc. degree in Mutlimedia Technologies at Faculty of Electronic Engineering, University of Niš at 2013. Currently he works as a lecturer of multimedia and 3D animation at Faculty of Electronics, University of Niš and a PhD student at department of Electronics at the same University.

Negovan STAMENKOVIĆ was born in 1979. He received the M.Sc. degree from the Department of Electronics and Telecommunication at the Faculty of Technical Sciences, University of Priština, Kosovska Mitrovica in 2006 and the Ph.D. degree in electrical and computer engineering from the Faculty of Electronic Engineering, Niš, Serbia, in 2011. He is assistant professor at Faculty of Natural Sciences and Mathematics, University of Priština. His current research interests lie in the area of analog and digital signal processing based on the residue number system.

Vidosav STOJANOVIĆ studied electrical engineering at the University of Niš, Serbia and he got his B.Sc. in 1964. The next year, he joined the Faculty of Electronic Engineering as a teaching and research assistant. He received the M.Sc. E.E. degree from the University of Belgrade, Serbia, in 1974. In 1977 he received Ph.D. in Electrical Engineering. 1981/82 he was a Humboldt Scholar at the University of Munich, working on the design of a high-speed digital transmission system. He joined Electronics industry of Niš, Serbia, in 1984. He was the director of the Institute for Research and Development and part-time professor for digital image processing at the Faculty of Electronic Engineering. After five years of working in the industry he became the full-time professor for analog and digital signal processing at the Faculty of Electronic Engineering in Niš, Serbia.

12

A Closed-Form Approximated Expression for the Residual ISI Obtained by Blind Adaptive Equalizers with Gain Equal or Less than One

Simon KUPCHAN, Monika PINCHAS

Dept. of Electrical and Electronic Engineering, Ariel University of Samaria, Ariel 40700, Israel

simonkupchan@gmail.com, monika.pinchas@gmail.com

Abstract. *In this paper we propose for the real and two independent quadrature carrier cases, a closed-form approximated expression for the achievable residual Inter-Symbol Interference (ISI). The expression depends on the step-size parameter, equalizer's tap length, equalized output gain, input signal statistics, channel power and SNR. This expression is valid for blind adaptive equalizers where the error fed into the adaptive mechanism, which updates the equalizer's taps, can be expressed as a polynomial function of order three of the equalized output, and where the gain between the input and equalized output signal is less than, or equal to one, as in the case of Godard ($gain = 1$) and WNEW ($gain < 1$) algorithm. Since the channel power is measurable, or can be calculated if the channel coefficients are given, there is no need for simulation with various step-size parameters to reach the required residual ISI. In addition, we show two new equalization methods (gain dependent) which have improved equalization performance compared to Godard and WNEW.*

Keywords

ISI distortion, residual ISI, blind equalizer.

1. Introduction

The wireless bandwidth limitation by government regulations, the large number of wireless applications sharing the limited bandwidth and constantly increasing communication speeds accentuate the ISI distortion. This is the main limitation factor for increasing communication speed. Today, wireless networks, such as GSM transmit training sequences, take up to 16% [1] of the channel capacity. This can be eliminated by using blind equalizers to retrieve transmitted data through the noisy channels by eliminating training sequence transmission. Thus preserving channel capacity for the data communication, which will increase speed. To develop a new blind equalizer we need to evaluate its performance using the achievable residual ISI. Developing a new blind equalizer involves choosing the equalizer's tap length and step-size parameter for a particular application or channel. Formerly, we used time consuming simulation for performance assessment. This part of the development process can be eliminated by using the closed-form approximated expression for the achievable residual ISI developed by Pinchas [2] for the noiseless case and expanded for the noisy environment by the same author in [3]. Both of the above mentioned expressions [2], [3] work well for equalizers with equalized output gain equal to one, as in Godard [5]. In [4] the WNEW algorithm was developed showing excellent equalization performance while having the same computational burden as the Godard algorithm. But its equalized output gain is lower than one. Thus, the expressions for the residual ISI developed in [2] and [3] are not applicable for the WNEW [4] algorithm as shown in Section V.

In this work, we develop a new closed-form approximated expression for the residual ISI for blind adaptive equalizers with equalized output gain lower or equal to one. As a by-product we present two new equalization methods (gain dependent) with improved equalization performance compared to the WNEW [4] and Godard [5] algorithm.

The paper is organized as follows: After describing the system under consideration in Section 2, the closed-form approximated expression for the achievable residual ISI is introduced in Section 3. In Section 4 two new equalization methods are introduced. Simulation results for the new closed-form approximated expression for the residual ISI and the new developed algorithms (gain dependent) are given in Section 5. The conclusion is presented in Section 6.

2. System Description

The system under consideration is the same system as used in [2], [3] and recalled here in Fig. 1.

We use in the following the same assumptions done in [2], [3]:

1. The transmitted sequence $x[n]$ is a Quadrature Amplitude Modulated (QAM) constellation, where x_r and x_i

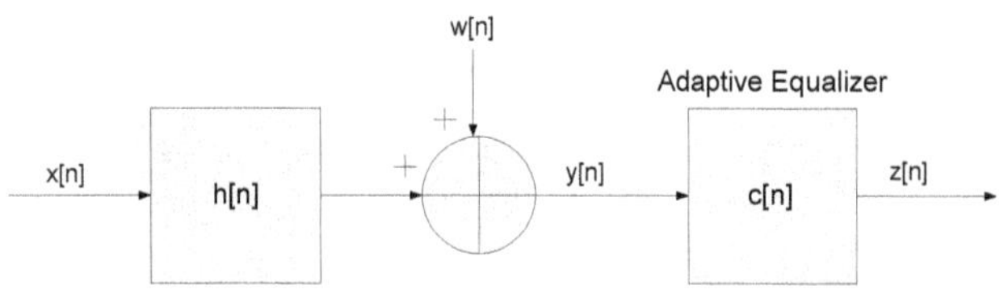

Fig. 1. Block diagram of the baseband communication system.

are the real and imaginary parts of $x[n]$ respectively and are independent. σ_x^2 is the variance of $x[n]$.

2. The unknown channel $h[n]$ is a possibly non-minimum phase linear time-invariant filter in which the transfer function has no "deep zeros", namely, the zeros lie sufficiently far from the unit circle.
3. The equalizer $c[n]$ is a tap-delay line.
4. The noise $w[n]$ is an additive Gaussian white noise with variance σ_w^2.

The sequence $x[n]$ is transmitted through the channel $h[n]$ and is corrupted with noise $w[n]$. Therefore, the equalizer's input sequence $y[n]$ may be written as:

$$y[n] = x[n] * h[n] + w[n] \tag{1}$$

where $*$ denotes the convolution operation. The equalized output signal is given by [2], [3]:

$$z[n] = y[n] * c[n] = x[n] * h[n] * c[n] + w[n] * c[n] \tag{2}$$

In the ideal case we have:

$$h[n] * c[n] = \alpha\delta[n-D]e^{j\theta} \tag{3}$$

where α is a constant gain between the input and equalized output signal, δ is the Kroneker delta function, D and θ are a constant delay and phase shift respectively. In the following, we denote $D = 0$ and $\theta = 0$ (as in [2], [3]). Thus, we write:

$$\tilde{s}[n] = h[n] * c_g[n] = \alpha\delta[n] + \xi[n] \tag{4}$$

where ξ stands for the difference (error) between the ideal value $c[n]$ and the guess $c_g[n]$ as stated in [3] and α is the equalizer's output gain. Substituting (4) into (2) yields:

$$z[n] = \alpha x[n] + p[n] + \beta\tilde{w}[n] \tag{5}$$

where β is a noise gain factor, $p[n]$ is the convolutional noise, produced from the difference between the initial guess $c_g[n]$ and the ideal values for $c[n]$, $\beta\tilde{w}[n] = w[n] * c_g[n]$ denotes the noise that passes through the equalizer.

The equalizer's update mechanism is defined by:

$$\underline{c}_{eq}[n+1] = \underline{c}_{eq}[n] - \mu \cdot \left(\frac{\partial F[n]}{\partial z[n]}\underline{y}^*[n]\right) \tag{6}$$

where μ is the equalizer's step size, $\underline{c}_{eq}[n]$ represents the current state of the equalizer's vector and $\underline{y}^*[n]$ is the input vector $y[n] = [y[n], \ldots, y[n-N+1]]^T$ where $()^*$ is the conjugate operator and N is the equalizer's tap length. In this paper the real part of $\frac{\partial F[n]}{\partial z[n]}$ is a polynomial function of order three of the equalized output defined (as in [2], [3]) by :

$$Re\left(\frac{\partial F[n]}{\partial z[n]}\right) = \left(a_1(z_r) + a_3(z_r)^3 + a_{12}(z_r)(z_i)^2\right) \tag{7}$$

where z_r , z_i are the real and imaginary parts of the equalized output $z[n]$ respectively and a_1, a_3, a_{12} are properties of the equalizer. The ISI is often used as a measure of performance in equalizer's applications, defined in [2]:

$$ISI = \frac{\sum_{\tilde{m}} |\tilde{s}(\tilde{m})|^2 - |\tilde{s}|^2_{\max}}{|\tilde{s}|^2_{\max}} \tag{8}$$

where $|s|_{max}$ is the component of $\tilde{s}$, given in (4), having the maximal absolute value.

In the next section we will develop a new closed-form approximated expression for the achievable residual ISI for blind adaptive equalizers where the gain of the equalized output is less than or equal to one.

2.1 ISI Performance

In this section we develop a closed-form approximated expression for the expected residual ISI as a function of the constellation input statistics, equalizer's tap length, equalized output gain, step-size parameter, channel power and SNR.

Theorem. For the following assumptions:

1. The convolutional noise p[n], is a zero mean, white Gaussian process with variance $\sigma_p^2 = E[p[n]p^*[n]]$, where $E[\,]$ stands for the expectation operator.
2. The variance and higher moments of the source signal $x[n]$ are known.
3. The convolutional noise p[n] and the source signal are independent.
4. α is the gain between the input and equalized output signal.
5. β is the noise gain factor for input noise.
6. $\max(|\tilde{s}|^2) = \alpha^2$.
7. $\frac{\partial F[n]}{\partial z[n]}$ can be expressed as a polynomial function of order three of the equalized output namely as $P(z)$.

The residual ISI expressed in dB units is defined as:

$$ISI = 10\log_{10}(m_p) - 10\log_{10}(\alpha^2) - 10\log_{10}(\sigma_{x_r}^2) \tag{9}$$

where:

$$m_p = \min[Sol_1^{mp_1}, Sol_2^{mp_1}] \text{ for } Sol_1^{mp_1} > 0 \text{ and } Sol_2^{mp_1} > 0$$

or

$$m_p = \max[Sol_1^{mp_1}, Sol_2^{mp_1}] \text{ for } Sol_1^{mp_1} \cdot Sol_2^{mp_1} < 0.$$

$Sol_1^{mp_1}$ and $Sol_2^{mp_1}$ are defined by:

$$Sol_1^{mp_1} = \frac{-B_1 + \sqrt{B_1^2 - 4A_1C_1B}}{2A_1},$$
$$Sol_2^{mp_1} = \frac{-B_1 - \sqrt{B_1^2 - 4A_1C_1B}}{2A_1},$$

$$\begin{aligned} A_1 =& B\left(45\alpha^2 m_2 a_3^2 + 18\alpha^2 m_2 a_3 a_{12} + 9\alpha^2 m_2 a_{12}^2 + \right.\\ & \left. 6a_1a_3 + 2a_1a_{12}\right) - 2\left(3a_3 + a_{12}\right) + \\ & B\left(45a_3^2 + 18a_3a_{12} + 9a_{12}^2\right)\beta^2\sigma_{\widetilde{w}_r}^2, \end{aligned}$$

$$\begin{aligned} B_1 =& \left(B\left(12\alpha^4 m_2^2 a_3 a_{12} + 6\alpha^4 m_2^2 a_{12}^2 + 12\alpha^2 m_2 a_1 a_3 + \right.\right.\\ & 4\alpha^2 m_2 a_1 a_{12} + a_1^2 + 15\alpha^4 m_4 a_3^2 + 2\alpha^4 m_4 a_3 a_{12} + \\ & \left.\left.\alpha^4 m_4 a_{12}^2\right) - 2\left(a_1 + 3\alpha^2 m_2 a_3 + \alpha^2 m_2 a_{12}\right)\right) + \\ & B\left(45a_3^2 + 18a_3a_{12} + 9a_{12}^2\right)\beta^4\sigma_{\widetilde{w}_r}^4 + \\ & \left(B\left(90\alpha^2 m_2 a_3^2 + 36\alpha^2 m_2 a_3 a_{12} + 12a_1a_3 + \right.\right.\\ & \left.\left.18\alpha^2 m_2 a_{12}^2 + 4a_1a_{12}\right) - 2a_{12} - 6a_3\right)\beta^2\sigma_{\widetilde{w}_r}^2, \end{aligned}$$

$$\begin{aligned} C_1 =& \left(2\alpha^4 m_2^2 a_1 a_{12} + \alpha^2 m_2 a_1^2 + 2\alpha^6 m_4 m_2 a_3 a_{12} + \right.\\ & \left. \alpha^6 m_4 m_2 a_{12}^2 + 2\alpha^4 m_4 a_1 a_3 + \alpha^6 m_6 a_3^2\right) + \\ & \left(15a_3^2 + 6a_3a_{12} + 3a_{12}^2\right)\beta^6\sigma_{\widetilde{w}_r}^6 + \\ & \left(45\alpha^2 m_2 a_3^2 + 18\alpha^2 m_2 a_3 a_{12} + 9\alpha^2 m_2 a_{12}^2 + 6a_1a_3 + \right.\\ & \left. 2a_1a_{12}\right)\beta^4\sigma_{\widetilde{w}_r}^4 + \\ & \left(a_1^2 + 12\alpha^2 m_2 a_1 a_3 + 4\alpha^2 m_2 a_1 a_{12} + 15\alpha^4 m_4 a_3^2 + \right.\\ & 12\alpha^4 m_2^2 a_3 a_{12} + 2\alpha^4 m_4 a_3 a_{12} + \\ & \left.\alpha^4 m_4 a_{12}^2 + 6\alpha^4 m_2^2 a_{12}^2\right)\beta^2\sigma_{\widetilde{w}_r}^2, \end{aligned} \quad (10)$$

$$B = \mu N \sigma_x^2 \left(\sum_{k=0}^{k=R-1} |h_k[n]|^2 + \frac{1}{SNR} \right) \quad (11)$$

where $m_\chi = E[x_r^\chi]$, $\sigma_{\widetilde{w}_r}^2 = \frac{\sigma_{x_r}^2}{SNR\sum_{k=0}^{k=R-1}|h_k[n]|^2}$, R is the channel's length, $SNR = \sigma_x^2/\sigma_w^2$ and a_1, a_3, a_{12} are the properties of the chosen equalizer and found via (7).

Proof. We begin our proof by recalling from [2] the expression for $E[\triangle(p_r^2)]$ (where p_r is the real part of $p[n]$ and $\triangle(p_r^2) = p_r^2[n+1] - p_r^2[n]$:

$$\begin{aligned} & E[\triangle(p_r^2)] = \\ & -2E\left[p_r\left(\mu P_r(z)\sum_{m=0}^{m=l} y[n-m]y^*[n-m]\right)\right] + \\ & E\left[\left(-\mu P_r(z)\sum_{m=0}^{m=l} y[n-m]y^*[n-m]\right)^2\right] \end{aligned} \quad (12)$$

where $P_r(z)$ is the real part of $P(z)$ and is given according to [2] as:

$$P_r(z) = \left(a_1(z_r) + a_3(z_r)^3 + a_{12}(z_r)(z_i)^2\right) \quad (13)$$

where z_r and z_i are the real and imaginary parts of (5) and equal to:

$$\begin{aligned} z_r &= \alpha x_r + p_r + \beta\widetilde{w}_r \\ z_i &= \alpha x_i + p_i + \beta\widetilde{w}_i \end{aligned} \quad (14)$$

where α and β may not be equal. In this paper, $p_r = p_r[n]$ where p_r and p_i are the real and imaginary parts of $p[n]$ respectively. Next, we calculate (12) in the same way as in [3]. We substitute (14) into (13) and evaluate (12) by using (14) and (13). Thus, we obtain for the latter stages of the convergence state:

$$E\left[\triangle\left(p_r^2\right)\right] \cong BA_1m_p^2 + BB_1m_p + B^2C_1 \quad (15)$$

where $E[p_r[n]^2] = m_p$ and B, B_1, A_1, C_1 are given in (10). Note that B, B_1, A_1, C_1 are different from those obtained in [3] due to the α and β parameters. In the latter stages of the deconvolution process, we may write: $E\left[\triangle\left(p_r^2\right)\right] \cong 0$. Thus, setting (15) to zero and solving the equation for m_p will lead to the solution for m_p given in (10).

Now to obtain the expression for the ISI given in (9) we use (2) and (5) thus we write for the noiseless case:

$$\begin{aligned} & E\left[z[n]z[n]^*\right] = \\ & E\left[(\widetilde{s}[n] * x[n])(\widetilde{s}[n] * x[n])^*\right] = \\ & E\left[x[n]x[n]^*\right]\sum_{\widetilde{m}}|\widetilde{s}[\widetilde{m}]|^2 = \sigma_x^2\sum_{\widetilde{m}}|\widetilde{s}[\widetilde{m}]|^2, \end{aligned} \quad (16)$$

$$\begin{aligned} & E\left[z[n]z[n]^*\right] = \\ & E\left[(\alpha x[n] + p[n] + \beta w[n])(\alpha x[n] + p[n] + \beta w[n])^*\right] = \\ & \alpha^2 E\left[x[n]x[n]^*\right] + E\left[p[n]p[n]^*\right] = \alpha^2\sigma_x^2 + \sigma_p^2. \end{aligned} \quad (17)$$

By comparing (16) with (17) we obtain:

$$\sigma_p^2 = \sigma_x^2\left[\sum_{\widetilde{m}}|\widetilde{s}[\widetilde{m}]|^2 - \alpha^2\right]. \quad (18)$$

Dividing (18) by a^2 leads to:

$$\frac{\sigma_p^2}{\alpha^2} = \sigma_x^2\frac{\left[\sum_{\widetilde{m}}|\widetilde{s}(\widetilde{m})|^2 - \alpha^2\right]}{\alpha^2}, \qquad |\widetilde{s}|_{max}^2 = \alpha^2 \quad (19)$$

which with the help of (8) can be written as:

$$ISI = \frac{\sigma_p^2}{\alpha^2 \sigma_x^2}. \tag{20}$$

Since $\sigma_p^2/\sigma_x^2 = \sigma_{p_r}^2/\sigma_{x_r}^2$ and the ISI is measured in the logarithmic scale we have (9).

Next, we turn to the various steps that lead to the expression for $\sigma_{\tilde{w}_r}^2$:

$$\beta \tilde{w}[n] = c_g[n] * w[n] \tag{21}$$

which with the help of (4) can be written as:

$$\begin{aligned}\beta \tilde{w}[n] * h[n] &= c_g[n] * w[n] * h[n] = \\ w[n] * \tilde{s}[n] &= w[n] * (\alpha \delta[n] + \xi[n])\end{aligned} \tag{22}$$

thus for $\xi[n] \to 0$ (for the ideal case) we have:

$$\beta^2 \sigma_{\tilde{w}_r}^2 \cong \alpha^2 \sigma_{w_r}^2 \frac{1}{\sum_{k=0}^{R-1} |h_k[n]|^2} \tag{23}$$

where $\sigma_{\tilde{w}_r}^2$ and $\sigma_{w_r}^2$ are the variances of the real part of $\tilde{w}[n]$ and $w[n]$ respectively. From (23) we may see that for $\xi[n] \to 0$ and $\alpha = \beta$ we return to the expression for $\sigma_{\tilde{w}_r}^2$ used in (10). This is the same expression for $\sigma_{\tilde{w}_r}^2$ used in [3] for the case where the equalized output gain is equal to one.

This completes our *Proof.*

2.2 The ANEW Equalizer

In this section we develop a new equalization method, namely we propose a new function for $Re\{\frac{\partial F[n]}{\partial z[n]}\}$. As already mentioned earlier in this paper, we write $E\left[\triangle\left(p_r^2\right)\right] \cong 0$ for the latter stages of the deconvolutional process. Therefore, by setting (15) to zero and dividing (15) by B (for $B \neq 0$), we may see that the convolutional noise power m_p does not converge in the steady state approximately to zero unless $C_1 = 0$ as stated in [3]. C_1 (10) depends on the constellation input statistics and on the algorithm itself via a_1, a_3 and a_{12}. By minimizing $C1$ with respect to the algorithm parameters (a_1, a_3, a_{12}) we may obtain a new equalizer.

Theorem. For the following assumptions:

1. The transmitted sequence $x[n]$ belongs to the square QAM constellation, thus a_{12} is set to zero.
2. No noise is added $\sigma_{\tilde{w}_r}^2 = 0$

The real part of $\frac{\partial F[n]}{\partial z[n]}$ can be written as:

$$Re\left(\frac{\partial F[n]}{\partial z[n]}\right) = \left(a_1(z_r) + a_3(z_r)^3\right) \tag{24}$$

with

$$a_1 = -\alpha; \quad a_3 = \frac{m_2}{\alpha m_4}; \quad a_{12} = 0. \tag{25}$$

Proof. We start the proof from recalling C_1 (10) and deleting there a_{12} and the noise component. Thus having:

$$C_1 = \alpha^2 m_2 a_1^2 + 2\alpha^4 m_4 a_1 a_3 + \alpha^6 m_6 a_3^2. \tag{26}$$

Now, minimizing (26) with respect to the coefficients a_1, a_3, α and then setting the relevant equations to zero, we obtain:

$$\frac{\partial C_1}{\partial \alpha} = 2\alpha m_2 a_1^2 + 8\alpha^3 m_4 a_1 a_3 + 6\alpha^5 m_6 a_3^2 = 0, \tag{27}$$

$$\frac{\partial C_1}{\partial a_1} = 2\alpha^2 m_2 a_1 + 2\alpha^4 m_4 a_3 = 0, \tag{28}$$

$$\frac{\partial C_1}{\partial a_3} = 2\alpha^4 m_4 a_1 + 2\alpha^6 m_6 a_3 = 0. \tag{29}$$

By solving (27), (28) and (29) the trivial solution ($a_1 = 0$, $a_3 = 0$) is obtained, which indicates that no equalizer exists. To find a non-trivial solution we set a_1 to $(-\alpha)$, which leads to two different solutions for a_3. The first solution is given in (25) obtained via (28) while the second solution obtained via (29) is given by:

$$a_1 = -\alpha; \quad a_3 = \frac{m_4}{\alpha m_6}; \quad a_{12} = 0. \tag{30}$$

Now, we turn to compare the two solutions (25) and (30), by substituting each of them into (26) and evaluate (26) for the 16QAM input case. For the 16QAM input case we have: $m_2 = 5$, $m_4 = 41$ and $m_6 = 365$. Thus we have:

$$C_1 \cong 0.43\alpha^4, \quad \text{for Case A}, \tag{31}$$

$$C_1 \cong 0.39\alpha^4, \quad \text{for Case B} \tag{32}$$

where Case A and Case B were obtained by substituting (25) and (30) into (26) respectively. According to (31) and (32), Case B may lead to a lower residual ISI in the steady state compared to Case A. However, the difference between the two cases (A, B) as appears in (31) and (32) is so small, that we may not see any difference in the equalization performance from the residual ISI point of view. In addition, we observe according to (31) and (32) that a smaller value for α may lead to a lower residual ISI. Note that for $\alpha = 0$ we obtain $C_1 = 0$ (perfect equalization), but for this case no equalizer exists (refer to (5)), thus no perfect equalization is obtained for $\alpha = 0$.
This completes our *Proof.*

2.3 Simulation

In this section we compare the usefulness of our new proposed expression for the residual ISI (9) with the expression for the residual ISI obtained in [3]. In the following we use (25) and (30) to define two new equalization methods, which we denote as ANEW and BNEW respectively. We compare the equalization performance obtained from the ANEW (with various values for α) and BNEW algorithm with Godard [5] and WNEW [4]. The equalizers were initialized by setting the center tap to one and all other taps to zero. The equalizer's taps according to Godard were updated by:

$$\underline{c}_G[n+1] = \underline{c}_G[n] - \mu_G G \underline{y}^*[n],$$
$$G = \left(|z|^2 - \frac{E[|x|^4]}{E[|x|^2]}\right) z \tag{33}$$

where $z = z[n]$, $x = x[n]$, $E[\,]$ is the expectation operator, $|\,|$ is the absolute operator, μ_G is the step-size parameter and a_1, a_3 and a_{12} were defined as a_1^G, a_3^G and a_{12}^G respectively and given by:

$$a_1^G = -\frac{E[|x|^4]}{E[|x|^2]}, \quad a_3^G = 1, \quad a_{12}^G = 1. \tag{34}$$

The equalizer's taps for WNEW [4] algorithm were updated according to:

$$\underline{c}_W[n+1] = \underline{c}_W[n] - \mu_W W_{new} \underline{y}^*[n],$$
$$W_{new} = \frac{1}{m_2}\left(z_r^3 + jz_i^3\right) - z \tag{35}$$

where μ_W is the step-size parameter and a_1, a_3 and a_{12} were defined as a_1^W, a_3^W and a_{12}^W respectively and given by:

$$a_1^W = -1, \quad a_3^W = -\frac{1}{m_2}, \quad a_{12}^W = 0. \tag{36}$$

The equalizer's taps for ANEW algorithm were updated according to:

$$\underline{c}_A[n+1] = \underline{c}_A[n] - \mu_A A_{new} \underline{y}^*[n],$$
$$A_{new} = \frac{m_2}{\alpha m_4}\left(z_r^3 + jz_i^3\right) - \alpha z \tag{37}$$

where μ_A is the step-size parameter and a_1, a_3 and a_{12} were substituted from (25). The equalizer taps for BNEW algorithm were updated according to:

$$\underline{c}_B[n+1] = \underline{c}_B[n] - \mu_B B_{new} \underline{y}^*[n],$$
$$B_{new} = \frac{m_4}{\alpha m_6}\left(z_r^3 + jz_i^3\right) - \alpha z \tag{38}$$

where μ_B is the step-size parameter and a_1, a_3 and a_{12} were substituted from (30). Two input sources were used: the 16QAM and 64QAM modulations with $\pm\{1,3\}$ and $\pm\{1,3,5,7\}$ levels respectively, for in-phase and quadrature components. Four different channels were considered.

Channel1 (initial ISI = 0.44): The channel parameters were determined according to Shalvi and Weinstein [10]:
h_n = (0 for $n < 0$; -0.4 for $n = 0$; $0.84 \cdot 0.4^{n-1}$ for $n > 0$).

Channel 2 (initial ISI = 0.5): The channel parameters were determined according to Fiori [11]:
h_n = (-0.0144, 0.0006, 0.0427, 0.0090, -0.4842, -0.0376, 0.8163, 0.0247, 0.2976, 0.0122, 0.0764, 0.0111, 0.0162, 0.0063).

Channel 3 (initial ISI = 0.88): The channel parameters were determined according to Pinchas [2]:
h_n = (0.4851, -0.72765, -0.4851)

Channel4 (initial ISI = 0.44): The channel parameters were determined according to Shalvi and Weinstein [10]:
h_n = (0 for $n < 0$; -0.4 for $n = 0$; $0.84 \cdot 0.4^{n-1}$ for $n > 0$) and normalized to $hh^T = 0.507$.

Let us start with the comparison of our closed-form approximated expression for the residual ISI (9) with the one obtained in [3]. Note the main difference between the two expressions (9) and [3], is in the α and β parameters. In [3] $\alpha = \beta = 1$, while in our case α and β receive various values.

Fig. 2 and Fig. 3 show the equalization performance comparison from the residual ISI point of view obtained by (9), [3] with the simulated results obtained by the WNEW algorithm, for the 16QAM input case and SNR values of 10 dB and 30 dB respectively. According to Fig. 2 and Fig. 3 the residual ISI obtained by (9) is very close to the simulated results, while this is not the case with the residual ISI obtained by [3].

Fig. 4 to Fig. 7 show the simulated performance of the WNEW equalization method for the 16QAM and 64QAM input case, namely the ISI as a function of iteration number for various step-size parameters, channel characteristics, equalizer's tap length and various SNR values, compared with the calculated residual ISI expression (9) used with $\alpha = 0.84$ and $\beta = 0.58$. Figs. 4 – 7 show a high correlation between the simulated results and those calculated with (9).

Fig. 8 and Fig. 9 show the simulated performance of the ANEW equalization method with $\alpha = 0.8$ for the 16QAM and 64QAM input case, namely the ISI as a function of iteration number for various step-size parameters, Channel characteristics and various SNR values, compared with the calculated residual ISI expression (9) used with $\alpha = 0.8$ and $\beta = 0.73$. Fig. 8 and Fig. 9, point to a high correlation between the simulated results and those calculated with (9).

Fig. 10 to Fig. 13 show the simulated performance of the ANEW equalization method with $\alpha = 1$ for the 16QAM and 64QAM input case, namely the ISI as a function of iteration number for various step-size parameters, channel characteristics, equalizer's tap length and various SNR values, compared with the calculated residual ISI expression (9) used with $\alpha = 1$ and $\beta = 0.93$. Fig. 10 to Fig. 13 indicate a high correlation between the simulated and calculated performance (9) of the achievable residual ISI.

Fig. 14 to Fig. 16 show the simulated performance of Godard's equalization method for the 16QAM and 64QAM input case, namely the ISI as a function of iteration number for various step-size parameters, channel characteristics, equalizer's tap length and various SNR values, compared with the calculated residual ISI expression (9) used with $\alpha = 1$ and $\beta = 1$. Fig. 14 to Fig. 16 show a high correlation between the simulated results and those calculated with (9).

Next, we turn to compare equalization performance obtained from the ANEW and BNEW algorithm. Fig. 17 and Fig. 18 show the ISI comparison between the simulated performance of the two equalization methods ANEW and

BNEW with $\alpha = 1$ (for both methods). The comparison was carried out for the 16QAM and 64QAM input case, for various step-size parameters, channel characteristics and SNR values of 10 dB and 30 dB. According to Fig. 17 and Fig. 18, ANEW and BNEW have approximately the same equalization performance.

Now, we turn to regarding the equalization performance comparison between ANEW with various gains ($\alpha = 0.8$, $\alpha = 1$), WNEW [4] and Godard [5].

Fig. 19 and Fig. 20 show the equalization performance comparison between the simulated performance obtained by the ANEW algorithm with two different values for α ($\alpha = 0.8$ and $\alpha = 1$), namely the ISI as a function of iteration number for Channel 1, 16QAM input case and SNR values of 10 dB and 30 dB. According to Fig. 19 and Fig. 20 and backed up by (31), a lower gain (α) leads to a lower residual ISI, hence to a better residual ISI performance.

Fig. 21 and Fig. 22 show the equalization performance comparison between the simulated performance obtained by the ANEW and WNEW equalization algorithm, namely the ISI as a function of iteration number for Channel 1, 16QAM input case and SNR values of 10 dB and 30 dB. According to Fig. 21 and Fig. 22 almost no difference is seen between the ANEW and WNEW equalizer. Note, that the ANEW equalizer was simulated with $\alpha = 1$. A lower value for α might have led the ANEW algorithm to a lower residual ISI compared to the WNEW method (please refer to Fig. 19 and Fig. 20).

Fig. 23 to Fig. 28 show the equalization performance comparison between the simulated performance obtained by the ANEW and Godard's equalization algorithm, namely the ISI as a function of iteration number for various channels, 16QAM and 64QAM input case, various step-size parameters, equalizer's tap length and SNR values of 10 dB and 30 dB. As seen from Fig. 23 to Fig. 28 the ANEW equalizer has improved equalization performance compared to Godard [5]. The improved equalization performance is seen in the residual ISI as well as in the convergence speed.

2.4 Conclusion

In this work, we have developed a new closed-form approximated expression for the achievable residual ISI valid for blind adaptive equalizers, where the gain between the input and equalized output signal is less than or equal to one, as is in the case of Godard [5], WNEW [4] and ANEW algorithm. Thus, the expressions for the residual ISI obtained in previous papers ([2], [3]) are special cases of our new proposed expression. In addition, we have developed two new equalization (α dependant) algorithms. The new algorithms were called ANEW and BNEW, shown to have improved equalization performance compared to Godard and WNEW. The new algorithms (ANEW and BNEW) have the same computational complexity as the classical Godard and WNEW algorithm.

Acknowledgements

We would like to thank the anonymous reviewers for their helpful comments.

References

[1] *Digital Cellular Telecommunications System (Phase 2+); Multiplexing and Multiple Access on the Radio Path.* ETSI European Telecommunications Standards Institute, ICS: 33.060.50, 1996.

[2] PINCHAS, M. A closed approximated formed expression for the achievable residual intersymbol interference obtained by blind equalizers. *Signal Processing (Eurasip)*, 2010, vol. 90, no. 6, p. 1940 - 1962.

[3] PINCHAS, M. A new closed approximated formed expression for the achievable residual intersymbol interference obtained by blind equalizers for the noisy case. In *IEEE International Conference on Wireless Communications, Networking and Information Security (WCNIS.* Beijing, (China), 2010, p. 26 - 30.

[4] PINCHAS, M., BOBROVSKY, B. Z. A maximum entropy approach for blind deconvolution. *Signal Processing (Eurasip)*, 2006, vol. 86, no. 10, p. 2913 - 2931.

[5] GODARD, D. Self-recovering equalization and carrier tracking in two- dimenional data communication system. *IEEE Transactions on Communications*, 1980, vol. 28, no. 11, p. 1867 - 1875.

[6] PINCHAS, M. A MSE optimized polynomial equalizer for 16QAM and 64QAM constellation. *Signal, Image and Video Processing*, 2011, vol. 5, no. 1, p. 29 - 37.

[7] PINCHAS, M. A novel expression for the achievable MSE performance obtained by blind adaptive equalizers. *Signal, Image and Video Processing*, 2013, vol. 7, no. 1, p. 67 - 74.

[8] GI-HONG IM, CHEOL-JIN PARK, HUI-CHUL WON A blind equalization with the sign algorithm for broadband access. *IEEE Communications Letters*, 2001, vol. 5, no. 2, p. 70 - 72.

[9] LAZARO, M., SANTAMARIA, I., ERDOGMUS, D., HILD, K. E., PANTALEON, C., PRINCIPE, J. C. Stochastic blind equalization based on PDF fitting using Parzen estimator. *IEEE Transactions on Signal Processing*, 2005, vol. 53, no. 2, p. 696 - 704.

[10] SHALVI O., WEINSTEIN, E. New criteria for blind deconvolution of nonminimum phase systems (channels). *IEEE Transactions on Information Theory*, 1990, vol. 36, no. 2, p. 312 - 321.

[11] SHALVI, O., WEINSTEIN, E. Super-exponential methods for blind deconvolution. *IEEE Transactions on Information Theory*, 1993, vol. 39, no. 2, p. 504 - 519.

[12] NIKIAS, C., PETROPULU A. P. (Eds.) *Higher-Order Spectra Analysis: A Non-linear Signal Processing Framework.* Prentice-Hall, 1993, p. 419 - 425.

About Authors ...

Monika PINCHAS is with the Ariel University. She is the Head of the graduate program at the faculty of Electrical and Electronic Engineering. Her research interests are in the area of blind equalization, frequency synchronization in OFDM systems and network synchronization where she has published several papers in leading journals and published two books. She was the CTO at Resolute Networks, managed the hardware group at Radiotel wireless transmission line of products, leading the development of modem technology. She worked at Tadiran Communication where she was recognized as an expert and worked for Scitex in the design and implementation of hardware systems.

Simon KUPCHAN received the B.Tech. degree in Electrical Engineering from Ariel University Center, Ariel, Israel, in 2006. From 2004, he worked in Infra-Com Ltd., Netanya, Israel, where he was HW digital designer in the R&D team. Since 2010, he has been with Orsan Medical Technologies Ltd. Netanya, Israel, where he is currently senior HW digital designer in the R&D team.

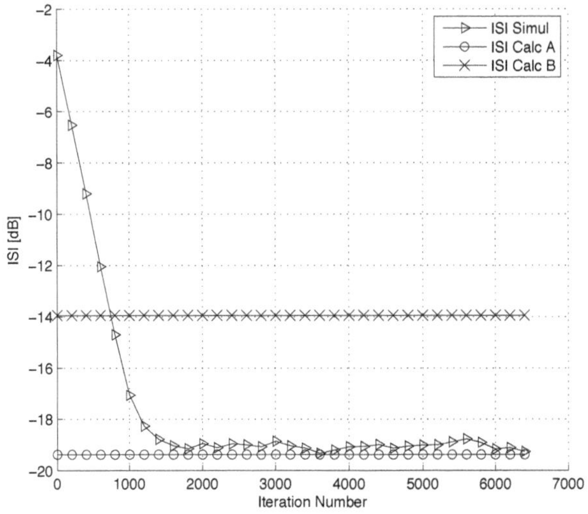

Fig. 2. A comparison between the simulated (with WNEW algorithm) and calculated residual ISI according to (9) (ISI Calc A) and [3] (ISI Calc B) for the 16QAM source going through Channel 1 and SNR of 10 dB. The averaged results were obtained in 100 Monte Carlo trials. The equalizer's tap length and step-size parameter were set to 13 and 0.0004 respectively. α and β gain parameters were set to 0.84 and 0.58 respectively.

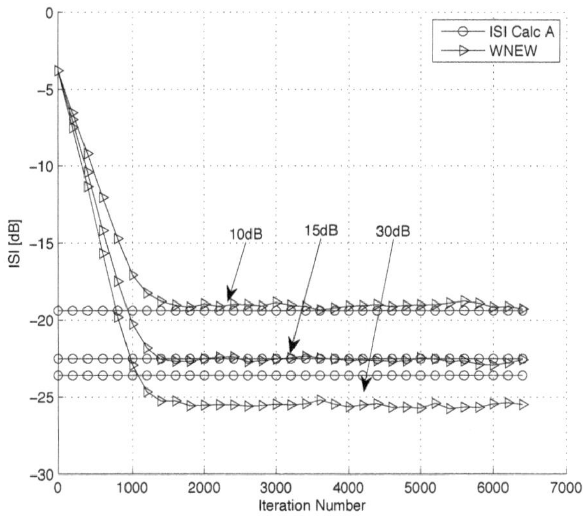

Fig. 4. A comparison between the simulated (with WNEW algorithm) and calculated residual ISI (9) (ISI Calc A) for the 16QAM source going through Channel 1 with various SNR values. The averaged results were obtained in 100 Monte Carlo trials. The equalizer's tap length and step-size parameter were set to 13 and 0.0004 respectively. α and β were set to 0.84 and 0.58 respectively.

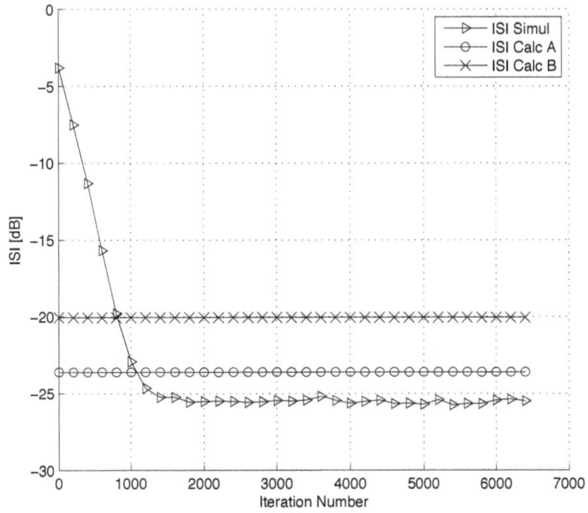

Fig. 3. A comparison between the simulated (with WNEW algorithm) and calculated residual ISI according to (9) (ISI Calc A) and [3] (ISI Calc B) for the 16QAM source going through Channel 1 and SNR of 30 dB. The averaged results were obtained in 100 Monte Carlo trials. The equalizer's tap length and step-size parameter were set to 13 and 0.0004 respectively. α and β gain parameters were set to 0.84 and 0.58 respectively.

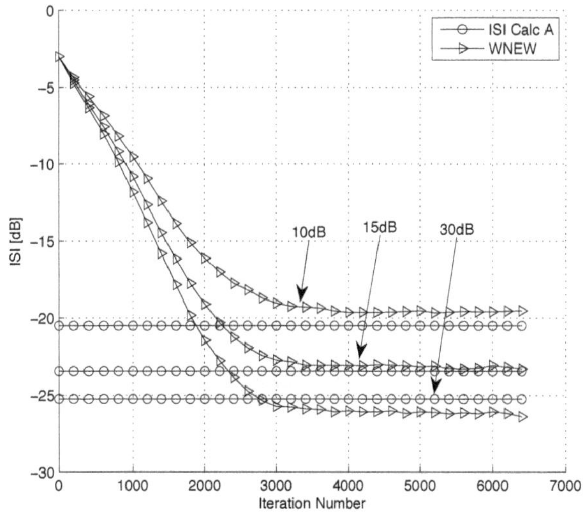

Fig. 5. A comparison between the simulated (with WNEW algorithm) and calculated residual ISI (9) (ISI Calc A) for the 16QAM source going through Channel 2 with various SNR values. The averaged results were obtained in 100 Monte Carlo trials. The equalizer's tap length and step-size parameter were set to 21 and 0.0002 respectively. α and β were set to 0.84 and 0.58 respectively.

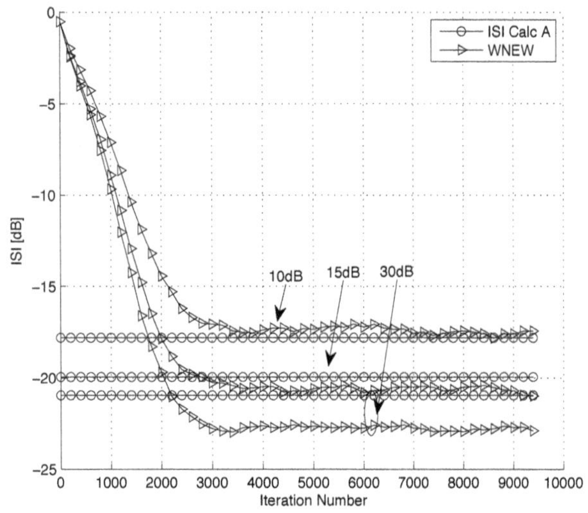

Fig. 6. A comparison between the simulated (with WNEW algorithm) and calculated residual ISI (9) (ISI Calc A) for the 64QAM source going through Channel 3 with various SNR values. The averaged results were obtained in 100 Monte Carlo trials. The equalizer's tap length and step-size parameter were set to 13 and 8e-5 respectively. α and β were set to 0.84 and 0.58 respectively.

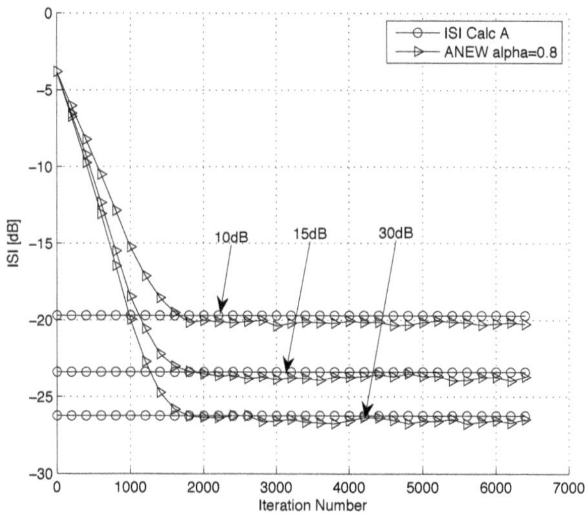

Fig. 8. A comparison between the simulated (with ANEW algorithm) and calculated residual ISI (9) (ISI Calc A) for the 16QAM source going through Channel 1 with various SNR values. The averaged results were obtained in 100 Monte Carlo trials. The equalizer's tap length and step-size parameter were set to 13 and 0.0004 respectively. α and β were set to 0.8 and 0.73 respectively.

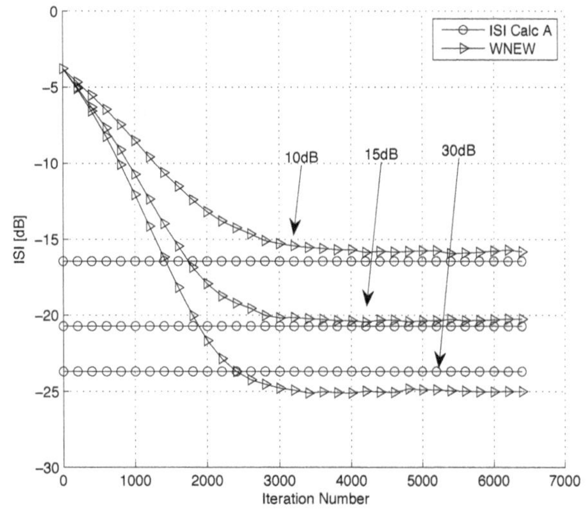

Fig. 7. A comparison between the simulated (with WNEW algorithm) and calculated residual ISI (9) (ISI Calc A) for the 16QAM source going through Channel 4 with various SNR values. The averaged results were obtained in 100 Monte Carlo trials. The equalizer's tap length and step-size parameter were set to 27 and 0.0004 respectively. α and β were set to 0.84 and 0.58 respectively.

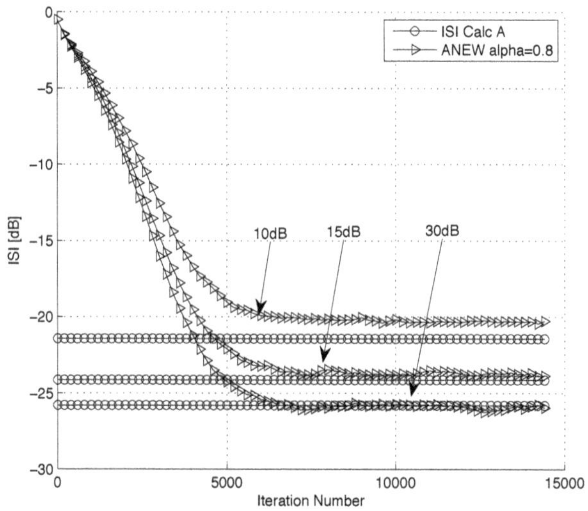

Fig. 9. A comparison between the simulated (with ANEW algorithm) and calculated residual ISI (9) (ISI Calc A) for the 64QAM source going through Channel 3 with various SNR values. The averaged results were obtained in 100 Monte Carlo trials. The equalizer's tap length and step-size parameter were set to 13 and 0.00005 respectively. α and β were set to 0.8 and 0.73 respectively.

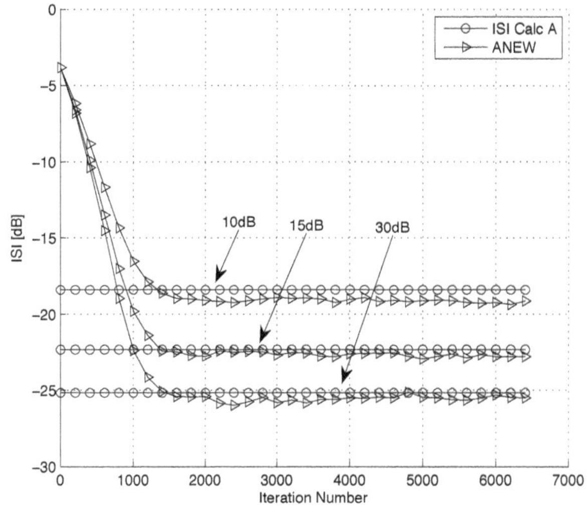

Fig. 10. A comparison between the simulated (with ANEW algorithm) and calculated residual ISI (9) (ISI Calc A) for the 16QAM source going through Channel 1 with various SNR values. The averaged results were obtained in 100 Monte Carlo trials. The equalizer's tap length and step-size parameter were set to 13 and 0.0004 respectively. α and β were set to 1 and 0.93 respectively.

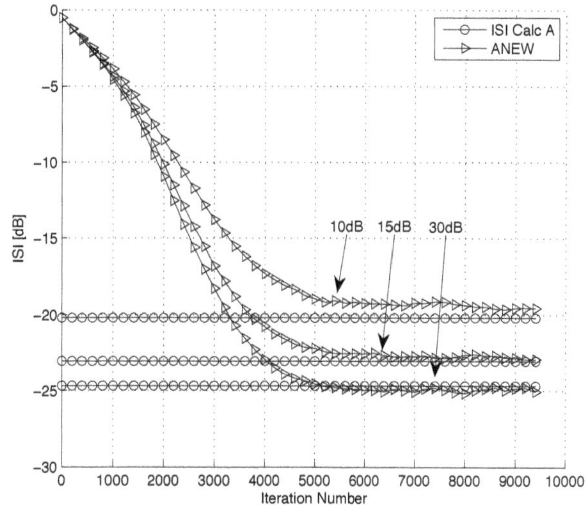

Fig. 12. A comparison between the simulated (with ANEW algorithm) and calculated residual ISI (9) (ISI Calc A) for the 64QAM source going through Channel 3 with various SNR values. The averaged results were obtained in 100 Monte Carlo trials. The equalizer's tap length and step-size parameter were set to 13 and 0.00005 respectively. α and β were set to 1 and 0.93 respectively.

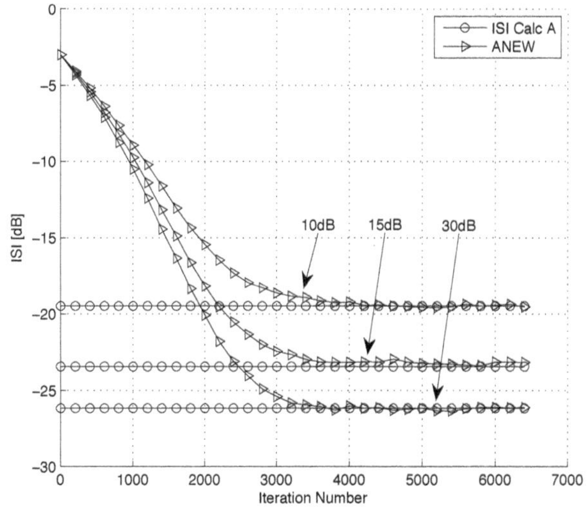

Fig. 11. A comparison between the simulated (with ANEW algorithm) and calculated residual ISI (9) (ISI Calc A) for the 16QAM source going through Channel 2 with various SNR values. The averaged results were obtained in 100 Monte Carlo trials. The equalizer's tap length and step-size parameter were set to 21 and 0.0002 respectively. α and β were set to 1 and 0.93 respectively.

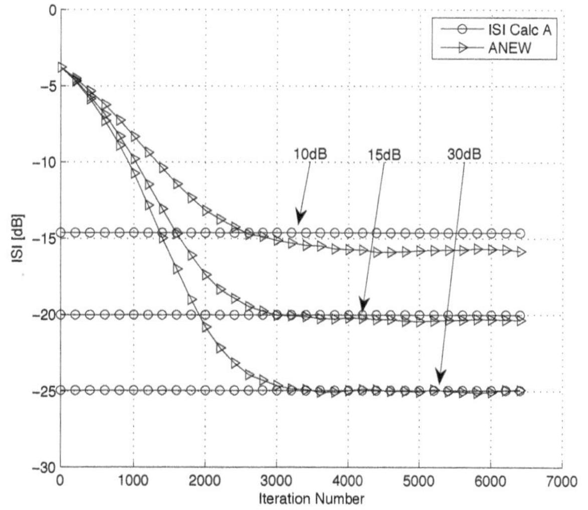

Fig. 13. A comparison between the simulated (with ANEW algorithm) and calculated residual ISI (9) (ISI Calc A) for the 16QAM source going through Channel 4 with various SNR values. The averaged results were obtained in 100 Monte Carlo trials. The equalizer's tap length and step-size parameter were set to 27 and 0.0004 respectively. α and β were set to 1 and 0.93 respectively.

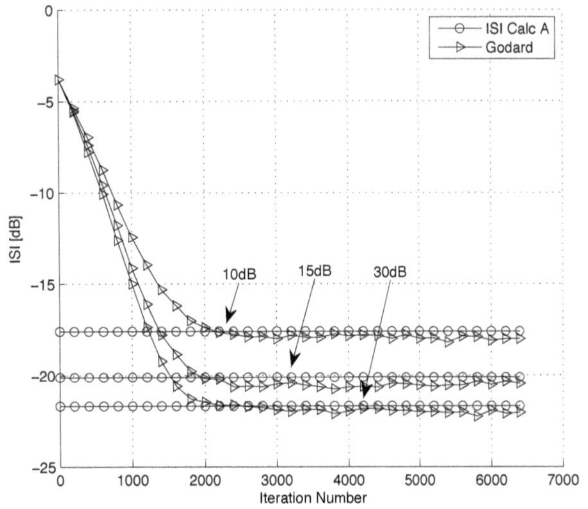

Fig. 14. A comparison between the simulated (with Godard algorithm) and calculated residual ISI (9) (ISI Calc A) for the 16QAM source going through Channel 1 with various SNR values. The averaged results were obtained in 100 Monte Carlo trials. The equalizer's tap length and step-size parameter were set to 13 and 5e-5 respectively. α and β were set both to 1.

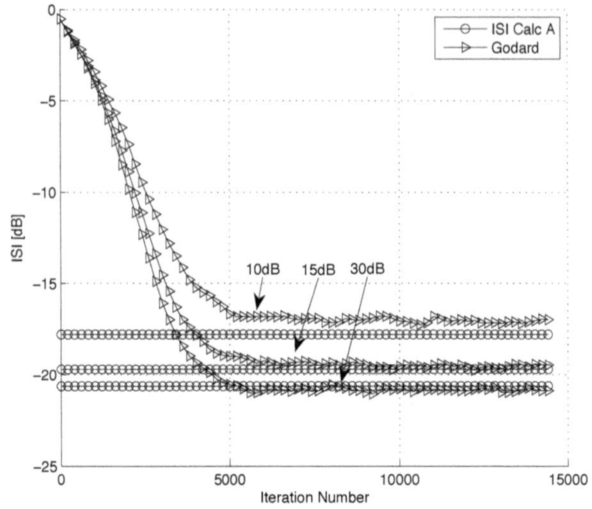

Fig. 16. A comparison between the simulated (with Godard algorithm) and calculated residual ISI (9) (ISI Calc A) for the 64QAM source going through Channel 3 with various SNR values. The averaged results were obtained in 100 Monte Carlo trials. The equalizer's tap length and step-size parameter were set to 13 and 1.2e-6 respectively. α and β were set both to 1.

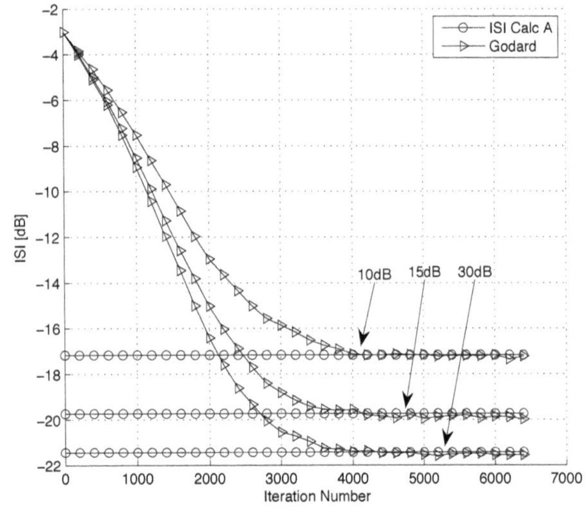

Fig. 15. A comparison between the simulated (with Godard algorithm) and calculated residual ISI (9) (ISI Calc A) for the 16QAM source going through Channel 2 with various SNR values. The averaged results were obtained in 100 Monte Carlo trials. The equalizer's tap length and step-size parameter were set to 21 and 2e-5 respectively. α and β were set both to 1.

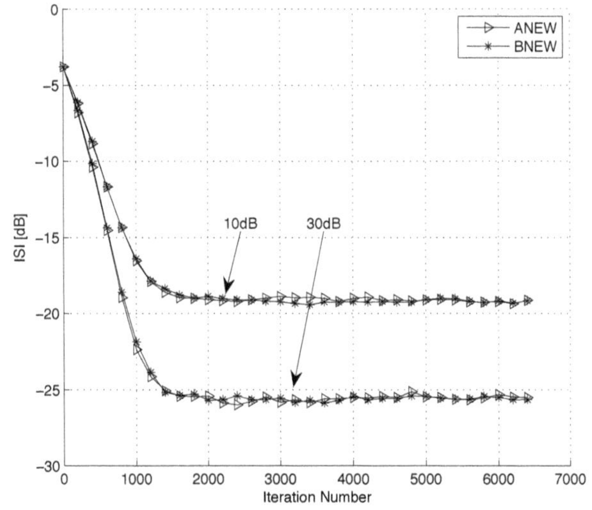

Fig. 17. A comparison between the ANEW and BNEW algorithm for the 16QAM source going through Channel 1 with SNR values of 10 dB and 30 dB. The averaged results were obtained in 100 Monte Carlo trials. The equalizer's tap length and step-size parameter were set to 13 and 0.0004 respectively. The parameter α was set to 1.

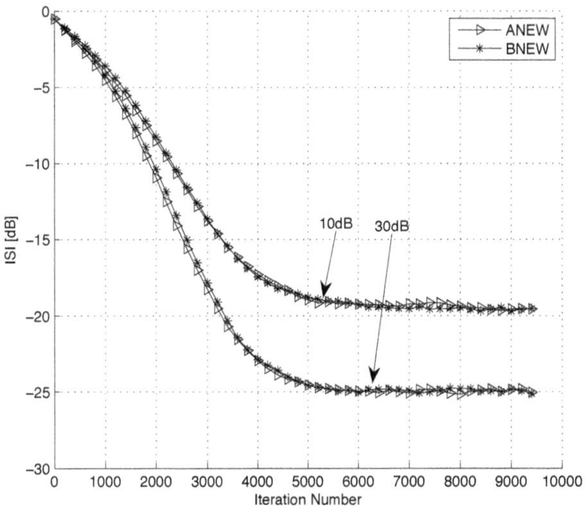

Fig. 18. A comparison between the ANEW and BNEW algorithm for the 64QAM source going through Channel 3 with SNR values of 10 dB and 30 dB. The averaged results were obtained in 100 Monte Carlo trials. The equalizer's tap length and step-size parameter were set to 13 and 0.00005 respectively. The parameter α was set to 1.

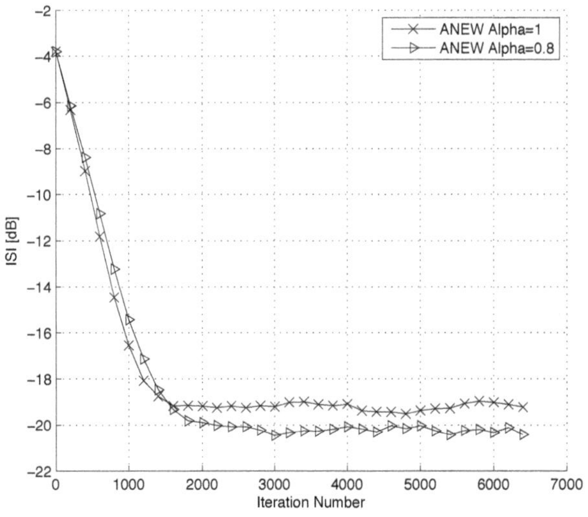

Fig. 19. A comparison between ANEW with gain $\alpha = 0.8$ and ANEW with gain $\alpha = 1$ for the 16QAM source going through Channel 1. The averaged results were obtained in 100 Monte Carlo trials. The equalizer's tap length and step-size parameter were set to 13 and 0.0004 respectively, SNR was set to 10 dB. The equalizer with gain $\alpha = 0.8$ achieves lower residual ISI by approximately 1 dB.

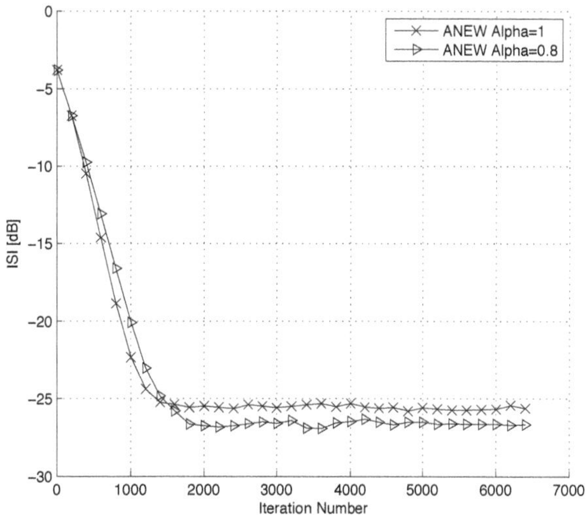

Fig. 20. A comparison between ANEW with gain $\alpha = 0.8$ and ANEW with gain $\alpha = 1$ for the 16QAM source going through Channel 1. The averaged results were obtained in 100 Monte Carlo trials. The equalizer's tap length and step-size parameter were set to 13 and 0.0004 respectively, SNR was set to 30 dB. The equalizer with gain $\alpha = 0.8$ achieves lower residual ISI by approximately 1 dB.

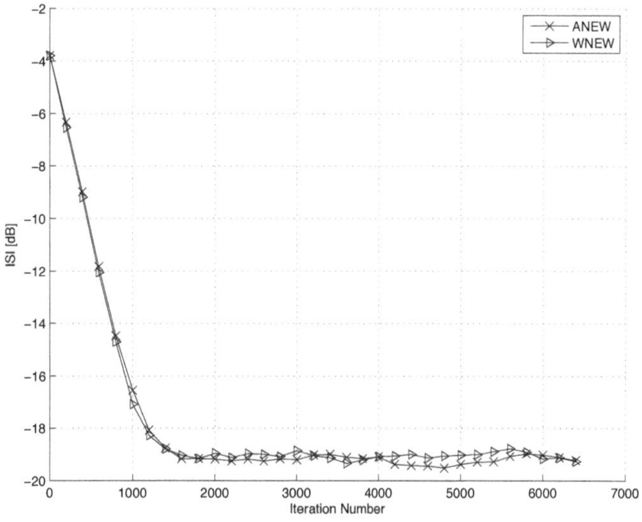

Fig. 21. A comparison between WNEW and ANEW algorithm for the 16QAM source going through Channel 1. The averaged results were obtained in 100 Monte Carlo trials. The equalizer's tap length was set to 13, the step-size parameter for WNEW and ANEW were set to $\mu_W = 0.0004$ and $\mu_A = 0.0004$ respectively, SNR was set to 10 dB. The gain α for ANEW equalizer was set to 1.

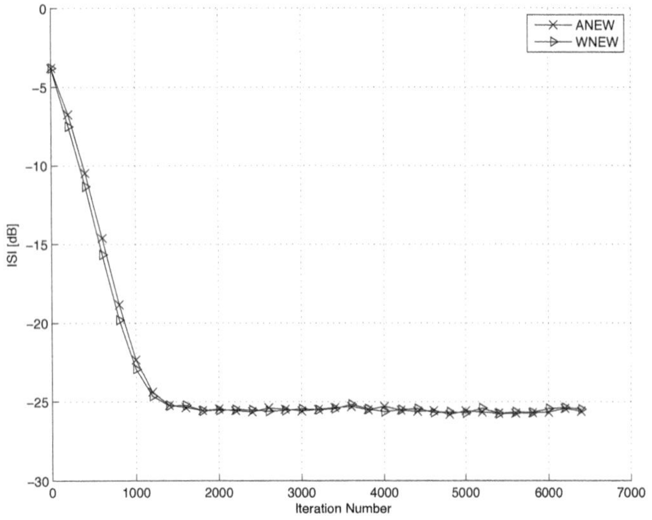

Fig. 22. A comparison between WNEW and ANEW algorithm for the 16QAM source going through Channel 1. The averaged results were obtained in 100 Monte Carlo trials. The equalizer's tap length was set to 13, the step-size parameter for WNEW and ANEW were set to $\mu_W = 0.0004$ and $\mu_A = 0.0004$ respectively, SNR was set to 30 dB. The gain α for ANEW equalizer was set to 1.

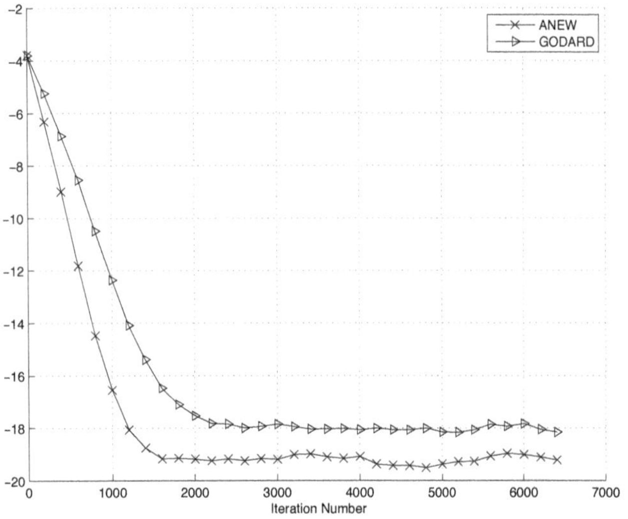

Fig. 23. A comparison between Godard and ANEW algorithm for the 16QAM source going through Channel 1. The averaged results were obtained in 100 Monte Carlo trials. The equalizer's tap length was set to 13, the step-size parameter for Godard and ANEW were set to $\mu_G = 0.00003$ and $\mu_A = 0.0004$ respectively, SNR was set to 10 dB. The gain α for ANEW equalizer was set to 1.

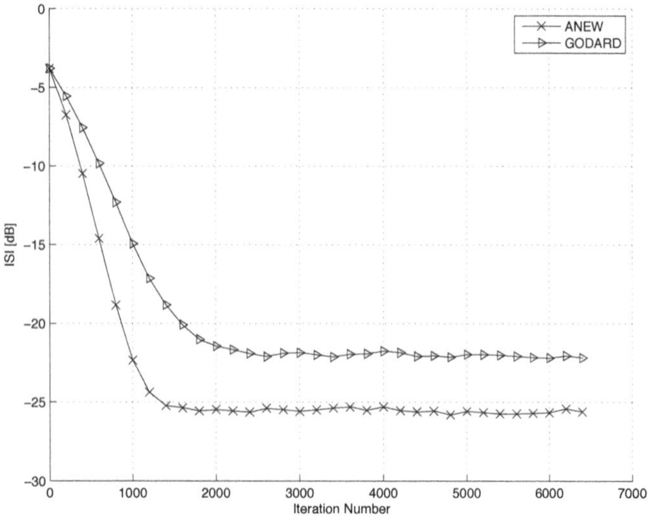

Fig. 24. A comparison between Godard and ANEW algorithm for 16QAM source input going through Channel 1. The averaged results were obtained in 100 Monte Carlo trials. The equalizer's tap length was set to 13, the step-size parameter for Godard and ANEW were set to $\mu_G = 0.00003$ and $\mu_A = 0.0004$ respectively, SNR was set to 10 dB. The gain α for ANEW equalizer was set to 1.

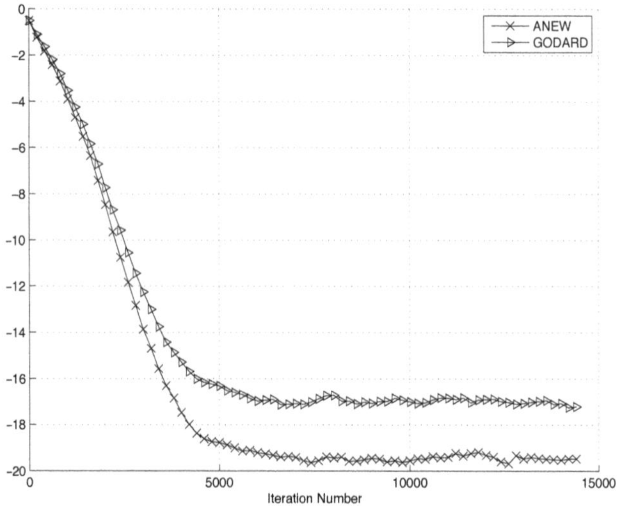

Fig. 25. A comparison between Godard and ANEW algorithms for the 64QAM source going through Channel 3. The averaged results were obtained in 100 Monte Carlo trials. The equalizer's tap length was set to 13, the step-size parameter for Godard and ANEW were set to $\mu_G = 0.0000012$ and $\mu_A = 0.00005$ respectively, SNR was set to 10 dB. The gain α for ANEW equalizer was set to 1.

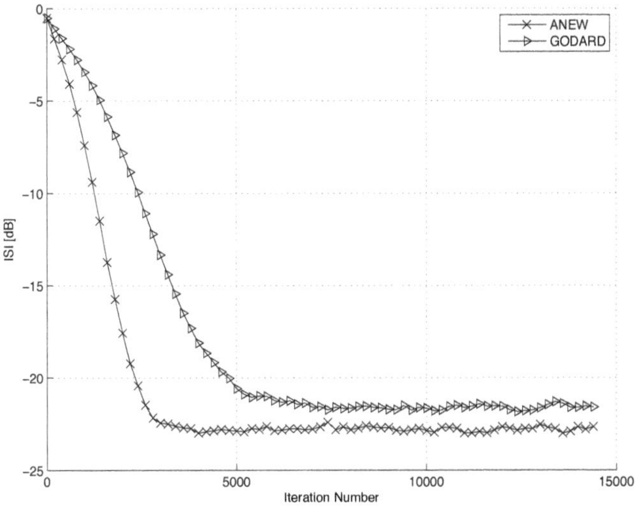

Fig. 26. A comparison between Godard and ANEW algorithm for the 64QAM source going through Channel 3. The averaged results were obtained in 100 Monte Carlo trials. The equalizer's tap length was set to 13, the step-size parameter for Godard and ANEW were set to $\mu_G = 0.000001$ and $\mu_A = 0.00008$ respectively, SNR was set to 30 dB. The gain α for ANEW equalizer was set to 1.

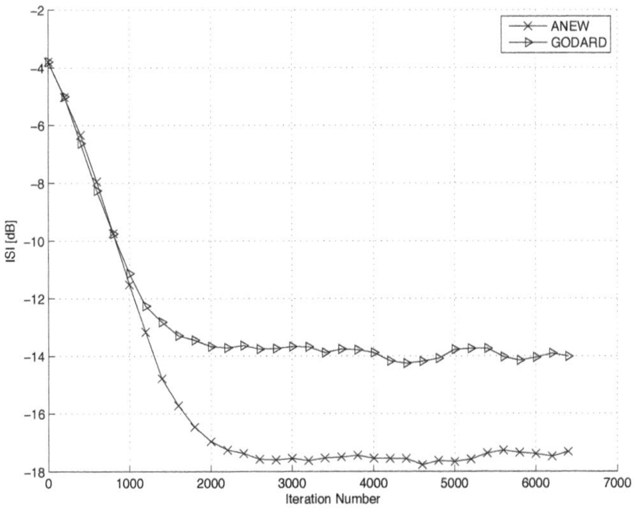

Fig. 27. A comparison between Godard and ANEW algorithms for the 16QAM source going through Channel 4. The averaged results were obtained in 100 Monte Carlo trials. The equalizer's tap length was set to 13, the step-size parameter for Godard and ANEW were set to $\mu_G = 0.00008$ and $\mu_A = 0.0006$ respectively, SNR was set to 10 dB. The gain α for ANEW equalizer was set to 1.

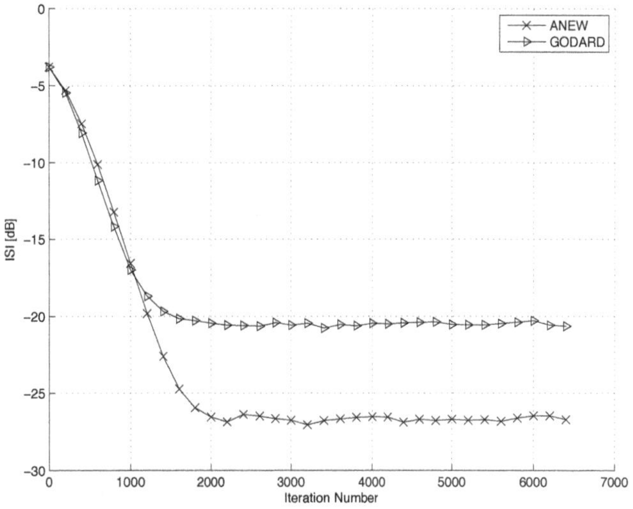

Fig. 28. A comparison between Godard and ANEW algorithms for the 16QAM source going through Channel 4. The averaged results were obtained in 100 Monte Carlo trials. The equalizer's tap length was set to 13, the step-size parameter for Godard and ANEW were set to $\mu_G = 0.00008$ and $\mu_A = 0.0006$ respectively, SNR was set to 30 dB. The gain α for ANEW equalizer was set to 1.

13

Design of Dual-Band Two-Branch-Line Couplers with Arbitrary Coupling Coefficients in Bands

Ivan PRUDYUS, Valeriy OBORZHYTSKYY

Institute of Telecommunications, Radioelectronics and Electronic Devices, Lviv Polytechnic National University, St. Bandery st., 12, 79013 Lviv, Ukraine

iprudyus@polynet.lviv.ua, oborzh@polynet.lviv.ua

Abstract. *A new approach to design dual-band two-branch couplers with arbitrary coupling coefficients at two operating frequency bands is proposed in this article. The method is based on the usage of equivalent subcircuits input reactances of the even-mode and odd-mode excitations. The exact design formulas for three options of the dual-band coupler with different location and number of stubs are received. These formulas permit to obtain the different variants for each structure in order to select the physically realizable solution and can be used in broad range of frequency ratio and power division ratio. For verification, three different dual-band couplers, which are operating at 2.4/3.9 GHz with different coupling coefficients (one with 3/6 dB, and 10/3 dB two others) are designed, simulated, fabricated and tested. The measured results are in good agreement with the simulated ones.*

Keywords

Dual band, branch-line coupler, even-odd-mode excitations, arbitrary coupling coefficient.

1. Introduction

Development of the modern wireless communication systems with various frequency standards is accompanied by the active usage of multiband microwave devices. As a directional coupler is one of the base components of Radio Frequency (RF) parts of these systems and can be used in structure of amplifiers, mixers, phase shifters and other devices, therefore in recent years much attention is given to new schemes for the couplers, operating at two arbitrary frequency bands. In the case of widely used branch-line couplers, a lot of dual-band schemes are proposed in the literature. In the majority of reports the structures of relatively simple two-branch-line couplers are considered, two-section topology (tri-branch-line construction) has been used with the object of bandwidth broadening. Among all variety of the offered options of a dual-band two-branch-line coupler realization one can note the next main approaches. It may be the usage of right/left-handed metamaterial transmission lines [1]–[3], the usage of stretched segments of line [4] or meander lines [5] for branches implementing, the usage of scheme in which all ports are extended through a transmission line section [6]. Very popular approach in obtaining the dual-band operation of the branch-line coupler is based on the use of stub lines. The structures in which open-circuit or short-circuit stubs are connected to all input ports are proposed in [7]. Further similar dual-band structures with some specialized functions were considered in [8]–[12]. In other options of coupler with dual-band response the loading of stub tapped to the center of through lines and branch lines [13]–[16], of branch lines [17], of through lines [18] is used. The difference of structures with stubs consists in a way of stubs realization for achievement of a definite purpose. In [8], for instance, the shunt stubs are folded for placement inside the coupler. The usage of stepped-impedance stubs for compactness and wide range of frequency ratio is proposed in [11], [15]. The multisection stubs, which are employed in [12], allow realizing any required value of shunt input susceptance. The structure, similar to [14] but with the shunt open-circuit dual composite right/left-handed cells, which provide the identical sign of phase difference of output ports within the two operating frequency bands is proposed in [16]. In most cases the principle of equivalent replacement was used for design of dual-band devices. According to this principle it is necessary to do a substitution of each branch of the conventional single-band device by a two-port structure. The π-type or T-type two-ports mostly are used. It is a transmission line segment loaded with shunt susceptance at its ends in the first case [7], [8], [10]–[12], and at its center in the second case [13]–[16]. The electrical parameters (characteristic impedance, electrical length, shunt impedance value) of equivalent structure have to provide the characteristics of removable branch at two frequencies.

Recently, the considerable attention is directed to couplers, which may have arbitrary output power division at the two operation bands. These couplers may be useful at the design of some devices such as antenna arrays, Doherty power amplifiers, mixers and others. In one of the first publications on this subject [8] it is offered to use the equivalent replacement of 90° section with different values of characteristic impedance for different frequencies by a two-port, which consists of a stepped-impedance section with open stubs attached to its ends. Similar approach but with replacement by conventional π-type two-port with single-section stubs is used in [10], and with multisection stubs in [12]. The other method of obtaining the required output power ratio at the dual frequencies is proposed in [19], [20]. It is based on the replacement of branches by shorted coupled line sections, application of which is obviously connected with certain difficulties because of different even and odd modes phase velocities and the complication of layout.

In this paper, a new method to design dual-band two-branch couplers with the desired coupling coefficients at two operating frequency bands is introduced. As distinct from above mentioned this method is based on the usage of input reactances of two-pole schemes which are obtained by means of the even-mode and odd-mode excitations. At such approach, exact design formulas are received for three options of the dual-band coupler with stub lines, which differ by number and location of stubs. Proposed calculation methods may be used in broad range of frequency ratio and power division ratio. They allows to obtain the different design variants with identical or opposite signs of phase differences within two frequency bands for the selection of variant with physically realizable values of characteristic impedances and shunt susceptances.

2. Design Methodology

It is known that the majority of directional couplers have the structure with one or two planes of symmetry. As a rule, for the analysis of a symmetrical four-port network the even-odd-mode decomposition method [21] is used, which is based on the implementation of magnetic and electric walls at this network. In case of bisymmetrical structure we can decompose the network into various single-port subcircuits (even-even, even-odd, odd-even, and odd-odd) by double using the superposition of an even-mode excitation and odd-mode excitation. The corresponding relations between the input resistances of these subcircuits and the scattering parameters of codirectional bisymmetrical coupler at the condition of ideal matching at its ports and ideal isolation are offered in [22]. From these relations it is possible to derive the following expressions for the transmission S-parameters of such lossless coupler:

$$S_{21} = \frac{x_{ee}^2 - x_{oo}^2 + j(x_{ee} - x_{oo} - x_{ee}^2 x_{oo} + x_{ee} x_{oo}^2)}{1 + x_{ee}^2 + x_{oo}^2 + x_{ee}^2 x_{oo}^2}, \quad (1.a)$$

$$S_{31} = \frac{x_{ee}^2 x_{oo}^2 - 1 + j(x_{ee} + x_{oo} + x_{ee}^2 x_{oo} + x_{ee} x_{oo}^2)}{1 + x_{ee}^2 + x_{oo}^2 + x_{ee}^2 x_{oo}^2}. \quad (1.b)$$

In the above, S_{21} is the transmission coefficient to direct port, S_{31} is the transmission coefficient to coupled port, and x_{ee}, x_{oo} are the input reactances for the even-even and odd-odd two-pole subcircuits, where the subscripts e and o denote the even and odd mode, respectively. The input reactances in (1) are normalized with respect to the reference (system) impedance Z_0 of the ports. The analysis of equations (1) showed such their peculiarities: 1) a preset combination of values of input reactances gives the required values of S-parameters, and consequently the required distribution of output power; 2) if to reverse the signs of preset values of input reactances, the values of magnitudes of scattering parameters will be not changed but the signs of their phases will be reversed. This also will provide the necessary distribution of output power and the quadrature of phase difference of coupler; 3) if to carry out a mutual exchange of preset values of input reactances, i. e. supply to reactance x_{ee} the value of reactance x_{oo}, and the value of reactance x_{ee} to reactance x_{oo}, it will provide the required value S_{31} and will provide the required value of the magnitude of parameter S_{21} but with a reverse sign of its phase. Such exchange will also give the necessary distribution of output power and the quadrature of phase difference; 4) if to carry out a mutual exchange of preset values of input reactances, as in point 3 but with reverse of their signs it will provide the required value of S-parameters magnitudes but with change of their phase. And such exchange will give the necessary power division ratio and 90 degrees phase difference.

The values of input reactances which provide the necessary value of coupling coefficient C of a bisymmetrical codirectional lossless coupler at the operating frequency may be calculated as is offered in [22]:

$$x_{ee} = \frac{-1}{x_{oe}} = \frac{|S_{21}|\sin\varphi_{21} + |S_{31}|\sin\varphi_{31}}{1 - |S_{21}|\cos\varphi_{21} - |S_{31}|\cos\varphi_{31}}, \quad (2.a)$$

$$x_{eo} = \frac{-1}{x_{oo}} = \frac{|S_{21}|\sin\varphi_{21} - |S_{31}|\sin\varphi_{31}}{1 - |S_{21}|\cos\varphi_{21} + |S_{31}|\cos\varphi_{31}}, \quad (2.b)$$

$$|S_{21}| = \sqrt{1 - |S_{31}|^2},\ |S_{31}| = 10^{-\frac{C}{20}},\ \varphi_{21} = \varphi_{31} - \frac{\pi}{2} \quad (2.c)$$

where φ_{21} and φ_{31} are the phases of S_{21} and S_{31}, coefficient C in dB. Unlike other methods, formulas (2) allow at the design of symmetrical directional couplers to preset a value of signal phase on the coupled port and in this way to influence the values of the electrical parameters in the process of their calculation.

2.1 Dual-Band Structure with Loaded Ports

Fig. 1(a) shows the structure of branch-line coupler with loaded ports. In [8], [12] for such dual-band structure with arbitrary division of power the calculation method only for $\varphi_{21} = -\pi$, $\varphi_{31} = -\pi/2$, was developed on the base of equivalent replacement of $\lambda/4$ sections. In offered approach by double application of even-odd-mode decomposition the initial structure can be divided into four reduced subcircuits with purely reactive input resistances (jX_{ee}, jX_{eo}, jX_{oe}, jX_{oo}) as shown in Fig. 1(b). The normalized input reactances of these reduced circuits can be expressed as follows:

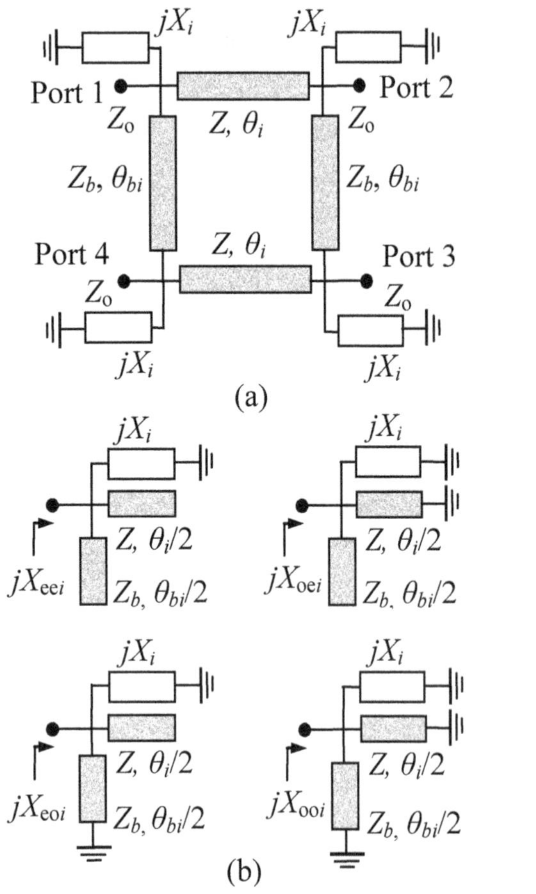

Fig. 1. (a) Dual-band coupler with loaded ports, (b) reduced subcircuits.

$$x_{eei} = \frac{x_i z z_b}{z z_b - x_i (z t_{bi} + z_b t_i)}, \tag{3.a}$$

$$x_{eoi} = \frac{x_i z z_b t_{bi}}{z z_b t_{bi} - x_i (z_b t_i t_{bi} - z)}, \tag{3.b}$$

$$x_{oei} = \frac{x_i z z_b t_i}{z z_b t_i - x_i (z t_i t_{bi} - z_b)}, \tag{3.c}$$

$$x_{ooi} = \frac{x_i z z_b t_i t_{bi}}{z z_b t_i t_{bi} + x_i (z t_i + z_b t_{bi})}. \tag{3.d}$$

In (3) subscript $i = 1,2$ points to number of frequency, all input reactances, addition reactance jX_i and characteristic impedances of through lines Z and branch lines Z_b are normalized to Z_o, $t_i = \tan(\theta_i/2)$, $t_{bi} = \tan(\theta_{bi}/2)$, where θ_i and θ_{bi} are the electrical lengths of lines. It follows from (3), that the functioning of such coupler with different values C_i at different frequencies can be ensured according to the first stated above feature of the equations (1), if to provide the values of input reactances as required for each frequency. If to use the second feature of the equations (1), when for the search of decision it is necessary to change the signs of input reactances, at that time it is necessary to change the signs of X_i, t_i, t_{bi}. The options, which conform to features of the equations (1) with a mutual exchange of the values of input reactances, may be used only to provide $C_1 = C_2$. To obtain formulas for calculation of unknown parameters Z, Z_b, X_i, we will express x_i from each of the equations (3), for example:

$$x_i = \frac{x_{eei} z z_b}{z z_b + x_{eei}(z t_{bi} + z_b t_i)}. \tag{4}$$

In the result of equating two pairs of these expressions for x_i, two formulas for normalized Z can be derived as

$$z = \frac{-x_{eei}}{x_{eei}^2 + 1} \cdot \frac{t_i^2 + 1}{t_i}, \quad z = \frac{-x_{eoi}}{x_{eoi}^2 + 1} \cdot \frac{t_i^2 + 1}{t_i}. \tag{5}$$

At the equating of formulas (4) with taking into account the interconnection of input reactances (2), the condition of solution of the set of equations (3) is established:

$$x_{eei} x_{eoi} = 1. \tag{6}$$

And the following formula for normalized Z_b is derived by the equating of two expressions for x_i with inserting (5) and by the taking into account (6):

$$z_b = \frac{x_{eei}}{x_{eei}^2 - 1} \cdot \frac{t_{bi}^2 + 1}{t_{bi}}. \tag{7}$$

In the result of equating of two expressions which have been written in terms of the first formula (5) for the center frequencies of the lower (f_1) and upper (f_2) bands (different values of i), the following relation is obtained :

$$\frac{x_{ee1}}{x_{ee2}} \cdot \frac{x_{ee2}^2 + 1}{x_{ee1}^2 + 1} = \frac{\sin\theta_1}{\sin(k_f \theta_1)} \tag{8}$$

where $k_f = f_2/f_1$ is the frequency ratio, θ_1 and $\theta_2 = k_f\theta_1$ are the electrical lengths at f_1 and f_2. At equating of z_b for different frequencies from (7) the similar relation can be derived as

$$\frac{x_{ee1}}{x_{ee2}} \cdot \frac{x_{ee2}^2 - 1}{x_{ee1}^2 - 1} = \frac{\sin\theta_{b1}}{\sin(k_f \theta_{b1})}. \tag{9}$$

It follows from (2) that the condition (6) can be carried out only when the phase φ_{31} at the coupled output will be equal 0 or $\pm\pi$. In this case signals at the outputs will be in a quadrature, if $\varphi_{21i} = \pm\pi/2$. Then all input reactances can be expressed by one of them, for example, by x_{eei}, which depending on a combination of phases φ_{31} and φ_{21} can be determined from (2) as

$$x_{eei} = \pm\sqrt{\frac{1 + |S_{31i}|}{1 - |S_{31i}|}}, \quad x_{eei} = \pm\sqrt{\frac{1 - |S_{31i}|}{1 + |S_{31i}|}} \tag{10}$$

where the first expression is used at $\varphi_{31i} = 0$ and the second expression is used at $\varphi_{31i} = \pm\pi$, the sign before the roots is identical to that for $\varphi_{21i} = \pm\pi/2$.

Thus, to define values of all electric parameters of the scheme in Fig. 1(a), it is necessary to set previously values of coupling coefficients and to choose combinations of phases φ_{31i} and φ_{21i} with required signs of phase difference.

By means of (8) and (9) with use of x_{eei} value calculated by (10) the values of θ_i and θ_{bi} can be defined. Values of X_i, Z, Z_b are calculate by formulas (4), (5) and (7). It should be noted that by a choice of values φ_{31i} and φ_{21i} it is possible to achieve physical realizable values of Z, Z_b and X_i.

2.2 Structure with Four Shunt Reactances

Fig. 2(a) shows the structure of dual-band branch-line coupler with four shunt reactances. Such coupler, but only with the same power division in both bands, was researched in [13]–[16] by using the method of equivalent replacement. In accordance with offered approach as a result of double application of even-odd-mode decomposition method this whole structure can be divided into reduced subcircuits as shown in Fig. 2(b). The normalized input reactances of such subcircuits can be evaluated by the following equations:

$$x_{eei} = \frac{zz_b p_i p_{bi}}{zp_i(z_b - 2x_{bi}t_i) + z_b p_{bi}(z - 2x_i t_i)}, \quad (11.a)$$

$$x_{eoi} = \frac{zz_b p_i t_i}{z_b t_i(z - 2x_i t_i) + zp_i}, \quad (11.b)$$

$$x_{oei} = \frac{zz_b p_{bi} t_i}{zt_i(z_b - 2x_{bi}t_i) + z_b p_{bi}}, \quad (11.c)$$

$$x_{ooi} = zz_b t_i /(z + z_b) \quad (11.d)$$

where $p_i = 2x_i + zt_i$, $p_{bi} = 2x_{bi} + z_b t_i$, $t_i = \tan(\theta_i/2)$. For the reduction of quantity of independent variables and also for the simple search of the solution of the electrical length of branch lines in Fig. 2(a) was accepted as equal to length of through sections. If to express x_i, x_{bi}, z_b from (11.b), (11.c), (11.d) and to insert these expressions in (11.a), then we reach the previously derived condition of solution (6) of the set of equations (11). Consequently, and for this structure the values of x_{eei} can be calculated by (10). In the result the expressions for the electrical parameters x_i, x_{bi}, and z_b can then be expressed in terms of x_{eei} as follows:

$$x_i = \frac{zt_i[z_b(1 - x_{eei}zt_i) + z]}{2z_b t_i(t_i + x_{eei}z) - z}, \quad (12)$$

$$x_{bi} = \frac{z_b t_i[z_b(1 + x_{eei}zt_i) + z]}{2zt_i(t_i - x_{eei}z_b) - z_b}, \quad (13)$$

$$z_b = -x_{eei}z /(x_{eei} + zt_i). \quad (14)$$

At the equating of x_{ooi} from formula (11.d) for different frequencies with taking into account that $x_{ooi} = -x_{eei}$ the following relation is established:

$$\frac{x_{ee1}}{x_{ee2}} = \frac{\tan(\theta_1/2)}{\tan(k_f\theta_1/2)}. \quad (15)$$

Thus, at the beginning of design of coupler with structure in Fig. 2(a) for pre-specified values of k_f, C_1, C_2 and the chosen combination of phases φ_{31i} and φ_{21i} it is necessary to define x_{eei} by (10), θ_1 by (15). Further, z_b is calculated by (14) and the values of x_i, x_{bi} are calculated by (12)–(13) for each frequency. Besides, for computations it is necessary to preset the value of Z as the number of independent variables exceeds the number of equations. If to assume that $Z_b = Z$, then the following formula for Z can be derived from (14):

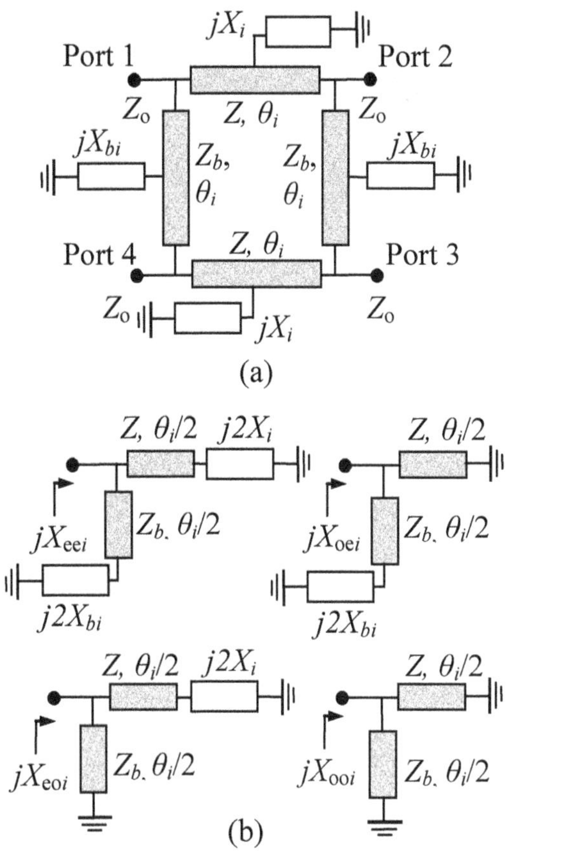

Fig. 2. (a) Structure with four reactances, (b) reduced subcircuits.

$$z = -2x_{eei}/t_i. \quad (16)$$

As in case of the previous structure due to a choice of phases' combinations it is possible as a result of calculation by (12)–(16) to receive different options of electrical parameters and consequently different features of scheme.

2.3 Structure with Two Shunt Reactances

Fig. 3 shows the dual-band structures, in which only two additional reactances are used. Such two-branch-line coupler in condition of dual-band operation but only with identical power division is investigated in [17] and [18]. In the first case reactances are attached to the center of branch lines as shown in Fig. 3 (a) and are implemented by open stubs with length equal to $\theta_i = \theta_{bi}$. In the second case the through lines of 3-dB coupler are loaded by stubs as in Fig. 3(b) and the length of branch lines is assumed $\theta_{bi} = 2\theta_i$.

In the structures under consideration the different length for the through sections and branches is accepted in

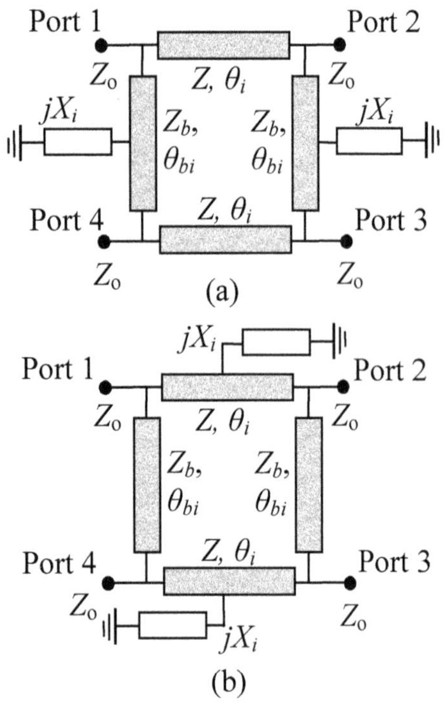

Fig. 3. Structures with two reactances, which (a) are attached to branches, (b) are attached to through lines.

order to receive the necessary number of independent variables. If to apply even-odd-mode decomposition twice to structure in Fig. 3(a), then we receive four two-pole subcircuits as in Fig. 2(b) but only without the reactances connected to segments of a through line. For input reactances of these subcircuits the following relations can be written:

$$x_{eei}=\frac{zz_b(2x_i+z_bt_{bi})}{z(z_b-2x_it_{bi})-z_bt_i(2x_i+z_bt_{bi})}, \tag{17.a}$$

$$x_{eoi}=zz_bt_{bi}/(z-z_bt_it_{bi}), \tag{17.b}$$

$$x_{oei}=\frac{zz_bt_i(2x_i+z_bt_{bi})}{z_b(2x_i+z_bt_{bi})+zt_i(z_b-2x_it_{bi})}, \tag{17.c}$$

$$x_{ooi}=zz_bt_it_{bi}/(zt_i+z_bt_{bi}). \tag{17.d}$$

If to express x_i from (17.a), (17.c) and z, z_b from (17.b), (17.d) and to equate inter se the expressions for x_i but with the insertion in these the expressions for z, z_b, then we reach the previously derived condition (6) of solution of the set of equations (17). Taking into account this condition the expressions for x_i, z, z_b in terms of x_{eei} can be written as

$$x_i=\frac{z_b[x_{eei}z-z_bt_{bi}(x_{eei}t_i+z)]}{2[x_{eei}zt_{bi}+z_b(x_{eei}t_i+z)]}, \tag{18}$$

$$z=-\frac{x_{eei}}{x_{eei}^2+1}\cdot\frac{t_i^2+1}{t_i}, \tag{19}$$

$$z_b=-\frac{x_{eei}}{t_i^2-x_{eei}^2}\cdot\frac{t_i^2+1}{t_{bi}}. \tag{20}$$

In the result of equating of two expressions (19), which have been written for two operation frequencies, the same relation (8) for evaluation of θ_i is obtained. In the result of equating of two expressions (20) for two frequencies, the following relation is established:

$$\frac{x_{ee1}}{x_{ee2}}\cdot\frac{t_2^2-x_{ee2}^2}{t_1^2-x_{ee1}^2}\cdot\frac{t_1^2+1}{t_2^2+1}=\frac{\tan(\theta_{b1}/2)}{\tan(k_f\theta_{b1}/2)}. \tag{21}$$

From (21) the values of θ_{b1} can be determined for required values of x_{eei} and θ_i value obtained from (8).

For structure in Fig. 3(b) the reduced subcircuits will be as in Fig. 2(b) but only without the reactances connected to segments of a branch line. The input reactances of these subcircuits can be evaluated by the following equations:

$$x_{eei}=\frac{zz_b(2x_i+zt_i)}{z_b(z-2x_it_i)-zt_{bi}(2x_i+zt_i)}, \tag{22.a}$$

$$x_{eoi}=\frac{zz_bt_{bi}(2x_i+zt_i)}{z(2x_i+zt_i)+z_bt_{bi}(z-2x_it_i)}, \tag{22.b}$$

$$x_{oei}=zz_bt_i/(z_b-zt_it_{bi}). \tag{22.c}$$

Equation for x_{ooi} is the same as (17.d). By the same procedure as in previous case, from these equations it is possible to receive the following expressions:

$$x_i=\frac{z[x_{eei}z_b-zt_i(x_{eei}t_{bi}+z_b)]}{2[x_{eei}zt_{bi}+z_b(x_{eei}t_i+z)]}, \tag{23}$$

$$z=-\frac{x_{eei}}{t_{bi}^2+x_{eei}^2}\cdot\frac{t_{bi}^2+1}{t_i}. \tag{24}$$

The equation for z_b is the same as (7), from which the relation (8) is obtained but for θ_{bi} determination. If to equate two expressions (24) for different frequencies, we will obtain the relation for θ_i determination:

$$\frac{x_{ee1}}{x_{ee2}}\cdot\frac{t_{b2}^2+x_{ee2}^2}{t_{b1}^2+x_{ee1}^2}\cdot\frac{t_{b1}^2+1}{t_{b2}^2+1}=\frac{\tan(\theta_1/2)}{\tan(k_f\theta_1/2)}. \tag{25}$$

In the same way as before at the design of structures in Fig. 3, for pre-specified values of k_f, C_1, C_2 and the chosen combination of phases φ_{31i} and φ_{21i} it is needed to define x_{eei} by (10), θ_i, θ_{bi} by (8), (21) or by (25), (8) and further to calculate Z, Z_b by (19), (20) or by (24), (7) and X_i values by (18) or by (23) for two frequencies. The results of calculations depend on a combination of φ_{31i}, φ_{21i}.

2.4 Realization of Dual-Frequency Reactances

Dual-band operation of above-mentioned structures requires the usage of additional reactances. For their realization it is possible to use both open or short circuit transmission line stubs and different types of multisection stubs formed by transmission line segments connections. Fig. 4 shows some options of such circuits. Their input reactances at frequencies f_1 and f_2 have to be respectively equal to X_1 and X_2 values. It can be provided by means of two independent electrical parameters (of two independent variables). The application with this purpose of an open-circuit stub or a short-circuit stub is a simplest. In this case their θ_1 electrical length can be defined from (15) where instead of the relation x_{ee1}/x_{ee2} it is necessary to use X_2/X_1 relation for the opened stub or X_1/X_2 for the shorted stub. Application of such stubs is limited to admissible X_i values and in many cases gives physically unrealizable results therefore more complex structure can be used.

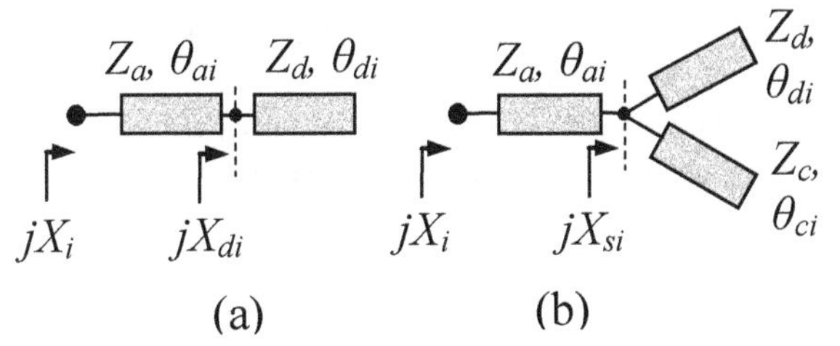

Fig. 4. (a) Stepped-impedance stub, (b) multisection stub.

The stepped-impedance-stub line in Fig. 4(a) is the series connection of two transmission line segments with different characteristic impedances Z_a, Z_d and electrical length θ_{ai}, θ_{di}. Opened or shorted at the end second segment must provide such X_{di} values of the input reactances, which can be transformed by the first segment to demanded X_i values. In this case there is a surplus of independent parameters therefore search of the solution can be carried out by various ways depending on basic data. For example, if Z_a of the first segment and its electrical length θ_{a1} (consequently $\theta_{a2} = k_f\theta_{a1}$) are given, the X_{di} can be defined as

$$X_{di} = Z_a(X_i - Z_a \cdot t_{ai})/(Z_a + X_i t_{ai}) \qquad (26)$$

where $t_{ai} = \tan(\theta_{ai})$, parameters Z_a, X_i, X_{di} are unrationed. Further, by method of calculation of single stub the electrical parameters of the second segment can be defined for the X_{di} values derived from (26). With a choice of other pair of independent variables the solution can be derived by iterative search of roots of the transcendental equations. Structure in Fig. 4(b) is formed by the connection of transmission line segment with a T-junction or Y-junction of transmission lines or with a segment of coupled lines. For the electrical parameters of this circuit the following relation is established:

$$(1/X_{di} + 1/X_{ci})(X_i - Z_a t_{ai})Z_a = Z_a + X_i t_{ai} \qquad (27)$$

where X_{di}, X_{ci} are the input reactances of segments with characteristic impedances Z_d, Z_c and electrical length θ_{di}, θ_{ci}. Whereas only two independent variables must be then values of four parameters need to be set, and so the calculations can be carried out by different ways. If to set, for example, Z_a, θ_{a1} provided that $Z_d = Z_c$, $\theta_{di} = \theta_{ci}$, then an input reactance X_{di} of each branch will be twice more than value calculated by (26). In this case instead of Y-junction it is possible to use a segment of coupled lines with joining of pair of its ends to the first segment. Thanks to even-mode excitation of coupled lines the characteristic impedance Z_e of even mode must be equal to Z_d, and the electrical length θ_{ei} of even mode must be equal to θ_{di}. At the same time the parameters of odd-mode excitation don't influence scheme characteristics. The values of Z_d, θ_{di} or Z_e, θ_{ei} can be determined from X_{di} values in way of the single stub. At other option of Fig. 4(b) structure calculation the values of Z_a, Z_d and θ_{ai}, θ_{di} can be preset. The X_{ci} values, which are necessary for definition Z_c, θ_{ci} can be calculated by (27). In accordance with approach which was considered in [12] one branch-line of Fig. 4(b) is assumed to be a quarter-wavelength long at one of the operating frequencies, and values of Z_a, Z_d, Z_c are preset. If, for example, $\theta_{c1} = 90°$, then $X_{c1} = 0$ (if the opened end) and therefore $t_{a1} = X_1/Z_a$. Further, from (27) for known Z_a, t_{a2}, X_{c2} X_{d2} can be defined, and then Z_c and θ_{ci} can be calculated. The other options of a choice of unknown parameters lead to the transcendental equations. Due to possibility of deriving the different results depending on initial data and consequently of deriving a different frequency dependence of jX in operating bands for structures in Fig. 4, thereby it is possible to influence characteristics of a branch-line coupler.

3. Simulated and Measured Results

To verify the offered design concept, and also for study of structures features, borders of their possible application each variant of a coupler was in detail investigated by calculations and simulation. In order to take into the account the effect of junction discontinuities and open-end effect, the tuning of layouts were carried out by using an electromagnetic solver. For experimental demonstration, the dual-band couplers were fabricated using the microstrip transmission lines on a Teflon substrate with dielectric constant of 2.68 and a thickness of 1.45 mm. Electrical parameters of these couplers, which are denominated further as A, B, C and values of geometrical sizes appropriate to them, which are received after correction by means of electromagnetic simulation, are listed in Tab. 1.

A. Coupler with loaded ports. Results of calculations of such a coupler by means of expressions given in Sec. 2.1, completely coincide with the results [8] (4.76/6.97 dB at 2.45/5.2 GHz), [12] (3/1.76 dB at 1.96/3.5 GHz), which are obtained only for one combination of phases $\varphi_{31i} = -\pi$, $\varphi_{21i} = -\pi/2$ at two frequencies.

The offered method allows designing the devices with smaller value of k_f thanks to possibility of a choice of different combinations of phase φ_{31i}, φ_{21i} at the operating frequencies. Tab. 2 gives the comparison of minimum values of frequency ratio k_f, which are permissible from point of view of the realizable values of characteristic impedances at various values of C_1 and various ratios C_1 to C_2 for proposed method at $\varphi_{311} = -\pi$, $\varphi_{211} = -\pi/2$, $\varphi_{212} = \pi/2$, for example, and methods [8], [12].

Coupl.	Z (Ω)	W (mm)	θ_1^o	l (mm)	Z_b (Ω)	W (mm)	θ_{b1}^o	l (mm)	X_i (Ω)	X_{bi} (Ω)
A	58.6	2.8	143	35.2	100.5	1.0	150.2	32.0	58.6/-46.5	–
B	54	3.4	129.4	30.4	83.3	1.8	–	–	416.1/-338.2	-26.5/-70.3
C	65.5	2.4	133.6	28.2	60.25	2.8	129.3	31.8	-15.1/ 45.35	–

Tab. 1. Circuit parameters and geometrical sizes of the three experimental couplers.

C_1(dB)	1.5		3		6		10	
C_1/C_2	[8]	Prop.	[8]	Prop.	[8]	Prop.	[8]	Prop.
0.3	2.1	1.2	-	1.4	-	-	-	-
0.4	1.8	1.15	2.2	1.2	-	1.7	-	-
0.6	1.3	1.1	1.5	1.1	1.9	1.35	2.3	2
0.8	1.1	-	1.1	-	1.3	1.3	1.4	1.6

Tab. 2. Permissible values of k_f for the coupler A with loaded ports.

The coupler of this kind for the coupling coefficients 3/6 dB at 2.45/3.9 GHz (k_f = 1.59, inadmissible value for methods [8], [12]) has been designed by the proposed method and then fabricated and measured. Influence of various combinations of phases upon the values of electrical parameters is evident from the results given for this coupler in Tab. 3. This affect can be used for choice of variant with smaller sizes (shorter segments) or with characteristic impedances, which are possible to implement.

φ_{311}	φ_{312}	φ_{211}	φ_{212}	Z (Ω)	θ_1^o	Z_b (Ω)	θ_{b1}^o
0	0	-π/2	-π/2	43.3	54.6	136.3	201.5
0	π	π/2	π/2	81	205.9	100.5	150.2
0	π	-π/2	π/2	58.6	143	136.1	201.5
-π	0	-π/2	-π/2	43.3	56.4	100.5	150.2
-π	-π	π/2	π/2	81	205.9	136.3	201.5
-π	-π	-π/2	π/2	58.6	143	100.5	150.2

Tab. 3. Dependence of electric parameters of the coupler with loaded ports A from combinations of phases.

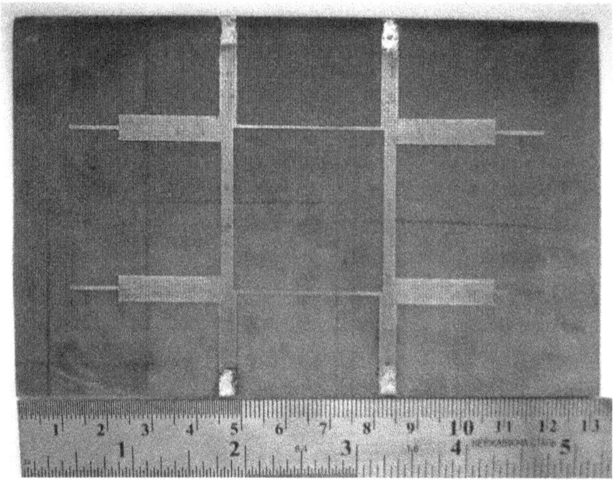

Fig. 5. Photograph of fabricated coupler A with loaded ports.

In Tab. 1, the values of electrical parameters of coupler A are given, which are calculated by using (4)-(10) for chosen combination of phases φ_{31i} = -π, φ_{211} = -π/2, φ_{212} = +π/2 and its sizes. Reactances of X_i were realized by structure as in Fig. 4(a) with electrical parameters Z_a = 38.2 Ω, Z_d = 102.7 Ω, θ_{a1} = 100°, θ_{d1} = 49° calculated by using (26). Fig. 5 shows the photograph of the fabricated dual-band coupler. Fig. 6 shows the simulated and measured scattering parameters of the coupler. The measured results indicate that the dual-band operation for small k_f has been achieved with a slight deviation of a coupling level at center frequencies from desired values. The insertion loss are $|S_{21}|$ = -3.12 dB and $|S_{31}|$ = -3.45 dB at 2.45 GHz and $|S_{21}|$ = -1.61 dB and $|S_{31}|$ = -6.73 dB at 3.9 GHz, respectively. The opposite sign of a phase difference φ_{21i} - φ_{31i} at f_1 and f_2 is caused by the choice at calculations of an opposite sign of φ_{21i} at these frequencies. Information on bandwidths of this coupler A is given in Tab. 4.

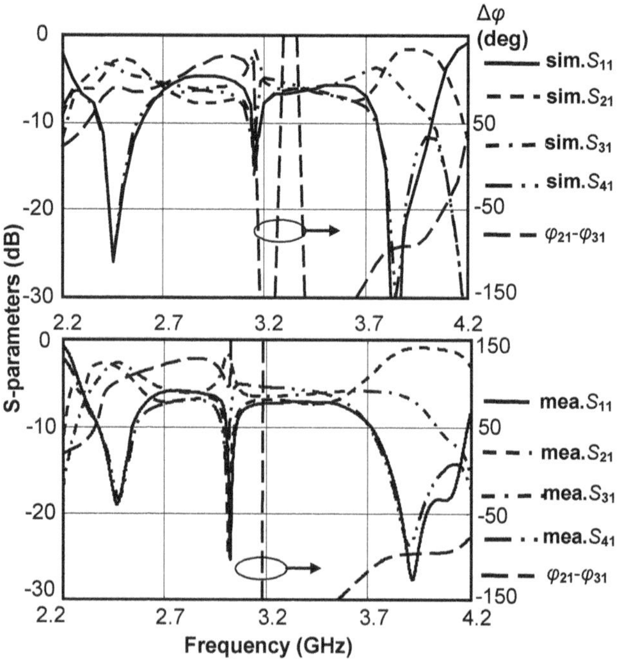

Fig. 6. Simulated and measured S-parameters of coupler A with loaded ports.

Coupler		$1/\|S_{11}\|$ (>15 dB)	$1/\|S_{41}\|$ (>15 dB)	$\varphi_{21} - \varphi_{31}$ (±90°±5°)
A	Sim.	4.0/4.4	4.0/3.6	5.2/4.4
	Meas.	3.9/8.2	3.9/5.6	5.0/8.1
B	Sim.	–/3.6	–/3.6	11.6/3.4
	Meas.	8.4/3.0	–/5.3	10.0/3.7
C	Sim.	–/3.6	–/3.4	11.1/3.4
	Meas.	11.8/3.0	–/3.9	11.5/3.8

Tab. 4. Bandwidths % of the three experimental dual-band couplers.

B. Coupler with four shunt reactances. Results of calculations of such coupler by an offered method coincide with the results [13] (0.9/2 GHz), [15] (2.4/5.8 GHz), which are obtained only for $C_1 = C_2$ at one combination of phases $\varphi_{31i} = -\pi$, $\varphi_{211} = -\pi/2$, $\varphi_{212} = +\pi/2$.

The identical signs of a phase difference at f_1 and f_2 of this structure with different power division can be achieved at short segments of lines only for ratio $C_1/C_2 = 0.2–0.8$, $k_f < 1.8$ and only at long segments for greater values of this ratio and k_f. The possibility to receive the values of impedances which are admissible for realization depends from the choice of phases combination and of Z value. Tab. 5 gives the examples of calculation of coupler parameters, which can serve as corroboration of the told.

$\varphi_{311}/\varphi_{312}$ $\varphi_{211}/\varphi_{212}$	C_1/C_2 (dB)	k_f	θ_1°	Z (Ω)	Z_b (Ω)
$-\pi/-\pi$ $-\frac{\pi}{2}/-\frac{\pi}{2}$	3/6	1.2	92.8	50	32.5
				30	57.3
	3/7.5	1.4	57.9	50	148
				100	59.7
$0/\pi$ $-\frac{\pi}{2}/\frac{\pi}{2}$	3/1	1.8	162.9	30	46.1
				50	28.6
	6/3	2.4	128.2	50	267.2
				100	72.8
$0/0$ $\frac{\pi}{2}/\frac{\pi}{2}$	6/3	2.4	245.7	100	127.3
				150	89.4

Tab. 5. Parameters of coupler with four shunt reactances B at the identical signs of phase difference.

The coupler with this structure can be used for dual-band work with a considerable difference of coupling coefficients C_i. This is demonstrated by the following example. The coupler for the coupling coefficients 10/3 dB at 2.4/3.9 GHz has been designed by the proposed method then has been fabricated and measured. The electrical parameters of this coupler B, which are calculated by using (12)-(15) for phases $\varphi_{31i} = 0$, $\varphi_{211} = -\pi/2$, $\varphi_{212} = +\pi/2$ and its

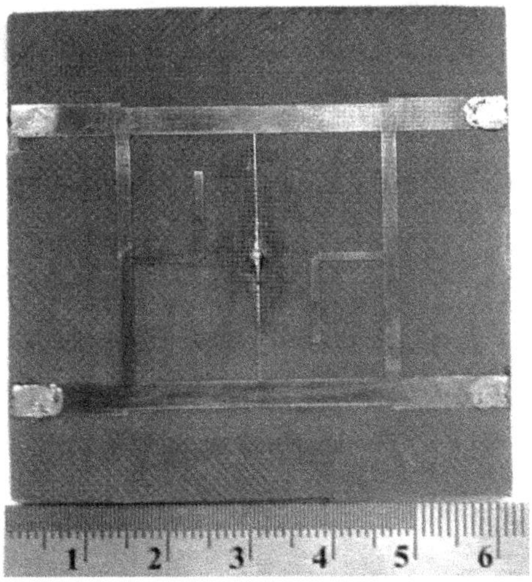

Fig. 7. Photograph of fabricated coupler B with four shunts.

sizes are given in Tab. 1.Reactances of X_i were realized by the short-circuit stub with electrical parameters $Z = 150.61$ Ω, $\theta_1 = 70.1$°, and reactances of X_{bi} were realized by the open-circuit stub with parameters $Z = 106.17$ Ω, $\theta_1 = 76$°. Fig. 7 shows the photograph of the fabricated coupler, and Fig. 8 shows the simulated and measured frequency responses of it, which agree closely with each other.

The measured data reveal that the required values of the insertion losses of a coupled port are obtained at 2.46/3.97 GHz with the phase differences of 95°/+86°. The small discrepancies between the simulated and measured results can be explained by deviation of a dielectric constant of substrate from a value used at calculations and by the limited accuracy of the prototypes fabrication. Tab. 4 gives the values of coupler B bandwidths.

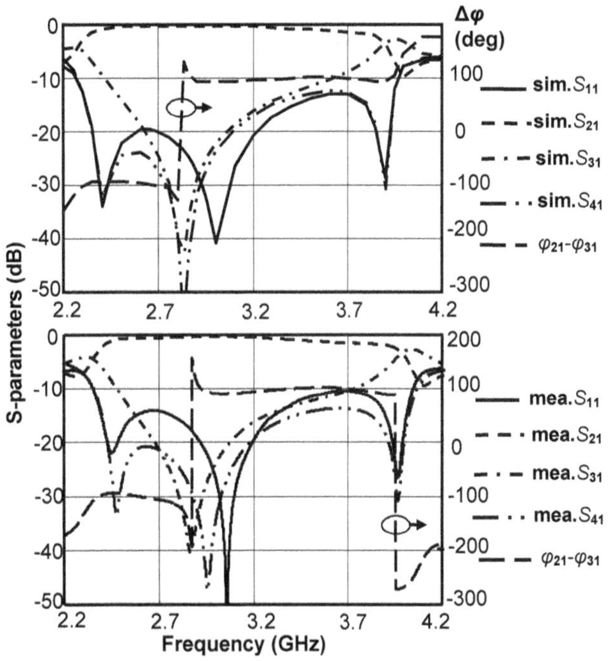

Fig. 8. Simulated and measured S-parameters of coupler B with four shunts.

C. Coupler with two shunt reactances. The calculations of such couplers showed that for structure in Fig. 3(b), physically admissible values of characteristic impedances issue for $k_f > 1.8$ and at different signs of φ_{21i}. The structure as in Fig. 3(a) gives more capabilities for use. Results of calculations of such coupler with $C_i = 3$ dB at 1/2.4 GHz and $\varphi_{31i} = \pi$, $\varphi_{211} = -\pi/2$, $\varphi_{212} = \pi/2$ by the offered method completely coincide with results of method [17], which was developed for $C_1 = C_2$ and opposite signs of phase difference at f_1 and f_2. The identical signs may be achieved by the proposed method in broad range of k_f thanks to selection of combination φ_{31i}, φ_{21i}. Tab. 6 gives the results, which corroborate that.

Results, acceptable for realization with the opposite signs of a phase difference may be obtained both at small difference of coupling coefficients and like the previous case at a big difference of them, which can be illustrated by

C_1/C_2 (dB)	$\varphi_{311}/\varphi_{312}$ $\varphi_{211}/\varphi_{212}$	k_f	θ_1°	Z (Ω)	θ_{b1}°	Z_b (Ω)
3/6	$0/\pi$	1.5	147.5	65.7	168.7	25.8
	$-\frac{\pi}{2}/\frac{\pi}{2}$	1.7	138.2	52.9	172.7	66.3
	$\pi/0$	1.8	134.2	49.3	99.2	21.4
	$-\frac{\pi}{2}/\frac{\pi}{2}$	2.2	126.0	43.7	35.4	85.4
6/3	$0/\pi$	1.6	134.6	60.8	168.0	22.6
	$-\frac{\pi}{2}/\frac{\pi}{2}$	1.8	123.8	52.1	172.3	52.8
	$0/0$	1.9	68.7	46.4	291.4	73.4
	$-\frac{\pi}{2}/-\frac{\pi}{2}$	2.1	63.4	48.4	258.5	37.3
	π/π	2.2	61.0	49.5	175.2	111.9
	$-\frac{\pi}{2}/-\frac{\pi}{2}$	2.4	57.0	51.6	183.4	29.3

Tab. 6. Parameters of coupler C with two shunt reactances at the identical signs of phase difference.

the following example. Calculation of coupler C with the same values 10/3 dB of coupling coefficients at frequencies 2.45/3.9 GHz for the same combination of phases $\varphi_{31i} = 0$, $\varphi_{211} = -\pi/2$, $\varphi_{212} = +\pi/2$ by means of (18)-(21) gave the values of electrical parameters, which are listed together with sizes in Tab. 1. Reactances of X_i were realized by the open-circuit stubs with parameters $Z = 68.45$ Ω, $\theta_1 = 77.58°$. Fig. 9 shows the photograph of dual-band coupler fabricated by sizes, specified as a result of electromagnetic simulation. Fig. 10 shows the simulated and measured coupler frequency responses which are very similar to the previous case. Required values of coupling coefficients are obtained at frequencies of 2.47/3.96 GHz with the phase differences of 95°/89°. Discrepancies between the simulated and measured results are caused by the same as stated above. The values of coupler bandwidths are given in Tab. 4. This type of a dual-band coupler is simpler in fabrication, has smaller quantity of discontinuities and the smaller sizes.

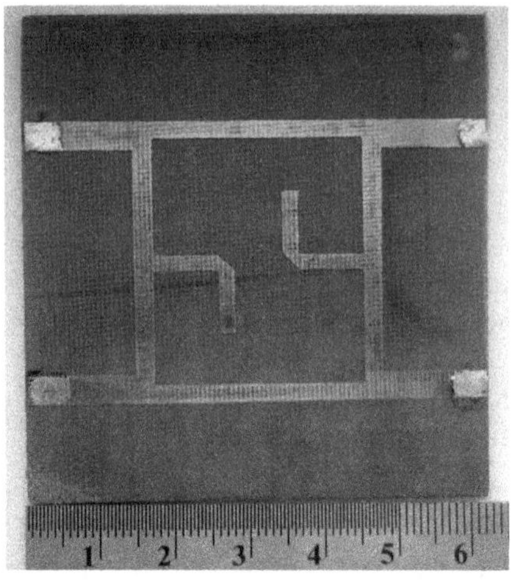

Fig. 9. Photograph of fabricated coupler C with two shunts.

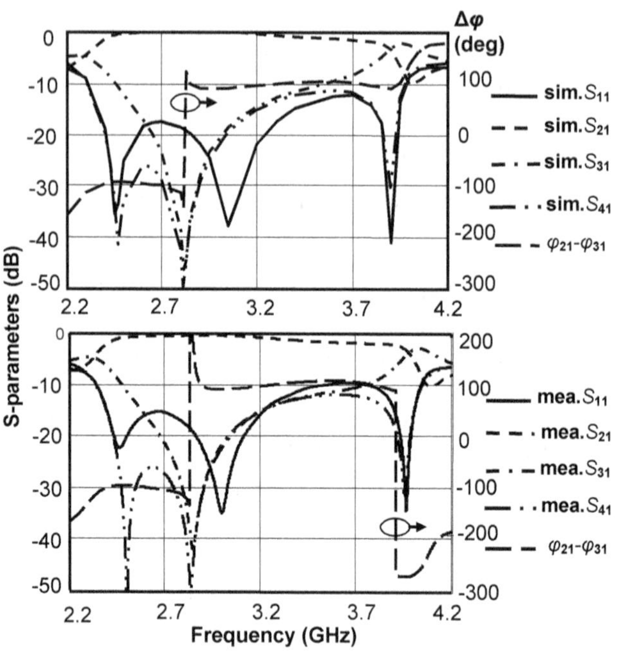

Fig. 10. Simulated and measured S-parameters of coupler C with two shunts.

4. Conclusions

The new approach to the design of dual-band two-branch-line directional couplers with arbitrary coupling coefficient at operating frequencies offered in this paper is based on the use of networks input impedances, which are obtained by even-odd-mode excitations. Three options of couplers with bisymmetrical structures, which contain four or two addition reactances are considered. Closed-form design equations have been formulated for the evaluation of parameters of their circuits. The proposed methods of calculation gives more design flexibility of coupler because of the possibility of choice of a suitable phases combination on outputs. For verification, three experimental dual-band circuits operating at 2.45/3/9 GHz with different coupling coefficients are demonstrated. Good agreement between the simulated and measured results has been observed. The proposed approach may easily be extended to the design of other directional couplers with two symmetry planes.

Acknowledgements

The authors would like to thank Prof. Yevhen Yashchyshyn and the research workers of the Institute of Radioelectronics, Warsaw University of Technology for their help in experimental investigations.

References

[1] LIN, X. Q., LUI, R. P., YANG, X. M., CHEN, J. X., YIN, X. X., CHENG, Q., CUI, T. J. Arbitrary dual-band components using

simplified structures of conventional CRLH TLs. *IEEE Transactions on Microwave Theory and Techniques,* 2006, vol. 54, no. 7, p. 2902–2909.

[2] CHI, I.-H., DeVINCENTIS, M., CALOZ, C., ITOH, T. Arbitrary dual-band components using composite right/left-handed transmission lines. *IEEE Transactions on Microwave Theory and Techniques,* 2004, vol. 52, no. 4, p. 1142–1149.

[3] LIN, P.-L., ITOH, T. Miniaturized dual-band directional couplers using composite right/left-handed transmission structures and their applications in beam pattern diversity systems. *IEEE Transactions on Microwave Theory and Techniques,* 2009, vol. 57, no. 5, p. 1207–1215.

[4] WONG, F.-L., CHENG, K.-K. M. A novel planar branch-line coupler design for dual-band applications. In *IEEE MTT-S Int. Microwave Symp. Dig.,* 2004, vol. 2, p. 903–906.

[5] JIZAT, N. M., RAHIM, S. K. A., RAHMAN, T. A., ABDULRAHMAN, A. Y., SABRAN, M. I., HALL, P. S. Miniaturized size of dual-band-meandered branch-line coupler for WLAN application. *Microwave and Optical Technology Letters,* 2011, vol. 53, no. 11, p. 2543–2547.

[6] KIM, H., LEE, B., PARK, M.-J. Dual-band branch-line coupler with port extensions. *IEEE Transactions on Microwave Theory and Techniques,* 2010, vol. 58, no. 3, p. 651–655.

[7] CHENG, K.-K., M., WONG, F.-L. A novel approach to the design and implementation of dual-band compact planar 90° branch-line coupler. *IEEE Transactions on Microwave Theory and Techniques,* 2004, vol. 52, no. 11, p. 2458–2463.

[8] HSU, C.-L., M., KUO, J.-T., CHANG, C.-W. Miniaturized dual-band hybrid couplers with arbitrary power division ratios. *IEEE Transactions on Microwave Theory and Techniques,* 2009, vol. 57, no. 1, p. 149–156.

[9] TANG, C.-W., CHEN, M.-G. Design of multipassband microstrip branch-line couplers with open stubs. *IEEE Transactions on Microwave Theory and Techniques,* 2009, vol. 57, no. 1, p. 196–203.

[10] KIM, K., LIM, J., KIM, K., AHN, D. A compact dual band branch line coupler with arbitrary power division ratio. *Microwave and Optical Technology Letters,* 2010, vol. 52, no. 7, p. 1476–1480.

[11] ZHENGL, N., ZHOU, L., YIN, W.-Y. A novel dual-band Π-shaped branch-line coupler with stepped-impedance stubs. *Progress In Electromagnetics Research Letters,* 2011, vol. 25, p. 11–20.

[12] RAWAT, K., RAWAT, M., HASHMI, M. S., GHANNOUCHI, F. M. Dual-band branch-line hybrid with distinct power division ratio over the two bands. *International Journal of RF and Microwave Computer-Aided Engineering,* 2013, vol. 23, no. 1, p. 90–98.

[13] ZHANG, H., CHEN, K. J. A stub tapped branch-line coupler for dual-band operations. *IEEE Microwave and Wireless Components Letters,* 2007, vol. 17, no. 2, p. 106–108.

[14] PARK, M.-J. Dual-band, unequal length branch-line coupler with center-tapped stubs. *IEEE Microwave and Wireless Components Letters,* 2009, vol. 19, no. 10, p. 617–619.

[15] CHIN, K.-S., LIN, K.-M., WEI, Y.-H., TSENG, T.-H., YANG, Y.-J. Compact dual-band branch-line and rat-race couplers with stepped-impedance-stub lines. *IEEE Transactions on Microwave Theory and Techniques,* 2010, vol. 58, no. 5, p. 1213–1221.

[16] LU, K., WANG, G.-M., TIAN, B. Design of dual-band branch-line coupler based on shunt open- circuit DCRLH cell. *Radioengineering,* 2013, vol. 22, no. 2, p. 618–623.

[17] KIM, T. G., LEE, B., PARK, M.-J. Dual-band branch-line coupler with two center-tapped stubs. *Microwave and Optical Technology Letters,* 2008, vol. 50, no. 12, p. 3136–3139.

[18] JIZAT, N. M., RAHIM, S. K. A., RAHMAN, T. A., KAMARUDIN, M. R. Miniaturize size of dual band branch-line coupler by implementing reduced series arm of coupler with stub loaded. *Microwave and Optical Technology Letters,* 2011, vol. 53, no. 4, p. 819–822.

[19] HSU, C.-L. Dual-band branch line coupler with large power division ratios. In *Proceedings of Asia Pacific Microwave Conference.* Singapore, 7-10 Dec. 2009, p. 2088–2091.

[20] YU, C.-H., PANG, Y.-H. Dual-band unequal-power quadrature branch-line coupler with coupled lines. *IEEE Microwave and Wireless Components Letters,* 2013, vol. 23, no. 1, p. 10–12.

[21] REED, J., WHEELER, G. J. A method of analysis of symmetrical four-port networks. *IEEE Transactions on Microwave Theory and Techniques,* 1956, vol. 4, no. 10, p. 246–252.

[22] PRUDYUS, I. N., OBORZHYTSKYY, V. I. A new approach to analytical calculation of microstrip directional couplers with full structure symmetry. *Radioelectronics and Communications Systems,* 2011, vol. 54, no. 9, p. 472–480.

About Authors ...

Ivan PRUDYUS was born in Ukraine in 1942. He received his M.Sc., Ph.D. and D.Sc. degree in Radioelectronics in Lviv Polytechnic Institute (Ukraine) in 1969, 1980, 2005, respectively. Since 2004 he is the head of the Institute of Telecommunications, Radioelectronics and Electronic Devices, Lviv Polytechnic National University. His research interests include antennas for communication systems, remote sensing systems.

Valeriy OBORZHYTSKYY was born in Ukraine, in 1949. He received the M.Sc. degree in Radioelectronics Engineering in Lviv Polytechnic Institute (Ukraine) in 1972, the Ph.D. degree in Microwave Devices and Antennas in Moscow Institute of the Electronic Techniques in 1983. His research interests are in the field of microwave network theory and methods of computer-aided design, microwave integrated multiband circuits, tunable circuits, matching circuits.

14

SIM-DSP: A DSP-Enhanced CAD Platform for Signal Integrity Macromodeling and Simulation

Chi-Un LEI

Department of Electrical and Electronic Engineering, The University of Hong Kong, Hong Kong

culei@eee.hku.hk

Abstract. *Macromodeling-Simulation process for signal integrity verifications has become necessary for the high speed circuit system design. This paper aims to introduce a "VLSI Signal Integrity Macromodeling and Simulation via Digital Signal Processing Techniques" framework (known as SIM-DSP framework), which applies digital signal processing techniques to facilitate the SI verification process in the pre-layout design phase. Core identification modules and peripheral (pre-/post-)processing modules have been developed and assembled to form a verification flow. In particular, a single-step discrete cosine transform truncation (DCTT) module has been developed for modeling-simulation process. In DCTT, the response modeling problem is classified as a signal compression problem, wherein the system response can be represented by a truncated set of non-pole-based DCT bases, and error can be analyzed through Parseval's theorem. Practical examples are given to show the applicability of our proposed framework.*

Keywords

Simulation, signal integrity, high-speed circuits, macromodeling.

1. Introduction

Radio-frequency systems and electronic systems, such as mobile communication systems and high-speed computers, have become essential to our daily lives. These electronic systems contain modules of integrated circuits (ICs). These modules are connected by signal and power distribution networks [1], [2]. With the increasing operation frequency and decreasing feature size of ICs, high-frequency effects of the network, such as signal delay, reflections, crosstalk and simultaneous switching noise, have significantly disturbed the original signals, and have become dominant factors that limit IC system performance [3], [4]. Therefore, signal integrity (SI) verification has become popular practices in the IC design process to address such limitations, so as to ensure consistent signal transmissions and reliable power distributions in high-speed electronic systems [5]. In the pre tape-out phase, detailed simulation and analysis are required to capture the high-frequency behaviors of systems. However, a full-wave electromagnetic (EM) analysis over the global system is impractical. Reduced (simplified) models, which preserve relevant properties from the original models, are therefore required for time-critical simulations. In particular, for structures with complicated geometry, such as packages, vias and radio-frequency (RF) objects, data-driven macromodeling is usually applied to generate reduced models. Responses of systems are usually approximated using s-domain Sanathanan-Koerner (SK) iteration [6] or equivalent discrete-time domain (z-domain) Steiglitz-McBride (SM) iteration [7].

Different stand-alone macromodeling algorithms [8]-[9] have been developed. However, peripheral processing techniques have not been developed. Therefore, they are not capable for a functional modeling-simulation-analysis process. Besides them, there are some comprehensive modeling-simulation methodologies. MATLAB RF toolbox [10] and IdEM toolbox [11] have adopted Vector Fitting (VF) [8] as the core identification algorithm for the modeling-simulation process. However, they suffer from ill-conditioned problems because of the iterative computation property. IBM AQUAIA [12], [13] provides a comprehensive simulation for on-chip interconnect analysis with an inbuilt EM field solver. However, it does not apply model order reduction techniques; therefore, often generate large scale models, which are not efficient for practical simulations. Another methodology BEMP [14], [15] constructs macromodels using a cascade of low-/band-/high-/all-pass filters, however it has a large computation complexity, which also makes it not practical for simulations. Meanwhile, discrete-time domain computations of transient simulation have been introduced to exploit its nice features [16].

Furthermore, besides improving the macromodeling process from a mathematical (mostly control-theoretic) perspective, pre-/post-processing techniques and improvement from non-control-theoretic perspectives have been less explored in the literature. Also, due to the emerging development of microelectronics features [3], [17], new considerations and new necessities have been raised for developments of the SI verification process. Existing standard macromodeling approaches have a room for improvement in modeling new interconnect features. Therefore, advanced macromodeling methodologies should be developed to meet the needs of SI simulations [18].

In this paper, a "VLSI Signal Integrity Macromodeling and Simulation via Digital Signal Processing Techniques" (SIM-DSP) framework is proposed for an effective interconnect macromodeling and simulation process [19]. This framework explores the feasibility and benefits of applying digital signal processing (DSP) techniques in the macromodeling process. We first give an introduction of SI (Section 2). Then we outline the macromodeling-simulation methodology flow and developed modules in the framework (Section 3). Various case studies are given to show the applicability and effectiveness of our proposed framework and algorithm (Section 4).

There are several contributions in this paper. First, a discrete cosine transform truncation (DCTT) approach is presented for the modeling-simulation process of time- and frequency-sampled responses. The response modeling is classified as a signal compression problem, such that DCTT provides a numerical-insensitive and simple computation with an exact *a priori* error analysis through Parseval's theorem. Second, the developed identification modules and peripheral processing modules have been assembled to form a functional modeling-simulation flow for SI verifications. Third, several benchmarks of industrial interest have been performed and are documented in this paper, with explanations of influences of macromodeling-simulation process in SI verifications.

2. Signal Integrity and Macromodeling

SI generally refers to problems that occur in the high speed circuit system due to physical interconnects. It happens as an EM phenomenon in nature. However, as ICs move toward the era of nano-scale and multi-GHz-operation, SI problems now become the bottleneck of the high speed system design. SI issues disturb the signal voltage, and make the transmitted signal become ambiguous. As a result, the quality of the signal transmission and the circuit performance are degraded by the disturbance, especially in high speed circuits. Discussions about SI can be found in [20], [21].

To reduce the design risk and shorten the design period of sophisticated circuits, modeling, simulation and verification process is developed, such that the electrical performance of the emerging inter-/intra-chip communication links in circuit systems can be simulated for *a priori* diagnosis.

To model a complicated geometry structure, generally, for a single-port system, macromodeling techniques intend to fit a linear time-invariant (LTI) system to the desired continuous-time frequency domain (s-domain) input/output (I/O) response $H(s)$ at a set of calculated/sampled points at the I/O ports. The model is usually a state-space system or a rational transfer function with a set of predefined basis $\{\phi_n\}$

$$H(s) \approx \frac{N(s)}{D(s)} = \frac{\sum_{n=1}^{N} b_n \phi_n(s)}{\sum_{n=1}^{N} \widetilde{b}_n \phi_n(s)} \tag{1}$$

where $\widetilde{b}_n, b_n \in \mathbb{R}$ and N is the macromodel order. Therefore, macromodeling can be regarded as a large-scale broadband system identification problem.

In the L_2-norm sense with N_s sampled points, the optimal model of a system can be obtained through minimizing the following objective function

$$\min \sum_{k=1}^{N_s} \left\| \frac{N(s_k)}{D(s_k)} - H(s_k) \right\|_2 . \tag{2}$$

However, this is a numerically sensitive non-linear problem with no prior information about the exact pole and zero locations of the system.

3. SIM-DSP Methodology Flow

SIM-DSP framework aims to explore the feasibility and benefits of applying DSP techniques in the macromodeling process, in order to improve the functionality and accuracy of the macromodeling-simulation process in a comprehensive manner, from the initial response characterization step to the final simulation step. The focus of this framework is on

1. Improving the functionality and automation of the identification process;
2. Increasing the fitting accuracy of the identification process;
3. Reducing the computation time of the macromodeling procedure.

A generic SIM-DSP macromodeling-simulation methodology flow in SIM-DSP is shown in Fig. 1., while Fig. 2 shows the data and process flow chat in SIM-DSP. The sampled structure responses can be obtained by exciting one input port at a time and then directly measuring the responses at the output ports by instruments [21] or computing the responses by the given structure geometries and the EM field solver [22] (**Response Characterization**). Sampled data are analyzed for *a priori* configuration in the later identification process, such as selecting the macromodel order and number of iteration. At the same time, sampled data are processed for the enhancement of identifications (**Data Pre-processing**). By approximating the processed system response data in a black-box approach with the determined configuration, a macromodel is generated to replace the original large-order system by a smaller-order one with similar input-output (I/O) behaviors (**Macromodeling**), in form of a rational function

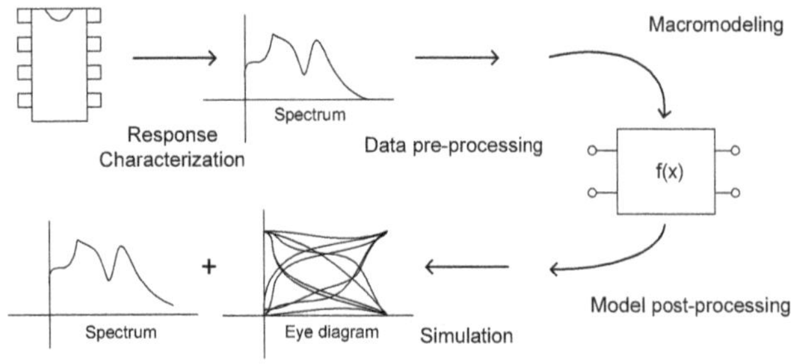

Fig. 1. Macromodeling-Simulation flow in SIM-DSP in a generic perspective.

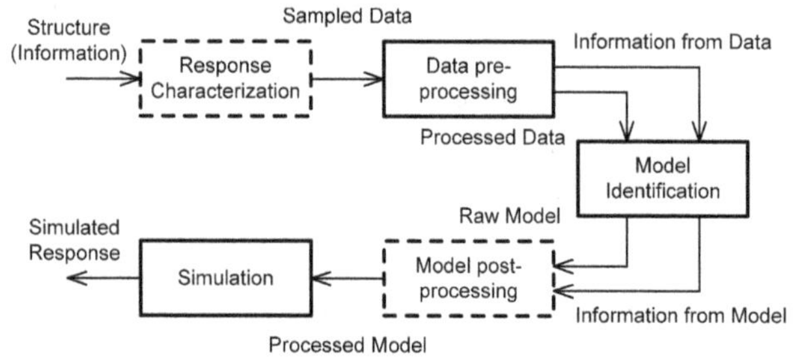

Fig. 2. Macromodeling-Simulation flow in SIM-DSP in a data-process perspective.

model, a pole-residue model, or a combination of specific bases. Then post-processing techniques are used to modify the macromodel characteristics and enhance the simulation performance (**Model Post-processing**), based on the macromodel and the peripheral information in the identification process. At last, the processed macromodel is used to produce spectra and waveforms numerically for SI analysis (e.g. eye diagram analysis and Bit Error Rate (BER) analysis), and/or converted into a circuit level model (e.g. SPICE equivalent circuit) or a mathematical model and coupled with other circuit model blocks for a global simulation (**Simulation**). In summary, SIM-DSP exploits nice features of discrete-time domain computations [16].

Currently, SIM-DSP framework has built up the following functional modules:

- Data pre-processing module for macromodeling configurations;
- System identification module for frequency-domain macromodeling;
- System identification module for time-domain macromodeling;
- Simulation module for real-time SI verifications;
- Modeling-Simulation module for time-domain SI verifications.

All these are captured and presented in the following subsections. Follow-up discussions are also shown in Section 3.6.

3.1 Data Pre-processing: Macromodeling Process Enhancement via Frequency Warping

To alleviate ill-conditioned computation problems in the linear-structure macromodeling process, we have proposed a spectral pre-processing scheme frequency warping. Frequency warping transforms the structure response, in order to introduce a numerically favorable fitting in the frequency domain, and to improve the fitting accuracy during the macromodeling process. The structure response is transformed by frequency warping which assigns an effective weighting in the frequency domain and gives a higher resolution (accuracy) in the desired region during computation.

3.2 Macromodeling: Macromodeling Framework Advancements via Discrete Time Domain Computation

To enhance the numerical robustness and functionality of the widely adopted, iterative based VF framework, we have proposed a family of improvements based on the root of the z-domain computation [23]. In the first part, we have shown some features and advantages of z-domain computation in the macromodeling process. Then, we have proposed the z-domain counterpart of VF (VFz), which uses z-domain partial fraction basis to seek a rational approximation to the desired response, namely,

$$\hat{f}(z) = \left(\sum_{b=1}^{B} \frac{c_b}{1 - z^{-1} a_b} \right) + d \approx f(z) \tag{3}$$

where B is the number of basis. This improves the numerical conditioning and convergence in broadband frequency-sampled system macromodeling.

In the second part, we have extended VFz as a functionality-oriented macromodeling algorithm. First, we have extended VFz to its time-domain (TD-VFz) [24] variants for the specific time domain macromodeling purpose. TD-VFz does not require discretizing the continuous-time convolution integrals for each iteration and it provides *a priori* model order selection through Hankel Singular Values (HSVs) computation as a pre-processing process. Second, we have also proposed the hybrid-domain (HD-VFz) variants, which model responses with a better hybrid-domain accuracy through providing extra informative data. Third, we have developed a versatile macromodeling adoption through a P-norm approximation expansion. We have modeled various structures, from chip-level to board-level, via VFz to demonstrate its excellent performance.

3.3 Macromodeling: Multiport Macromodeling Technique without Eigenvalue Computation and Initial Guess

To avoid numerically sensitive initial guess and expensive computation in time-domain macromodeling process, we have developed rational function macromodeling algorithms (VISA [25] and WISE [26]) for accuracy-oriented macromodeling. The idea is to regard the system response as the impulse response of a finite-impulse-response (FIR) filter, and then apply infinite-impulse-response (IIR) filter approximation techniques to generate the macromodel. By applying the idea of Walsh theorem and complementary signal, this approach can be interpreted as an interpolation problem. This approach can be interpreted as a numerically simple, non-pole-based Steiglitz-McBride (SM) iteration without initial guess and eigenvalue computation.

3.4 Simulation: Real-time Simulation for Emerging Interconnect System SI Verifications

Computed macromodels are used to generate responses for later-stage simulations and verifications. A visual simulation platform is realized using Matlab Simulink, for multilevel SI simulations and analysis, because of the following reasons,

1. Simulations can be easily coupled with other physical-domain blocks and simulation platforms through a simple interface, in order to model mechanical, thermal and other physical characteristics of nano-scale circuits and give a more realistic transient simulation.
2. Complex models can be segmented into hierarchies of design components (blocks), with levels of abstractions. Thus, this helps the division of labor and shortens the design cycle.
3. Tested signal in the simulation process can be generated and monitored in a real-time manner. The simulation process becomes convenient to examine simulation results, analysis the performance and diagnose unexpected and inconsistent behaviors in the simulation. Thus, the design cycle can be shortened.
4. Signal properties, such as clock jitter, signal delay and signal noise, can be modeled with supplemental components and a quick configuration, for a more realistic simulation.

3.5 Modeling-Simulation: Single-Step Modeling-Simulation via Discrete Cosine Transform Truncation (DCTT)

Different macromodeling techniques have been proposed. However, most of them are based on iterative methods with fundamental drawbacks. On the other hand, impulse response models have been introduced for SI simulation with the help of efficient convolution methods [27], [28]. However, impulse response models contain lots of parameters, which are not effective for model storing and response computation. Therefore, a compacted model formulation is proposed in this section.

Discrete Cosine Transform (DCT), which is a generalization of the Discrete Fourier Transform (DFT), has been proposed to represent the (discrete) time-sampled sequences using a set of energy-compacted, orthogonal, real-valued basis sequences [29]. In general, DCT is superior to all other discrete transforms in terms of accuracy [29]. In the mean time, DCT inherits some properties in DFT, such as basis orthogonality and unity, while the symmetrical extension property in DCT reduces the abruptness of truncation, when compared to the DFT. Therefore, it has been used in audio/visual signal compression and coding [29], as well as combinational circuit current simulation [30].

In this section, a discrete cosine transform truncation (DCTT) method is proposed for time- and frequency-sampled modeling-simulation in the discrete-time domain, in which the system response is modeled by a small number of DCT bases instead of a rational model. Two main features of DCTT are:

1. DCTT handles the macromodeling problem as a single-step signal compression problem instead of a numerical-sensitive system identification problem, which avoids many drawbacks in the iterative approach.
2. DCTT provides an exact *a priori* error analysis through Parseval's theorem for model size selection.

3.5.1 Decomposition and Reconstruction of Sampled Responses using DCT

Assuming a N-point (discrete) time-sampled output signal sequence is given ($x[n]$ for $n = 0, 1, \ldots, N-1$), and the input signal sequence is a normalized impulse response ($i[0] = 1$ and $i[n] = 0$ for $n = 1, \ldots, N-1$), we would approximate the finite-length time-sampled signal sequences using a truncated set of orthogonal spectral bases $\phi_k[n]$

$$x[n] = \sum_{k=0}^{N-1} X[k]\,\phi_k[n]. \tag{4}$$

Having assumptions of both periodicity ($x[n] = x[nT]$ for $n = 0, 1, \ldots, N-1$, where T is the sampling period) and even symmetry (i.e., a symmetrically extended sequence x_2 exists, where $x_2[n] = x[n]$ for $n = 0, 1, \ldots, N-1$ and $x_2[n] = x[2N-1-n]$ for $n = N, N+1, \ldots, 2N-1$), DCT-II (the most commonly used DCT) uses periodic cosine function as basis. DCT decomposes the sampled response signal into a summation of DCT bases

$$X_{DCT}[k] = \sqrt{\frac{2}{N}}\beta[k] \sum_{n=0}^{N-1} x[n] \cos\left(\frac{\pi k(2n+1)}{2N}\right), \tag{5}$$

for $k = 0, 1, \ldots, N-1$, where β is a normalization factor, and $\beta[k] = \frac{1}{\sqrt{2}}$ for $k = 0$ and 1 for $k = 1, 2, \ldots, N-1$. The response can be reconstructed using the inverse discrete cosine transform (IDCT)

$$x[n] = \sqrt{\frac{2}{N}}\beta[k]\sum_{k=0}^{N-1} X_{DCT}[k]\cos\left(\frac{\pi k(2n+1)}{2N}\right), \quad (6)$$

for $n = 0, 1, \ldots, N-1$.

Meanwhile, an N-point DCT X_{DCT} is closely related to a $2N$-point DFT X_{DFT} of the N-point signal sequence $x[n]$,

$$X_{DCT}[k] = 2\Re\left\{X_{DFT}[k]e^{-j\pi k/(2N)}\right\}, \quad (7)$$

for $k = 0, 1, \ldots, N-1$. The relation from $X_{DCT}[k]$ to $X_{DFT}[k]$ is not as straightforward as from $X_{DFT}[k]$ to $X_{DCT}[k]$, but a relationship can be derived,

$$X_{2-DFT}[k] = \begin{cases} X_{DCT}[0] & k = 0, \\ e^{j\pi k/2N} X_{DCT}[k] & k = 1, \ldots, N-1, \\ 0 & k = N, \\ -e^{j\pi k/2N} X_{DCT}[2N-k] & k = N+1, \ldots, 2N-1 \end{cases} \quad (8)$$

where $X_{2-DFT}[k]$ is the $2N$-point DFT of the $2N$-point symmetrically extended sequence $x_2[n]$ of the response sequence $x[n]$. If inverse DFT is applied to X_{2-DFT}, we obtain

$$x_2[n] = \frac{1}{2N}\sum_{k=0}^{2N-1} X_{2-DFT}[k]e^{j2\pi kn/(2N)} \quad (9)$$

and $x[n] = x_2[n]$ for $n = 0, 1, \ldots, N-1$, then we can apply DFT to obtain X_{DFT} (frequency-sampled response).

3.5.2 Response Decomposition and Basis Truncation in DCTT

As discussed in the previous section, we can obtain DCT coefficients from arbitrary frequency-sampled data (DFT coefficients) or arbitrary time-sampled impulse response through (7) or (5), respectively. Similar to DFT, according to the Parseval's theorem, the total energy contained in the time-sampled signal is equal to the total energy of the DCT bases. Therefore, the signal energy can be separated into the energy of an approximant signal E_{DCT} and the energy of an error signal E_{error}, namely,

$$\sum_{n=0}^{N-1}|x(n)|^2 = \frac{1}{N}\sum_{n=0}^{N-1}\beta(k)|X_{DCT}(k)|^2 = \underbrace{\frac{1}{N}\sum_{n=0}^{N_{pr}-1}\beta(k)|X_{DCT}(k)|^2}_{E_{DCT}} + \underbrace{\frac{1}{N}\sum_{n=N_{pr}}^{N-1}\beta(k)|X_{DCT}(k)|^2}_{E_{error}} \quad (10)$$

where N_{pr} is the number of the preserved DCT bases. Therefore, the exact error E_{error} can be calculated from the truncated bases, which allows *a priori* error analysis and a perceptual model size selection to facilitate the macromodeling process. Furthermore, as DCTT uses real-coefficient bases, the algorithm generates real-valued output signal for real-valued input signal with truncation of arbitrary number of bases. Also, DCTT avoids extra considerations (and computations) in rational-function fitting approaches to handle response with complex conjugate poles as DCTT does not model the response using its pole information.

In general, DCTT reduces the system order significantly for lowpass-response signals, which is common for many physical systems. It is also shown that DCT is nearly optimum in the sense of minimum mean-squared truncation error for sequences with exponential correlation functions [29], which makes DCTT superior in modeling exponentially decaying physical system responses.

3.5.3 Response Reconstruction in DCTT

As discussed in the previous section, the response signal can be perfectly reconstructed by (6). By truncating insignificant bases, i.e., $X_{DCT}[k] := 0$ for k with small $|X_{DCT}[k]|^2$, the time-sampled signal can be almost fully reconstructed, whereas insignificant bases are not calculated and stored. Since all computations in DCTT involve only multiplications (no root-finding nor eigenvalue calculation), it is easy to apply in the high-level simulation environment (e.g., Verilog-A and Matlab). The continuous-time system response can be obtained from the discrete-time system response through a stability- and passivity-preserving bilinear transform ($z = e^{sT} \approx (1+sT/2)/(1-sT/2)$).

For electrically long structures (e.g., broad-level transmission line), time delay may exist and cause difficulties in modeling and simulation. In DCTT, to model responses with time delay, the peripheral time delay can be included artificially during the simulation process, or adopted into the DCT basis parameters. The adoption with a unit time delay is described by

$$X_{shift}[m] = \cos\left(\tfrac{m\pi}{N}\right)X_{DCT}[m] + \sin\left(\tfrac{m\pi}{N}\right)X_{DST}[m] + \tfrac{2}{N}\beta[k]\left[(-1)^m x[n] - x(0)\right]\cos\left(\tfrac{m\pi}{2N}\right) \quad (11)$$

where $X_{DST}[m]$ is the Discrete Sine Transform (DST) of the signal $x(n)$, which can be calculated and adopted into DCT parameters during the decomposition stage. An arbitrary time shift can be similarly adopted through the recursion of (11).

3.5.4 Computation Complexity

Similar to DFT, by exploiting the Fast Fourier Transform (FFT) approach, the relationship between DCT and DFT, factorization and the real-valued coefficient property, the computation can be reduced to $N\log N - 3N/2$ real multiplications for a complete DCT. Different implementations have been proposed for different tradeoff [29]. The computation is further reduced in IDCT as most of the DCT bases in the reduced macromodel have been truncated to zero.

Comparing to conventional approaches, DCTT is a non-iterative approach, which is numerically simple (i.e., accurate and efficient) in computation. On the other hand, the computation complexity of the iterative macromodeling approach (e.g. VF(z) and WISE) is $O\left(kM^2N\right)$, where M is the model order size, and k is the number of the iterations, which is the computation bottleneck in the iterative approach. Furthermore, DCTT avoids expensive root or eigenvalue calculation, overhead computation (e.g. pole stabilization) and manual effort for selection of model order and number of iterations.

3.5.5 Remarks in Using DCT Basis

Some remarks are in order:

1. Karhunen-Loeve transform (KLT) is shown to be an ideal transform [31], where the most signal energy is contained in the fewest number of bases. However, KLT involves expensive computations to determine the basis. Therefore, discrete cosine transform (DCT) has been proposed. DCT is shown to be the best approximation to KLT among different kinds of transforms in most situations [29].
2. By considering the transform as a generalized Wiener filtering problem, DCT is shown to be less disturbed by additive noise, comparing to other transforms [29]. The uncorrelated white noise signal only affects small-magnitude DCT bases, which will be truncated during macromodeling. The numerically robust performance of DCTT is shown in the numerical examples.
3. To achieve a more accurate modeling of highpass or arbitrary response, warped DCT (WDCT) can be applied for response compensation [32], which bilinearly maps the response to a weighted response for numerically favorable computations. This further enhances the universal performance of DCTT.

3.6 Remarks in SIM-DSP

Some remarks are in order:

1. The purpose of introducing VISA is for accuracy-oriented macromodeling. Its compact formation and specialized time domain properties adoption makes VISA a numerically simple (and thus efficient and accurate) time-domain macromodeling algorithm. These specialties also make VISA difficult for generalizations, such as developing its frequency domain variants. On the other hand, VFz are purposed for the functionality-oriented macromodeling development. As a result, VFz and its variants provide different generalizations for various applications. Meanwhile, DCTT is also introduced for a non-iterative, numerical robust modeling-simulation of arbitrary port-to-port responses. DCTT requires only a single-step computation without knowing the location of poles, which is much less numerical sensitive to iterative-based computation. However, DCTT is needed to introduce a more direct and convenient interface for ease of simulations.
2. Developed modules may not generate a passivity-guaranteed macromodel. However, passivity check/enforcement techniques in continuous-time domain (as shown in [33]) can be used to rectify the model, since the bilinear continuous-to/from-discrete transformation is passivity-preserving [34]. Furthermore, discussions about passivity in z-domain system can be found in [34].
3. The purpose of applying digital signal processing (DSP) is to represent, transform and manipulate digitized (discretized) signals and information in an organized and meaningful manner [35]. In SIM-DSP, we have leveraged on features of (pre-)processing (Section 3.1), computation in a transformed domain (from Section 3.2 to Section 3.5), sampling (Section 3.1) and architectures (Section 3.3 and 3.5) in the DSP area to benefit VLSI macromodeling and simulation process in a well-rounded perspective.

4. Case Studies Using SIM-DSP Framework

To illustrate the performance of SIM-DSP framework, we model responses of several practical testbenches using SIM-DSP. The proposed framework is coded in Matlab m-script (text) files and runs in the Matlab 7.4 environment on a 1GB-RAM 3.4GHz PC. Table 1 summarizes the implementation analysis of all the examples. A detailed macromodeling-simulation process is shown in the first example. Other examples show the performance of the core macromodeling module using VFz and DCTT.

4.1 A Detailed Macromodeling-Simulation Process of a Four-port Backplane using VFz

A 520 mm differential transmission channel (including Tyco line cards) on a full mesh ATCA Kaparel backplane is designed using the Tyco Quadroute technique [36]. The standard 4-port scattering parameters (4-port) are measured using a vector network analyzer (VNA) over 1497 frequencies ranging from 50 MHz to 15 GHz. The frequency-sampled response of some typical time-delayed port responses are extracted. Each response is fitted using VFz with a 130-pole approximant. Each response takes about 10 iterations and 12 seconds for VFz to reach convergence. Fig. 3 plots the fitted responses. VFz gives an accurate identification result. The implementation analysis is summarized in Tab. 1.

This example shows the numerical difficulties of approximating time-delayed responses using ordinary system

	No. of points per response	No. of responses	Frequency range (Hz)	Types of sampled response	No. of poles	Average relative error	CPU time (sec.)
4-port (1,1)	1497	1	50 M - 15 G	Scattering with delay	130	0.1123	10
4-port (2,1)	1497	1	50 M - 15 G	Scattering with delay	130	0.0928	10
4-port (4,1)	1497	1	50 M - 15 G	Scattering with delay	130	0.0755	10
IEEE_802 (3,1)	1496	1	50 M - 15 G	Scattering	136	0.0177	16
IEEE_802 (4,1)	1496	1	50 M - 15 G	Scattering	136	0.0127	26
INC	1286	196/54	10 k - 9 G	Admittance	100	0.0189	271

Tab. 1. Implementation summary of different macromodeling examples.

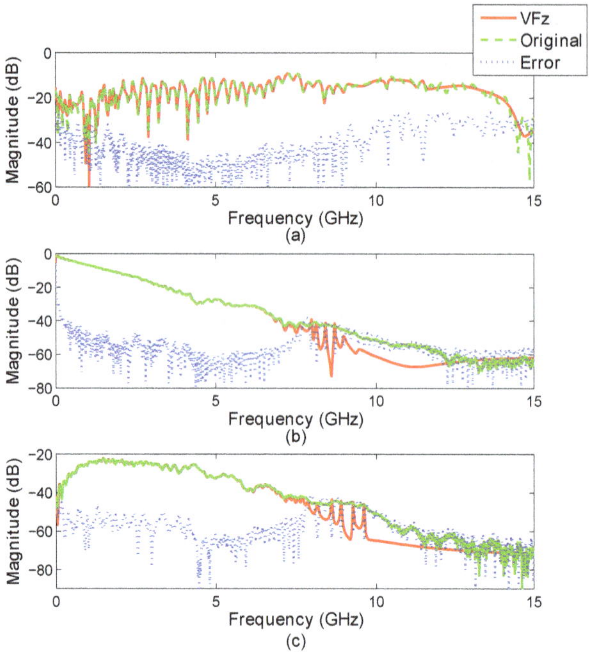

Fig. 3. Magnitude responses of the four-port backplane (4-port) macromodeling: (a) Port (1,1), (b) Port (2,1) and (c) Port (4,1).

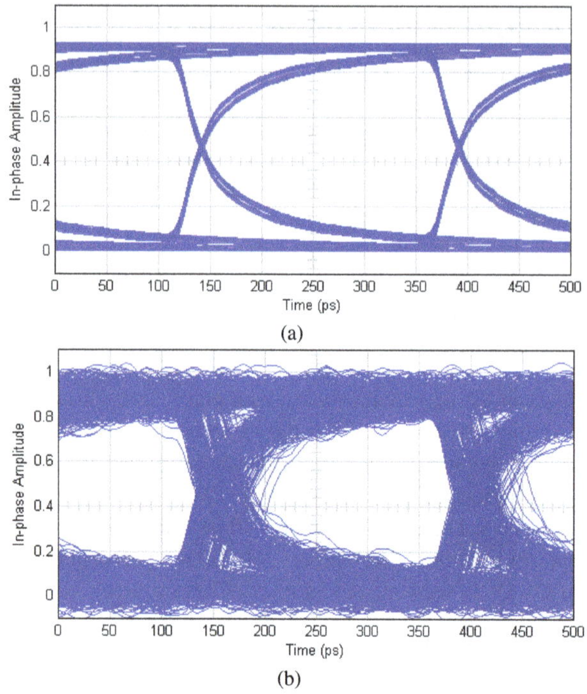

Fig. 4. Eye diagram of the channel: (a) uncorrupted input bit sequence, and (b) input bit sequence with Gaussian noise and Gaussian signal delay.

identification techniques and rational function macromodel. In some cases, interconnects are considered as electrically long for high-frequency circuits with time delay in the propagation of the signals. To handle this situation, delay extraction pre-processing is needed, or the response should be modeled using delay-based macromodels.

Fig. 3 shows the characteristics of the transmission channel. In Fig. 3(a) (Port (1,1)), we can observe a smaller (-20 dB) and then a larger return loss (-9 dB) in the lower- and higher-frequency reflection, respectively. An accurate identification is needed for the higher-frequency reflection because the upper envelopes characterize the overall worst-case reflection and the channel performance. In Fig. 3(b) (Port (2,1)), we can observe a visually smooth response with many small ripples, which is a typical measured insertion loss response. Fig. 3(c) (Port (4,1)) shows the crosstalk between channels, i.e., the coupling of signal from the aggressor to the victim. There is a -40 dB (1 %) coupling in the low frequency operation, and the coupling becomes -22 dB (8 %) when the frequency becomes 1.5 GHz. In high-speed digital transmission, 1 % crosstalk can cause a significant deterioration and even a failure. Therefore, accurate channel identification and differential signaling are common practices in the high-speed electronic system design.

The obtained macromodel is used for a transient simulation on the Matlab-Simulink platform, as shown in Fig. 5. The model is excited by an uncorrupted 4000-bits random bit sequence (Bernoulli binary signal) with a pulse period of 5 ps (i.e., 20 Gbps bit rate). Also, the model is excited by the same configured bit sequence with noise (Gaussian distribution with zero mean and 0.03 variance) and signal delay (Gaussian distribution with zero mean and 15 variance). Eye diagrams are generated from the obtained signal. The eye diagram is updated simultaneously during the computation, and the final eye diagram is computed within 20 seconds. The eye diagram result in Fig. 5 shows that the digital signal is deteriorated during the transmission, and the corrupted signal can probably lead to an accidental signal mis-triggering. This means that SI analysis is necessary for modern high-speed circuit system design.

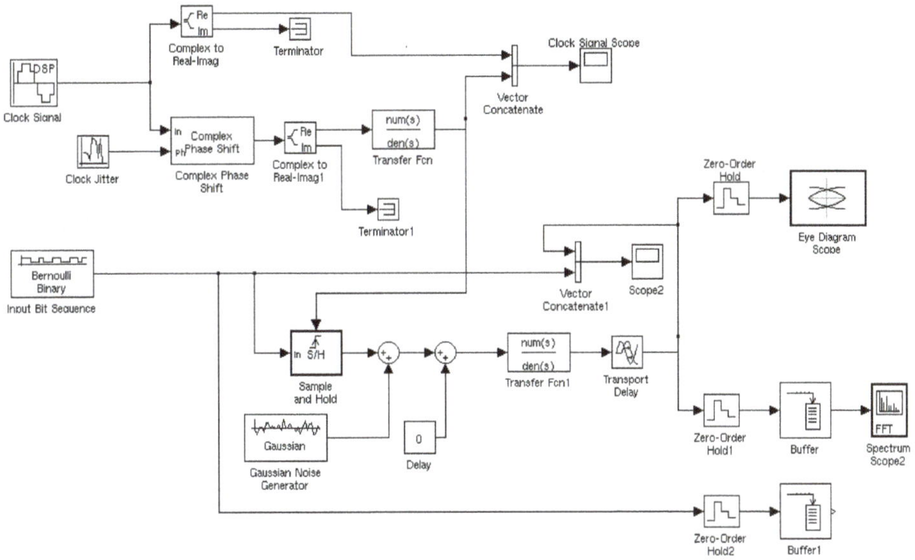

Fig. 5. Configuration of the backplane transient simulation in Matlab-Simulink, with clock jitter and transmission delay.

4.2 Macromodeling of the Crosstalk of a Four-port Backplane Using VFz

An Intel ATCA Ethernet backplane test system is designed, fabricated and measured for the IEEE 802.3 standardization. The backplane system contains line cards with via stubs, backplane, connectors and AC coupling capacitors. There are eight channels from three signaling layers in the backplane. The 4-port scattering parameters of the top signal signaling layers (IEEE_802) are measured at 1496 frequencies ranging from 50 MHz to 15 GHz. Frequency-sampled signals of some typical port responses are extracted. Each response is fitted using VFz with a 136-pole approximant. It takes about 12 iterations (16 seconds) and 20 iterations (26 seconds) for VFz to reach convergence of port response (3,1) and port response (4,1), respectively. Fig. 6 plots the fitted responses. The implementation analysis is summarized in Table 1. These promising results demonstrate the superiority of VFz in modeling board-level interconnect systems.

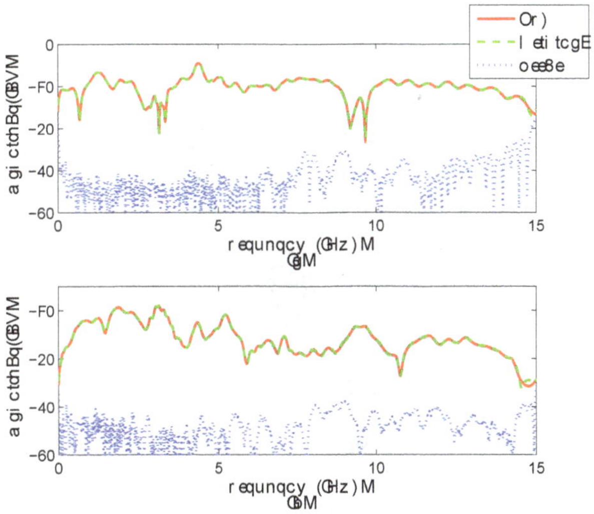

Fig. 6. Magnitude responses of the four-port Intel backplane (IEEE_802) macromodeling: (a) Port (3,1) and (b) Port (4,1).

4.3 Macromodeling of a Power Distribution Network of a System-In-Package Board Using VFz

The tested System-In-Package (SIP) intelligent network communicator (INC) board contains digital, radio frequency (RF) and optoelectronic sections on a single 83 mm × 65 mm test bed [15]. The board has two FPGA chips, one multiplexer (MUX), one RF amplifier, one signal layer, one irregularly shaped ground layer and one irregularly shaped power plane layer. A 14 × 14 admittance parameter matrix is generated. Fig. 7 plots the signal energy distribution of the approximated system, from which we can see that some pairs of ports are not coupled with each other. By extracting the coupled responses from all the responses, the approxi-

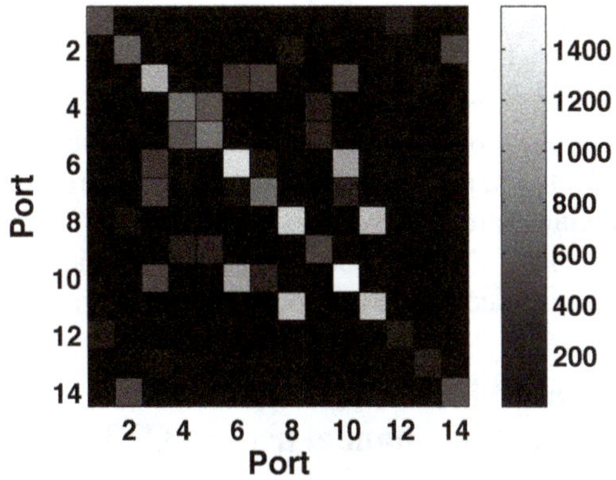

Fig. 7. Signal energy distribution of the power distribution network (INC) model.

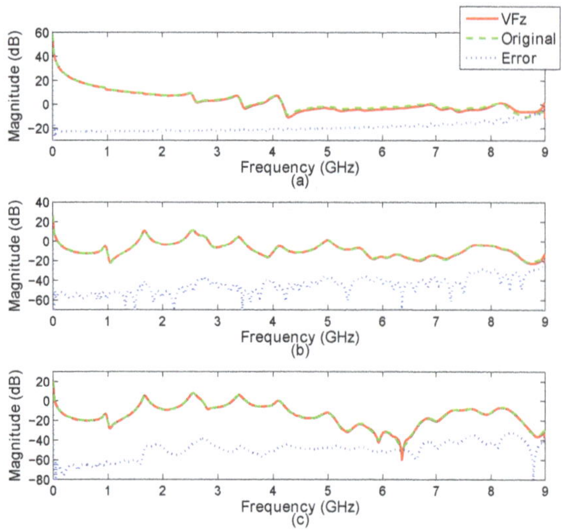

Fig. 8. Magnitude responses of the power distribution network (INC) macromodeling: (a) Port (6,10), (b) Port (7,13), and (c) Port (10,13).

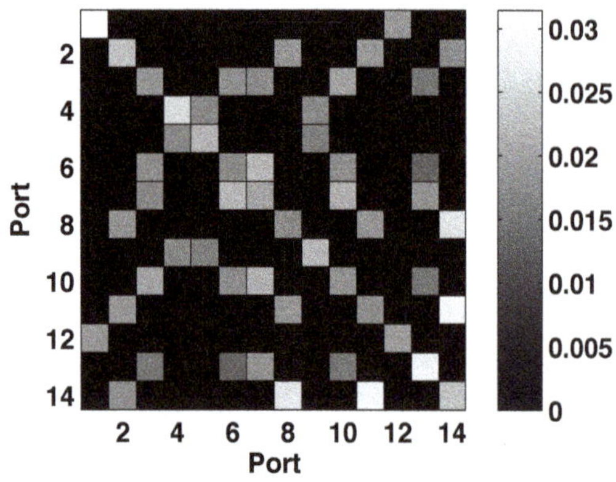

Fig. 9. Relative error energy distribution of the power distribution network (INC) macromodel.

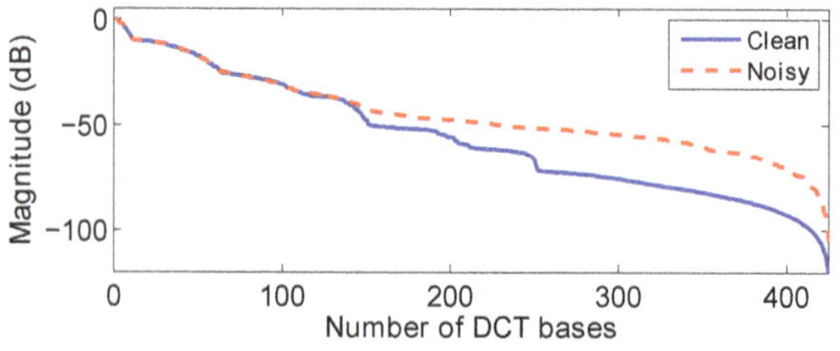

Fig. 10. Magnitude of the DCT bases of clean and noisy channels in the descending order.

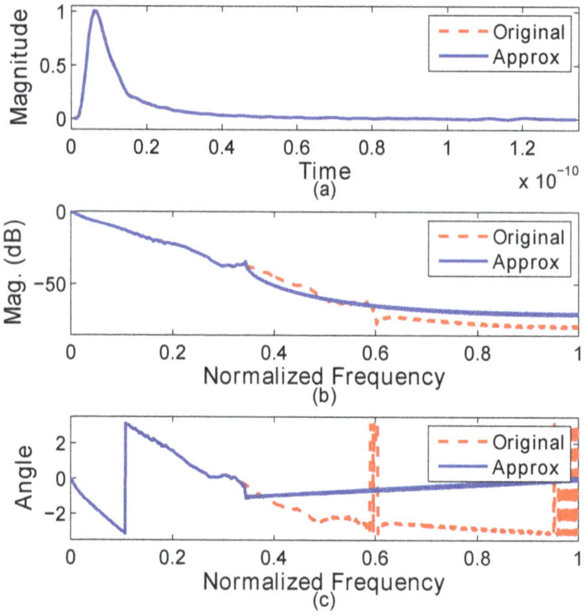

Fig. 11. (a) Time impulse responses, (b) magnitude responses and (c) phase responses of the clean differential channel example in the discrete-time frequency.

mation is significantly reduced from fitting 196 responses to fitting 54 responses. The frequency-sampled responses are fitted using VFz with a 100-pole approximant. It takes 271 seconds and 5 iterations for VFz to converge, with the average relative error being 0.0189. The implementation analysis is summarized in Table 1. Fig. 8 plots approximation of some typical responses. Fig. 9 plots the distribution of the relative error of energy, and shows that the error distribution is independent to the energy distribution. This large-scale macromodeling example has a large amount of response data to be fitted, but VFz is computationally well-conditioned and gives an accurate approximation within a few iterations.

4.4 Modeling-Simulation Process of a Differential Channel using DCTT

The backplane example in Section 4.1 is used to show the efficiency and accuracy of DCTT. The example arises from modeling a differential transmission channel. The time-domain response is generated, normalized and fitted using DCTT. Time samples are taken at 0.67 ps intervals for the first 425 points (0.28 ns). The algorithm requires 0.023 seconds to decompose the signal into its DCT bases. As shown by the DCT basis energy distribution in Fig. 10, most energy is compacted by a small amount of DCT bases, therefore the least significant DCT bases can be truncated without loss of accuracy. DCTT requires 141 and 109 DCT bases to model the signal with a 1 % and 3 % relative error $\left(\|\text{error energy}\|_2 / \|\text{signal energy}\|_2\right)$, respectively. Fig. 11 plots the normalized discrete-time frequency-domain responses and the normalized time-domain responses of the approximant with a 1 % relative error, demonstrating the excellent fitting accuracy in both time and frequency (magnitude and phase) domains. Given a 2^{17}-points input signal sequences, DCTT requires 0.08 seconds to calculate the output response sequence, and the transient response and the eye diagram of the channel are shown in Fig. 12. The system is also modeled using an eigenvalue-calculation-free algorithm (VISA [25]) using a 45th-order rational-function macromodel and 20 (converged) algorithm iterations. The quantitative data are shown in Tab. 2, which shows a significant ($\sim 9.94\times$) speed-up using DCTT due to its efficient single-step calculation.

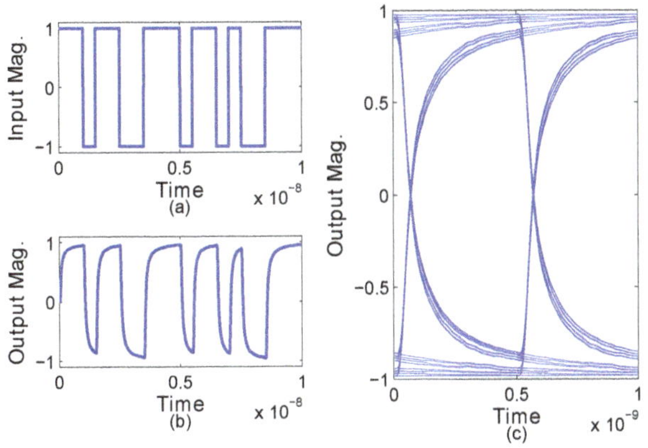

Fig. 12. (a) Time-domain random high-frequency input signal (the first 2000-points), (b) Output response of the differential channel using DCTT modeling (the first 2000 points) and (c) eye diagram of the modeled system using DCTT (2^{17}-points).

Next, we study the robustness of DCTT. First, we repeat the example with the response corrupted by white noise under a signal-to-noise ratio (SNR) of -30 dB ($\sim$ 3.14 % error). DCTT requires 116 DCT bases (7 additional DCT bases) to model the response with a 3 % relative error. The magnitude of DCT bases are shown in Fig. 10. As shown in figure, the noise signals are decomposed and spread in the small-magnitude basis region, and most of them are truncated. Second, the differential channel example is repeated with different signal sequence lengths (longer impulse response tail). The quantitative result is shown in Tab. 3, which shows that there is a slight increase in computation time, and an increase of the required bases for longer sequences due to a wider spread of the DCT basis distribution.

	Channel			Crosstalk		Serial
	WISE	DCTT	DCTT (noisy)	WISE	DCTT	DCTT
Var #	45	141	118	55	251	664
Rel. err.	0.011	0.010	0.030	0.028	0.0298	0.030
Time (s)	0.348	0.035	0.036	0.417	0.062	0.077

Tab. 2. Comparison between DCTT and other algorithms in the examples.

Sequence length	200	425	600	800	1000
Number of var.	66	141	198	264	330
CPU Time (sec.)	0.031	0.035	0.039	0.042	0.046

Tab. 3. Comparison of DCTT in the differential channel example with different response sequence lengths.

4.5 Modeling-Simulation Process of a Crosstalk Response and a Serial-Link Type Response using DCTT

The example arises from modeling the bandpass-response crosstalk between transmission channels on the backplane in Section 4.1. The result is shown in Fig. 13 and Tab. 2. Fig. 13 shows that DCTT accurately approximates the response with a higher magnitude (region near $0.7 f_s$, where f_s is the normalized sampling frequency).

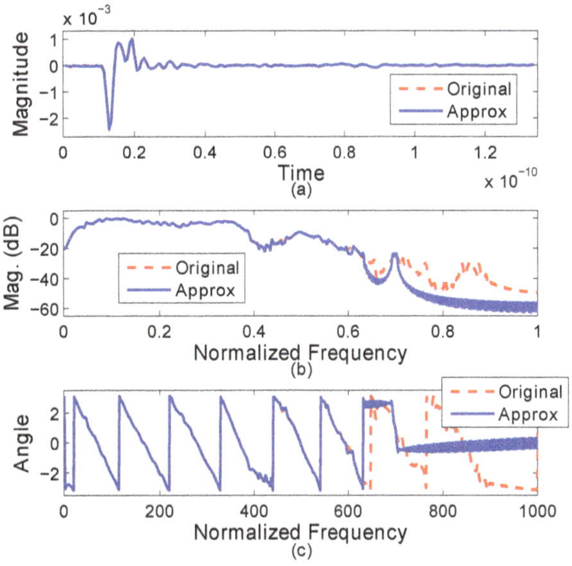

Fig. 13. (a) Time impulse responses, (b) magnitude responses and (c) phase responses of the crosstalk example in the discrete-time frequency.

Lastly, a complicated 2000-point impulse response of a serial-link type response is created and imported for macromodeling, which cannot be approximated effectively by iterative pole-finding calculation approaches. The modeling results are shown in Fig. 14 and Tab. 2, DCTT can accurately model complicated and long responses. In summary, it is a good choice to use non-iterative, numerically simple DCTT for the macromodeling process.

5. Conclusions

SIM-DSP framework has been presented for the macromodeling-simulation process. We have developed a digital signal processing (DSP) enhanced macromodeling-simulation process for signal integrity (SI) verifications. In particular, SIM-DSP framework, which is assembled by our core identification modules and peripheral modules, has improved the accuracy and functionality of the macromodeling process, through the validation from numerical examples. Furthermore, DCTT is also introduced for a single-step, numerical robust modeling-simulation of arbitrary port-to-port responses. Non-linear macromod-

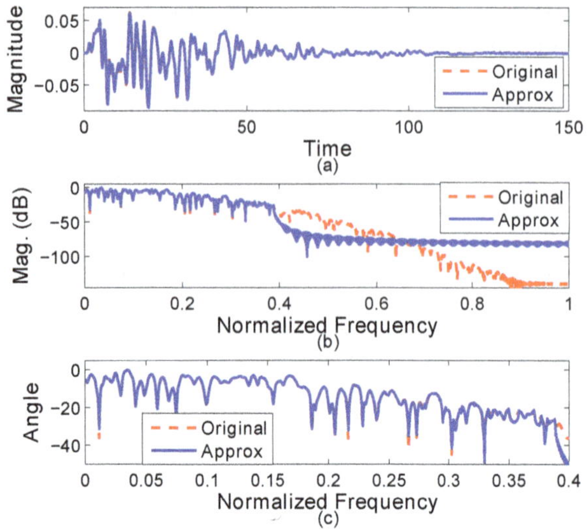

Fig. 14. (a) Time impulse responses, (b) magnitude responses and (c) low-frequency magnitude responses of the serial link-type example in the discrete-time frequency.

eling algorithms and multi-rate macromodeling-simulation process can be developed in the future to meet the simulation requirements of emerging circuit techniques.

References

[1] DAVIS, J. A., VENKATESAN, R., KALOYEROS, A., et al. Interconnect limits on gigascale integration (GSI) in the 21st century. *Proceedings of the IEEE*, 2001, vol. 89, no. 3, p. 305 - 324.

[2] GOUDOS, S. Calculation and modeling of EMI from integrated circuits inside high-speed network devices. *Radioengineering*, 2006, vol. 15, no. 4, p. 2 - 8.

[3] LI, E.-P., WEI, X.-C., CANGELLARIS, A. C., LIU, E.-X., ZHANG, Y.-J., D'AMORE, M., KIM, J., SUDO, T. Progress review of electromagnetic compatibility analysis technologies for packages, printed circuit boards, and novel interconnects. *IEEE Transactions on Electromagnetic Compatibility*, 2010, vol. 52, no. 2, p. 248 - 265.

[4] ARITRA, A., BANERJEE, S., BANERJEE, J. P. Large-signal simulation of 94 GHz pulsed silicon DDR IMPATTs including the temperature transient effect. *Radioengineering*, 2012, vol. 21, no. 4, p. 1218 - 1225.

[5] HANY, F., CHEN, W., PISSOORT, D., BADESHA, A. Virtual-EMI lab: Removing mysteries from black-magic to a successful front-end design. In *Proceedings of IEEE Workshop on Signal Propagation on Interconnects*. Hildesheim (Germany), 2010, p. 105 - 108.

[6] SANATHANAN, C., KOERNER, J. Transfer function synthesis as a ratio of two complex polynomials. *IEEE Transactions on Automatic Control*, 1963, vol. 8, no. 1, p. 56 - 58.

[7] STEIGLITZ, K., MCBRIDE, L. E.A technique for the identification of linear systems. *IEEE Transactions on Automatic Control*, 1965, vol. 10, no. 4, p. 461 - 464.

[8] GUSTAVSEN, B., SEMLYEN, A. Rational approximation of frequency domain responses by vector fitting. *IEEE Transactions on Power Delivery*, 1999, vol. 14, no. 3, p. 1052 - 1061.

[9] CERNY, D., DOBES, J. Common lisp as simulation program (CLASP) of electronic circuits. *Radioengineering*, 2011, vol. 20, no. 4, p. 880 - 889.

[10] *Official website of MATLAB RF toolbox.* [Online]. Available at: http://www.mathworks.com/products/rftoolbox/

[11] *Official website of E-System Design.* [Online]. Available at: http://www.e-systemdesign.com/

[12] ELFADEL, I. M., ANAND, M. B., DEUTSCH, A., et al. AQUAIA: A CAD tool for on-chip interconnect modeling, analysis, and optimization. In *Proceedings of IEEE Electrical Performance of Electronic Packaging Conference*. Monterey (CA, USA), 2002, p. 337 - 340.

[13] ELFADEL, I. M., DEUTSCH, A., KOPCSAY, G. V., RUBIN, B. J., SMITH, H. H. A CAD methodology and tool for the characterization of wide on chip buses. *IEEE Transactions on Advanced Packaging*, 2005, vol. 28, no. 1, p. 63 - 70.

[14] MIN, S. H., SWAMINATHAN, M. Construction of broadband passive macromodels from frequency data for simulation of distributed interconnect networks. *IEEE Transactions on Electromagnetic Compatibility*, 2004, vol. 46, no. 4, p. 544 - 558.

[15] MIN, S. H. *Automated Construction of Macromodels from Frequency Data for Simulation of Distributed Interconnect Networks*, Ph.D. dissertation. Georgia Institute of Technology (USA), 2004.

[16] NAREDO, L., RAMIREZ, A., AMETANI, A., et al. Z-transform-based methods for electromagnetic transient simulations. *IEEE Transactions on Power Delivery*, 2007, vol. 22, no. 3, p. 1799 - 1805.

[17] BRENNER, P. A general modeling approach for linear circuit blocks and passive multiport components. In *Proceedings of IEEE European Packaging Workshop*, 2007.

[18] DENK, G., FELDMANN, U. Circuit simulation for nanoelectronics. In *From Nano to Space*. Springer, 2008, p. 11 - 26.

[19] LEI, C. U. *VLSI Macromodeling and Signal Integrity Analysis via Digital Signal Processing Techniques*, Ph.D. dissertation. Hong Kong: University of Hong Kong, 2011.

[20] BOGATIN, E. *Signal and Power Integrity – Simplified*. Upper Saddle River (NJ, USA): Prentice Hall, 2010.

[21] RESSO, M., BOGATIN, E. *Signal Integrity Characterization Techniques*. International Engineering Consortium, 2009.

[22] YU, W. *Electromagnetic Simulation Techniques Based on the FDTD Method*. Wiley, 2009.

[23] LEI, C. U. Exploiting implicit information from data for linear macromodeling. *IEEE Transactions on Components, Packaging and Manufacturing Technology*, 2013, vol. 3, no. 9, p. 1570 - 1577.

[24] LEI, C. U., WONG, N. Efficient linear macromodeling via discrete-time time-domain vector fitting. In *Proceedings of International Conference on VLSI Design*. 2008, p. 469 - 474.

[25] LEI, C. U., WONG, N. VISA: Versatile Impulse Structure Approximation for time-domain linear macromodeling. In *Proceedings of Asia and South Pacific Design Automation Conference*. Taipei (Taiwan), 2010, p. 37 - 42.

[26] LEI, C. U., WONG, N. WISE: Warped Impulse Structure Estimation for time-domain linear macromodeling, *IEEE Transactions on Components, Packaging and Manufacturing Technology*, 2012, vol. 2, no. 1, p. 131 - 139.

[27] BASEL, M., STEER, M., FRANZON, P. Simulation of high speed interconnects using a convolution-based hierarchical packaging simulator. *IEEE Transactions on Components, Packaging, and Manufacturing Technology, Part B: Advanced Packaging*, 1995, vol. 18, no. 1, p. 74 - 82.

[28] ROY, S., DOUNAVIS, A. Transient simulation of distributed networks using delay extraction based numerical convolution. *IEEE Transactions on Components, Packaging, and Manufacturing Technology, Part B: Advanced Packaging*, 2011, vol. 30, no. 3, p. 364 - 373.

[29] RAO, K. R., YIP, P. *Discrete cosine transform: algorithms, advantages, applications*. Academic Press Professional, 1990.

[30] BODAPATI, S., NAJM, F. N. High-level current macro model for logic blocks. *IEEE Transactions on Computer-Aided Design of Integrated Circuits and Systems*, 2006, vol. 25, no. 5, p. 837 - 855.

[31] AHMED, N., RAO, K. R. *Orthogonal Transforms for Digital Signal Processing*. Springer, 1975.

[32] CHO, N. I., MITRA, S. K. Warped discrete cosine transform and its application in image compression. *IEEE Transactions on Circuits and Systems for Video Technology*, 2000, vol. 10, no. 8, p. 1364 - 1373.

[33] GRIVET-TALOCIA, S., UBOLLI, A. A comparative study of passivity enforcement schemes for linear lumped macromodels. *IEEE Transactions on Advanced Packaging*, 2008, vol. 31, no. 4, p. 673 - 683.

[34] SMITH, J. O. *Physical Audio Signal Processing for Virtual Musical Instruments and Audio Effects*. W3K Publishing, 2010.

[35] VENKATARAMANI, B., BHASKAR, M. *Digital Signal Processors: Architecture, Programming and Applications*. Boston (MA, USA): McGraw-Hill Higher Education, 2002.

[36] ZENG, R. X., SINSKY, J. H. Modified rational function modeling technique for high speed circuits. In *Proceedings of MTT-S International Microwave Symposium Digest*. 2006, p. 1951 - 1954.

About Authors...

Chi-Un LEI received his B.Eng. (first class honors) and Ph.D. in Electrical and Electronics Engineering from the University of Hong Kong in 2006 and 2011, respectively. He is now a Honorary Assistant Professor and a Research Scientist at the same university. His research interests include VLSI signal integrity analysis, cyber physical systems, learning analytics and engineering education.

15

A Versatile Active Block: DXCCCII and Tunable Applications

Sezai Alper TEKİN[1], *Hamdi ERCAN*[2], *Mustafa ALÇI*[3]

[1] Dept. of Industrial Design Engineering, Erciyes University, 38039, Kayseri, Turkey
[2] Dept. of Avionics, Erciyes University, 38039, Kayseri, Turkey
[3] Dept. of Electrical and Electronics Engineering, Erciyes University, 38039, Kayseri, Turkey

satekin@erciyes.edu.tr, hamdiercan@erciyes.edu.tr, malci@erciyes.edu.tr

Abstract. *The study describes dual-X controlled current conveyor (DXCCCII) as a versatile active block and its application to inductance simulators for testing. Moreover, the high pass filter application using with DXCCCII based inductance simulator and oscillator with flexible tunable oscillation frequency have been presented and simulated to confirm the theoretical validity. The proposed circuit which has a simple circuit design requires the low-voltage and the DXCCCII can also be tuned in the wide range by the biasing current. The proposed DXCCCII provides a good linearity, high output impedance at Z terminals, and a reasonable current and voltage transfer gain accuracy. The proposed DXCCCII and its applications have been simulated using the CMOS 0.18 μm technology.*

Keywords

Dual-X current conveyor, low voltage, tunable circuit, controlled oscillator.

1. Introduction

The second-generation current conveyor (CCII) which is the different versions of the current conveyors has achieved to be a functionally usable and accomplished building block for the realization of the analog circuits. Owing to having a variable gain-bandwidth product, high slew rate, wide dynamic range, higher bandwidth and good linearity, these structures have recently become attractive [1].

The current controlled conveyor (CCCII) which is an electronically tunable type of the CCII has been used in many electronic circuit applications as a part of oscillators, inductance simulators, active resistors, filters and multipliers [2-4]. Parasitic resistance seen at the port X of the CCII is fundamentally seen as a drawback in the analog circuit design. Parasitic resistance can be easily tuned by the biasing current. This resistance providing to obtain numerous tunable functions is used to advantage in current controlled conveyors. Furthermore, it provides to reduce the use of passive components in the design.

An active building block combining the main advantages of CCII and inverting second-generation current conveyor (ICCII) is dual-X second generation current conveyor (DXCCII). The dual-X structure of the DXCCII lately popularized by circuit designers helps reducing the number of components used in the same applications. Although the tunability of the DXCCII can be feasible with MOSFET operating triode region, an extra MOS manufacturing process is required for tuning [5]. On the other hand, there are a lot of applications realized using DXCCII such as inductance simulators, oscillators and filters, etc. Some of these circuits suffer from using passive resistors [6-10].

In this work, dual-X controlled current conveyor (DXCCCII) is presented as an active block for tunable applications. The proposed circuit is a new controllable version of the conventional DXCCII. Parasitic resistances of the conveyors, seen at ports X, are controlled by biasing current. However, the proposed circuit does not require external passive or active elements except for DXCCCII to be controlled. The adjustable range of the DXCCIIs is restricted for generally utilized the gate voltage of the MOS transistor as a control argument. The control voltage is limited by supply voltage. The current as a control argument provides facility to tune wider range [2], [11]. Therefore, the current control is more useful than the voltage control for low voltage and low power circuits. Also, DXCCCII operates at low voltage as ±0.75 V. Considering these advantages, the inductance simulators, adjustable oscillator and tunable high pass filter circuits are presented as applications in this work employing the only active elements and the grounded capacitors. The obvious advantage of the proposed oscillator is using less active and passive elements. Also, the grounded and floating active resistors are presented as other applications. Finally, PSpice simulation results are given to validate the theory.

2. Proposed Dual-X Controlled Current Conveyor

The DXCCCII is implemented by using floating gate MOS transistors (FGMOS). The symbol and the equivalent circuit of an n-type FGMOS transistor with two inputs are shown in Fig. 1. There are several simulation models for the FGMOS transistors in [12-14]. In this proposed circuit, the model is based on connecting capacitors in parallel with the resistors as given in [12].

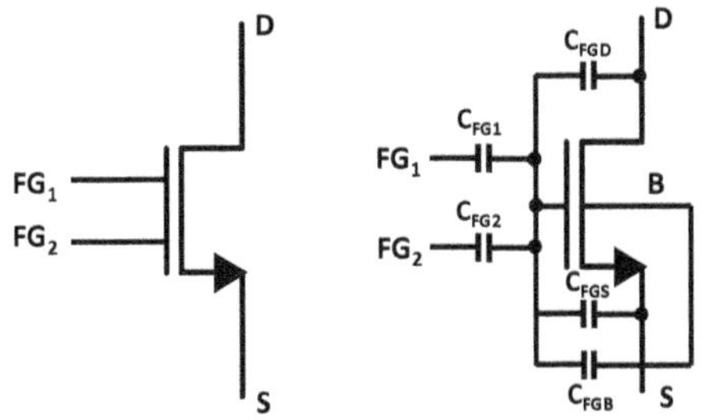

Fig. 1. The n-type FGMOS transistor with three inputs: a) symbol, b) equivalent circuit.

As shown in Fig. 1, FG_1 and FG_2 are the input gate terminals of the FGMOS transistor. The input capacitances are C_{FG1} and C_{FG2} and the input gates are coupled to the floating gate of the FGMOS transistor. C_{FGD}, C_{FGS} and C_{FGB} are the parasitic capacitances between the drain, source, bulk and gate, respectively. Input gate voltages and drain, source and bulk voltages affect an effective floating gate voltage in proportion to the value of the capacitances. When the relation among the capacitances are assumed that $C_{FGD} + C_{FGS} + C_{FGB} << C_{FG1} + C_{FG2}$, the total capacitance C_T is approximately equal to $C_{FG1} + C_{FG2}$. V_{FG} is the effective floating gate voltage and it can be defined as

$$V_{FG} = \frac{C_{FG1} V_{FG1} + C_{FG2} V_{FG2}}{C_T}. \quad (1)$$

The drain current of the FGMOS in saturation region can be calculated as

$$I_D = \frac{k_n}{2}\left[V_{FG} - V_S - V_{TH}\right]^2 \quad (2)$$

where V_S is the source voltage, V_{FG} is the effective floating gate voltage, V_{TH} is the threshold voltage and I_D is the drain current of the FGMOS transistor. In addition, k_n known as the transconductance parameter is $\mu_n . C_{ox} . (W/L)$ where W/L is the aspect ratio of the FGMOS transistor.

The block diagram of the dual-X second generation controlled current conveyor as a versatile active element is demonstrated in Fig. 2. The proposed circuit has six terminals and principle of the operation is similar to the conventional DXCCII [15]. Y and Z terminals of the DXCCCII have high impedances.

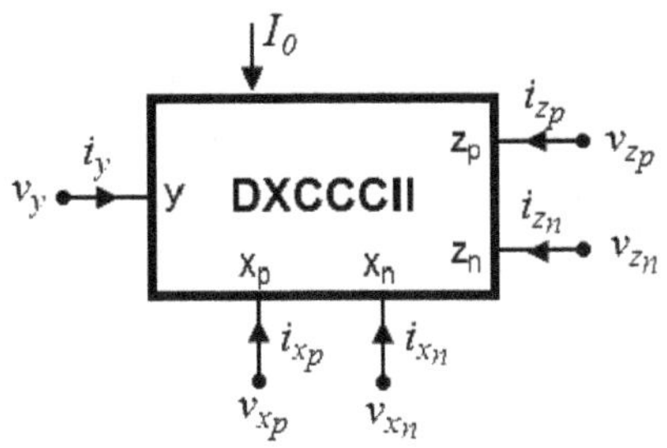

Fig. 2. The symbolic representation of the DXCCCII.

The impedances of the X terminals exhibit a resistance behavior known as a parasitic resistance and its value can be tuned by bias current I_0 of the DXCCCII. The matrix equations of the DXCCCII can be characterized in the following form:

$$\begin{bmatrix} I_Y \\ V_{X_p} \\ V_{X_n} \\ I_{Z_p} \\ I_{Z_n} \end{bmatrix} = \begin{bmatrix} 0 & 0 & 0 \\ 1 & R_{X_p} & 0 \\ -1 & R_{X_n} & 0 \\ 0 & 1 & 0 \\ 0 & 0 & 1 \end{bmatrix} \cdot \begin{bmatrix} V_Y \\ I_{X_p} \\ I_{X_n} \end{bmatrix}. \quad (3)$$

The effective floating gate voltages of M_1 and M_2 transistors are V_{FG1} and V_{FG2}. A loop equation written from floating gate of M_2 to floating gate of M_1 transistor can be expressed as

$$V_{FG2} - V_{FGS2} + V_{FGS1} - V_{FG1} = 0. \quad (4)$$

If it is supposed that $C_{FG1} = C_{FG2} = C_{FG}$, the total capacitance C_T is equal to $2C_{FG}$. The gate-source voltages in (5) can be given as $V_{FG1} = (1/2)V_Y$ and $V_{FG2} = (1/2)V_{Xp}$. If the related equation is arranged, it can be written as below.

$$V_{FGS2} - V_{FGS1} = \frac{1}{2}\left(V_{X_p} - V_Y\right). \quad (5)$$

The drain currents of the transistors M_1 and M_2 can be defined as

$$I_{D1} = \frac{1}{2} k_n \left(\frac{W}{L}\right)\left(\frac{1}{2} V_Y - V_{TH}\right)^2. \quad (6.a)$$

$$I_{D2} = \frac{1}{2} k_n \left(\frac{W}{L}\right)\left(\frac{1}{2} V_{X_p} - V_{TH}\right)^2. \quad (6.b)$$

I_{D1} and I_{D2} are the drain currents of the transistors M_1 and M_2, respectively. The expression belonging to the difference voltage of the terminals X_p and Y can be defined as $V_{XYp} = V_{Xp} - V_Y$.

The relationship between input voltages is given below.

$$V_{XY_p} = 2 \cdot \left[\sqrt{\frac{I_0 + I_{X_p}}{k_n (W/L)}} - \sqrt{\frac{I_0 - I_{X_p}}{k_n (W/L)}}\right]. \quad (7)$$

where I_0 is the biasing current of the differential pair structures. The current I_{Xp} shown in Fig. 3 can be calculated as shown in (8)

$$I_{X_p} = \frac{1}{2} V_{XY_p} \sqrt{k_n (W/L)} \sqrt{2I_0 - \frac{1}{2} k_n (W/L)(V_{XY_p})^2}. \quad (8)$$

From (8), it is assumed that $2I_0 >> k_n(W/L)(V_{XYp})/2$ for small input voltages. Using this approximation, the output current of the differential pair I_{Xp} is obtained as

$$I_{X_p} \cong \frac{1}{2} V_{XY_p} \sqrt{\frac{1}{2} k_n (W/L)} \sqrt{2I_0}. \quad (9)$$

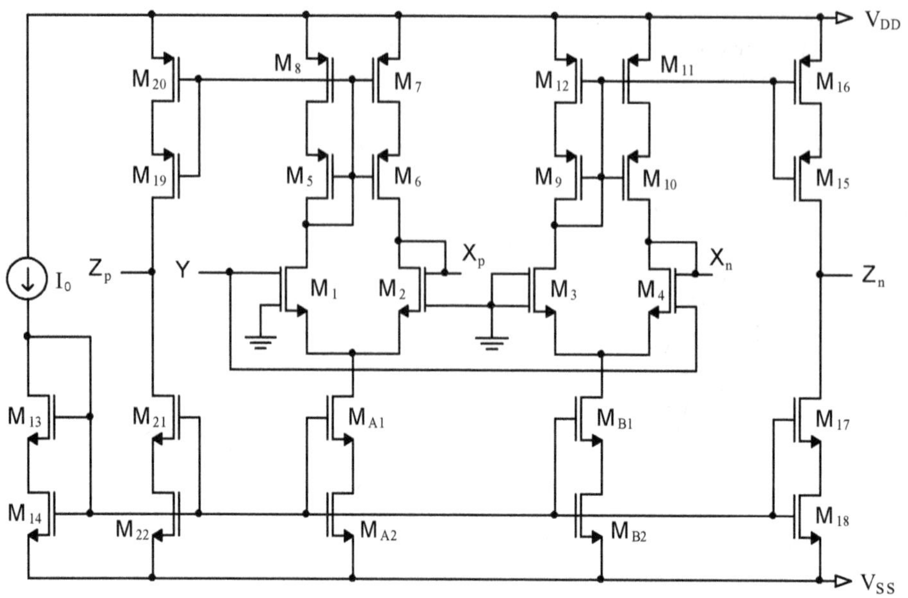

Fig. 3. The circuit structure of the DXCCCII.

From (9), parasitic resistance at terminal X_p of the circuit will be expressed as

$$R_{X_p} \cong \frac{V_{XY_p}}{I_{X_p}} = \frac{2}{\sqrt{I_0 k_n (W/L)}}. \tag{10}$$

As shown in (10), the parasitic resistance of the proposed circuit is easily controlled by biasing current. It is obvious that the electronic adjustability of this resistance is provided by the useful structure. Similarly, the other parasitic resistance at terminal X_n can be described as

$$R_{X_n} \cong \frac{V_{XY_n}}{I_{X_n}} = \frac{2}{\sqrt{I_0 k_n (W/L)}}. \tag{11}$$

In addition, it can be seen that the values of R_{Xp} and R_{Xn} depend on the aspect ratios of M_A and M_B transistors, respectively. If it is desired, each intrinsic resistance value can be changed by the aspect ratio of these transistors or the FGMOS transistors.

The non-ideal model of the DXCCCII is shown in Fig. 4. The real DXCCCII has parasitic resistors and capacitors at the terminal y and z to the ground, and a serial resistor at the terminal x.

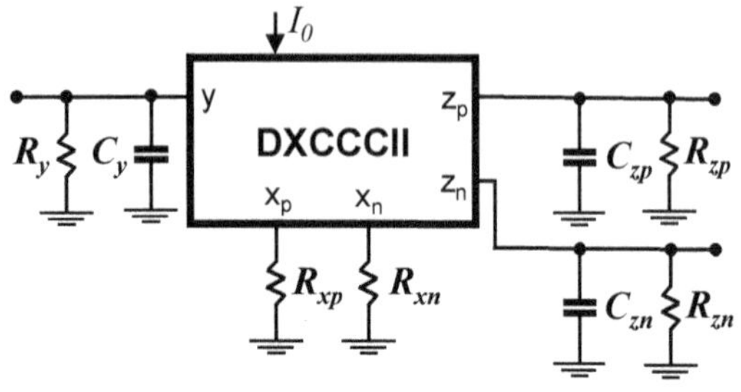

Fig. 4. The non-ideal model of the DXCCCII.

3. Simulation Results

The proposed DXCCCII is simulated using the schematic implementation shown in Fig. 3 with low supply voltages ± 0.75 V. The simulations are based on 0.18 μm level 7 TSMC CMOS process parameters. The dimensions of the transistors used in the DXCCCII implementation are demonstrated in Tab. 1. C_{FG1} and C_{FG2} is selected as 0.02 pF. These capacitances have been known as parasitic capacitances selected for each simulation model of the FGMOS transistor.

Transistor	W/L
$M_1 - M_4$	1.1/0.36
M_{17},M_{18}, M_{21},M_{22},M_{A1},M_{A2},M_{B1},M_{B2}	3.6/0.36
$M_5 - M_{14}$	3.6/0.36
M_{15},M_{16},M_{19},M_{20}	7.2/0.36

Tab. 1. The aspect ratio of the MOS transistors.

Figure 5 shows the changing of the input voltage V_Y versus voltages V_{Xn} and V_{Xp} for the proposed DXCCCII. The curve exhibits a linear characterization approximately between -400 mV and +380 mV. Also, the voltage transfer gain of the DXCCCII is equal to 0.985.

The changing of the input currents I_{Xn} and I_{Xp} versus output current I_Z for the DXCCCII is depicted in Fig. 6. The curve exhibits a highly linear characterization between -60 μA and +60 μA. The current transfer gain of the DXCCCII is equal to 0.99. Considering these findings, this value is highly acceptable and good enough.

Also, the proposed circuit is observed that how the voltage and current ranges depend on the bias current I_0. Voltage and current ranges have been almost decreased as

±200 mV and ±2 μA, respectively. In addition, the voltage and current errors are investigated for the changing biasing current. The findings have proved that voltage and current errors are equal to 0.97 and 0.98, respectively. The changing of I_0 versus R_X has been illustrated in Fig. 7.

Parasitic resistances of the DXCCCII can be approximately tuned between 6.5 kΩ and 400 kΩ by biasing current.

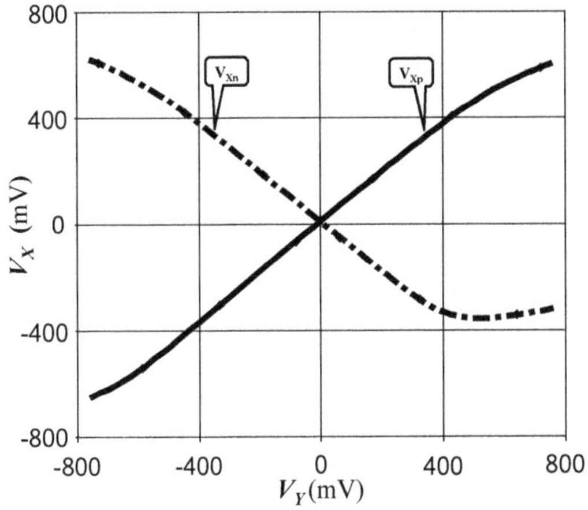

Fig. 5. The voltage transfer curve for the DXCCCII.

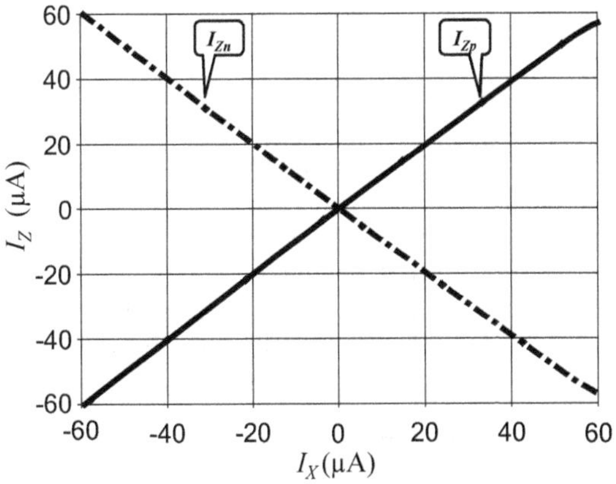

Fig. 6. The current transfer curve for the DXCCCII.

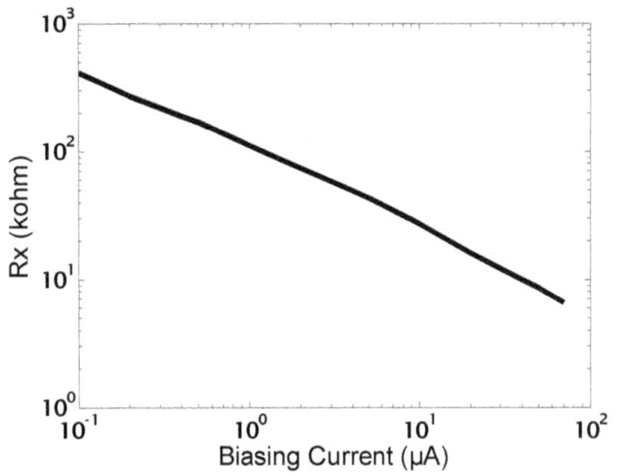

Fig. 7. The parasitic resistance of the proposed DXCCCII for different biasing currents.

The current gains between terminals X and terminals Z are almost 1. The transfer of current is linear from X to Z node. Figure 8 displays the frequency response for the voltage transfer gains and current transfer gains of the DXCCCII. At the same time, Fig. 8 shows the frequency responses of current transfer gains I_{Zp}/I_{Xp}, I_{Zn}/I_{Xn} and voltage gain V_{Xp}/V_Y, the 3 dB cut-off frequencies are 290 MHz, 908 MHz and 265 MHz, respectively. It is investigated that the bandwidth of the voltage and current gains are depended on the bias current. When the biasing

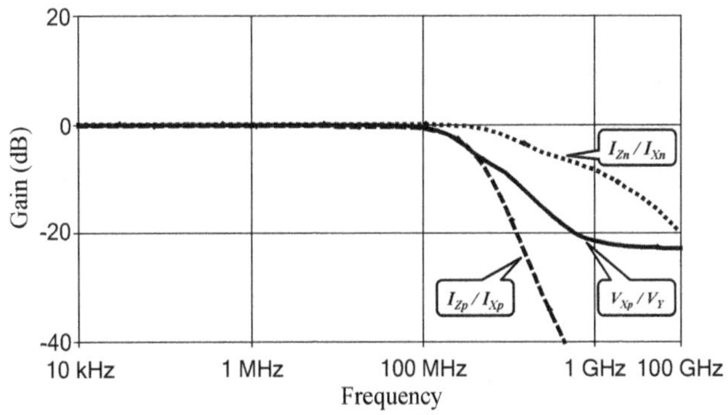

Fig. 8. The frequency response of the voltage transfer gains and current transfer gains of the proposed DXCCCII.

Parameters	**DXCCII [14]**	**DXCCII [16]**	**Proposed DXCCCII**
Supply voltage	±2.5V	±1.5V	±0.75V
Input voltage range	−360 mV, +400mV	−500 mV, +600mV	−400 mV, +380mV
Input voltage / Supply voltage (%)	15,2	36	52
Input current range (I_X)	−60 μA, +60 μA From Fig. 5 in Ref. [16]	−70 μA, +90 μA From Fig. 5 in Ref. [16]	±60μA
Output current range (I_Z)	−90 μA, +110 μA	−100 μA,+125 μA	±60μA
Voltage transfer gain	0.95	0.95	0.985
Current transfer gain	0.97	0.98	0.99
-3 dB bandwidth (I_{Zp}/I_{Xp}, I_{Zn}/I_{Xn})	-	10.35 GHz	290 MHz, 908 MHz
-3 dB bandwidth (V_{Xp}/V_Y, V_{Xn}/V_Y)	580 MHz	1.05 GHz	265 MHz, 350 MHz
R_X (adjustable range) (I_0=0.1μA-70μA)	-	-	400 kΩ-6.5 kΩ
Y input resistance	-	-	10 GΩ
Z output resistance	0.18 GΩ	5.83 GΩ	9.2 GΩ
Parasitic capacitance (C_y)	-	-	0.02 pF
Parasitic capacitance (C_z)	-	-	0.017 pF
Power dissipation	-	-	200 μW
Tunability	No	No	Yes
Technology / Number of transistors	0.35 μm CMOS / 20	0.35 μm CMOS / 48	0.18 μm CMOS / 20
* for I_0 = 35 μA			

Tab. 2. The parametric characteristics of the DXCCCII.

current is increased from 0.1 μA to 70 μA, the bandwidth of the voltage and current gains has approximately varied from 8.5 MHz to 800 MHz and 13 MHz to 3 GHz, respectively. As shown in Fig. 3, while terminal X_n is composed of gate and drain of the transistor M_4, terminal Y is connected to the another gate of the transistor M_4. Therefore, taking into consideration Fig. 1, it can be easily seen that frequency performance only depends on the input capacitances C_{FG1} and C_{FG2}, in addition, the frequency behavior of the voltage gain V_{Xn}/V_Y is rather reasonable. Phase difference for gain values below 350 MHz can be accepted zero. Also, the proposed circuit consumes 200 μW for $I_0 = 35$ μA.

The performance parameters of the proposed circuit are depicted in Tab. 2. The proposed circuit offers some advantages. For instance, low-voltage power supply has been required about ±750 mV. Likewise, the circuit which has a simple circuit design consumes power about 200 μW. The main feature of the DXCCCII is that the intrinsic resistance can be usefully tuned in the wide range by biasing current. Also, the output (port Z) and input (port Y) of the DXCCCII has very high resistance and the proposed circuit has reasonable number of transistors. Voltage and current transfer gain of the proposed circuit is more adorable than the other references in Tab. 2

4. Tunable Applications of the DXCCCII

In order to reveal the performance and the usability of the DXCCCII, application examples such as active inductors and tunable oscillator have been introduced using PSpice simulations.

4.1 The Grounded Active Inductor

The proposed circuits for accomplishing grounded inductors are shown in Fig. 9. All of the circuits are constructed with two DXCCCII and one grounded capacitor. It is obviously known that a circuit with grounded capacitors has appreciable advantages in analog integrated circuit implementations.

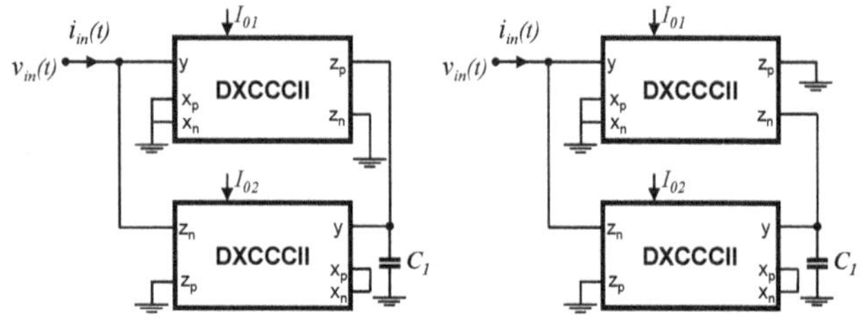

(a) (b)

Fig. 9. Inductance simulators realized using DXCCCII: a) positive, b) negative.

The circuits can simulate diverse combinations of the inductances as shown in Fig. 9. These circuits can operate as positive and negative inductances. The previously presented grounded inductance simulators suffer from some disadvantages such as passive component mismatch, use of capacitor that connected in series to the port X of the current conveyor, and eventually operating in lower frequency, use of two or more passive elements [10], [17-20]. Whereas, these drawbacks have partly been overcame by the proposed circuit which has only two grounded capacitors and two DXCCCIIs. The equivalent inductance value of the both positive and negative simulators is given by the following equation

$$L_{eq} \cong \frac{v_{in}}{i_{in}} = |R_{X1}R_{X2}C_1| \tag{12}$$

where R_{X1} and R_{X2} are the parasitic resistances of the first and second DXCCCII, respectively. Considering both the voltage and the current tracking errors of the DXCCCII, $\beta_p = 1 - \varepsilon_{Vp}$ and $\beta_n = 1 - \varepsilon_{Vn}$ define the voltage tracking errors from Y terminal to X_p and X_n terminals; $\alpha_p = 1 - \varepsilon_{Ip}$ and $\alpha_n = 1 - \varepsilon_{In}$ define the current tracking errors from X_p and X_n terminals to Z_p and Z_n terminals, respectively. β_p, β_n and α_p, α_n are the voltage and the current transfer gains ; ε_{Vp}, ε_{Vn} and ε_{Ip}, ε_{In} are the voltage and the current transfer errors of the DXCCCII, respectively. Taking into consideration both the voltage and current tracking errors of the DXCCCII, the input current of the inductance simulator as shown in Fig. 9 can be calculated as,

$$i_{in} = \frac{v_{in}\beta_{p1}\alpha_{p1}\alpha_{n2}(\beta_{p2} + \beta_{n2})}{sC_1(2R_{X1}R_{X2})} \tag{13}$$

where β_{p1}, β_{n1} and α_{p1}, α_{n1} are the voltage and current transfer gains of the first current conveyor (DXCCCII 1), respectively, and, β_{p2}, β_{n2} and α_{p2}, α_{n2} are the voltage and current transfer gains of the second current conveyor (DXCCCII 2).

Considering tracking errors of the DXCCCII, positive and negative inductance of the simulator can be described respectively as

$$L_{eq}^{+} = \frac{2R_{X1}R_{X2}C_1}{\beta_{p1}\alpha_{p1}\alpha_{n2}(\beta_{p2} + \beta_{n2})}, \tag{14.a}$$

$$L_{eq}^{-} = -\frac{2R_{X1}R_{X2}C_1}{\beta_{n1}\alpha_{n1}\alpha_{n2}(\beta_{p2} + \beta_{n2})}. \tag{14.b}$$

The frequency response of the impedance value for the inductance simulator is displayed in Fig. 10. The graph is drawn by using different biasing currents. The capacitance value of the C_1 shown in Fig. 9 is equal to 15 pF. When the graph is investigated, it can be seen that the curve exhibits approximately a linear behavior between 100 kHz–7 MHz, 100 kHz–8.5 MHz and 100 kHz–10 MHz for $I_0 = 20$, 30, 40 μA, respectively.

To prove the theoretical validity of the inductance simulator given in Fig. 9, the classical high pass filter shown in Fig. 11 was employed as an application.

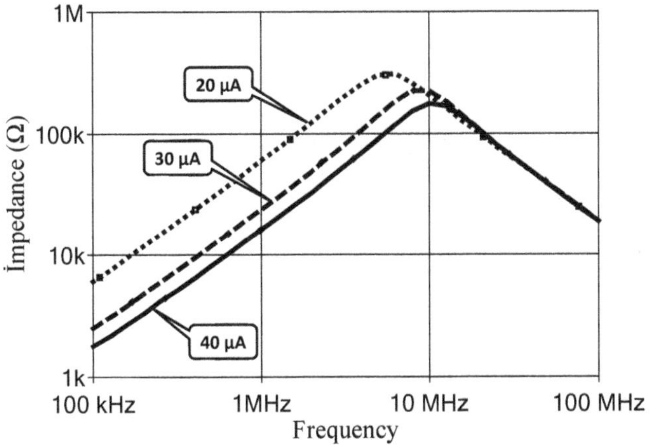

Fig. 10. The impedance values of the simulators for different biasing currents.

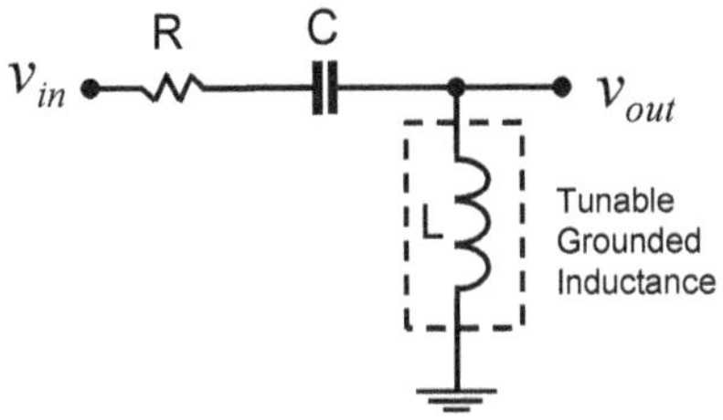

Fig. 11. The classical high pass filter.

Frequency response of the filter is illustrated in Fig. 12. The passive elements are selected as $R = 4$ kΩ, $C = 35$ pF, and $C_1 = 15$ pF, which results in a 3dB frequencies of 267 kHz, 390 kHz, 530 kHz for $I_0 = 20$, 30, 40 µA, respectively. In addition, the 3dB frequencies of the high pass filter is theoretically calculated as 277 kHz, 405 kHz and 569 kHz. The reason of the difference between the calculated and simulated values is the voltage and current tracking errors of the DXCCCII.

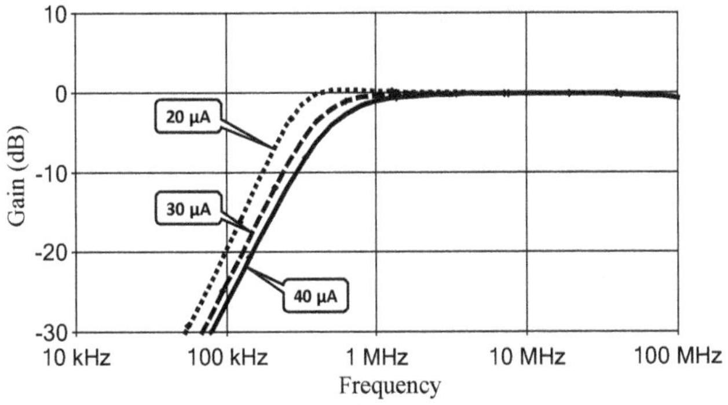

Fig. 12. Frequency responses of the filter.

4.2 Tunable Oscillator

As another application, the tunable oscillator circuit using only two DXCCCII shown in Fig. 13 has been presented. Also, it consists of two grounded capacitors.

Characteristic equation and oscillation frequency of the tunable oscillator shown in Fig. 13 can be obtained

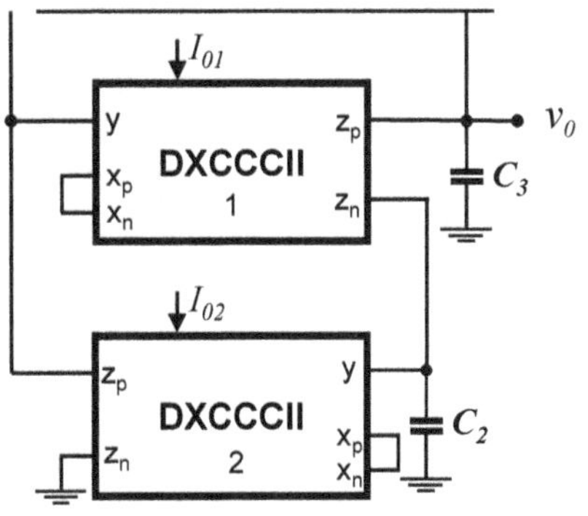

Fig. 13. The tunable oscillator circuit.

doing routine circuit analysis following characteristic equations.

$$s^2 - \frac{1}{R_{X1}C_3}s + \frac{1}{R_{X1}R_{X2}C_2C_3} = 0, \quad (15.a)$$

$$\omega_0 = \frac{1}{\sqrt{R_{X1}R_{X2}C_2C_3}} \quad (15.b)$$

where R_{X1} and R_{X2} are the parasitic resistances of the first and second DXCCCII, respectively. As shown in (15.a), oscillation condition of the oscillator is always provided when the non-ideal effects are ignored. Considering tracking errors of the DXCCCII, characteristic equation and oscillation frequency of the tunable oscillator can be described respectively as

$$s^2 - \frac{\alpha_{p1}(\beta_{p1}+\beta_{n1})}{2R_{X1}C_3}s + \frac{\alpha_{n1}\alpha_{p2}(\beta_{p1}+\beta_{n1})(\beta_{p2}+\beta_{n2})}{4R_{X1}R_{X2}C_2C_3} = 0, \quad (16.a)$$

$$\omega_0 = \sqrt{\frac{\alpha_{n1}\alpha_{p2}(\beta_{p1}+\beta_{n1})(\beta_{p2}+\beta_{n2})}{4R_{X1}R_{X2}C_2C_3}}. \quad (16.b)$$

Oscillation waveform of the oscillator is displayed in Fig. 14. In order to obtain the frequency responses of the oscillator, C_2 and C_3 are set to 1 pF and 3 pF, respectively. The biasing currents for the DXCCCIIs I_{01} and I_{02} are equal to 1 µA and 40 µA, respectively.

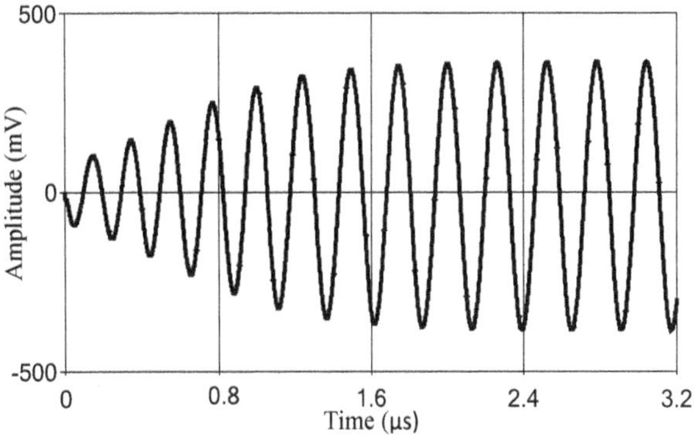

Fig. 14. Voltage waveforms of the oscillator in time domain.

Fig. 15 shows that the calculated results have a good agreement with the simulated results. This figure depicts that the oscillation frequency of the oscillator can be tuned by I_{02}. The oscillation frequency of the oscillator can be varied from 1.83 MHz to 4.79 MHz if the biasing current is tuned from 10 μA to 70 μA.

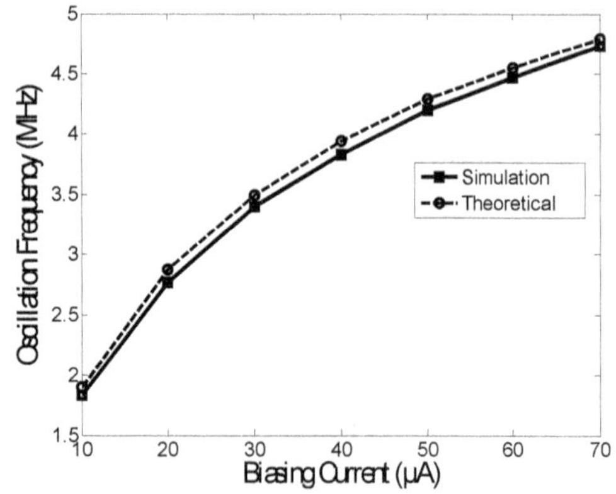

Fig. 15. Oscillation frequency versus biasing current of the second DXCCCII.

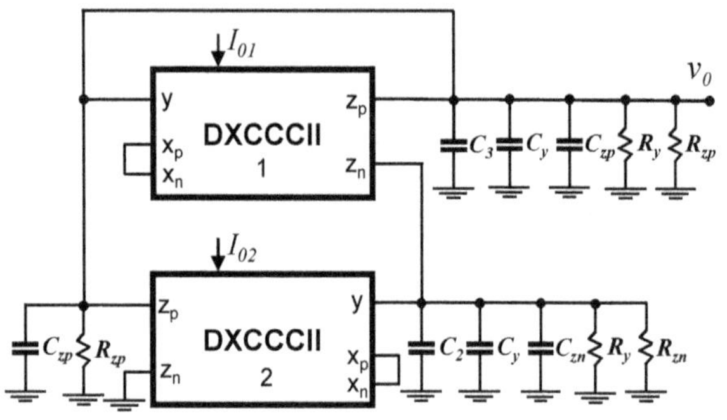

Fig. 16. The non-ideal model of the oscillator.

The non-ideal model of the oscillator is shown in Fig. 16. The real DXCCCII has parasitic resistors and capacitors at the terminal y and z to the ground, and a serial resistor at the terminal x. R_{zp1}, R_{zp2}, R_{zn1}, C_{zp1}, C_{zp2}, C_{zn1} can be defined as the parasitic resistors and capacitors of the z terminals of DXCCII 1 and DXCCII 2 in Fig. 16. R_{y1}, R_{y2}, C_{y1}, C_{y2} are the parasitic resistors and capacitors at the y terminals of DXCCII 1 and DXCCII 2. Because of $C_3 >> C_{y1} + C_{zp1} + C_{zp2}$ and $C_2 >> C_{y2} + C_{zn1}$, the effect of the parasitic capacitances to the total capacitances can be neglected. Considering non-ideal effects of the DXCCCII, oscillation frequency and condition of the tunable oscillator can be expressed respectively as

$$\omega_0 = \sqrt{\frac{1}{R_{X1}R_{X2}C_2C_3} + \frac{1}{R_AR_BC_2C_3} - \frac{1}{R_BR_{X1}R_{X2}C_2C_3}} \text{,(17.a)}$$

$$\text{CO: } \frac{R_AC_3 + R_BC_2}{R_AR_BC_2C_3} - \frac{1}{R_{X1}C_3} \leq 0 . \qquad (17.b)$$

where $R_A = R_{y1} // R_{zp1} // R_{zp2}$ and $R_B = R_{y2} // R_{zn1}$. It can be seen that values of the R_A and R_B are approximately GΩs. Thus, the effects of the parasitic resistance to the frequency are highly poor. In (17.b), the condition of oscillation is given as the formulation including non-ideal effects.

Comparison between various oscillators using active element is shown in Tab. 3. Some circuits have no electronical tunability as shown in Tab. 3. Also, the proposed circuit has low power consumption compared to these circuits due to the fact that DXCCCII used in the proposed oscillator has simple structure and low supply voltage. The obvious advantage of the proposed oscillator is using less active and passive elements.

Ref	Supply voltage	Active element	Number of active / passive elements	FO range (MHz)	THD (%)	Technology	Electronical tunability	Power consumption (mW)
21	± 2.5 V	CCCDTA	2 / 2	0.1 – 5	1.14	BJT	Yes	12.1
22	± 5 V	ECCII -, CCII	3 / 5	0.26 – 1.25	0.2 – 1.5	BJT	Yes	N/A
23	± 1.25 V	CDTA	3 / 3	0.4 – 0.8	10	CMOS	Yes	2.87
24	± 1.25 V	DDCC, OTA	2 / 3	1.69	1.75	CMOS	No	1.86
25	± 2 V	OTA	4 / 4	0.2 – 21.5	1	CMOS	Yes	1.52
26	± 1.25 V	DX-MOCCII, MOS	2 / 5	1.59	3.5	CMOS	No	N/A
27	± 2.5 V	DVCC	1 / 5	0.096	3.76	CMOS	No	N/A
28	± 2.5 V	OTA, CDTA	2 / 4	0.053	1.17	CMOS	No	N/A
29	± 1.5 V	MCCCDTA	1 / 6	0.076	0.13	BJT	No	N/A
30	± 5 V	CCII	2 / 2	0.035	3	BJT	No	N/A
Prop.	± 0.75 V	DXCCCII	2 / 2	1.83 – 4.79	2.4 – 4.4	CMOS	Yes	0.28

Tab. 3. Comparison between various oscillators.

4.3 The Grounded and Floating Active Resistors

DXCCCII based grounded positive and negative active resistor structures have been displayed in Fig. 17.

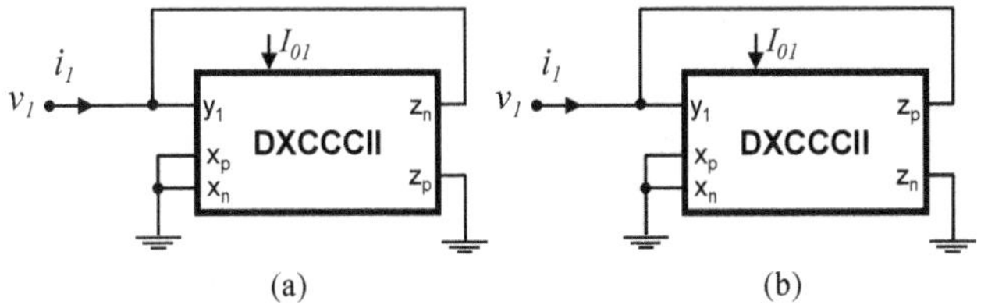

Fig. 17. DXCCCII based grounded resistors a) positive resistor, b) negative resistor.

From (10) and (11), it can be seen that R_{Xn} is equal to R_{Xp}. So, all parasitic resistances can be represented as R_X. The resistance value of the DXCCCII based grounded resistors will be expressed as,

$$R_1 \cong \frac{v_1}{i_1} = |R_X| \cdot \tag{18}$$

Floating positive and negative active resistor structure based on DXCCCII have been depicted in Fig. 18. This figure demonstrates two types of an active resistor designed by using only two DXCCCIIs. The proposed active resistors might be tuned by the biasing currents. So, the resistance value of the resistors is calculated as,

$$R_{12} = \frac{v_1 - v_2}{i_1} = 2 \cdot |R_X| \tag{19}$$

where all biasing currents are equal to I_0 and $i_2 = -i_1$.

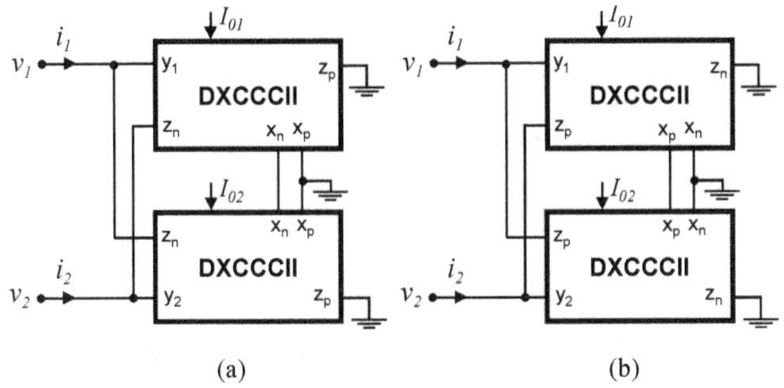

Fig. 18. DXCCCII based floating resistors a) positive resistor, b) negative resistor.

The I-V characteristics of the proposed resistors are shown in Fig. 19. The behaviors of the resistors are fairly linear between -400 mV and +400 mV. Additionally, the estimation of the proposed circuit's dynamic-range is calculated as

$$|v_1 - v_2| \le 2\left(\frac{I_0}{k_n W / L}\right)^{1/2} . \tag{20}$$

The dynamic range highly depends on the biasing current as shown in (20). Thus, the dynamic range will be expanded for the high values of the biasing current. Considering both the voltage and the current tracking errors of the DXCCCII, the resistance values of the floating active resistors can be calculated as

$$R_{12}^{+} = \frac{v_1 - v_2}{i_1} = 2\alpha_p \beta_p R_X , \tag{21.a}$$

$$R_{12}^{-} = \frac{v_1 - v_2}{i_1} = -2\alpha_n \beta_n R_X . \tag{21.b}$$

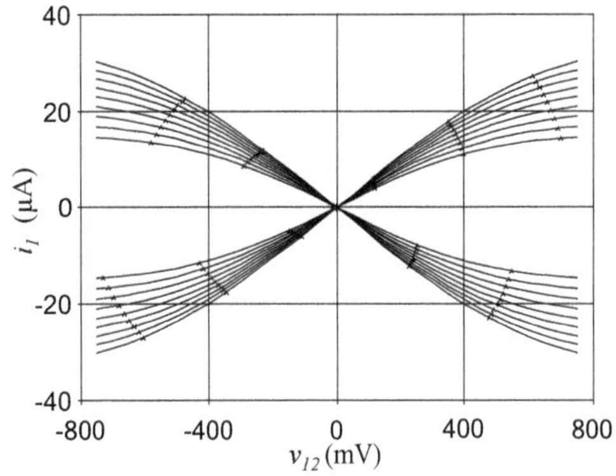

Fig. 19. The I-V characteristics of the floating positive and negative active resistor.

5. Conclusion

In this paper, dual-X controlled current conveyor (DXCCCII) as a versatile active block is presented and its applications to pure inductance simulators has been tested. Also, high pass filter application using DXCCCII based inductance simulator has been simulated to prove the theoretical validity. In addition, DXCCCII based oscillator with flexible tunable oscillation frequency and active resistors are presented as other applications. Only grounded capacitor and DXCCCII have been employed in the all designed applications. The adjustment capability of the proposed circuits is the functional feature in electronic circuit designs. In this context, the proposed circuits are rather convenient for IC realizations. The proposed circuits have been simulated using a PSpice simulation program, and its simulation results were compared with the theoretical approaches and the other DXCCIIs. Theoretical analyses of these circuits were achieved, and the performances of the proposed circuits have been verified by the simulation results. For the proposed DXCCCII, the parasitic resistance value can be tuned from 6.5 kΩ to 400 kΩ if the biasing current is changed from 0.1 µA to 70 µA with a good coherence between the theoretical and simulation results. Besides this good coherence, the proposed circuit is required a low voltage as well as ±0.75 V. As a consequence, we believe that it is absolutely an admirable design because of having low power dissipation.

References

[1] CHANG, C. M., CHEN P. C. Realization of current-mode transfer function using second-generation current conveyors. *International Journal of Electronics*, 1991, vol. 71, no. 5, p. 809–15.

[2] FABRE, A., SAAID, O., WIEST, F., BOUCHERON, C. Current controlled bandpass filter based on translinear conveyors. *Electronic Letters,* 1995, vol. 31, no. 20, p. 1727–28.

[3] ERCAN, H., ALÇI, M. A new design for a BiCMOS controlled current conveyor. *Elektronika Ir Elektrotechnika (Journal Electronics and Electrical Engineering),* 2013, vol. 19, no. 1, p. 56–60.

[4] FANI, R., FARSHIDI, E. A FG-MOS based fully differential current controlled conveyor and its applications. *Circuits, Systems and Signal Processing,* 2012, vol. 32, no. 3, p. 993–1011.

[5] MINAEI, S., YÜCE, E. A new full-wave rectifier circuit employing single dual-X current conveyor. *International Journal of Electronics*, 2008, vol. 95, no. 8, p. 777–784.

[6] METIN, B. Supplementary inductance simulator topologies employing single DXCCII. *Radioengineering,* 2011, vol. 20, no. 3, p. 614–618.

[7] MAHESHWARI, S., ANSARI, M. S. Catalog of realizations for DXCCII using commercially available ICs and application. *Radioengineering*, 2012, vol. 21, no. 1, p. 281–289.

[8] MINAEI, S., YÜCE, E. Unity/variable-gain voltage-mode/current-mode first-order all-pass filters using single dual-X second-generation current conveyor. *IETE Journal of Research*, 2010, vol. 56, no. 6, p. 305–312.

[9] ANSARI, M. S., KHAN, I. A., BEG, P., NAHHAS, A. M. Three phase mixed-mode CMOS VCO with grounded passive components. *SAP publishing, Electrical and Electronic Engineering,* 2013, vol. 3, no. 6, p. 149–155.

[10] MYDERRIZIA, I., MINAEI, S., YÜCE, E. DXCCII-based grounded inductance simulators and filter applications. *Microelectronics Journal,* 2011, vol. 42, no. 9, p. 1074–1081.

[11] TEKİN, S. A., ERCAN, H., ALÇI, M. Novel low voltage CMOS current controlled floating resistor using differential pair. *Radioengineering,* 2013, vol. 22, no. 2, p. 428–433.

[12] RODRIGUEZ-VILLEGAS, E. *Low Power and Low Voltage Circuit Design with the FGMOS Transistor.* Institution of Engineering and Technology, London, UK, 2006.

[13] KHATEB, F., KHATIB, N., KOTON, J. Novel low-voltage ultra-low-power DVCC based on floating-gate folded cascode OTA. *Microelectronics Journal,* 2011, vol. 42, no. 8, p. 1010–1017.

[14] TEKIN, S. A. Voltage summing current conveyor (VSCC) for oscillator and summing amplifier applications. *Informacije MIDEM-Journal of Microelectronics, Electronic Components and Materials*, 2014, vol. 44, no. 2, p. 159–167.

[15] ZEKI, A., TOKER, A. The dual-X current conveyor (DXCCII): A new active device for tunable continuous-time filters. *International Journal of Electronics*, 2003, vol. 89, no. 13, p. 913–923.

[16] KAÇAR, F., METİN, B., KUNTMAN, H. A new CMOS dual-X second generation current conveyor (DXCCII) with an FDNR circuit application. *International Journal of Electronics and Communications (AEU),* 2010, vol. 64, no. 8, p. 774–778.

[17] KUMAR, P., SENANI, R. New grounded simulated inductance circuit using a single PFTFN. *Analog Integrated Circuits and. Signal Processing,* 2010, vol. 62, no. 1, p. 105–112.

[18] YUCE, E. Inductor implementation using a canonical number of active and passive elements. *International Journal of Electronics,* 2007, vol. 94, no. 4, p. 317–326.

[19] YUCE, E., MINAEI, S., ÇİÇEKOĞLU, O. A novel grounded inductor realization using a minimum number of active and passive components. *ETRI Journal,* 2005, vol. 27, no. 4, p. 427–432.

[20] KAÇAR, F., YEŞİL, A., MINAEI, S., KUNTMAN, H. Positive/negative lossy/lossless grounded inductance simulators employing single VDCC and only two passive elements. *International Journal of Electronics and Communications (AEU),* 2014, vol. 68, no. 1, p. 73–78.

[21] JAIKLA, W., LAHIRI, A. Resistor–less current–mode four–phase quadrature oscillator using CCCDTAs and grounded capacitors. *International Journal of Electronics and Communications (AEU),* 2011, vol. 66, no. 3, p. 214–218.

[22] SOTNER, R., HRUBOS, Z., SEVCIK, B., SLEZAK, J., PETRZELA, J., DOSTAL, T. An example of easy synthesis of active filter and oscillator using signal flow graph modification and controllable current conveyors. *Journal of Electrical Engineering,* 2011, vol. 62, no. 5, p. 258–266.

[23] HORNG, J.W., LEE, H., WU, J. Electronically tunable third–order quadrature oscillator using CDTAs. *Radioengineering,* 2010, vol. 19, no. 2, p. 326–330.

[24] KWAWSIBSAM, A., SREEWIROTE, B., JAIKLA, W. Third-order voltage mode quadratrue oscillator using DDCC and OTAs. In *International Conference on Circuits, System and Simulation IPCSIT.* Singapore, 2011, vol. 7, p. 317–321.

[25] GALAN, J., CARVAJAL, R. G., MUNOZ, F., TORRALBA, A., RAMIREZ-ANGULO, J. A low-power low-voltage OTA-C sinusoidal oscillator with more than two decades of linear tuning range. In *Proc. of the 2003 International Symposium on Circuits and Systems.* Bangkok (Thailand), 2003, vol. 1, p. 677–680.

[26] BEG, P., SIDDIQI, M. A., ANSARI, M. S. Multi output filter and four phase sinusoidal oscillator using CMOS DX-MOCCII. *International Journal of Electronics*, vol. 98, no. 9, p. 1185 – 1198.

[27] CHIEN, H. C. Voltage- and current-modes sinusoidal oscillator using a single differential voltage current conveyor. *Journal of Applied Science and Engineering,* 2013, vol. 16, no. 4, p. 395 to 404.

[28] BIOLEK, D., KESKIN, A. Ü., BIOLKOVA, V. Grounded capacitor current mode single resistance-controlled oscillator using single modified current differencing transconductance amplifier. *IET Circuits, Devices & Systems,* 2010, vol. 4, no. 6, p. 496–502.

[29] LI, Y. A new single MCCCDTA based Wien-bridge oscillator with AGC. *International Journal of Electronics and Communications (AEU),* 2012, vol. 66, no. 2, p. 153–156.

[30] HORNG, J. W. A sinusoidal oscillator using current-controlled current conveyors. *International Journal of Electronics,* 2001, vol. 88, no.6, p. 659–664.

About Authors ...

Sezai Alper TEKİN was born in Kayseri, Turkey. He received the B. Engineering degree from Erciyes University, Kayseri, Turkey, in 2000, the M.S. degree from Niğde University, Niğde, Turkey, in 2005 and the Ph.D. in Electrical and Electronics Engineering from Erciyes University, Kayseri, Turkey in 2012. From 2006 to 2012 he was a member of academic staff with the Erciyes University, Engineering Faculty, Electrical and Electronics Dept., Kayseri, Turkey. Since 2012, he has been a member of academic staff with the Erciyes University, Engineering Faculty, Industrial Design Dept., Kayseri, Turkey. His current research interests include free space optics, current mode circuits, current conveyors, electronic circuit design and solar cells.

Hamdi ERCAN received the B. Engineering and M.S. degrees from Erciyes University, Kayseri, Turkey, in 2004 and 2007, respectively, all in Electronic Engineering. He received the Ph.D. in Electrical and Electronics Engineering from Erciyes University, Kayseri, Turkey in 2012. Since 2005, he has been a member of academic staff with the Erciyes University, Civil Aviation School, Avionics Dept., Kayseri, Turkey. His current research interests include current mode circuits, current conveyors and electronic circuit design.

Mustafa ALÇI was born in Kayseri, Turkey, in 1957. He graduated Electronic Dept. of Technology Faculty from Gazi University, Ankara in 1979. Then he received the B. Sc. degree from Erciyes University, Kayseri, in 1983, M. Sc. degree from Middle East Technical University, Ankara, in 1986 and Ph.D. degree from Erciyes University, Kayseri, in 1989, respectively, all in Electrical & Electronics Engineering. Since 1979, he is a member of academic staff with Erciyes University Engineering Faculty, Electronic Engineering Dept., Kayseri, Turkey. His current research interests include image processing, noise and coding artifacts suppression, fuzzy systems, medical electronics, chaotic systems and circuit design.

16

Low Voltage Floating Gate MOS Transistor Based Four-Quadrant Multiplier

Richa SRIVASTAVA, Maneesha GUPTA, Urvashi SINGH

Dept. of Electronics and Communication Engineering, NSIT, New Delhi, India

richa_ec@yahoo.co.in, maneeshapub@gmail.com, urvashi.singh27@gmail.com

Abstract. *This paper presents a four-quadrant multiplier based on square-law characteristic of floating gate MOSFET (FGMOS) in saturation region. The proposed circuit uses square-difference identity and the differential voltage squarer proposed by Gupta et al. to implement the multiplication function. The proposed multiplier employs eight FGMOS transistors and two resistors only. The FGMOS implementation of the multiplier allows low voltage operation, reduced power consumption and minimum transistor count. The second order effects caused due to mobility degradation, component mismatch and temperature variations are discussed. Performance of the proposed circuit is verified at ±0.75 V in TSMC 0.18 μm CMOS, BSIM3 and Level 49 technology by using Cadence Spectre simulator.*

Keywords

FGMOS, low-voltage, low-power, four-quadrant, multiplier.

1. Introduction

Increasing density of components on chip and growing demand of battery-powered portable equipments have directed the research towards the development of low-voltage low-power analog signal processing circuits. Such research involves finding new and promising design techniques so that the complete system could meet the specified design constraints. Threshold voltage is one of the most important design parameter for low voltage analog circuit designers. Threshold voltage reduction not only helps in reducing the supply voltage requirement but also the power consumption in many analog applications. Various low-voltage low-power techniques reported in literature include sub-threshold MOSFETs, level shifters, self cascode, bulk-driven and FGMOS techniques [1–10]. In the last few years, FGMOS transistor has gained wide popularity because of its ability to reduce or remove the threshold voltage requirement of the circuit. Recently number of publications showing the application of FGMOS in various analog signal processing circuits such as voltage squarer and multiplier have been reported [7–11].

Four-quadrant multiplier is an important and very useful building block in many analog signal processing circuits such as modulators, frequency doublers, adaptive filters etc. The Gilbert six transistor cell based on variable transconductance technique is very popular in bipolar technology [12], [13]. The proposed configuration uses exponential characteristic of bipolar transistor and is useful for the frequency ranging from DC to unity gain frequency of BJT. Although the Gilbert multiplier has been the most widely used bipolar multiplier, it is not suitable for low voltage applications. Kimura [14] has proposed a bipolar multiplier which can operate at lower supply voltage (< 3 V) and can replace Gilbert multiplier for low voltage operation. The results of the multipliers based on MOS square-law characteristic are not as good as that produced by using exponential characteristic of BJT. Hence, several linearization techniques have been used to improve the performance of MOS based multipliers [15–19]. The multiplier proposed by Qin and Geiger [16] is based upon MOS version of Gilbert cell and uses differential active attenuators to increase the input signal swing. Soo and Meyer [17] have used cascaded MOS differential pair to implement four-quadrant multiplier with linearity comparable to bipolar circuits over a wide input range. The simplest form of four-quadrant analog multiplier consists of a pair of MOSFETs, a pair of current/voltage convertors and a subtractor. The multiplier based on this concept employs various resistive components and leads to serious problems such as large power consumption and large offset error. To overcome these issues, the four-quadrant multiplier based on switched capacitor technique has been proposed by Yasumoto and Enomoto [20]. The multiplier exhibits excellent characteristics such as low THD, large dynamic range and high-speed operation. Most of the multipliers based on MOS square law characteristics have low input range and small bandwidth. To overcome this issue, various BiCMOS based multipliers have been reported in [21–23]. These multipliers combine the high input impedance property of MOS transistor along with the linearity and high speed of bipolar transistors. Another technique to implement analog multiplier is the use of active blocks [24], [25] such as operational amplifiers, current conveyors, and second generation current-controlled conveyors. Recently, the multipliers based on FGMOS transistors [26–29] have gained wide popularity because of their high input range. But most of these designs operate at high supply voltage and have complex structure. This paper presents a very simple four-quadrant multiplier based on square law characteristic of FGMOS, consisting of eight FGMOS transistors and two resistors only.

This paper is organized as follows: The basic structure and operation of FGMOS transistor is described in Section 2. The principle of operation of the proposed multiplier is presented in Section 3. Second order effects over the proposed configuration are discussed in Section 4. Section 5 deals with the simulation results. Finally the conclusions are drawn in the last section.

2. FGMOS Transistor

FGMOS is a multiple-input floating-gate transistor whose floating gate is formed by the first polysilicon layer and multiple-input gates are formed by the second polysilicon layer. Symbol of N-input FGMOS with input voltages V_1, V_2.....V_N and its equivalent circuit are shown in Fig. 1a and 1b respectively.

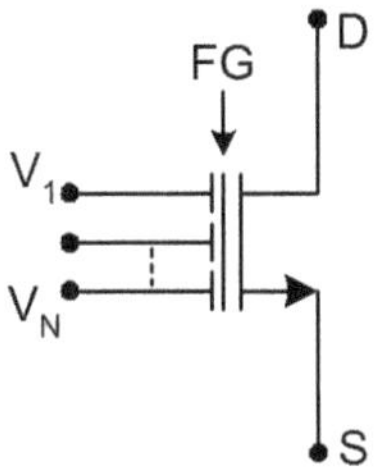

Fig. 1a. Symbol of N-input FGMOS.

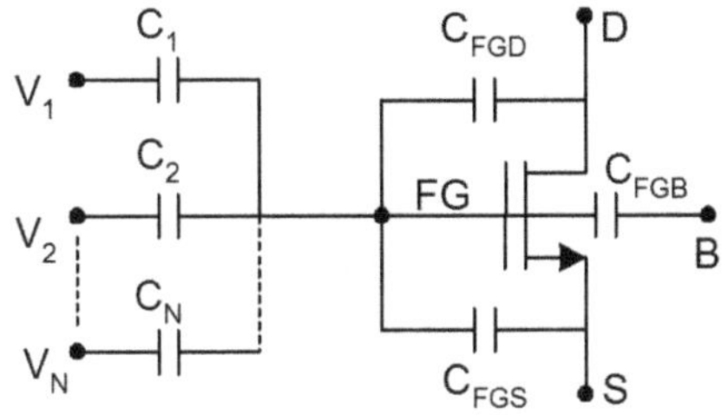

Fig. 1b. Equivalent circuit of FGMOS.

The drain current (I_D) of the FGMOS transistor operating in saturation region is given by [10]

$$I_D = \frac{\beta}{2}(\sum_{i=1}^{N}\frac{C_i}{C_T}V_{is} + \frac{C_{GD}}{C_T}V_{DS} + \frac{C_{GB}}{C_T}V_{BS} + \frac{Q_{FG}}{C_T} - V_T)^2 \quad (1)$$

where C_i is the set of input capacitors associated with effective inputs and the floating gate; C_{FGD}, C_{FGS} and C_{FGB} are the parasitic capacitances of floating gate with drain, source, and bulk respectively; V_D, V_S and V_B denote the drain, source and bulk voltages respectively. Q_{FG} is the residual charge trapped in the oxide-silicon interface during fabrication process, β is the transconductance parameter and V_T stands for the threshold voltage.

$C_T(=\sum_{i=1}^{N}C_i + C_{FGS} + C_{FGD} + C_{FGB})$ is the total floating gate capacitance. Assuming $C_i >> C_{FGD}$, C_{FGB} and $Q_{FG} = 0$ [28], the drain current of FGMOS transistor in saturation region can be expressed as

$$I_D = \frac{\beta}{2}(\sum_{i=1}^{N}k_i V_{iS} - V_T)^2 \quad (2)$$

where $k_i = C_i/C_T$. It can be seen from (2) that the multiple input voltages along with capacitance ratio can be used to cancel the threshold voltage term so as to get the perfect squaring function which in turn is used to realize the proposed multiplier.

2.1 Analysis of Two-Input FGMOS Transistor

Two-input FGMOS transistor with inputs V_b (bias voltage) and V_{in} (input signal) is shown in Fig. 2.

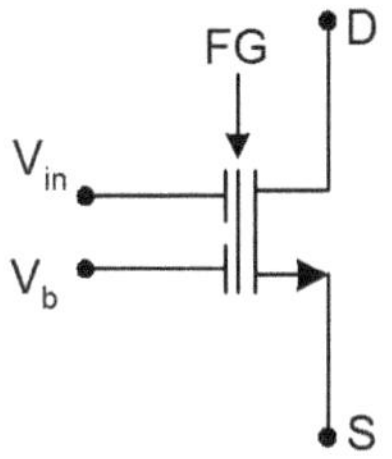

Fig. 2. Two input FGMOS transistor.

For 2-input FGMOS (with source grounded), the drain current equation (2) is modified as

$$I_D = \frac{\beta}{2}\left(\frac{C_1}{C_T}V_b + \frac{C_2}{C_T}V_{in} - V_T\right)^2. \quad (3)$$

From (3) the effective threshold voltage is given as [30]

$$V_{Teff} = \frac{V_T - k_1 V_b}{k_2} \quad (4)$$

where $k_1 = C_1/C_T$, $k_2 = C_2/C_T$, are the capacitive coupling ratios and $C_T = C_1 + C_2$ is the total capacitance after neglecting the parasitic capacitances.

Equation (4) shows that the threshold voltage of the FGMOS transistor can be tuned and made much lower than the threshold voltage of the standard MOSFET by choosing proper values of V_b, k_1, k_2.

Fig. 3 shows the DC transfer characteristic of FGMOS and standard MOS transistor. For the FGMOS transistor, the value of bias voltage $V_b = 0.75$ V and the capacitive coupling ratio $k_1 = k_2 = 0.5$. It is observed that in case of FGMOS transistor the threshold voltage is completely removed from signal path by choosing proper value of bias voltage.

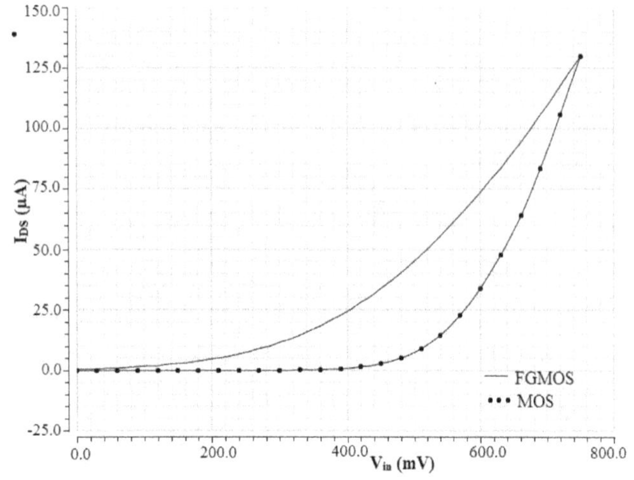

Fig. 3. Transfer characteristic of standard MOS and FGMOS transistor.

3. Proposed Four-Quadrant Multiplier

The principle of operation of the proposed multiplier is based on well known square-difference identity and is given by

$$V_{out} = (V_1 + V_2)^2 - (V_1 - V_2)^2 = 4V_1V_2 . \quad (5)$$

It is evident from (5) that in order to implement the multiplier function, square of the sum and difference of two voltages V_1 and V_2 are needed. The squarer circuit (proposed by Gupta et al.) used to implement the squared sum and difference of two inputs V_1 and V_2 is shown in Fig. 4 [31].

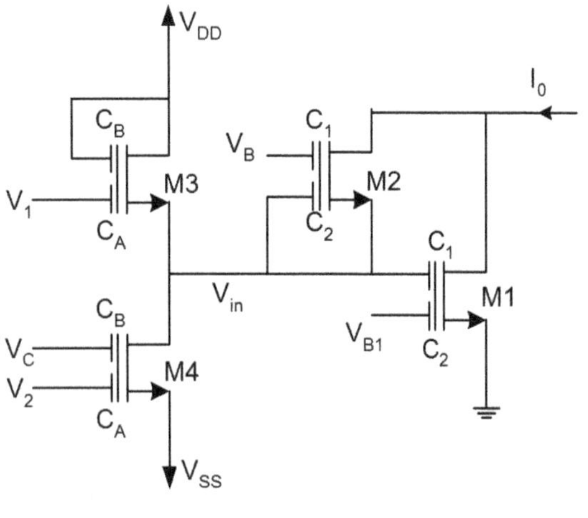

Fig. 4. Squarer circuit [31].

The output current of the basic squarer [32] (formed by transistors M1 and M2) after neglecting the parasitic capacitances, channel-length modulation, mobility degradation and the body effect is given by

$$I_O = \frac{\beta_1}{2}\left(k_1 V_{BS1} + k_2 V_{inS1} - V_{T1}\right)^2 \;\; if\; V_{in} > 0 , \quad (6)$$

$$I_O = \frac{\beta_2}{2}\left(k_1 V_{BS2} + k_2 V_{inS2} - V_{T2}\right)^2 \;\; if\; V_{in} < 0 . \quad (7)$$

If $\beta_1 = \beta_2 = \beta$, $V_{T1} = V_{T2} = V_T$, $k_1 = k_2 = k$ and $kV_B = V_T$ then the output current of the squarer can be approximated as [32]

$$I_O = \frac{\beta}{2}\left(kV_{in}\right)^2 . \quad (8)$$

The input voltage V_{in} of the squarer is generated by voltage attenuator formed by FGMOS transistor M3 and M4. The output voltage of the attenuator is given by [26],

$$V_{in} = \alpha\left(V_1 - V_2\right) + \frac{k_A V_{SS} - k_B V_C}{k_A + k_B} \quad (9)$$

where $k_A = C_A/(C_A + C_B)$, $k_B = C_B/(C_A + C_B)$ are the capacitive coupling ratios and $\alpha = k_A/(k_A + k_B)$, is the attenuation factor which can be tuned by choosing proper values of k_A and k_B. From (9) it can be seen that the offset voltage term $(k_A V_{ss} - k_B V_C) / (k_A + k_B)$ can be cancelled by the proper choice of the bias voltage V_C. For offset cancellation, the bias voltage V_C must be equal to

$$V_C = \frac{k_A}{k_B} V_{SS} . \quad (10)$$

Assuming zero-output offset for the voltage attenuator, the output current of the squarer using (8) and (9) is given as

$$I_O = \frac{\beta}{2}\left\{k\alpha\left(V_1 - V_2\right)\right\}^2 . \quad (11)\backslash$$

If $k\alpha = k_{eq}$, (11) can be written as

$$I_O = \frac{\beta}{2}\left\{k_{eq}\left(V_1 - V_2\right)\right\}^2 . \quad (12)$$

It can be seen from (12) that the squarer (Fig. 4) gives the output current proportional to the difference of input voltages V_1 and V_2 and the voltage range of the squarer can be determined by the factor k_{eq}.

The proposed multiplier (Fig. 5) has been implemented by using two squarers shown in Fig. 4. The first squarer (Fig. 5) has been formed by transistors M1, M2, M3, M4 with inputs V_1 and V_2; the second squarer has been formed by transistors M1d, M2d, M3d, M4d with inputs V_1 and $-V_2$. Since the inputs are applied at the gates of FGMOS transistor, so the proposed configuration has very high input impedance. The output impedance of the proposed multiplier can be given as

$$R_{out} = R_1 \| r_{Ox} = R_2 \| r_{Oy} \quad (13)$$

where R_1 and R_2 are the load resistances and $r_{ox,y}$ is the small signal output resistance of the transistor (where $x = 1$ or 2 and $y = 1$d or 2d).

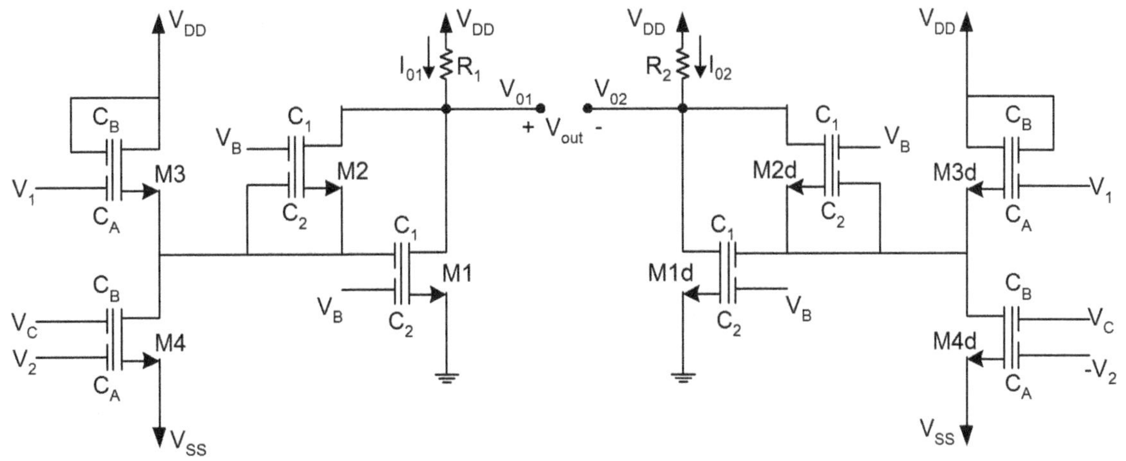

Fig. 5. Proposed four-quadrant multiplier.

The output of the squarer with inputs V_1 and V_2 is given by (assuming $R_1 = R_2 = R$)

$$V_{o1} = V_{DD} - R\frac{\beta}{2}\{k_{eq}(V_1 - V_2)\}^2. \quad (14)$$

The output of the squarer with inputs V_1 and $-V_2$ is given by

$$V_{o2} = V_{DD} - R\frac{\beta}{2}\{k_{eq}(V_1 + V_2)\}^2. \quad (15)$$

Thus the output of the multiplier using (14) and (15) is given as

$$V_{out} = V_{o1} - V_{o2} = 2\beta R k_{eq}^2 V_1 V_2. \quad (16)$$

It is evident from (16) that the output voltage V_{out} is equal to the four quadrant multiplication of input voltages V_1 and V_2 and the voltage range being determined by the factor k_{eq}. Therefore, the factor k_{eq} should be chosen properly so as to maximize the input range.

4. Second Order Effects

The operation of the multiplier has been analyzed by neglecting the deviations from ideal square-law characteristics due to component mismatch, mobility degradation and temperature dependence. These non-ideal effects are the basic source of discrepancy between the ideal and simulated output voltage of the proposed multiplier.

4.1 Component Mismatch

Assuming that the load resistors of multiplier are R_1, R_2 and the mismatch between these resistors is ΔR_L, the output voltage of the multiplier considering this mismatch is given by

$$V_{out} = R_1\frac{\beta}{2}\{k_{eq}(V_1 + V_2)\}^2 - R_2\frac{\beta}{2}\{k_{eq}(V_1 - V_2)\}^2. (17)$$

Equation (17) can be simplified as

$$V_{out} = \frac{\beta}{2}k_{eq}^{\ 2}\{\Delta R_L(V_1^2 + V_2^2) + 2V_1V_2(R_1 + R_2)\}. \quad (18)$$

It can be seen from (18) that the mismatch in load resistors R_1 and R_2 produces an error voltage proportional to the square of the two input voltages V_1 and V_2.

4.2 Mobility Degradation

Drain current equation including the mobility degradation effect can be modeled as

$$I_{DS} = \frac{1}{2}\left\{\frac{\mu_O}{1 + \theta(V_{GS} - V_T)}\right\}C_{OX}\left(\frac{W}{L}\right)(V_{GS} - V_T)^2 \quad (19)$$

where θ is the mobility degradation parameter whose value is 0.1~0.001 V^{-1}. Using Taylor series expansion equation (19) can be rewritten as

$$I_{DS} = \frac{1}{2}K_n\left(\frac{W}{L}\right)(V_{GS} - V_T)^2 \quad (20)$$
$$\{1 - \theta(V_{GS} - V_T) + \theta^2(V_{GS} - V_T)^2 - \theta^3(V_{GS} - V_T)^3 + \ldots\}$$

where $K_n = \mu_o C_{ox}$. Considering mobility degradation effect and using (20) the output voltage of the multiplier can be modeled as

$$V_{out} = R_1 k_{eq}^{\ 2}\beta V_1 V_2 + \varepsilon \quad (21)$$

where the error in output voltage is given by

$$\varepsilon = -\theta\beta R_1 k_{eq}^{\ 2} V_2 (V_2^2 + 3V_1^2). \quad (22)$$

Due to extremely small value of θ the output voltage of the multiplier will be slightly affected by the errors introduced due to mobility degradations.

4.3 Temperature Effect

The output of the proposed multiplier will vary because of the temperature dependence of its gain factor (βR), as can be seen from (16).

The temperature dependence of resistance R for any crystalline metal is given as

$$R_T = R_{ref}\left[1+\alpha\left(T-T_{ref}\right)\right] \quad (23)$$

where R_T is the resistance at any temperature T, R_{ref} is the resistance at reference temperature (usually 20°C but sometimes it is 0°C) and α is the temperature coefficient of resistance.

Assuming $R = 20\ \text{k}\Omega$ at $20°C$, $\alpha = .004$ (copper), from (23) the value of resistance R at $30°C$ is 20.8 kΩ and the percentage increase in resistance is approximately 4 %.

The transconductance parameter β depends upon mobility of carriers; the temperature dependence of the mobility is given as

$$\mu(T) = \mu(T_r)\left(\frac{T}{T_r}\right)^{\gamma} \quad (24)$$

where T is the absolute temperature (300 K), T_r is the temperature at which the mobility is to be calculated, and γ is a constant between 1.5 and 2.0. According to (24), for temperature variation of 10°C around room temperature the percentage decrease in mobility is 4.7 % (for $\gamma = 1.5$) and 6.3 % (for $\gamma = 2$).

For 10°C variation in temperature, increase in resistance R is 4 % and decrease in mobility is 4.7 % or 6.3 % for $\gamma = 1.5$ and 2 respectively.

According to (16), the output voltage variations caused due to temperature changes will be cancelled up to some extent. But for wider temperature range gain variations can be compensated to large extent by implementing load resistor using diode connected PMOS transistor. The resistance of diode connected MOS transistor is given by

$$R_{MOS} \propto \left\{\mu_0 C_{ox}\left(V_{GS} - V_T\right)\right\}^{-1}. \quad (25)$$

It can be seen from (25) that for large value of the gate to source voltage the variation in threshold voltage with temperature can be neglected [18]. Above equation shows that for the temperature variation of 10°C around room temperature if decrease in mobility is 4.7 % as discussed above, the variation in resistance R_{MOS} is just inverse and therefore, the variation in output voltage is cancelled out.

5. Simulation Results

The designed circuits are simulated using Cadence Spectre simulator in TSMC 0.18 µm CMOS technology using ±0.75 V power supply. The design parameters of the proposed circuit are given in Tab. 1. The approximate value of V_B (Tab. 1) is chosen by using equation $kV_B = V_T$. The value of threshold voltage V_T has been taken from the TSMC 180 nm CMOS model file specified in Spectre simulator of Cadence and $k = C_1/C_2 = 1/2$. At $V_B = 750$ mV, the threshold voltage is being cancelled out in simulations. By using equation $V_C = (k_A/k_B)V_{SS}$ the value of V_C (Tab. 1) is chosen, where, $k_A/k_B = C_A/C_B = 1/3$ and $V_{SS} = -750$ mV.

Design parameters	Values
Transistor sizes (µm)	M1,M2,M1d,M2d : W=4.4, L=0.18
	M3,M4,M3d,M4d: W=0.54, L=0.18
$R_1 = R_2$ (kΩ)	20
FGMOS capacitances(fF)	$C_B = 432$, $C_A = 144$ $C_1 = C_2 = 100$
Bias voltages (V)	$V_C = -0.25$, $V_B = 0.75$

Tab. 1. Design parameters of the proposed multiplier.

Since the floating gate (FG) of FGMOS does not have any connection to ground, the simulator cannot understand the floating gate and reports dc convergence problem during simulation. To avoid dc convergence error during simulation, model suggested in [10] has been used in this work. This model is based on connecting large value resistors in parallel with the input capacitors as shown in Fig. 6.

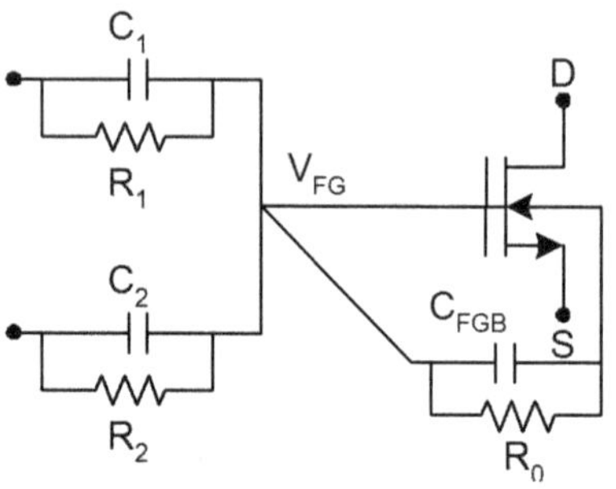

Fig. 6. Simulation model of two input FGMOS.

In this model the relation between resistances and capacitances can be given as follows: $R_i = 1/(kC_i) = 1000\ \text{G}\Omega$.

The DC response of the multiplier is shown in Fig. 7.

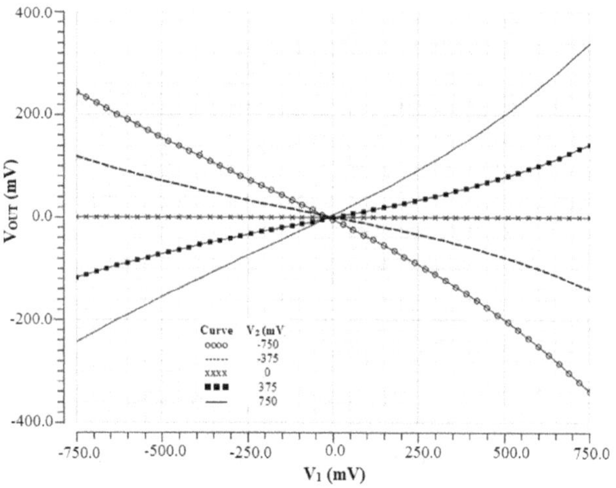

Fig. 7. DC response of the multiplier.

The figure shows that for input varying from -750 mV to 750 mV the output of the multiplier varies linearly up to ±250 mV but significant nonlinearities occur for input voltages higher than ±250 mV. The proposed multiplier operates at low supply voltage (±0.75 V) with total quiescent power consumption of only 35.28 μW. The linearity error corresponding to V_1 = 750 mV and V_2 = 500 mV is 4.1 %.

The input and output impedance plots for the proposed multiplier are shown in Fig. 8a and 8b respectively. It is observed from the figures 8a and 8b that the suggested topology has the high input impedance of 192 GΩ and moderate output impedance of 19 kΩ respectively.

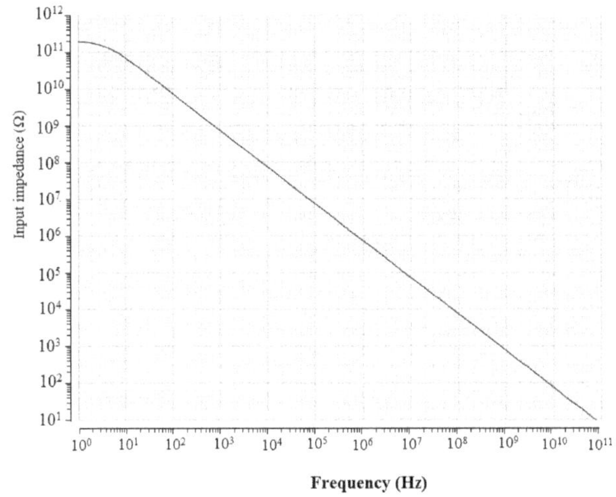

Fig. 8a. Input impedance.

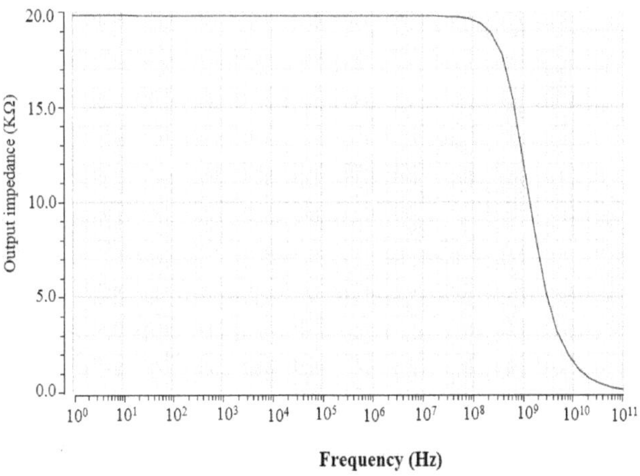

Fig. 8b. Output impedance.

The Total Harmonic Distortion (THD) for the proposed multiplier with V_1 as 1 kHz sinusoid while V_2 as 750 mV is shown in Fig. 9. The maximum THD due to second order effects, such as component mismatch, mobility degradation and temperature variation is about 6.8 %. In the proposed circuit for input voltage up to 90 mV, THD is less than / equal to 1% and the input noise is 457.5 μ$V_{rms.}$ Using these values the input dynamic range is found to be 45.8 dB.

The dynamic range of the circuit can be increased by varying both the capacitive coupling ratios and the aspect ratios of the transistors. Fig. 10a shows the DC response of the proposed multiplier for the capacitor values of C_B = 432 fF, C_A = 54 fF and the corresponding THD plot is shown in Fig. 10b. In the proposed circuit for input voltage up to 130 mV, THD is less than / equal to 1 % and input noise is 389.4 μV_{rms}. Using these values the input dynamic range is found to be 50.4 dB. The dynamic range of the proposed circuit increases at the cost of reduced output voltage range.

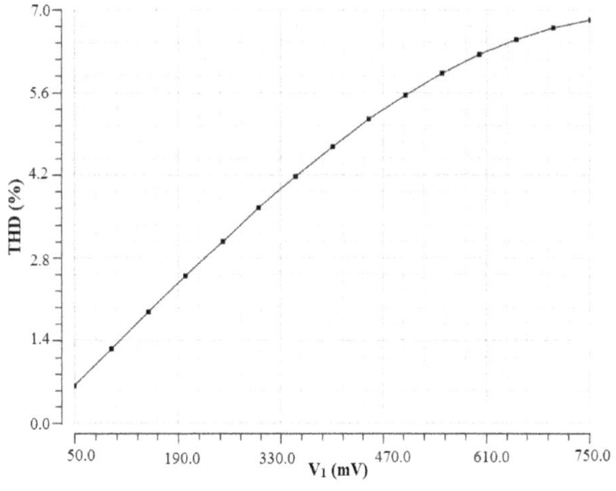

Fig. 9. THD plot.

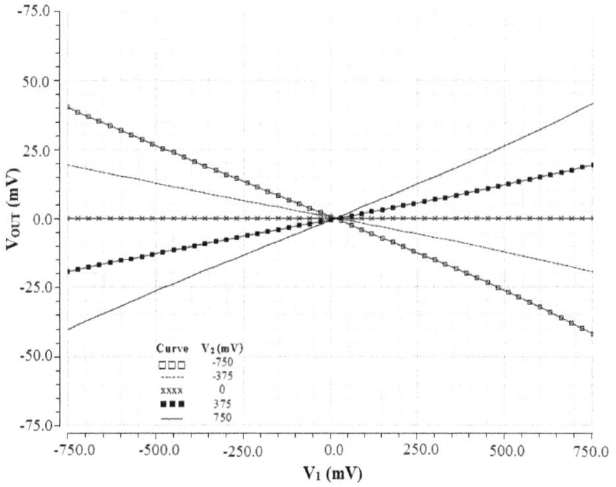

Fig. 10a. DC response of the multiplier for C_B = 432 fF and C_A = 54 fF.

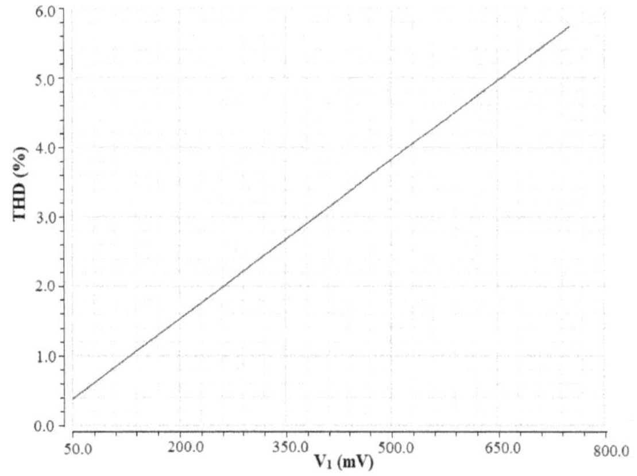

Fig. 10b. Percentage THD for C_B = 432 fF and C_A = 54 fF.

Fig. 11a shows the DC response of the proposed multiplier for the aspect ratios $(W/L)_1 = (W/L)_2 = (W/L)_{1d} = (W/L)_{2d} = (24\ \mu m / 6\ \mu m)$ and the corresponding THD plot is shown in Fig. 11b. In the proposed circuit for input voltage range up to 170 mV, THD is less than or equal to 1 % and input noise is 484.8 μV_{rms}. Using these values the input dynamic range is found to be 50.8 dB.

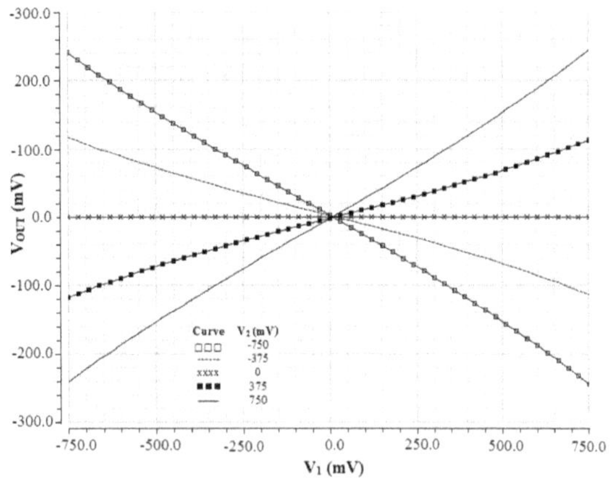

Fig. 11a. DC response of the multiplier for $(W/L)_1 = (W/L)_2 = (W/L)_{1d} = (W/L)_{2d} = (24\ \mu m / 6\ \mu m)$.

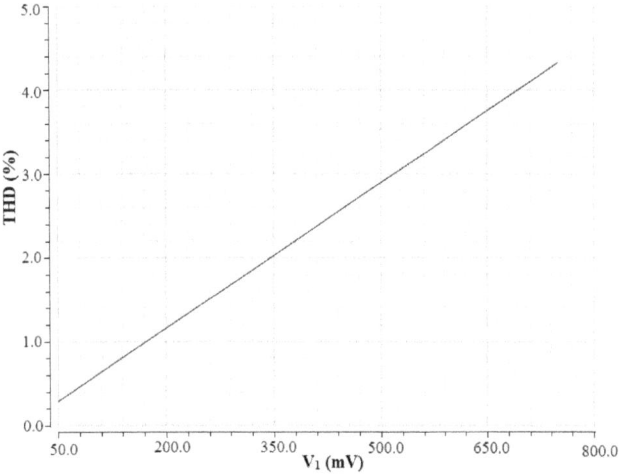

Fig. 11b. Percentage THD for $(W/L)_1 = (W/L)_2 = (W/L)_{1d} = (W/L)_{2d} = (24\ \mu m / 6\ \mu m)$.

The frequency response of the multiplier at different values of temperature is shown in Fig. 12a. It can be seen that as the temperature varies from -50°C to 50°C the gain of the multiplier varies from -25.2 dB to -13.9 dB and the pole frequency varies from 205.6 MHz to 168.9 MHz respectively.

The frequency response of the multiplier with temperature compensated gain is shown in Fig. 12b. For temperature compensation the load resistors R_1, R_2 are replaced by diode connected PMOS transistors with $W = 10.8\ \mu m$ and $L = 0.18\ \mu m$. The frequency response of the temperature compensated multiplier shows that as the temperature varies from -50°C to 50°C the variation of gain is very small (-19.0 to -19.5 dB) and the pole frequency varies from 77.7 MHz to 155.5 MHz respectively.

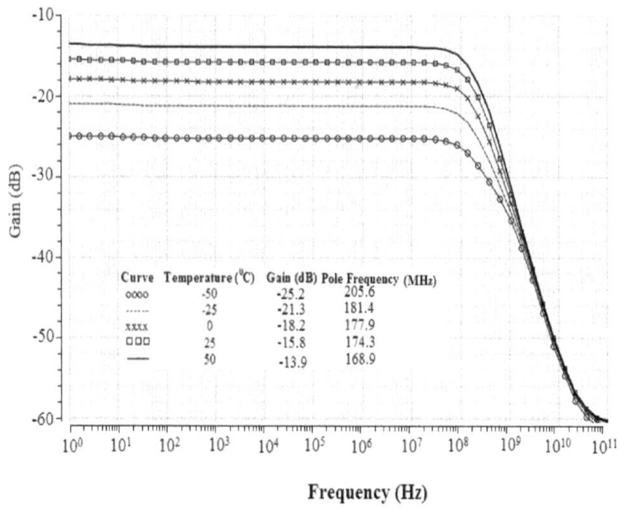

Fig. 12a. Frequency response of the multiplier at different temperatures.

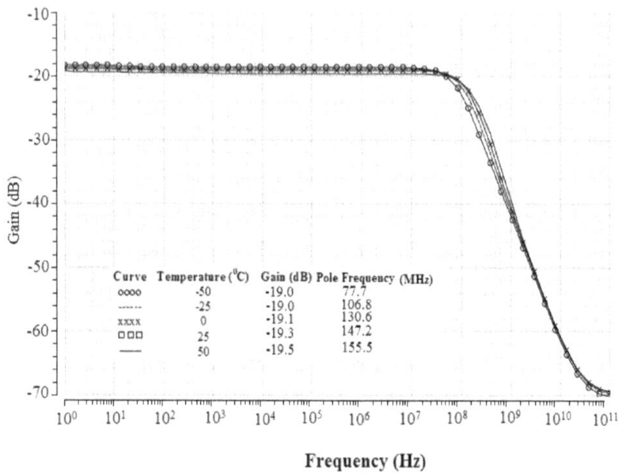

Fig. 12b. Frequency response with temperature compensated gain.

Finally, to demonstrate the effectiveness of the proposed configuration the circuit has been employed as an amplitude modulator, as shown in Fig. 13. The carrier

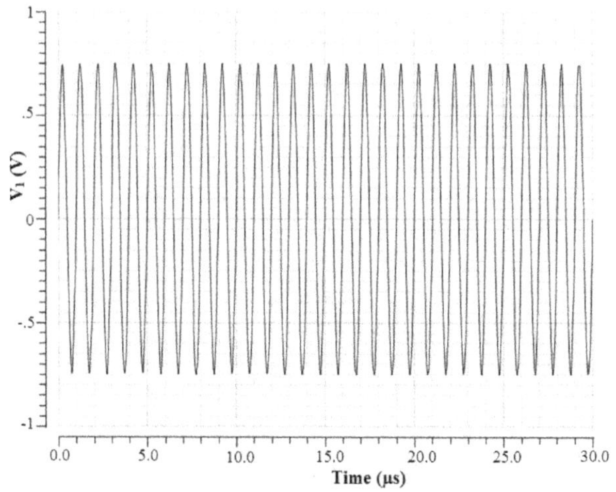

Fig. 13a. Carrier input signal V_1.

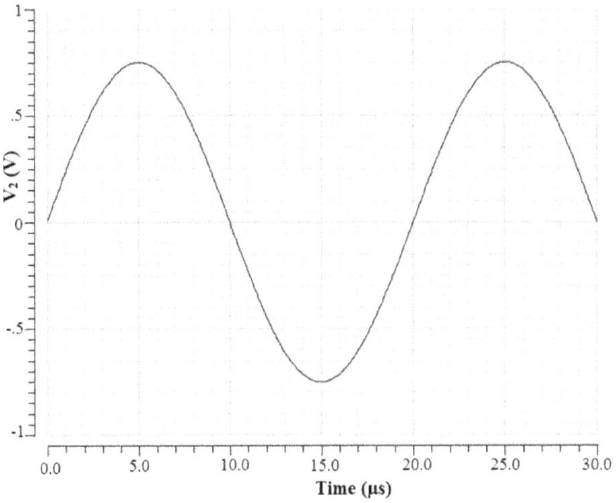

Fig. 13b. Modulating input signal V_2.

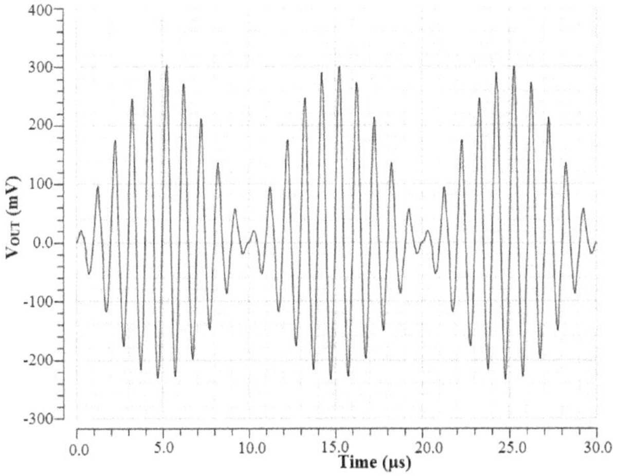

Fig. 13c. Output of the amplitude modulator.

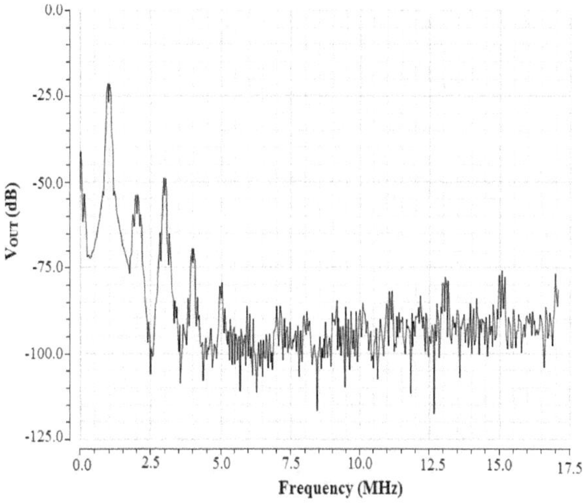

Fig. 13d. Spectrum of the output of the amplitude modulator.

input V_1 with frequency 1 MHz, amplitude 750 mV and the modulating input V_2 with frequency 50 kHz, amplitude 750 mV are shown in Fig. 13a and 13b respectively. The modulated output waveform is shown in Fig. 13c and its spectrum is shown in Fig. 13d. From Fig. 13d it can be seen that the maximum distortion peak is 20 dB below the desired modulation

Tab. 2 compares the performance parameters of the proposed circuit with various commercially available multipliers (AD633, AD734, AD835) and the multipliers based on CMOS/BJT/FGMOS technology.

It is observed that among all the multipliers mentioned in Tab. 2, the proposed configuration has the highest input impedance of 192 GΩ which makes the signal source loading effect negligible in the circuit. Apart from this, the

Parameters	AD633	AD734	AD835	[22]	[23]	[27]	[28]	[29]	Proposed
Technology (μm)	NA	NA	NA	0.35	1	2	0.8	0.35	0.18
Maximum Supply Voltage (V)	±18	±16.5	±5.5	1.5	±1.5	±2.5	±1	2	±0.75
Input range	±10 V	±12.5 V	±1 V	±60 mV	±0.8 V	100 % of supply voltage	100 % of supply voltage	100 % of supply voltage	100 % of supply voltage
Linearity error (%)	±1 (X=±10 V) (Y =+10 V)	0.05 (X=±10 V) (Y =+10 V)	0.5 (X=±1V, Y=1V)	< 3.2 (for input range ±60 mV)	< 2 (for input range ±0.8 V)	< 0.5 (for input range ±2.5 V)	NA	0.0081	4.1 (V_1=750mV, V_2=500 mV)
No. of Components	NA	NA	NA	8 PMOS, 2 Resistors, 4 Biasing current sources	8 NMOS, 4 BJT, 2 Resistors	9 FGMOS, 2 Resistors, 3 biasing current sources	4 FGMOS, 2 MOS, 1 OTA	6 FGMOS, 6 MOS, 3 biasing current sources	8 FGMOS, 2 Resistors
Power consumption	NA	NA	NA	6.7 μW (Maximum)	50 μW (Static)	NA	Large power consumption	NA	35.28 μW (Static)
Input resistance	10 MΩ	50 kΩ	60 kΩ	NA	NA	NA	NA	NA	192 GΩ
Output resistance	NA	NA	NA	NA	NA	NA	NA	NA	19 kΩ
Bandwidth (MHz)	1	10	250	0.26	10	NA	NA	1400	173
THD (%)	NA	NA	NA	4.2	NA	NA	1.4	2.6	6.8

Tab. 2. Comparison of various conventional and proposed multipliers.

proposed multiplier has the simplest structure, operates at the lowest supply voltage of ±0.75 V and has low quiescent power consumption of 35.28 µW at the cost of low linearity.

The layout of the proposed multiplier is shown in Fig. 14 and the associated layout-level simulations are given in Figs. 15–17. Fig. 14 shows that the designed multiplier occupies an area of 69.9 × 26.4 µm^2. All the DC responses (Fig. 15a, 16a, 17a) have been obtained by varying V_1 from -750 mV to 750 mV and V_2 = -750 mV.

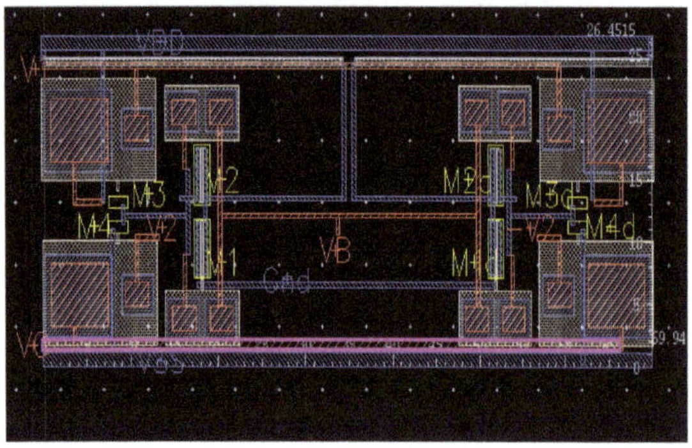

Fig. 14. Layout of the proposed multiplier.

The proposed multiplier is based on only NMOS transistors; therefore it has three process corners i.e. Typical (T), Fast (F) and Slow (S). The simulated DC and AC responses of the proposed circuit (Fig. 5) at different process corners are shown in Figs. 15a and 15b respectively. Fig. 15a shows that at V_1 = -750 mV and V_2 =-750 mV the output voltages are 243.9 mV, 399.5 mV and 152 mV for the process corners T, F and S respectively. Similarly, at V_1 = 750 mV and V_2 = -750 mV the output voltages are -334 mV, -419 mV and -138 mV at process corners T, F and S respectively. The AC response (Fig. 15b) shows that at process corners T, F and S the gain varies from -15.6 dB to -4.9 dB and the corresponding pole frequency varies from 173.3 MHz to 282.4 MHz.

The simulated DC and AC responses of the proposed circuit (Fig. 5) while considering the mismatch between the load resistances are shown in Figs. 16a and 16b respectively. The DC response (Fig. 16a) shows that as one of the load resistances varies from 18 kΩ to 22 kΩ (keeping other load resistance constant at 20 kΩ); the output voltage of the multiplier varies from -362 mV to -300 mV respectively. The frequency response (Fig. 16b) shows that as one of the load resistance varies from 18 kΩ to 22 kΩ the gain of the multiplier varies from -16.25 dB to -14.9 dB and the pole frequency varies from 163 MHz to 173 MHz respectively.

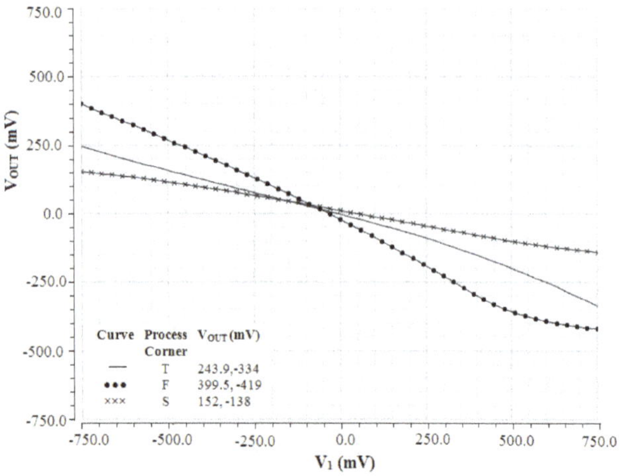

Fig. 15a. DC response at different process corners.

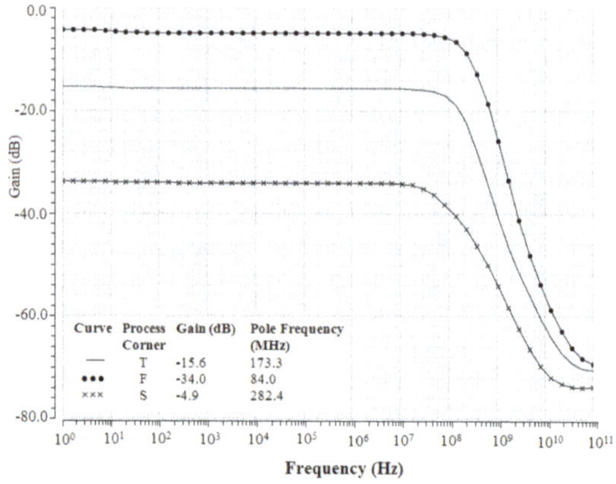

Fig. 15b. Frequency response at different process corners.

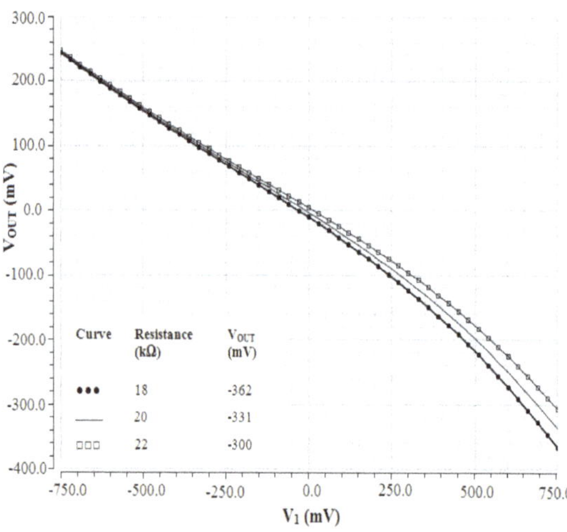

Fig. 16a. DC response at different values of load resistance.

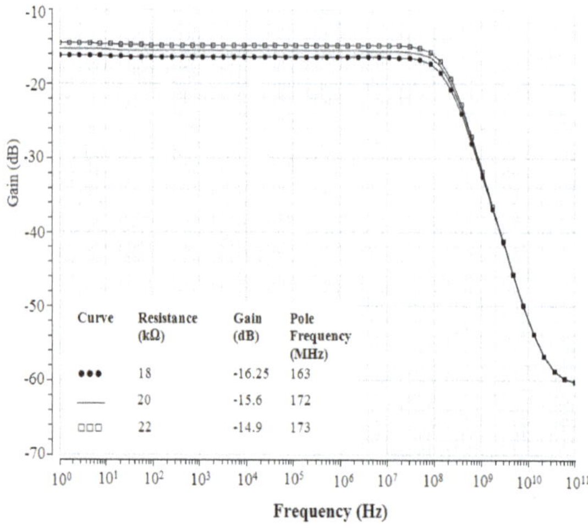

Fig. 16b. Frequency response at different values of load resistance.

The simulated DC and AC responses of the proposed circuit at different temperatures are shown in Figs. 17a and 17b respectively. The DC response (Fig. 17a) shows that as the temperature varies from -50°C to 50°C the output voltage of the multiplier varies from 109 mV to 277 mV for V_1 = -750 mV and V_2 = -750 mV. Similarly, as the temperature varies from -50°C to 50°C the output voltage of the multiplier varies from -113 mV to -395 mV for V_1 = 750 mV and V_2 = -750 mV. The frequency response (Fig. 17b) shows that as the temperature varies from -50°C to 50°C the gain of the multiplier varies from -27.4 dB to -13.4 dB and the pole frequency varies from 155.2 MHz to 125.8 MHz respectively.

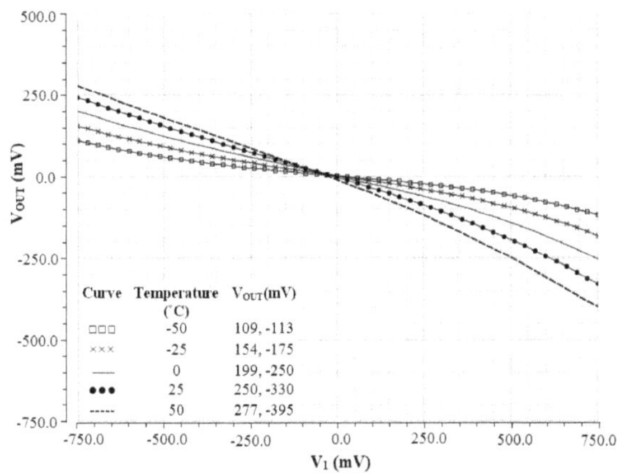

Fig. 17a. DC response at different temperatures.

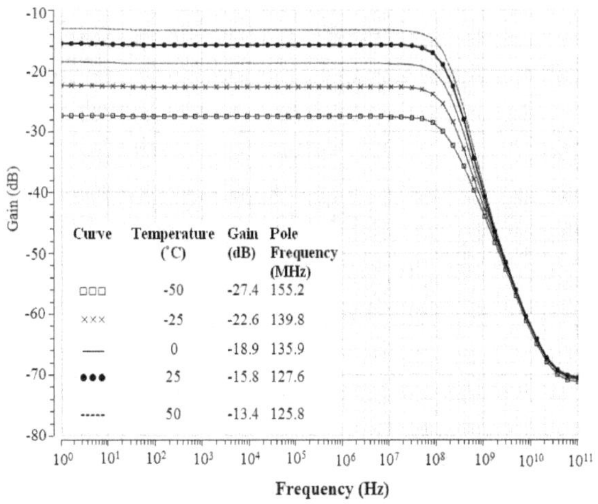

Fig. 17b. Frequency response at different temperatures.

6. Conclusions

In this paper, a simple FGMOS transistor based four-quadrant multiplier has been proposed. First the sum and difference of the input voltages are squared and then their difference is taken so as to generate the four quadrant multiplier function. The proposed circuit operates at ±0.75 V with quiescent power consumption of only 35.28 µW and the bandwidth of approximately 173 MHz. Thus the newly developed four-quadrant multiplier is one of the best choices for low voltage/low power analog signal processing applications.

References

[1] RAJPUT, S. S., JAMUAR, S. S. Low voltage analog circuit design techniques. *IEEE Circuits and Systems Magazine*, 2002, vol. 2, no. 1, p. 24–42. DOI: 10.1109/MCAS.2002.999703.

[2] HAGA, Y., ZARE-HOSEINI, H., BERKOVI, L., KALE, I. Design of a 0.8 Volt fully differential CMOS OTA using the bulk- driven technique. In *IEEE International Symposium on Circuits and Systems*. Kobe (Japan), 2005, vol. 1, p. 220–223.

[3] AGGARWAL, B., GUPTA, M., GUPTA, A. K. Analysis of low voltage bulk-driven self-biased high swing cascode current mirror. *Microelectronics Journal*, 2013. vol. 44, no. 3, p. 225–235. DOI: 10.1016/j.mejo.2012.12.006.

[4] BERG, Y., LANDE, T. S., NAESS, S. Low-voltage floating gate current mirrors. In *10th Annual IEEE International ASIC Conference and Exhibit*. Portland (USA), 1997, p. 21–24. DOI: 10.1109/ASIC.1997.616971.

[5] LANDE, T. S., WISLAND, D. T., SOETHER, T., BERG, Y. FLOGIC-Floating gate logic for low-power operation. In *3rd IEEE International Conference on Electronics, Circuits and Systems*. Rhodos (Greece), 1996, vol. 2, p. 1041–1044. DOI: 10.1109/ICECS.1996.584565.

[6] MINAEI, S., YUCE, E. New squarer circuits and a current-mode full-wave rectifier topology suitable for integration. *Radioengineering*, 2010, vol. 19, no. 4, p. 657–661.

[7] GUPTA, M., PANDEY, R. FGMOS based voltage-controlled resistor and its applications. *Microelectronics Journal*, 2010, vol. 41, no. 1, p. 25–32. DOI: 10.1016/j.mejo.2009.12.001.

[8] GUPTA, M., PANDEY, R. Low-voltage FGMOS based analog building blocks. *Microelectronics Journal*, 2011, vol. 42, no. 6, p. 903-912. DOI: 10.1016/j.mejo.2011.03.013.

[9] PANDEY, R., GUPTA, M. FGMOS based tunable grounded resistor. *Analog Integrated Circuits & Signal Processing*, 2010, vol. 65, no. 3, p. 437–443. DOI: 10.1007/s10470-010-9500-x.

[10] RODRIGUEZ-VILLEGAS, E. Low power and low voltage circuit design with the FGMOS transistor. *IET Circuits, Devices and Systems Series 20*. The Institution of Engineering and Technology, London, United Kingdom, 2006.

[11] LOPEZ-MARTIN, A. J., RAMIREZ-ANGULO, J., CHINTHAM, R., CARVAJAL, R. G. Class AB CMOS analogue squarer circuit. *Electronics Letters*, 2007, vol. 43, no. 20, p. 1059–1060.

[12] GILBERT, B. A high-performance monolithic multiplier using active feedback. *IEEE Journal of Solid-State Circuits*, 1974, vol. 9, no. 6, p. 364–373.

[13] GILBERT, B. A precise four-quadrant multiplier with subnanosecond response. *IEEE Journal of Solid-State Circuits*, 1968, vol. 3, no. 4, p. 365–373.

[14] KIMURA, K. A bipolar four-quadrant analog quarter-square multiplier consisting of unbalanced emitter-coupled pairs and expansions of its input ranges. *IEEE Journal of Solid-State Circuits*, 1994, vol. 29, no. 1, p. 46–55.

[15] BABANEZHAD, J. N., TEMES, G. C. 20-V four-quadrant CMOS analog multiplier. *IEEE Journal of Solid-State Circuits*, 1985, vol. 20, no. 6, p. 1158–1168.

[16] QIN, S. C., GEIGER, R. L. A ±5-V CMOS analog multiplier. *IEEE Journal of Solid-State Circuits*, 1987, vol. 22, no. 6, p. 1143 to 1146.

[17] SOO, D. C., MEYER, R. G. A four-quadrant NMOS analog multiplier. *IEEE Journal of Solid-State Circuits*, 1982, vol. 17, no. 6, p. 1174–1178.

[18] SONG, H. J., KIM, C. K. An MOS four-quadrant analog multiplier using simple two-input squaring circuits with source followers. *IEEE Journal of Solid-State Circuits*, 1990, vol. 25, no. 3, p. 841 to 848. DOI: 10.1109/4.102683.

[19] LIU, S. I., HWANG, Y. S. CMOS squarer and four-quadrant multiplier. *IEEE Transactions on Circuits and Systems-I: Fundamental Theory and Applications*, 1995, vol. 42, no. 2, p. 119–122. DOI: 10.1109/81.372853.

[20] YASUMOTO, M., ENOMOTO, T. Integrated MOS four-quadrant analogue multiplier using switched-capacitor technique. *Electronics Letters*, 1982, vol. 18, no. 18, p. 769–771.

[21] RAMÍREZ-ANGULO, J. Highly linear four quadrant analog BiCMOS multiplier for ±1.5 V supply operation. *Electronics Letters*, 1992, vol. 28, no. 19, p. 1783–1785.

[22] LIU, W., LIU, S. I. Design of a CMOS low-power and low-voltage four-quadrant analog multiplier. *Analog Integrated Circuits & Signal Processing*, 2010, vol. 63, no. 2, p. 307–312.

[23] LIU, S. I, LEE, J. L., CHANG, C. C. Low-voltage BiCMOS four-quadrant multiplier and squarer. *Analog Integrated Circuits and Signal Processing*, 1999, vol. 20, no. 1, p. 25–29.

[24] YUCE, E. Design of a simple current-mode multiplier topology using a single CCCII+. *IEEE Transactions on Instrumentation and Measurement*, 2008, vol. 57, no. 3, p. 631–637.

[25] KESKIN, A. U. A four quadrant analog multiplier employing single CDBA. *Analog Integrated Circuits & Signal Processing*, 2004, vol. 40, no. 1, p. 99–101.

[26] VLASSIS, S., SISKOS, S. Differential-voltage attenuator based on floating-gate MOS transistors and its applications. *IEEE Transactions on Circuits and Systems-I: Fundamental Theory and Applications*, 2001, vol. 48, no. 11, p. 1372–1378.

[27] VLASSIS, S., SISKOS, S. Analogue squarer and multiplier based on floating-gate MOS transistors. *Electronics Letters*, 1998, vol. 34, no. 9, p. 825–826. DOI: 10.1049/el:19980639.

[28] NAVARRO, I., LOPEZ-MARTIN, A. J., DE LA CRUZ, C. A., CARLOSENA, A. A compact four-quadrant floating-gate MOS multiplier. *Analog Integrated Circuits & Signal Processing*, 2004, vol. 41, no. 2–3, p. 159–166.

[29] KELEŞ, S., KUNTMAN, H. H. Four quadrant FGMOS analog multiplier. *Turkish Journal of Electrical Engineering & Computer Sciences*, 2011, vol. 19, no. 2, p. 291–301.

[30] SHARMA, S., RAJPUT, S. S., MANGOTRA, L. K., JAMUAR, S. S. FGMOS current mirror: behaviour and bandwidth enhancement. *Analog Integrated Circuits & Signal Processing*, 2006, vol. 46, no. 3, p. 281–286.

[31] GUPTA, M, SRIVASTAVA, R., SINGH, U. Low voltage floating gate MOS transistor based differential voltage squarer. *ISRN Electronics*, vol. 2014, article ID 357184, 6 pages. DOI:10.1155/2014/357184.

[32] SRIVASTAVA, R., GUPTA, M., SINGH, U. FGMOS transistor based low voltage and low power fully programmable Gaussian function generator. *Analog Integrated Circuits and Signal Processing*, 2014, vol. 78, no. 1, p. 245–252. DOI: 10.1007/s10470-013-0207-7.

About Authors ...

Richa SRIVASTAVA received the B. E. degree in Electronics and Communication from Dr. B. R. Ambedkar University, Agra, and M.Tech degree in VLSI Design from Banasthali Vidyapeeth, India in 2003 and 2006 respectively. During 2006-10, she was a lecturer in A. K. G. E. C, Ghaziabad, India. She is currently working towards the Ph.D. degree at Netaji Subhas Institute of Technology (NSIT), New Delhi, India. Since August 2010, she has been with NSIT, where her research focuses on design of analog integrated circuits for low voltage/low power applications. She has thorough experience on working with various industry-standard VLSI design tools (Tanner EDA; Cadence Virtuoso).

Maneesha GUPTA received B.E. in Electronics and Communication Engineering from Government Engineering College, Jabalpur in 1981, M.E. in Electronics and Communication Engineering from Government Engineering College, Jabalpur in 1983 and Ph.D. in Electronics Engineering (Analysis, Synthesis and Applications of Switched Capacitor Circuits) from Indian Institute of Technology, Delhi in 1990. She is working as Professor in the Division of Electronics and Communication Engineering of the Netaji Subhas Institute of Technology, New Delhi from 2008. She is working in the areas of switched capacitors circuits and low voltage design techniques. She has co-authored over 60 research papers in the above areas in various international/national journals and conferences.

Urvashi SINGH received the M.Sc degree in Electronics and Communication from CSJM University and M.Tech degree in VLSI design from MITS University, India in 2007 and 2009 respectively. During 2009-10, she was a lecturer in GLA University, Mathura, India. She is currently working towards the Ph.D. degree at Netaji Subhas Institute of Technology (NSIT), New Delhi, India. Since August 2010, she has been with NSIT, where her research focuses on design of analog integrated circuits for broadband applications. She is also working on analog circuit characterization at both schematic and layout level. She has thorough experience on working with various industry-standard VLSI design tools (Mentor Graphics TC status; Cadence Virtuoso).

Digital Demodulator for BFSK Waveform Based Upon Correlator and Differentiator Systems

Jorge TORRES[1], *Fidel HERNANDEZ*[2], *Joachim HABERMANN*[3]

[1]Dept. of Telecommunication and Telematic, CUJAE University, 114 Street, 11901, La Habana, Cuba
[2]Dept. of Telecommunication and Electronics, UPR University, Marti 270, Pinar del Río, Cuba
[3]Department for ICT, Electrical Engineering & Mechatronics, TH Mittelhessen University of Applied Sciences,D-61169 Friedberg, Germany

[1] jorge.tg@electrica.cujae.edu.cu, [2] fidel@tele.upr.edu.cu, [3] joachim.habermann@iem.thm.de

Abstract. *The present article relates in general to digital demodulation of Binary Frequency Shift Keying (BFSK waveform) . New processing methods for demodulating the BFSK-signals are proposed here. Based on Sampler Correlator, the hardware consumption for the proposed techniques is reduced in comparison with other reported. Theoretical details concerning limits of applicability are also given by closed-form expressions. Simulation experiments are illustrated to validate the overall performance.*

Keywords

BFSK, correlator, digital receiver.

1. Introduction

Frequency Shift Keying (FSK) is a digital modulation with applications in wireless technologies [1] and also in satellite communications [2]. Various configurations of circuitry have previously been reported for BFSK demodulators. As a case in point, there are systems with the Envelope Detector [3] and self-tuning systems [4].

Some of them, specially those based upon correlator receiver employment, are particularly interesting because they perform signal to noise ratio maximization [5] and do not use symbol synchronization blocks. As a case in point, such detectors are Quadricorrelator [6], [7], Balanced Quadricorrelator [7], [8], [9], [10], [11], [12] and Quotient Detector [13].

These types of detectors are implemented in a two-part structure; the first one is an analogue correlator [14]. In the second part, the instantaneous frequency is extracted by performing proper derivation, addition and multiplication procedures, Fig. 1 depicts the case of the Balanced Quadricorrelator. However, due to the large number of adders and multipliers needed, such digital implementations demand a high cost. For instance, the number of floating adders and multipliers exceeds 47 in the case of the Balanced Quadricorrelator. In this case an elliptic filter is employed for the digital lowpass filter, and a FIR filter of fifth order as a discrete time differentiator [15], both of them employed by the detectors mentioned above.

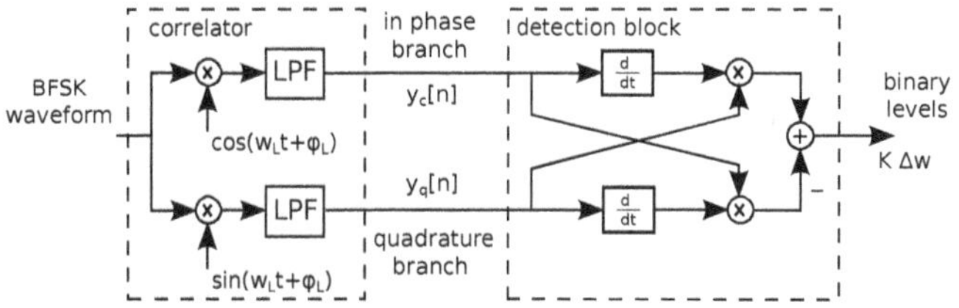

Fig. 1. Block scheme of the balanced quadricorrelator.

Although the low pass filter and the multiplication procedure could be obtained through a flip-flop element, as in the digital logic quadricorrelator scheme [16], and the differentiator can be accomplished by flip-flop blocks followed by AND-gates, this scheme is prone to instabilities under conditions of low signal to noise ratio, causing total interruption of the counting process at the system output, despite of its low hardware consumption.

The method discussed in this paper outperforms all previous schemes, based upon correlator, in a hardware consumption sense. New BFSK detection schemes based upon correlator are devised, which consider the recognition of slope instead of instantaneous frequency at the output of correlator. In this paper, new slope detection algorithm are derived.

2. Receiver Conception

2.1 Correlator Block

The main idea of the proposed receiver is to identify the approximately linear ramp observed at the output of a correlator for a sine-wave input [5]. The BFSK waveform is conformed with only two symbols, comprised by a radio-frequency pulse with two different frequencies, and they are orthogonal to each other, so if a tone is correlated with itself a linear increase in time is obtained at the output. Conversely, if such a tone is correlated with the other one, approximately a constant output is derived. Figure 2 b) shows the correlation operation for the BFSK signal shown in Fig. 2 a) when the reference is the lower frequency tone.

This behavior is obtained at the output of the correlator when the frequency w_L is the same as the received frequency w_0. Thus, techniques for recovering the average slope leads to new schemes for BFSK demodulators in the detection block. For instance, Fig. 2 c) shows high and low levels in accordance with the signal received in a) when substraction between consecutive points in b) is performed. This aspect will be considered in the next Section in detail.

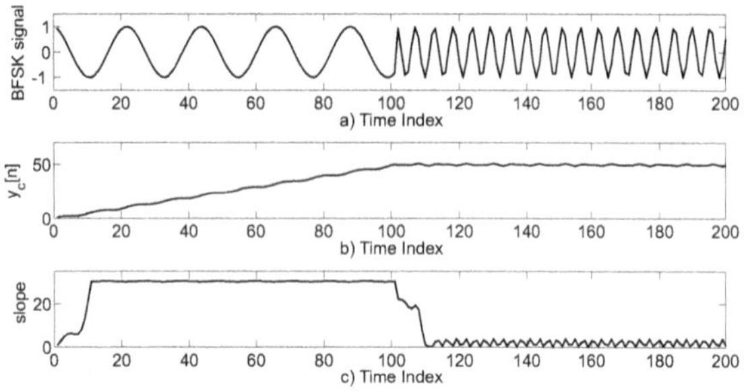

Fig. 2. Output of a correlator for a BFSK-wave input and a sine-wave reference.

In order to reduce complexity a Sampler Correlator [17] is employed instead of other types reported in scientific papers. It is preferred here since it has the same Signal to Noise Ratio (SNR) as the Analogue Correlator [18], which has higher SNR than other types such as the Digital [19], [20] and the Stieltjets [21], [14] receiver. Besides, modified versions of Stieljets [21] or Relay Correlator [19], [22], [23] are implemented with additional sources of random waves which add complexity to the receiver.

Figure 3 depicts the block diagram of the discrete correlator. It is comprised of an IIR filter acting as an accumulator. Besides, the analytic expressions for the output of the in phase branch are described in (1) when a symbol of the type $A\cos(w_i n+\varphi_i)$ is received. At the in-phase branch, in Fig. 3, the summation of $A\cos(w_i n+\varphi_i)\cdot\cos(w_L n+\varphi_L)$ from 0 to n is carried out using the well-known trigonometric identity called Euler's theorem and the sum formula for the geometric progression.

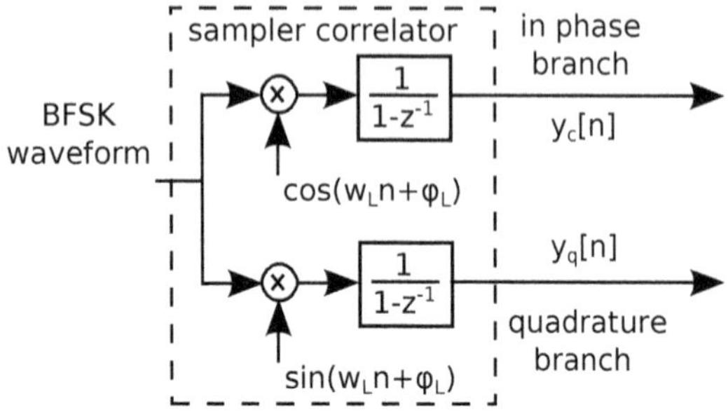

Fig. 3. Sampler correlator scheme.

$$y_c[n] = \frac{A}{4}\left\{\frac{1}{\sin\left(\frac{w_i-w_L}{2}\right)}\left[\sin\left(\varphi_L-\varphi_i+\frac{w_i-w_L}{2}\right)+\right.\right. \quad (1)$$
$$\left.+\sin\left(\varphi_i-\varphi_L+(w_i-w_L)\left(n+\frac{1}{2}\right)\right)\right]+$$
$$+\frac{1}{\sin\left(\frac{w_i+w_L}{2}\right)}\left[\sin\left(\frac{w_i+w_L}{2}-\varphi_L-\varphi_i\right)\right.$$
$$\left.\left.+\sin\left(\varphi_i+\varphi_L+(w_i+w_L)\left(n+\frac{1}{2}\right)\right)\right]\right\}.$$

On the other hand, expression (2) describes the case of a received symbol with frequency and phase (w_0, φ_0) and considering also that $w_L = w_0$. This is the case depicted in Fig. 2 b) when the symbol with the lower frequency is being transmitted, where the term $(n+1)\cos(\varphi_0-\varphi_1)$ is related to the ramp behavior. The expression (1) also describes the output of the correlator when the high frequency $w_i = w_1$ is received and $w_L = w_0$. The description for the quadrature branch is simply obtained by substituting φ_L by $\varphi_L - \frac{\pi}{2}$ in both expressions (1) and (2)

$$y_c[n] = \frac{A}{2}\left((n+1)\cos(\varphi_0-\varphi_L)+\right. \quad (2)$$
$$+\frac{1}{2\sin(w_0)}\left(\sin(w_0-\varphi_0-\varphi_L)+\right.$$
$$\left.\left.+\sin\left(2w_0\left(n+\frac{1}{2}\right)\varphi_0+\varphi_L\right)\right)\right).$$

2.2 Detection Block

Towards detecting the average slope in Fig. 2 the use of a differentiator systems is considered. This is a point of view different to that of the Balanced Quadricorrelator and the others scheme mentioned above. In case of the Balanced Quadricorrelator, the output is directly related to $\Delta w = w_i - w_L$. In the algorithm proposed in this paper, instead of extracting the instantaneous frequency we consider the methods of slope recognition at the correlator output. This is a problem that is also considered in applications for QRS detection [24].

Differentiator systems reported in scientific papers are either based on a FIR discrete-time differentiator [15], or simply perform substraction of consecutive points in the temporal input series. The FIR discrete-time differentiator is equivalent to the ideal derivator $H_d(e^{jw}) = jwe^{-jw\frac{M}{2}}$ in the frequency domain, except by additional windowing with Kaiser's window. M represents the filter order and for $M = 5$ a small amplitude approximation error is reported [15].

The differentiator systems without floating multiplier are also described by $H_1(z) = 1 - z^{-1}$, $H_2(z) = 1 - z^{-2}$, $H_4(z) = 2 + z^{-1} - z^{-3} - 2z^{-4}$ or $H_4(z) = 1 - z^{-4}$, first order [25], second order [26], and fourth order [27] respectively. However, a more general case can be considered when the substraction is performed between points at the distance of L samples; in this case, $H_L(z) = 1 - z^{-L}$. Performance analyzes and selection criteria regarding this system are discussed in the Result Section.

The differentiator systems, described above, extract correctly the term $cos(\varphi_0 - \varphi_1)$ from (2), identifying with a high level the interval in wich the symbol of low frequency is received, as in Fig. 2 c). By the other hand, the quadrature branch will detect the term $sin(\varphi_0 - \varphi_1)$ in (2) when φ_L is replaced by $\varphi_L - \frac{\pi}{2}$. After this, the squaring and adding of both branches will neglect the effects of phase in detection.

Figure 4 depicts the general scheme of the receiver with replications of the structure shown in Fig. 3 with different values of w_L; the upper pair of branches are configured for detecting the low frequency, the lower one are configured for the high frequency. The squaring procedures are needed for recovering the energy in both the in phase and quadrature branches. Nevertheless, spurious frequencies must be suppressed in order to obtain the average slope at the output. This can be accomplished through lowpass filters.

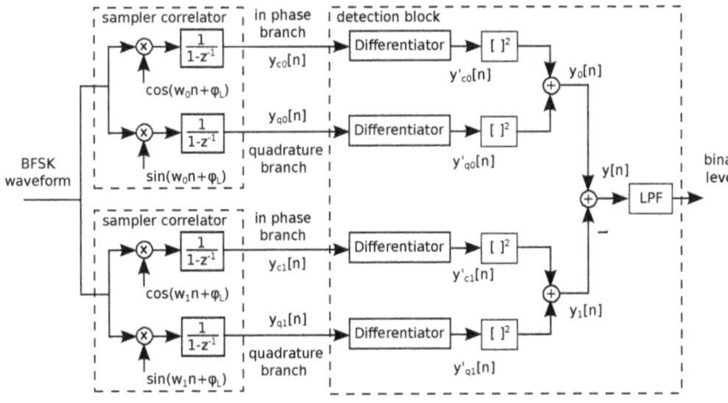

Fig. 4. Block diagram of the receiver.

The differentiator block in Fig. 4 is implemented by means of the transfer function $H_1(z)$, $H_2(z)$, $H_3(z)$ or $H_4(z)$ described above. However, the response function for $H_L(e^{jw}) = -2j\sin(w\frac{L}{2})e^{jw\frac{L}{2}}$ can also be employed to suppress the high frequencies at the accumulator output. Indeed, the term $\sin(w\frac{L}{2})$ has nulls in the frequency domain when $w\frac{L}{2} = k\pi$, $k \in N$, thus L is a parameter for selecting the null position at frequency w_i by the expression $L = \left\lfloor \frac{2k\pi}{w_i} \right\rfloor$.

In this case the implementation of only one branch, the upper or the lower in Fig. 4, is only needed if a proper expression for the threshold is obtained. Indeed, considering the upper branch for detecting the low frequency, the threshold in (3) is obtained for identifying the high and low levels at $y_0[n]$ in Fig. 4. For further details see the derivation in the Appendix. However, a loss of performance, regarding the noise margin, is also obtained

$$y_{th} = \frac{1}{2}\left(\frac{A^2}{4}L^2 - |R_{00}| + |R_{01}|\right). \tag{3}$$

In (3) the term $\frac{A^2}{4}L^2$ is related with the average slope when the symbol of low frequency is detected, R_{00} is related with the amplitude of oscillations around this average slope, and R_{01} represents an upper bound for the oscillations when the symbol of high frequency is received. By way of example, Fig. 5 depicts this values when a BFSK waveform is received with the parameters $w_0 = 0.2856$ rad/s, $w_1 = 0.6347$ rad/s, $f_m = 44100$ Hz, $T_s f_m = 882$ samples, and $L = \left\lfloor \frac{2\pi}{w_0} \right\rfloor = 21$ samples. The value for L was chosen in order to attenuate the term R_{00} in (16) on the Appendix, since it yields the largest oscillation regarding R_{01}. The threshold in (3) is selected as the average between the lines conformed by $\frac{A^2}{4}L^2 - |R_{00}|$ and $|R_{01}|$.

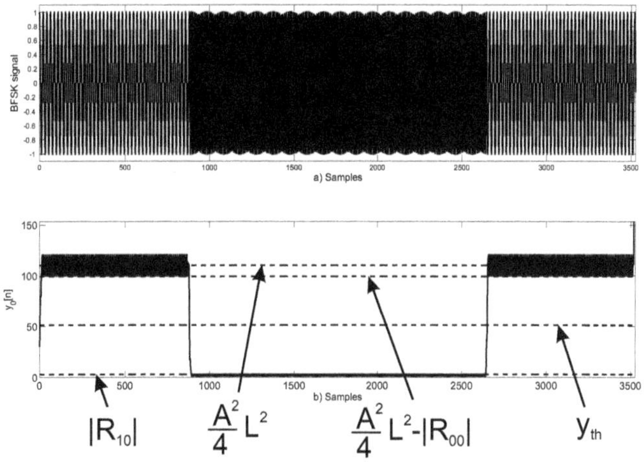

Fig. 5. Performance for the receiver in Figure 4 by means of the $H_L(z)$ differentiator and without the use of a lowpass filter at the output.

2.3 Data Recovery

The recovery of the binary levels from a BFSK waveform using the scheme in Fig. 4 has been described. In this Section the discussion about the performance of digital information associated with the binary data is presented.

After the high and low levels are recovered, as indicated in Figure 6, it is needed to obtain the quantity of "1's" and "0's" that are represented under each level. The length in time of a "1" or a "0" is given by the time-symbol T_s, the comparison of the length of the level and T_s will introduce the quantity of "1's" and "0's" that represents each one. However, this comparison produces errors since the transition of each levels are not abrupt, in this case there is an upper bound on the total of bit to be analyzed. The present section analyzes this situation bringing a closed form expression for determining the quality of the reception.

The algorithm for recovery the data is as follows:

1. Sketching a histogram where the abscissa represents the length of transition in samples; the amplitude is given by the number of occurrences of these lengths. An example is given in Figure 6 b), the abscissa represents the duration of the interval in a). This histogram is used for the estimation of the time-symbol duration.

2. The histogram is comprised of maximal separated by the same distance because the information transmitted consists of successive "1's" or "0's"; thus, in Figure 6 c), the number of samples of each level at the output are multiples of each other. The maximum closer to zero is related to the time-symbol duration, leading to the T_s estimation. The estimation of T_s is obtained with (4), in which N_1 and N_2 represents the interval of the x-axis (e.g. $N_1 = 200$ and $N_2 = 250$ in Figure 6 c), where the histogram has values unequal to zero), T_m represents the sample time, and $h[k]$ represents the values of the histogram obtained

$$\widehat{T_s} = \frac{\sum_{k=N_1}^{N_2} k T_m \cdot h[k]}{\sum_{k=N_1}^{N_2} h[k]}. \tag{4}$$

The length of the high levels in Fig. 6 are established by the intersection of the output of the system with the threshold indicated in (3). In this case, the length of this measure is modified from symbol to symbol by the smooth transition between levels. The accuracy in the determination of T_s, denoted by Δt, could be estimated by calculating the deviation as indicated in Fig. 6 c). The parameters T_s and Δt are related to the binary levels just as indicated in Figure 6 a)

$$\widehat{\Delta t} = \frac{\sum_{k=N_1}^{N_2} (kTm - \widehat{T_s})^2 \cdot h[k]}{\sum_{k=N_1}^{N_2} h[k]}. \tag{5}$$

3. Once $\widehat{T_s}$ is obtained, the transmitted digital information is recovered by dividing the length of each level with the time-symbol parameter, rounding the result towards the nearest integer.

The accuracy to be obtained on the third step depends on the variance Δt. If k identical symbols are supposed to be transmitted sequentially, then, after many symbols the output in Fig. 4 will have an average duration between symbols of $k \cdot T_s$ plus the variance Δt. If this resulting duration is divided by T_s plus the same Δt, then the total bits to be recovered can be described by the expression (6). Using Laurent expansion and considering $\Delta t << T_s$, we obtain:

$$\widehat{k} = \frac{k \cdot T_s + \Delta t}{T_s + \Delta t} \approx k + (k+1)\frac{\Delta t}{T_s}. \tag{6}$$

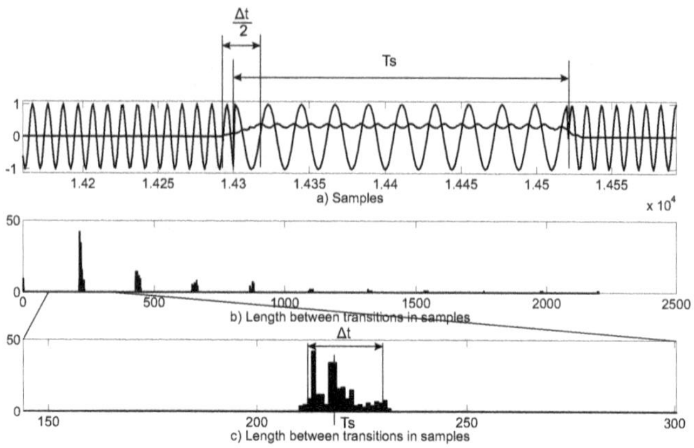

Fig. 6. Details regarding the transition between consecutive symbols. a) Signal received and binary levels. b) Histogram of the length of transitions. c) Horizontal zoom of b).

Expression (6) considers the addition of the variance instead of substraction because the linear system presented in Fig.4 tends to expand the transitions and not to contract. Besides, the same variance is also considered on each transition as a simplification in the determination of accuracy. This considers that the system responds in the same way no matter the total symbols received.

An error in the estimation of k occurs when the second term in (6) is higher than 0.5. In such a case, the estimated value will be superior to $\hat{k} = k+1$ and the correct value is k. Hence, in the absence of noise the occurrence of consecutive symbols is limited in order to perform a reception free of error.

Equation (7) represents the upper bound for k when T_s and Δt are substituted by $\widehat{T_s}$ and $\widehat{\Delta t}$ respectively, since both of them are obtained via the histogram. Expression (7) gives an idea about how many bits the system in Fig. 2 might receive free of errors in the absence of noise. Indeed, the probability of transmitting k bits comprised of repetitive sequences of "1's" or "0's" is $2\frac{1}{2^k}$, so an error could happen once in 2^{k-1} transmitted bits. This is why the receiver is upper bounded in the total of bits to be processed. The relation (7) is a closed form expression that represents a figure of merit for the receivers analyzed, and its values are analyzed in Sec. 3.2

$$k < 0.5\frac{\widehat{T_s}}{\widehat{\Delta t}} - 1. \tag{7}$$

3. Results

Three key aspects are considered for comparing the proposed solution with the systems reported:

1. Hardware Complexity
2. Precision through relation (7).
3. Bit Error Rate (BER) performance.

The Balanced Quadricorrelator is less expensive than the Quadricorrelator and the Quotient Detector, and mainly has the best accuracy. Thus, comparison of the system in Figure 4 will be made with regard to the Balanced Quadricorrelator.

3.1 Hardware Complexity

In this Section the complexity is measured with regard to the total of floating adders and multipliers. The total of adders and multipliers to implement the system proposed in Fig. 4 depends upon the employed differentiator. The lowest complexity is obtained when the transfer functions $H_1(z)$, $H_2(z)$, $H_L(z)$ or $H_4(z) = 1 - z^{-4}$, all these systems employ just 1 floating adder. Taking into account one branch in Fig. 4 the total of adders and multipliers is 9, and in cases that 17 is not fulfilled an additional digital filter is also needed.

On the other hand, the Balanced Quadricorrelator employs 2 digital filters, two discrete-time differentiators and 5 additional floating elements for adding and multiplying. The differentiator can be implemented with a FIR of fifth order [15], which requires another 5 adders and 6 multipliers. Summarizing, the total of floating elements is 27 and also 2 digital filters.

In contrast with the Balanced Quadricorrelator, the proposed scheme reduces complexity to 18 floating elements and 1 digital filter. However, in case that the proposed system in Figure 4 is employed with the differentiator $H_L(z)$, without the use of the low pass filter and with only one branch, then in total 7 adders and multipliers are only needed. Which represents a considerable reduction in hardware consumption. This scheme can be employed if relation (17) is fulfilled, this means that oscillations of low levels are not higher than the oscillations of high levels of the binary signal at the output. Besides, relation (18) in the Appendix establishes an upper bound for the total bits to be demodulated by the system, without additional bits and without the presence of noise.

3.2 Precision

Regarding precision, obtained by means of (7), the quantity depends on the parameters of the modulation format. The BFSK waveform is driven by three parameters, f_0, f_1 and T_s, which are related between each other by $f_1 = f_0 + \frac{m}{T_s}$. The simulation is made by holding $f_0 = 1000$ Hz and varying T_s and m as indicated in (8). The parameter r in (8) controls the symbol duration, while m the distance between frequencies

$$\begin{aligned} f_0 &= 1000 \\ T_s &= \frac{r}{f0} \\ f_1 &= f_0 + \frac{m}{T_s} = f_0\left(1 + \frac{m}{r}\right). \end{aligned} \tag{8}$$

A family of curves for the k parameter can be obtained when r is the independent variable and m is evaluated for several values. The simulations show that precision is not to much affected with changes in m but with changes in r, these curves are almost the same when $m \geq 4$. Figure 7 shows in average the results for both systems, the Balanced Quadricorrelator and the proposed scheme with one branch and the differentiator $H_L(z)$. The graphic is conformed by averaging the different values given by the variation of m when r is fixed, m is considered in the interval 2 to 20.

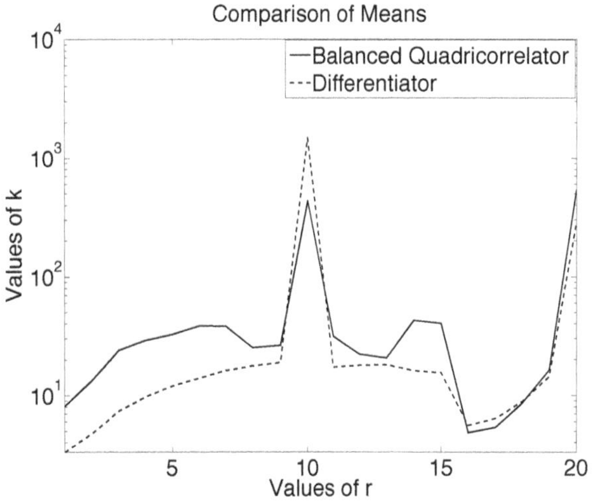

Fig. 7. Precision for different values of r when the different values given by the variation of m when r is fixed, m is considered in the interval 2 to 20.

The curves depicted in Fig. 7 show that the Balanced Quadricorrelator is superior for almost every value of r. This can be explained by the fact the Quadricorrelator performs the operations locally, mainly with the use of discrete-time differentiators, while the proposed scheme estimates slope with samples separated by L points, which further smooth the transitions between levels making it more inaccurate.

3.3 BER Parameter

Figure 7 depicts the measured bit error rate (BER) of the proposed receiver with and without the use of a lowpass filter as a function of the bit energy to power density (SNR), in comparison with that of the Balanced Quadricorrelator. The parameters employed are $w_0 = 0.2856$ rad/s, $w_1 = 0.6347$ rad/s, $f_m = 44100$ Hz, $T_s \cdot f_m = 882$ samples, and $L = \left\lfloor \frac{2\pi}{w_0} \right\rfloor =$

21 samples, the simulation was performed with a total number of 10^6 bits in steps of 0.25 dB on the SNR axis. The range analyzed is below an $\frac{E_b}{N_0}$ of 5 dB since up to this value error correcting codes are usually employed [28].

The proposed receiver without LPF has the worst performance regarding the P_b parameter, and the best performance when the LPF is employed. The lowpass filter reduces the amount of noise in the binary levels, and the Balanced Quadricorrelator in Figure 1 implements both of them on each branch. In contrast, the proposed receiver with the use of only one LPF, saves approximately 1.5 dB of energy regarding the Balanced Quadricorrelator. In both cases, the proposed receiver is used with just one branch.

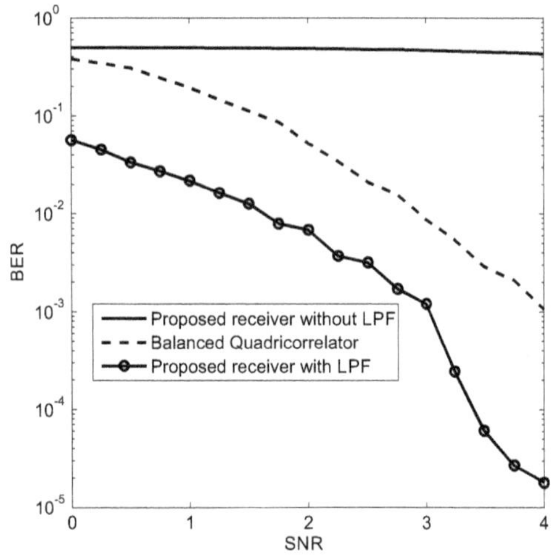

Fig. 8. Bit Error Rate.

At first glance, some of the curves in Fig. 8 suggest a better performance with regard to the ideal receiver. However, the use of lowpass filters in the structures of Fig.1 and Fig. 4 reduces the noise density for the decision rule and this is not the case for the ideal receiver [5].

4. Conclusions

The solution presented proposes a different detection scheme for the Balanced Quadricorrelator regarding to the detection scheme. High-order filters are avoided thanks to one important matter: the symbol detection is accomplished through the constant slope recognition verified at the output of Sampler Correlator. Although this simple scheme exhibits fluctuations at the output that restrict proper symbol identification, this could be avoided by means of a lowpass filter. However, the entire solution has less complexity than other reported methods. The proposed solution based upon correlator maximizes signal to noise ratio at the input detector like previous solutions, but with a major advantage: it reduces complexity for hardware implementation. Accordingly, advantages of lower energy consumption and hardware saving are also provided.

Appendix

In the present Section it is derived a closed-form expression for the output of system in Fig. 4, when a differentiator with transfer function $H_L(z) = 1 - z^{-L}$ is considered. Consider that a symbol $A\cos(w_0 n + \varphi_0)$ is received; the expression (9), corresponding to the output of the differentiator of the in phase branch that detects the frequency w_1, Fig. 4, is obtained by grouping terms with the trignometric relations in (10). The upper bound for (9) is obtained in (11) taking into account (12). The upper bound for the quadrature branch $y'_{s1}[n]$ is the same, since the phase is not present in the expression for the upper bound

$$\begin{aligned} y'_{c1}[n] &= y_{c1}[n] - y_{c1}[n-L] \\ &= a\cos\left(\varphi_0 - \varphi_L + (w_0 - w_1)\left(n - \frac{L}{2} + \frac{1}{2}\right)\right) + \\ &\quad + b\cos\left(\varphi_0 + \varphi_L + (w_0 + w_1)\left(n - \frac{L}{2} + \frac{1}{2}\right)\right) \\ a &= \frac{A}{2}\frac{\sin\left(\frac{L(w_0 - w_L)}{2}\right)}{\sin\left(\frac{w_0 - w_L}{2}\right)} \\ b &= \frac{A}{2}\frac{\sin\left(\frac{L(w_0 + w_1)}{2}\right)}{\sin\left(\frac{w_0 + w_1}{2}\right)}, \end{aligned} \tag{9}$$

$$\begin{aligned} \cos(\mathrm{A}) - \cos(\mathrm{B}) &= 2\sin\left(\frac{B+A}{2}\right)\sin\left(\frac{B-A}{2}\right) \\ \sin(A) - \sin(B) &= 2\cos\left(\frac{B+A}{2}\right)\sin\left(\frac{A-B}{2}\right), \end{aligned} \tag{10}$$

$$|y'_{c1}[n]| \le \frac{A}{2}\left|\frac{\sin\left(\frac{L(w_0 - w_1)}{2}\right)}{\sin\left(\frac{w_0 - w_1}{2}\right)}\right| + \frac{A}{2}\left|\frac{\sin\left(\frac{L(w_0 + w_1)}{2}\right)}{\sin\left(\frac{w_0 + w_1}{2}\right)}\right|, \tag{11}$$

$$\begin{aligned} A\cos(\theta + \alpha) + B\cos(\theta + \beta) &\le C \\ C &= \sqrt{A^2 + B^2 + 2AB\cos(\alpha - \beta)} \\ A\cos(\theta + \alpha) + B\cos(\theta + \beta) &\le |A| + |B|. \end{aligned} \tag{12}$$

Since $y'_{c1}[n]$ is in quadrature respect to $y'_{q1}[n]$, the upper bound for $y'_{c1}[n]^2 + y'_{q1}[n]^2$ is just the square of the upper bound of $y'_{c1}[n]^2$ or $y'_{q1}[n]^2$. Thus, a symbol with frequency w_0 produces, at the output of the branch, an upper bound determined by relation (14)

$$|y_1[n]| \leq R_{01} \tag{13}$$
$$R_{01} = \frac{A^2}{4}\left[\left|\frac{\sin\left(\frac{L(w_0-w_1)}{2}\right)}{\sin\left(\frac{w_0-w_1}{2}\right)}\right| + \left|\frac{\sin\left(\frac{L(w_0+w_1)}{2}\right)}{\sin\left(\frac{w_0+w_1}{2}\right)}\right|\right]^2.$$

The case for the branch that detects w_0 is shown in (14). Through the operation $y'_{c0}[n] = \lim_{w_1 \to w_0} y'_{c1}[n]$, the term $\cos(\varphi_0 - \varphi_L)\frac{A}{2}L$ is related to the constant behavior in Figure 2 c)

$$y'_{c0}[n] = \cos(\varphi_0 - \varphi_L)\frac{A}{2}L + R_{c0} \tag{14}$$
$$R_{c0} = \frac{A}{2}\frac{\sin(w_0 L)}{\sin(w_0)}\cos\left(\varphi_0 + \varphi_L + 2w_0\left(n - \frac{L}{2} + \frac{1}{2}\right)\right)$$
$$|R_{c0}[n]| \leq \frac{A}{2}\left|\frac{\sin(w_0 L)}{\sin(w_0)}\right|.$$

The differentiator output for the in quadrature branch is obtained in (15) by changing φ_L by $\varphi_L - \pi$ in (14)

$$y'_{s0}[n] = -\sin(\varphi_0 - \varphi_L)\frac{A}{2}L + R_{s0}[n] \tag{15}$$
$$R_{s0}[n] = \frac{A}{2}\frac{\sin(w_0 L)}{\sin(w_0)}\sin\left(\varphi_0 + \varphi_L + 2w_0\left(n - \frac{L}{2} + \frac{1}{2}\right)\right)$$
$$|R_{s0}[n]| \leq \frac{A}{2}\left|\frac{\sin(w_0 L)}{\sin(w_0)}\right|.$$

Finally, the output of the branch that detects the frequency w_0 is obtained in (16) by straightforward procedures

$$y_0[n] = {y'_{c0}}^2[n] + {y'_{s0}}^2[n] \tag{16}$$
$$= \frac{A^2}{4}L^2 + R_{00}[n]$$
$$|R_{00}[n]| \leq \frac{A^2 L}{2}\left|\frac{\sin(w_0 L)}{\sin(w_0)}\right| + \frac{A^2}{4}\frac{\sin^2(w_0 L)}{\sin^2(w_0)}.$$

A proper detection of the symbol with frequency w_0 is obtained if the condition $y_0[n] > y_1[n]$ is fulfilled during the interval duration of such a symbol. Upon substituting (12) and (16) in the condition above, and considering the worst case, that is a negative value for $R_{00}[n]$, (17) is obtained

$$\frac{A^2}{4}L^2 - |R_{00}| - |R_{01}| > 0. \tag{17}$$

Even further, the L parameter is directly related to the precision of the system. The longer the L value is, the longer will be the transition between symbols in Fig. 6. Thus, in relation (7), the L parameter is directly related to the variance in the determination of the symbol transition Δt and (18) gives a measure of the total bits, in average, to be analyzed without errors; N_{samp} represents the number of samples per symbol

$$k < 0.5\frac{N_{samp}}{L} - 1. \tag{18}$$

Expressions (18) and (17) give the rules for the proper performance of the system given in Figure 4 with a differentiator $H_L(z) = 1 - z^{-L}$. The latter evaluate L for a given precision, the former determines if a filter at the output is needed.

References

[1] PENG, K.-C., LIN, C.-C., CHAO, C.-H.. A novel three-point modulation technique for fractional-N frequency synthesizer applications, *Radioengineering*. 2013, vol. 22, no. 1, p. 269–275.

[2] VERTAT, I., MRAZ, J. Hybrid M-FSK/DQPSK modulations for CubeSat picosatellites, *Radioengineering*. 2013, vol. 22, nol. 1, p. 389–393.

[3] THOMPSON, A. C., HUSSAIN, Z. M., O'SHEA P. A single-bit digital non-coherent baseband BFSK demodulator, *In IEEE Region 10 Conference TENCON*. 2004, vol. 1, p. 515–518.

[4] TERVO, R., ZHOU, K. DSP based self-tuning BFSK demodulation, *In IEEE Pacific Rim Conference on Communications, Computers and Signal Processing*, 1993, vol. 1, p. 68–71.

[5] SKLAR, B. *Digital Communications, Fundamental and Applications*, 2nd ed. New Jersey: Prentice Hall, 2001.

[6] RICHMAN, D. Color-carrier reference phase synchronization accuracy in NTSC color television. *Proceedings of the IRE*, 1954, vol.42, no.1, p. 106–133.

[7] GARDNER, F., Properties of Frequency Difference Detectors. *In IEEE Transactions on Communications*,1985, vol. 33, no. 2, p. 131–138.

[8] PARK, J., An FM detector for low S/N. *IEEE Transactions on Communication Technology*, 1970, vol. 18, no. 2, p. 110–118.

[9] NATALI, F., AFC tracking algorithms. *IEEE Transactions on Communications*, 1984, vol. 32, no. 8, p. 935–947.

[10] FARRELL, K., A., McLANE, P., J. Performance of the cross-correlator receiver for binary digital frequency modulation. *IEEE Transactions on Communications*, 1997, vol. 45, no. 5, p. 573–582.

[11] KANG, H., KIM, D., PARK, S.C. Coarse frequency offset estimation using a delayed auto-quadricorrelator in OFDM-based WLANs, In *3rd International Congress on Ultra Modern Telecommunications and Control Systems and Workshops (ICUMT)*, 2011, p. 1–4.

[12] ORDU, G., KRUTH, A., SAPPOK, S., WUNDERLICH, R., HEINEN. S. A quadricorrelator demodulator for a Bluetooth low-IF receiver. In *IEEE Radio Frequency Integrated Circuits (RFIC) Symposium. Digest of Papers*. 2004, p. 351–354.

[13] KREUZGRUBER, P. A class of binary FSK direct conversion receivers. In *IEEE 44th Vehicular Technology Conference*, . 1994, vol. 1, p. 457–461.

[14] EGAU, PC. Correlation systems in radio astronomy and related fields.In *IEEE Proceedings F Communications, Radar and Signal Processing*. 1984, vol. 131, no. 1, p. 32–39.

[15] OPPENHEIM, A., V., SCHAFER, R., W., BUCK, J., R. *Discrete-Time Signal Processing*. 2nd ed. New Jersey: Prentice Hall, 1998.

[16] AHN, T., YOON, C.-G., MOON, Y. An adaptive frequency calibration technique for fast locking wideband frequency synthesizers, In *48th Midwest Symposium on Circuits and Systems*. 2005, vol. 2, p. 1899–1902.

[17] LEE, Y., W., CHEATHAM, T., P., WIESNER, J., B. Application of correlation analysis to the detection of periodic signals in noise, *Proceedings of the IRE*, 1950, vol. 38, no. 10, p. 1165–1171.

[18] PEEK, J., B., H. The measurement of correlation functions in correlators using "shift-invariant independent functions", *Philips Res. Rep. Suppl.* 1968, vol. 1.

[19] CHANG, K.-Y., MOORE, A. Modified digital correlator and its estimation errors, *IEEE Transactions on Information Theory*, 1970, vol. 16, no. 6, p. 699–706.

[20] VAN VLECK, J., H. *The spectrum of clipped noise*, Tech. Rep. No. 51. Cambridge, Mass., Radio Res. Lab., Harvard University, 1943.

[21] WATTS, D., G. A general theory of amplitude quantization with applications to correlation determination, *Proceedings of the IEEE - Part C: Monographs*. 1962, vol. 109, no. 15, p. 209–218.

[22] JESPER, P., CHU, P., T., FETTWEIS, A: A new method to compute correlation functions, In *Int. Symp. Inform. Theory, and IRE Trans. Inform. Theory*. 1962, p. 106–107.

[23] BERNDT, H., Correlation function estimation by a polarity method using stochastic reference signals, *IEEETransactions* 1968, IT-14, p.796–801.

[24] ARZENO, N. M., DENG, Z. D., POON C. S. Analysis of first-derivative based QRS detection algorithms, *IEEE Transactions on Biomedical Engineering*. 2008, vol. 55, no. 2, p.478–84.

[25] MUKHOPADHYAY, S., MITRA, M., MITRA, S. Time plane ECG feature extraction using hilbert transform, variable threshold and slope reversal approach, In *2011 International Conference on Communication and Industrial Application (ICCIA)*. 2011, p. 1–4.

[26] FRIESEN, G., JANNET, T., JADALLAH, M., YATES, S., QUINT, S., Nagle, H. A comparison of the noise sensitivity of nine QRS detection algorithms, *IEEE Transactions on Biomedical Engineering*, 1990, vol. 37, no. 1, p. 85–98, 1990.

[27] ARZENO, N., DENG, Z., D., POON, C., S. Analysis of first-derivative based QRS detection algorithms, *IEEE Transactions on Biomedical Engineering*, 2008, vol. 55, no. 2, p. 478–484.

[28] CARLSON, A., B., Crilly, P., B., RUTLEDGE, J., C. Communication systems: an introduction to signals and noise in electrical communication, 4th ed .McGraw-Hill, 2002.

About Authors...

Jorge TORRES received his M.Sc. from CUJAE university in 2010. His research interests include digital demodulation, digital signal processing, estimation theory. He is currently a full assistant professor at CUJAE University, Habana, Cuba.

Fidel HERNANDEZ M.Sc. in Digital Systems from CUJAE in 2000, Ph.D. degree in Electronics and Industrial Automation from University of Mondragon, Spain, in 2006. Currently he works as full professor in University of Pinar del Rio, Cuba. Head of the Research Group for Advanced Machine Diagnosis (GIDAM) since 2000. He has directed several international and national projects involving eye image processing for medical diagnostic applications, higher-order statistical signal processing applied on mechanical vibrations, classification and demodulation of communication signals, and so on. He is a CYTED expert, and is associate editor and reviewer of various international journals. He is member of the administration committee of the Cuban Association of Pattern Recognition.

Joachim HABERMANN is currently a full professor in the Department of Electrical Engineering and Information Technology of the Technische Hochschule Mittelhessen (THM), University of Applied Sciences Giessen, Germany. He obtained his Diploma and Dr.-Ing. (Ph.D.) degrees in Electrical Engineering both with highest honors from the Technical University of Darmstadt, Germany. Prior to joining THM he worked in the Asea Brown Boveri (ABB) research center, Switzerland in the Telecommunications department as a researcher and project leader. He was part of the working groups defining the global mobile communication systems GSM and UMTS and is now a contributor to the development of the enhanced LTE system. He has authored and coauthored more than 70 technical papers in journals and international conferences and one textbook. He is a senior member of the Institute of Electrical and Electronics Engineers (SM IEEE) and served as Technical Chair and member of the program committee of many international conferences. He is also reviewer of several research institutions, such as the German Research Council (DFG).

18

Efficient Raw Signal Generation Based on Equivalent Scatterer and Subaperture Processing for SAR with Arbitrary Motion

Hongtu XIE, Daoxiang AN, Xiaotao HUANG, Zhimin ZHOU

College of Electronic Science and Engineering, National University of Defense Technology, Changsha, Hunan, P. R. China, 410073

xht20041623@163.com

Abstract. *An efficient SAR raw signal generation method based on equivalent scatterer and subaperture processing is proposed in this paper. It considers the radar's motion track, which can obtain the precise raw signal for the real SAR. First, the imaging geometry with arbitrary motion is established, and then the scene is divided into several equidistant rings. Based on the equivalent scatterer model, the approximate expression of the SAR system transfer function is derived, thus each pulse's raw signal can be generated by the convolution of the transmitted signal and system transfer function, performed by the fast Fourier transform (FFT). To further improve the simulation efficiency, the subaperture and polar subscene processing is used. The system transfer function of pulses for the same subaperture is calculated simultaneously by the weighted sum of all subscenes' equivalent backscattering coefficient in the same equidistant ring, performed by the nonuniform FFT (NUFFT). The method only involves the FFT, NUFFT and complex multiplication operations, which means the easier implementation and higher efficiency. Simulation results are given to prove the validity of this method.*

Keywords

Arbitrary motion, equivalent scatterer, raw signal generation, synthetic aperture radar, subaperture processing.

1. Introduction

Synthetic aperture radar (SAR) plays a significant role in the remote sensing, geosciences, and surveillance applications. It can effectively operate in severe weather condition and achieve high resolution images [1]-[3].

SAR raw signal generation technology is a key ring in the research and development of SAR system and takes an important role in many aspects, such as the system design, parameter optimization, meeting different requirements, testing of imaging algorithms, analyzing the jamming and noises, etc. [4]-[5]. To ensure that designed SAR systems satisfy the specific requirements of the different users, raw signal generation and imaging processing should be carried out before the development of the SAR systems. Therefore, the SAR raw signal generation is a significant component of the SAR simulation technology.

Generally, the SAR raw signal generation of discrete scattering targets is suitable for testing the performance of imaging algorithms and for testing system parameters such as the resolution, peak sidelobe ratio (PSLR), and integral sidelobe ratio, etc. When the number of targets is small, the raw signal can be calculated pulse by pulse and target by target (TBT) according to the SAR geometry. It works in the time domain and can generate precise raw signal [6]. However, as the jamming, thermal noise, clutter, multipath, analog-to-digital quantization errors and image ambiguity are concerned, the natural scene raw signal has to be generated. For the natural scene raw signal generation, there are some difficulties from two aspects. One is the simulation of the natural scene's backscattering coefficient, and the other is the efficient raw signal generation method while a large number of scatterers are contained. As to the former problem, the simulation of the natural scene's backscattering coefficient has been investigated by several researchers, which has been partially indicated in [7]-[14]. In this paper, we focus on the latter problem, which means that the natural scene's backscattering coefficient is known and considered as the input in the raw signal generation.

In general, the SAR raw signal generation algorithms include the TBT [15] method in the time domain, the one-dimensional fast Fourier transform (1D FFT) algorithm, the two-dimensional FFT (2D FFT) algorithm in frequency-domain and imaging inverse processing (IMIP) algorithm. TBT algorithm calculates the raw signal by adding all distributed point targets' echo coherently, according to the imaging geometry, so it works pulse by pulse and target by target. It doesn't make any approximation, thus it has the highest accuracy. However, it costs huge calculation and time, thus it is only suitable for small number of discrete scattering targets' raw signal generation. 2D FFT method is first proposed by Franceschetti [16]-[19]. He developed a series of SAR raw signal simulators for terrain scenes, such as bare land ground, sea surface, and buildings, and then provided a series of improved versions. It was realized by multiplying the 2D spectrum of the targets' scattering characteristic and 2D frequency-domain expression of the SAR system transfer function, and the time domain raw signal is obtained using 2D inverse FFT (IFFT). Compared with the TBT algorithm, the simulation efficiency of 2D FFT algorithm is improved obviously. However, when the radar's motion errors exist, it is very difficult to simulate these errors for the SAR raw signal generation. Therefore, the 2D FFT algorithm is restricted by the radar's motion errors. IMIP [20]-[22] is to transform the image data into signal space by the inverse imaging processing, and then the simulated raw signal is obtained. Its efficiency is also higher than that of the TBT method. However, it is also limited by the radar's motion errors. When SAR system contains these errors, the simulated raw signal by the 2D FFT or IMIP methods is imprecise. 1D FFT algorithm [23]-[24] generates the SAR raw signal by the convolution of the radar signal and SAR system impulse response function, which is performed by the FFT. It also works pulse by pulse, so it is especially useful for the raw signal generation of the natural scene when the radar's motion errors exist. Its efficiency is higher than that of the TBT method, but lower than that of 2D FFT and IMIP methods.

To make a compromise between the simulation efficiency and precision, the reference [25] proposed a subaperture based echo simulation algorithm. It used the FFT to simultaneously calculate the SAR system transfer function for the same subaperture pulses, then subaperture pulses' raw signal is generated by the convolution of the transmitted signal and system transfer function using the FFT, which can further improve the simulation efficiency. But, it doesn't consider the radar's motion error, which can't obtain the precise raw signal of the nature scene for the real SAR system. Reference [26] presented a raw signal simulation algorithm based on subaperture processing. First, the FFT algorithm in range dimension is used to get the central pulse's echo signal of a sub-aperture. Then, the 1st-order approximate model of slant range in azimuth dimension is utilized to generate other pulses' echo signal in a sub-aperture. However, it also doesn't consider the radar's motion error in the simulation.

Based on these previous works, this paper explores an efficient SAR raw signal generation algorithm based on the equivalent scatterer and subaperture processing. It considers the radar's motion track to generate the high-precision raw signal. The subaperture processing and nonuniform FFT (NUFFT) are used to simultaneously calculate the SAR system transfer function for the same subaperture pulses, then subaperture pulses' raw signal is generated by the convolution of the transmitted signal and system transfer function using the FFT. The utilization of the equivalent scatterer, subaperture processing and FFT (NUFFT) operation can greatly improve the SAR raw signal simulation efficiency.

This paper is organized as follows. Section 2 presents the SAR geometry mode, and the raw signal generation based on equivalent scatterer is introduced. Section 3 gives a detail description of the proposed raw signal generation method. First, the SAR geometry mode in polar coordinate is established; and then the subaperture processing in this method is derived; finally implementation of the proposed method is discussed. The validity of the proposed approach is proved in Section 4 based on the simulation data. Section 5 provides the conclusion.

2. SAR Raw Signal Generation based on Equivalent Scatterer

2.1 Geometry Model

Because the radar speed is much less than the speed of light, the stop and go assumption is commonly used in SAR [1]. In other words, the radar's antenna is assumed to be at a standstill during ranging signal propagation and it moves on to the next position only after the echo signal is received. According to this hypothesis, the signal can be modeled as a function of two independent variables, i.e., the fast time and slow time. However, its validity needs to be evaluated in each particular SAR case. Here, we assumed that the stop and go assumption is valid for the SAR signal generation.

SAR geometry model with arbitrary motion is shown in Fig. 1. The dashed straight line parallel to the Y axis is the radar's nominal track, while the solid curve is its actual track. The radar coordinates are $(x(\eta), y(\eta), z(\eta))$ at the slow time η. The simulated scene is composed of large numbers of point targets distributed on the rectangle grid. Provided that the scene grid has dimensions $M \times N$, and range and azimuth intervals between the adjacent grids are Δx and Δy, respectively. $x_{\min}$ is the minimum ground range of the scene, and its azimuth centre line is the X axis. So the range between the radar and the grid (m,n) at η is

$$R(\eta,m,n) = [\left(x(\eta)-(x_{\min}+m\cdot\Delta x)\right)^2 + \left(y(\eta)-(n-N/2)\cdot\Delta y\right)^2 + \left(z(\eta)\right)^2]^{1/2} \quad (1)$$

where $1 \le m \le M$, $1 \le n \le N$

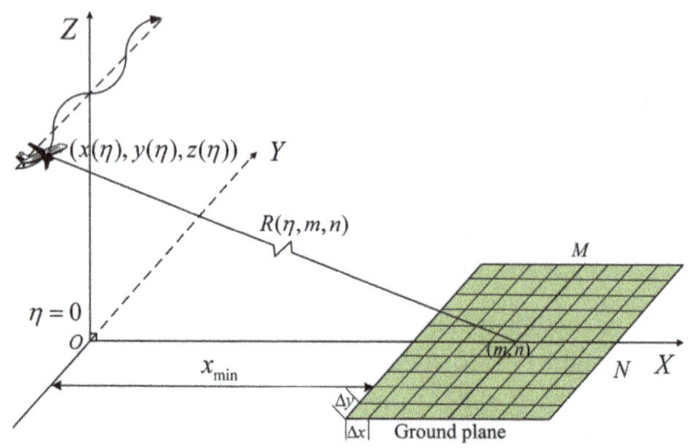

Fig. 1. SAR geometry model with arbitrary motion.

Suppose that the transmitted signal is a chirp signal

$$p(\tau) = w_r(\tau)\cdot \exp(j2\pi f_c \tau + j\pi\gamma\tau^2) \tag{2}$$

where τ is the fast time, γ is the chirp rate, $w_r(\cdot)$ is the range envelope and f_c is the center frequency. Because the scene can be separated into large numbers of point targets, the scene's raw signal can be represented as

$$\begin{aligned} s(\tau,\eta) = &\sum_{m=1}^{M}\sum_{n=1}^{N}\sigma_{mn} w_r\left[\tau - 2R(\eta,m,n)/c\right] \\ &\cdot \exp\left[-j4\pi f_c R(\eta,m,n)/c + j\pi\gamma\left(\tau - 2R(\eta,m,n)/c\right)^2\right] \end{aligned} \tag{3}$$

where σ_{mn} is target's backscattering coefficient of the grid (m,n), and c is the speed of light.

2.2 Calculate Equivalent Scatterer's Backscattering Coefficients

From (3), we can find that it is easy to generate the raw signal in time domain, but it has high computational load. To improve the simulation efficiency, this paper first introduces a fast raw signal generation method based on the equivalent scatterer. Its basic idea is to divide the scene into several equidistant rings, and targets on rectangle grids can be considered as distributing in those equidistant rings that nominally exist, which is shown in Fig. 2. Therefore, backscattering coefficient of targets in the same equidistant ring can be substituted for an equivalent scatterer.

Assume that Δs is the interval between the adjacent equidistant rings. $R_{\min}(\eta)$ and $R_{\max}(\eta)$ are the minimum and maximum ranges from the radar to the scene at η, respectively. Then, the number of equidistant rings is

$$N_p(\eta) = round((R_{\max}(\eta) - R_{\min}(\eta))/\Delta s) + 1\,. \tag{4}$$

Then, the range of grid (m, n) meets the following inequality

$$R_{\min}(\eta) + (p-1)\Delta s \le R(\eta,m,n) \le R_{\min}(\eta) + p\Delta s \quad 1 \le p \le N_p(\eta)\,. \tag{5}$$

Targets' distribution between two adjacent equidistant rings is shown in Fig. 3. p-1 and p denote two adjacent equidistant rings, and the ranges from them to the radar are $R_{\min}(\eta) + (p-1)\Delta s$ and $R_{\min}(\eta) + p\Delta s$, respectively.

The spacing between the (p-1)-th and p-th equidistant rings is defined as the p-th equidistant ring domain. The dashed ring is the center of the equidistant ring domain, and the range from it to the radar is

$$R(\eta, p) = R_{\min}(\eta) + (p - 1/2)\Delta s\,. \tag{6}$$

To calculate the targets' backscattering coefficients in the equidistant ring domain's center, the range between the radar and the targets in the equidistant ring domain is approximately replaced by the range from the radar to the equidistant ring domain's center. Therefore, the equivalent scatterer is substituted for all targets in this domain, whose backscattering coefficient is computed by

$$\sigma(\eta, p) = \sum_{i=1}^{I_p}\sigma_i \exp\left[-j4\pi f_c \Delta R_i(\eta,p)/c\right] \tag{7}$$

where I_p is the number of targets in the p-th equidistant ring domain, σ_i is the backscattering coefficient of the i-th target in this domain. $\Delta R_i(\eta,p)$ is the difference between the range from the i-th target to the radar (i.e., $R_i(\eta)$) and the range from the p-th equidistant ring domain's center to the radar, which can be given by

$$\Delta R_i(\eta,p) = R_i(\eta) - \left(R_{\min}(\eta) + (p-1/2)\Delta s\right). \tag{8}$$

So the raw signal of the scene can be also expressed as

$$\begin{aligned} s(\tau,\eta) = &\sum_{p=1}^{N_p(\eta)}\sigma(\eta,p)\exp\left[-j4\pi f_c R(\eta,p)/c\right] \\ &\cdot w_r\left[\tau - 2R(\eta,p)/c\right]\exp\left[j\pi\gamma\left(\tau - 2R(\eta,p)/c\right)^2\right] \end{aligned} \tag{9}$$

We can find that the latter two items of (9) are only related to the fast time τ. So the raw signal of the scene can also be written in the following convolution format

$$\begin{aligned} s(\tau,\eta) = &w_r(\tau)\exp(j\pi\gamma\tau^2)\otimes_\tau \sum_{p=1}^{N_p}\sigma(\eta,p) \\ &\cdot \exp\left[-j4\pi f_c R(\eta,p)/c\right]\cdot\delta\left[\tau - 2R(\eta,p)/c\right] \end{aligned} \tag{10}$$

where $\otimes_\tau$ is the convolution versus the fast time τ. The physical meanings of (10) is that the received signal of the radar is the response of a linear time-invariant system, the system's input is the transmitted base-band chirp signal, and its output is the result of the convolution between the transmitted signal and the impulse response function with a certain amplitude, phase and delay.

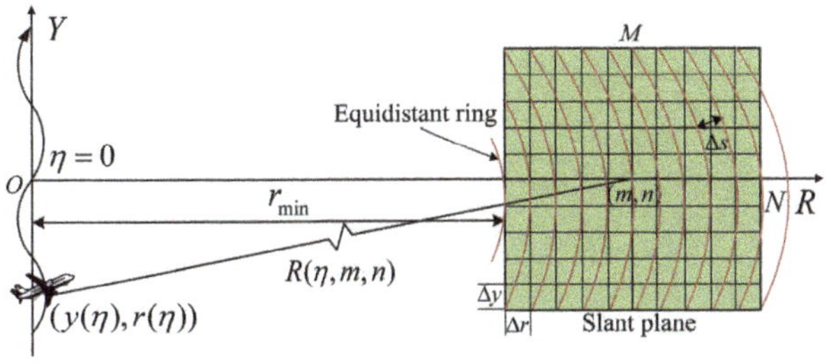

Fig. 2. Geometry model in slant plane.

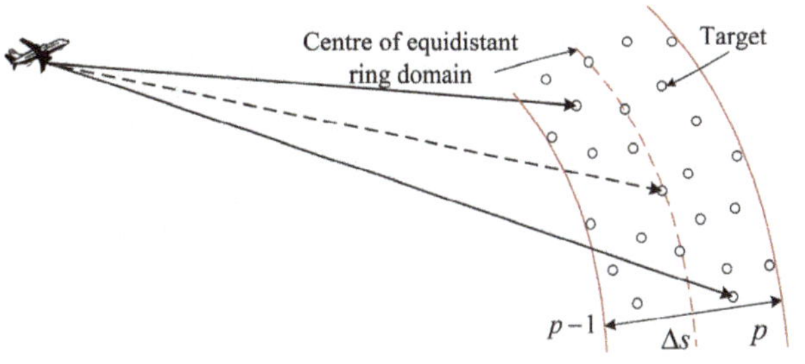

Fig. 3. Targets' distribution between adjacent equidistant rings

It is known that the convolution can be performed by the FFT, so the raw signal of the scene is given by

$$s(\tau,\eta) = \mathrm{IFFT}\left\{\mathrm{FFT}\left[w_r(\tau)\exp(j\pi\gamma\tau^2)\right]\right.$$
$$\left.\cdot\mathrm{FFT}\left[\sum_{p=1}^{N_p}\sigma(\eta,p)\exp(-j4\pi f_c R(\eta,p)/c)\cdot\delta(\tau-2R(\eta,p)/c)\right]\right\} \quad (11)$$

3. SAR Raw Signal Generation Based on Subaperture Processing

SAR raw signal generation method in Section 2 can improve the simulation efficiency to a certain extent. Since the raw signal is calculated pulse by pulse and equivalent scatterer by equivalent scatterer, thus generation with the large number of pulses is not suitable for the raw signal.

In order to further improve the simulation efficiency, the subaperture and polar subscene processing is used. Based on the above method, the system transfer function of pulses for the same subaperture can be calculated by the weighted sum of all subscenes' equivalent backscattering coefficient in the same equidistant ring.

3.1 Geometry Model in Polar Coordinate

SAR geometry model in polar coordinate is shown in Fig. 4. Assume that A_c is the synthetic aperture center at aperture center time η_c, A_η is the aperture position at η. Let ρ be the range between A_c and the scene and θ be the angle between ρ and the Y positive axis, respectively. $\Delta\theta$ is the angular sampling spacing of the scene. (p,k) is the k-th polar subscene in the p-th equidistant ring domain, and $R(\eta,p,k)$ is the range between it and A_η at η. Targets in the scene can be considered as distributing in a large number of small subscenes depicted as red arc and dashed line in Fig. 4. So, the scene's raw signal can be rewritten as

$$s(\tau,\eta) = w_r(\tau)\exp(j\pi\gamma\tau^2)\otimes_\tau \sum_{p=1}^{N_p(\eta)}\sum_{k=1}^{N_k(\eta)}\sigma(\eta,\rho_p,\theta_k)$$
$$\cdot\exp\left[-j4\pi f_c R(\eta,\rho_p,\theta_k)/c\right]\cdot\delta\left[\tau-2R(\eta,\rho_p,\theta_k)/c\right] \quad (12)$$

where $N_k(\eta)$ is the number of the polar subscenes at η. $R(\eta,\rho_p,\theta_k)$ is the range from the radar to the k-th subscene in the p-th equidistant ring domain. $\sigma(\eta,\rho_p,\theta_k)$ stands for the backscattering coefficient of this equivalent scatterer (polar subscene), which can be computed as

$$\sigma(\eta,\rho_p,\theta_k) = \sum_{i=1}^{I_{p,k}}\sigma_i\exp\left[-j4\pi f_c\left(R_i(\eta)-R(\eta,\rho_p,\theta_k)\right)/c\right]. \quad (13)$$

where $I_{p,k}$ is the number of targets in the k-th subscene of the p-th equidistant ring domain.

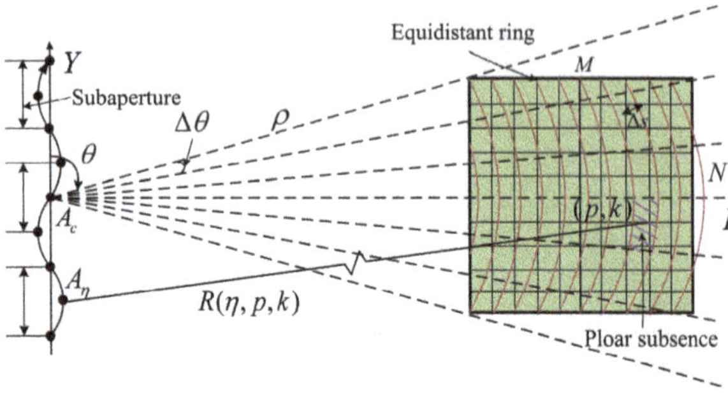

Fig. 4. Geometry model in polar coordinate.

3.2 Subaperture Processing

In the following, SAR raw signal generation method based subaperture processing is derived in detail.

Radar pulses are split into several small subapertures (see Fig. 4). Each subaperture contains the same number of radar pulses. We assume that each subaperture time is very short, thus all pulses in the same subaperture can be considered as sharing the same scene scope. The SAR raw signal generation based on the subaperture processing is shown in Fig. 5. Taking the n-th subaperture for example, $A_{\eta_{nc}}$ is the center pulse position, A_{η_n} is the radar's current position. ρ_n and θ_n are the corresponding scene polar coordinates, and $\Delta\theta_n$ is the corresponding scene angular sampling spacing. d_{η_n} is the length of the straight line $\overline{A_{\eta_{nc}}A_{\eta_n}}$, and β_{η_n} is the angle between $\overline{A_{\eta_{nc}}A_{\eta_n}}$ and the Y axis. $T(\rho_p,\theta_{n0})$ and $T'(\rho_p,\theta_{n0}+k\Delta\theta_n)$ are two polar subscenes in the p-th equidistant ring domain.

Taking into account the central pulse position $A_{\eta_{nc}}$ of the n-th subaperture, $R(\eta,\rho_p,\theta_k)$ in (12) is the same to each polar subscene in the p-th equidistant ring domain. Thus, (12) can be written as

$$s(\tau,\eta_{nc}) = w_r(\tau)\exp(j\pi\gamma\tau^2)\otimes_\tau \sum_{p=1}^{N_p(\eta_{nc})}\sum_{k=-N_k(\eta_{nc})/2}^{N_k(\eta_{nc})/2}\sigma(\eta_{nc},\rho_{np},\theta_{nk})$$
$$\cdot\exp\left[-j4\pi f_c R(\eta_{nc},\rho_{np})/c\right]\cdot\delta\left[\tau-2R(\eta_{nc},\rho_{np})/c\right] \quad (14)$$

where $R(\eta_{nc},\rho_{np})$ denotes the range from the central pulse position $A_{\eta_{nc}}$ to the p-th equidistant ring domain.

As taking into account other pulse position A_{η_n} rather than the central pulse position $A_{\eta_{nc}}$, the range $R(\eta_n,\rho_{np},\theta_{nk})$ is different from the range $R(\eta_{nc},\rho_{np})$. However, the range

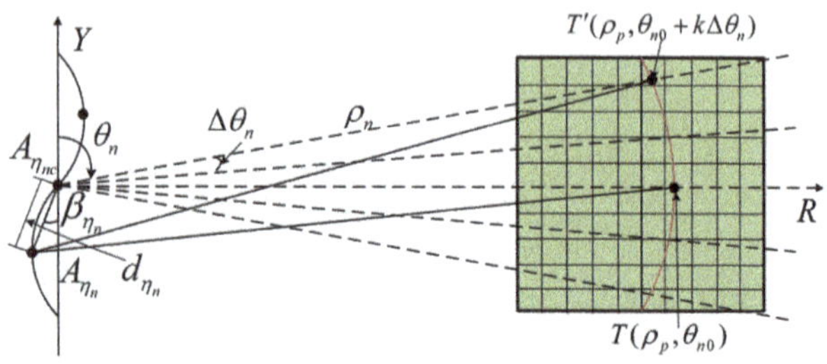

Fig. 5. Subaperture processing.

$R(\eta_n, \rho_{np}, \theta_{nk})$ can be computed from the range $R(\eta_{nc}, \rho_{np})$ according to the subaperture geometry model in Fig. 5.

Then, we can get the following relationship

$$\begin{aligned} R(\eta_n, \rho_{np}, \theta_{nk}) &= |A_{n\eta}T'| \\ &= |A_{nc}T'| + \left(|A_{n\eta}T'| - |A_{nc}T'|\right) \\ &= |A_{nc}T'| + \left(|A_{n\eta}T'| - |A_{n\eta}T|\right) + \left(|A_{n\eta}T| - |A_{nc}T'|\right) \\ &= |A_{nc}T'| + \left(|A_{n\eta}T'| - |A_{n\eta}T|\right) + \left(|A_{n\eta}T| - |A_{nc}T|\right) \\ &= R(\eta_{nc}, \rho_{np}) + \Delta R_1 + \Delta R_2 \end{aligned} \quad (15)$$

From (15), it is found that both $R(\eta_{nc}, \rho_{np})$ and ΔR_2 are independent of the variable *k*. Thus, (12) can be written as

$$\begin{aligned} s(\tau, \eta_n) = w_r(\tau)\exp(j\pi\gamma\tau^2) \otimes_\tau \sum_{p=1}^{N_p(\eta_{nc})} \sum_{k=-N_k(\eta_{nc})/2}^{N_k(\eta_{nc})/2} \sigma(\eta_n, \rho_{np}, \theta_{nk}) \\ \cdot \exp\left[-j4\pi f_c\left(R(\eta_{nc}, \rho_{np}) + \Delta R_1 + \Delta R_2\right)/c\right] \\ \cdot \delta\left[\tau - 2R(\eta_n, \rho_{np}, \theta_{nk})/c\right] \end{aligned} \quad (16)$$

Provided that each subaperture time is very short, it is reasonable that the backscattering coefficient $\sigma(\eta_n, \rho_{np}, \theta_{nk})$ is approximately substituted by $\sigma(\eta_{nc}, \rho_{np}, \theta_{nk})$. We further assume that radar's beamwidth is not very wide, so $R(\eta_n, \rho_{np}, \theta_{nk})$ is also approximated as $R(\eta_n, \rho_{np}, \theta_{n0})$, which hardly impacts the value of $\delta\left[\tau - 2R(\eta_n, \rho_{np}, \theta_{nk})/c\right]$ in the digital signal processing. Thus, (16) is approximated as

$$\begin{aligned} s(\tau, \eta_n) \approx w_r(\tau)\exp(j\pi\gamma\tau^2) \otimes_\tau \sum_{p=1}^{N_p(\eta_{nc})} \Big\{ \delta\left[\tau - 2R(\eta_n, \rho_{np}, \theta_{n0})/c\right] \\ \left[\sum_{k=-N_k(\eta_{nc})/2}^{N_k(\eta_{nc})/2} \sigma(\eta_{nc}, \rho_{np}, \theta_{nk}) \cdot \exp[-j4\pi f_c \Delta R_1/c]\right] \\ \cdot \exp\left[-j4\pi f_c\left(R(\eta_{nc}, \rho_{np}) + \Delta R_2\right)/c\right] \Big\} \end{aligned} \quad (17)$$

Here, we define the backscattering coefficient of the *p*-th equidistant ring domain at the pulse position A_{η_n} as

$$\sigma(\eta_n, \rho_{np}) = \sum_{k=-N_k(\eta_{nc})/2}^{N_k(\eta_{nc})/2} \sigma(\eta_{nc}, \rho_{np}, \theta_{nk}) \cdot \exp[-j4\pi f_c \Delta R_1/c]. \quad (18)$$

We use R_p to denote the range $R(\eta_{nc}, \rho_{np})$, then we have

$$\begin{aligned} |A_{n\eta}T| &= \sqrt{R_p^2 + d_{\eta_n}^2 + 2R_p d_{\eta_n}\cos(\theta_{n0} - \beta_{\eta_n})} \\ &= R_p\sqrt{1 + d_{\eta_n}^2/R_p^2 + 2d_{\eta_n}\cos(\theta_{n0} - \beta_{\eta_n})/R_p} \end{aligned} \quad (19)$$

It is known that (19) can be expanded by the Taylor series

$$\begin{aligned} |A_{n\eta}T| &= R_p\left[\begin{aligned} &1 + \frac{1}{2}\left(d_{\eta_n}^2/R_p^2 + 2d_{\eta_n}\cos(\theta_{n0} - \beta_{\eta_n})/R_p\right) \\ &- \frac{1}{8}\left(d_{\eta_n}^2/R_p^2 + 2d_{\eta_n}\cos(\theta_{n0} - \beta_{\eta_n})/R_p\right)^2 + \cdots \end{aligned}\right] \\ &= R_p + d_{\eta_n}\cos(\theta_{n0} - \beta_{\eta_n}) + \frac{d_{\eta_n}^2}{2R_p}\sin^2(\theta_{n0} - \beta_{\eta_n}) + \cdots \end{aligned} \quad (20)$$

Due to the fact that $d_{\eta_n} << R_p$, terms excepted the first two terms of (20) are approximated to zeros. Only considering the first two terms of (20), $|A_{n\eta}T|$ is approximated as

$$|A_{n\eta}T| \approx R_p + d_{\eta_n}\cos(\theta_{n0} - \beta_{\eta_n}) . \quad (21)$$

In a similar way, the range $|A_{n\eta}T'|$ is given by

$$|A_{n\eta}T'| = R_p\sqrt{1 + d_{\eta_n}^2/R_p^2 + 2d_{\eta_n}\cos(\theta_{n0} + k\Delta\theta_n - \beta_{\eta_n})/R_p} \quad (22)$$

and then approximated as

$$|A_{n\eta}T'| \approx R_p + d_{\eta_n}\cos(\theta_{n0} + k\Delta\theta_n - \beta_{\eta_n}) . \quad (23)$$

Then the range difference between $|A_{n\eta}T|$ and $|A_{n\eta}T'|$ is

$$\begin{aligned} \Delta R_1 &= |A_{n\eta}T'| - |A_{n\eta}T| \\ &\approx d_{\eta_n}\left(\cos(\theta_{n0} + k\Delta\theta_n - \beta_{\eta_n}) - \cos(\theta_{n0} - \beta_{\eta_n})\right) \end{aligned} \quad (24)$$

Under the above assumption that radar's beamwidth is not very wide, then the value $k\Delta\theta_n$ is usually small enough. So $\cos(\theta_{n0} + k\Delta\theta_n - \beta_{\eta_n}) - \cos(\theta_{n0} - \beta_{\eta_n}) \approx -k\Delta\theta_n\sin(\theta_{n0} - \beta_{\eta_n})$ is reasonable. Then the range difference ΔR_1 becomes

$$\Delta R_1 \approx -d_{\eta_n}k\Delta\theta_n\sin(\theta_{n0} - \beta_{\eta_n}) . \quad (25)$$

Then, (18) can be written as

$$\begin{aligned} \sigma(\eta_n, \rho_{np}) = \sum_{k=-N_k(\eta_{nc})/2}^{N_k(\eta_{nc})/2} \sigma(\eta_{nc}, \rho_{np}, \theta_{nk}) \\ \cdot \exp\left[j4\pi f_c d_{\eta_n}k\Delta\theta_n\sin(\theta_{n0} - \beta_{\eta_n})/c\right] \end{aligned} \quad (26)$$

If the angular sampling spacing of the scene for the *n*-th subaperture satisfies the following relationship

$$\Delta\theta_n = c/\left(2f_c d_{T_n}\right). \quad (27)$$

where d_{T_n} is the range from the center pulse position $A_{\eta_{nc}}$ to the edge pulse position in the *n*-th subaperture, then (27) can be written as

$$\begin{aligned} \sigma(\eta_n, \rho_{np}) = \sum_{k=-N_k(\eta_{nc})/2}^{N_k(\eta_{nc})/2} \sigma(\eta_{nc}, \rho_{np}, \theta_{nk}) \\ \cdot \exp\left[j2\pi k d_{\eta_n}\sin(\theta_{n0} - \beta_{\eta_n})/d_{T_n}\right] \end{aligned} \quad (28)$$

Assume that the size of the *n*-th subaperture is L_n, and its index is $l_n = -L_n/2 \cdots L_n/2$, then (28) can be rewritten as

$$\sigma(l_n, \rho_{np}) = \sum_{k=-N_k(\eta_{nc})/2}^{N_k(\eta_{nc})/2} \sigma(\eta_{nc}, \rho_{np}, \theta_{nk}) \cdot \exp\left[j2\pi k W_{l_n}\right]. \quad (29)$$

$W_{l_n} = d_{l_n}\sin(\theta_{n0} - \beta_{l_n})/d_{L_n}$. d_{l_n}, β_{l_n} and d_{L_n} correspond to d_{η_n}, β_{η_n} and d_{T_n}, respectively. Compared with the formula of the NUFFT [27]-[28], we can find that (28) can be calculated by the NUFFT from the scattering coefficient $\sigma(\eta_{nc}, \rho_{np}, \theta_{nk})$ of the *k*-th subscene in the *p*-th equidistant ring domain at the subaperture central pulse position $A_{\eta_{nc}}$.

For the pulse l_n in the n-th subaperture, the equivalent scatterer's backscattering coefficient $\sigma(l_n,\rho_{np})$ of the p-th equidistant ring domain will be stored in the index l_n of $\sigma(\eta_{nc},\rho_{np},\theta_{nk})$ after NUFFT. It means that all the scattering coefficient $\sigma(l_n,\rho_{np})$ of the p-th equidistant ring for the n-th subaperture can be calculated by the NUFFT at the same time, which can improve the efficiency of the raw signal generation. Therefore, the scene's raw signal for the n-th subaperture can be computed by

$$s(\tau,\eta_n) \approx w_r(\tau)\exp(j\pi\gamma\tau^2)\otimes_\tau \sum_{p=1}^{N_p(\eta_{nc})}\Big\{\mathrm{FFT}\big[\sigma(\eta_{nc},\rho_{np},\theta_{nk})\big] \cdot\exp\big[-j4\pi f_c\big(R(\eta_{nc},\rho_{np})+\Delta R_2\big)/c\big] \cdot\delta\big[\tau-2R(\eta_n,\rho_{np},\theta_{n0})/c\big]\Big\} \quad (30)$$

3.3 Implementation

According to the principle of the proposed raw signal generation algorithm, its implementation is given as follows. Fig. 6 shows the flowchart of the implementation. Then, a brief description of the flowchart:

a. Set system simulation parameters and input the targets' backscattering coefficients of the scene.

b. Generate the transmitted signal and compute its FFT.

c. Divide the synthetic aperture into several subapertures.

d. Start subaperture processing, separate the scene into several equidistant ring domain and polar subscenes.

e. Calculate the equivalent scattering coefficient of polar subscenes at the subaperture center pulse. Then, based on the subscenes' scattering coefficient, calculate equidistant rings' scattering coefficient for all subaperture pulses by the NUFFT at the same time.

f. Calculate system's impulse response function and its FFT for all subaperture pulses.

g. Multiply the results of b and f, and then transform their product by IFFT to obtain subaperture pulses' echo signal.

h. In terms of subaperture order repeat (d)-(g) procedure, then the scene's echo signal can be generated.

4. Simulation Results

In order to verify the validity of the proposed method, simulation results are shown in this section. The raw signal generated by the TBT algorithm is used as the reference for the comparison, because this method has no approximation in theory. Besides, to further prove the correctness of the proposed algorithm by the evaluation of the imaging results, raw signals simulated by different methods are processed by backprojection algorithm (BPA), since it is the most accurate imaging algorithm. Here, SAR raw signals of the discrete scattering targets and nature scene are simulated.

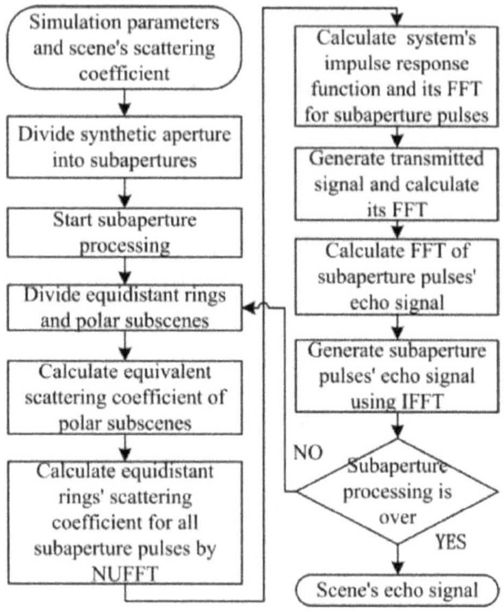

Fig. 6. Flowchart of the implementation of proposed method.

4.1 Discrete Scattering Targets Simulation

The simulation parameters are shown in Tab. 1. Based on the simulation parameters, it is easy to compute that the synthetic aperture time T_a is about 3.8 s. The motion errors are added to the nominal track of the radar. The azimuth error in Y axis is $\delta y = 2\cdot\sin\big(2\pi\cdot(0.3/T_a)\eta\big)$, the range error in X axis is $\delta x = 5\cdot\sin\big(2\pi\cdot(1/T_a)\eta\big)+0.3\eta$, and the altitude error in Z axis is $\delta z = 3\cdot\sin\big(2\pi\cdot(0.5/T_a)\eta\big)$. 81 discrete scattering targets are located in scene, which are arranged in 9 rows and 9 columns. Range and azimuth spacing between the targets are 20.98 m and 22.44 m (in ground plane), respectively. The center scattering target's position is (1100 m, 0 m, 0 m). The backscattering coefficient of all scattering targets is assumed to be 1. Note that the jamming, thermal noise and clutter aren't considered in simulation.

Fig. 7 shows the simulated raw signals by the TBT and proposed algorithms. We can find that both amplitude and phase of the two raw signals look quite like each other. The amplitude relative error and phase error between two raw signals are given in Fig. 8, which is shown that the maximum amplitude relative error between two raw signals

Center frequency	400 MHz
Signal bandwidth	230 MHz
Sampling frequency	250 MHz
Pulse duration	1 μs
Pulse repetition frequency	100 Hz
Azimuth beamwidth	9.5°
Radar's nominal altitude	100 m
Radar's nominal velocity	45 m/s
Size of the scene	200 m × 200 m (range×azimuth)
Ground range of the scene centre	1100 m

Tab. 1. Simulation parameters for discrete scattering targets.

caused by the approximation of the equivalent scatterer and subaperture processing is about 0.01. The maximum phase error between two raw signals is about 0.201 radian, and its mean value and standard deviation are -4.079×10^{-7} radian and 0.0147 radian, respectively. These results prove the raw signal generated by the proposed method is precise.

To further verify the validity of the proposed method, raw signals generated by different methods are processed by the BPA, and the imaging results are shown in Fig. 9. From Fig. 9 (a) and (b), it is seen that all discrete scattering targets are full focused. Fig. 9 (c) and (d) show contours of the center target in the range [-30, 0] dB with a contour step of -3 dB, which are extracted from Fig. 9 (a) and (b), respectively. It is seen that the imaging result of the center target in Fig. 9 (d) is very close to that in Fig. 9 (c), but its focusing quality is slightly degraded due to approximations in raw signal generation. However, the effect is invisible at the high contour levels and some small influences can be observed at the lower contour levels.

The amplitude and phase profiles in the azimuth and range directions are shown in Fig. 10. From these figures, we can see clearly that both amplitude and phase profiles are almost identical. In the mainlobes of the amplitude profiles, it is impossible to see any difference between them. Whereas in the sidelobes of the amplitude profiles, there are slight differences between them. The similar performance can also be observed in the phase profiles, both in the range and azimuth directions.

The center target's resolution and PSLR are measured utilizing the amplitude profiles. The measured results are shown in Tab. 2. From Tab. 2, it is found that the measured parameters of the center target obtained from different raw signals are almost identical, and nearly theoretical results are achieved, which indicates that the presented raw signal generation algorithm is valid.

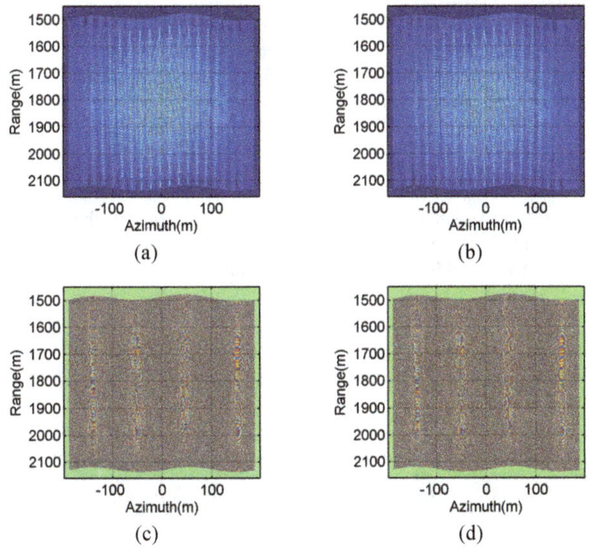

Fig. 7. Simulated raw signals of the discrete scattering targets. (a), (b) Amplitude and phase of the raw signal by the TBT algorithm; (c), (d) Amplitude and phase of the raw signal by the proposed algorithm.

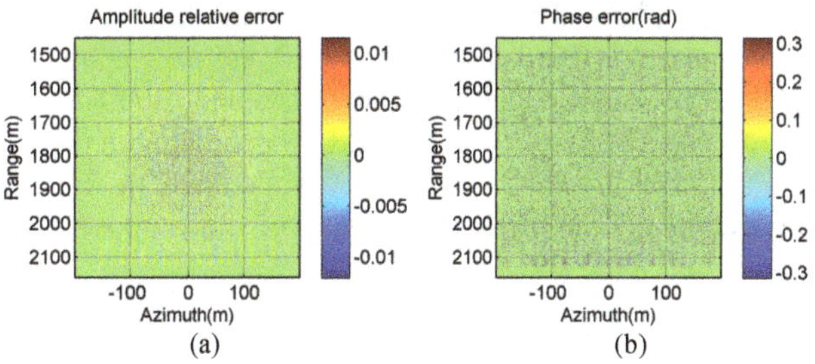

Fig. 8. Comparison between the simulated raw signals by different algorithms. (a) Amplitude relative error; (b) Phase error.

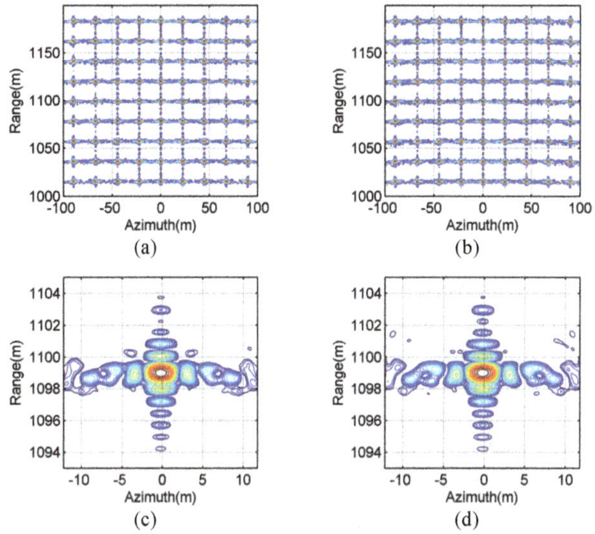

Fig. 9. Imaging results of the raw signals processed by the BPA. (a) TBT algorithm; (b) Proposed algorithm; (c) Imaging result of the center target in (a); (d) Imaging result of the center target in (b).

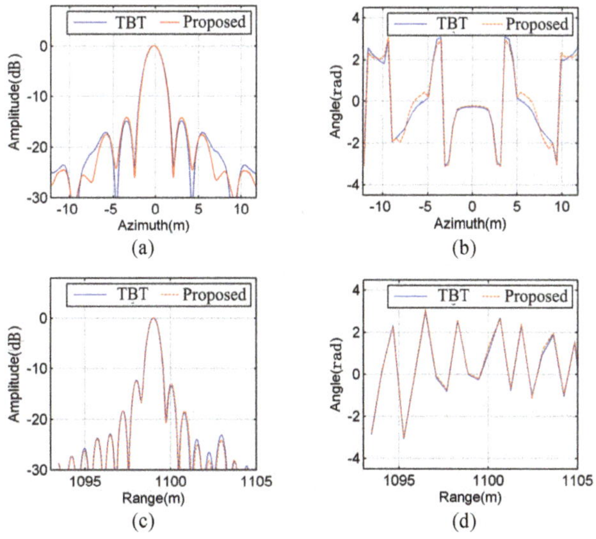

Fig. 10. Comparison between the results in Fig. 9(c) and (d). (a) Azimuth amplitude; (b) Azimuth phase; (c) Range amplitude; (d) Range phase.

To prove the simulation efficiency of the proposed method, the processing time of two algorithms is measured on the same simulation condition. The two algorithms have been implemented in Matlab version 7.10.0 and on a computer with a 2.93 GHz Dual-Core processor and 2.00 GB RAM. Processing time of the TBT and proposed methods are 2243 s and 312.1 s, respectively. Compared with the TBT algorithm, the simulation efficiency of the proposed algorithm is improved about 7.18 times.

Measured parameters of the center target		Raw signal obtained by TBT method	Raw signal obtained by proposed method
Resolution	Azimuth	1.9575 m	1.9993 m
	Range	0.6415 m	0.6437 m
PSLR	Azimuth	13.47 dB	13.85 dB
	Range	13.35 dB	13.54 dB

Tab. 2. Measured parameters for the center target in the scene.

4.2 Nature Scene Raw Signal Simulation

To prove the feasibility of the proposed algorithm, the raw signal generation of the natural scene is performed.

The system simulation parameters for the natural scene are shown in Tab. 3. Fig. 11 is a high-resolution airborne SAR image with 1051 × 900 pixels. According to the point scattering model, single look complex data of the airborne SAR image can be directly regarded as the natural scene target's backscattering coefficient. And, TBT and proposed algorithms are used to simulate the natural scene's raw signal, which are also reconstructed by the BPA.

The simulated raw signals of the natural scene by the different methods are shown in Fig. 12. Fig. 13 gives the amplitude relative error and phase error between two raw signals in Fig. 12. It can be seen that two raw signals are very similar, and both amplitude relative error and phase error are very small, so the proposed method can satisfy the precise scene raw signal generation.

Fig. 14 shows the imaging results from the simulated raw signals in Fig. 12 by BPA, we can find that the nature scenes are reconstructed well. From the angle of sight, the simulated SAR images in Fig. 12 are similar to the real airborne SAR image in Fig. 11, and the only difference is that the former's resolution is lower than that of the latter. The reason is that the resolution scheduled in the simulated system is lower than that of the airborne SAR image, and the effects of the jamming, noise, clutter, multipath and incidence angle on target's backscattering coefficient aren't considered in simulation. The contours and the amplitude and phase profiles of the focused point target in the red rings in Fig. 14 are given in Fig. 15 and Fig. 16. We can find that the selected target is well focused, the focusing performance of the two imaging results are almost identical. Besides, the measured resolutions of this target are about 1.24 m in azimuth and 1.5 m in range, respectively.

Center frequency	400 MHz
Signal bandwidth	100 MHz
Sampling frequency	120 MHz
Pulse duration	1 μs
Pulse repetition frequency	100 Hz
Azimuth beamwidth	15.4°
Radar's nominal altitude	100 m
Radar's nominal velocity	45 m/s
Size of the scene	120 m × 195 m (range×azimuth)
Ground range of the scene centre	1060 m

Tab. 3. Simulation parameters for the natural scene.

Finally, the running time of the proposed algorithm is 1547.5 s comprised with the TBT method with 11204 s, which is acceptable for the scene simulation experiment.

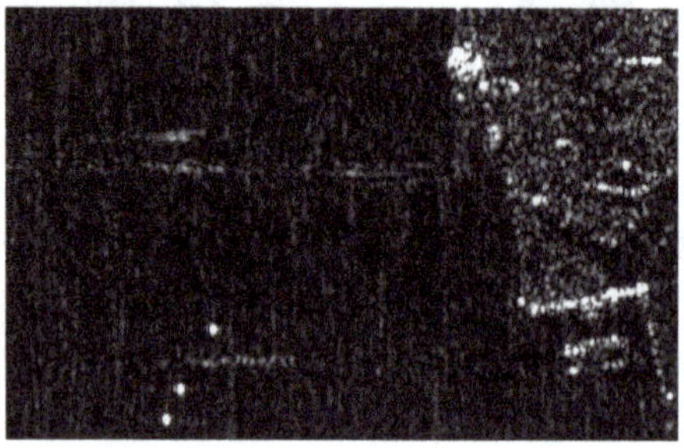

Fig. 11. Real airborne SAR image.

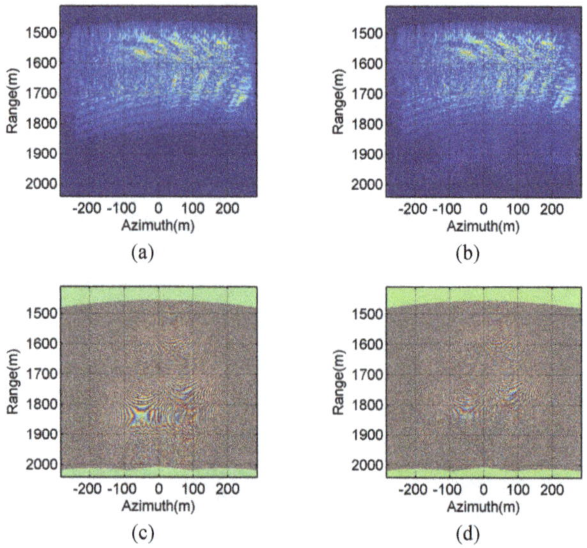

Fig. 12. Simulated raw signals of the natural scene. (a), (b) Amplitude and phase of the raw signal by the TBT algorithm; (c), (d) Amplitude and phase of the raw signal by the proposed algorithm.

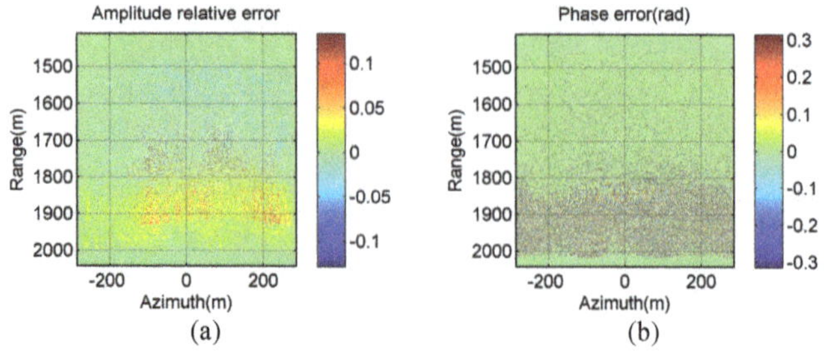

Fig. 13. Comparison between the simulated raw signals by different algorithms. (a) Amplitude relative error; (b) Phase error.

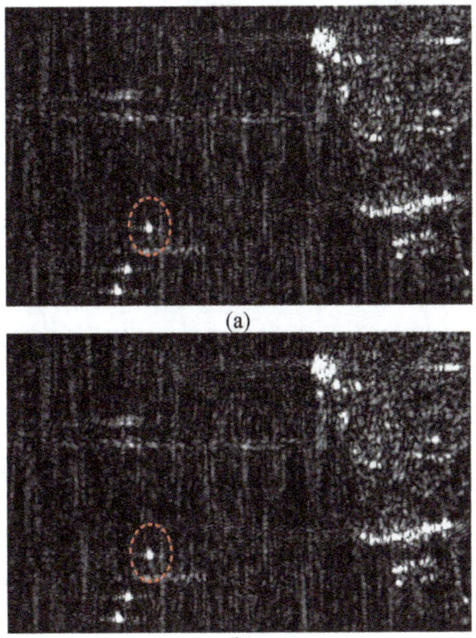

(a)

(b)

Fig. 14. Imaging results of the raw signals by the BPA. (a) TBT algorithm; (b) Proposed algorithm.

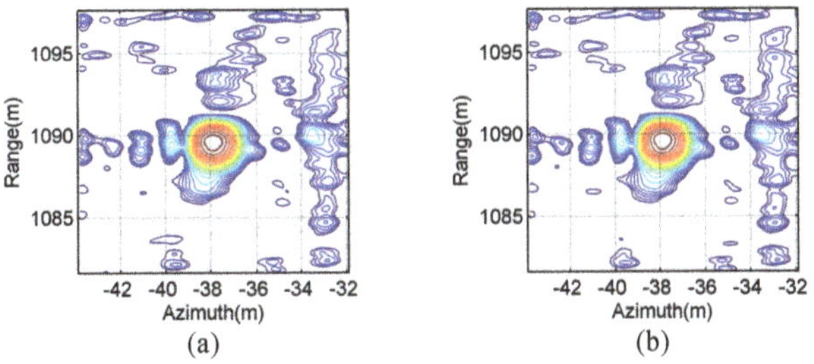

(a) (b)

Fig. 15. Imaging results of the point target in the red rings in Fig. 14: (a) TBT algorithm; (b) Proposed algorithm.

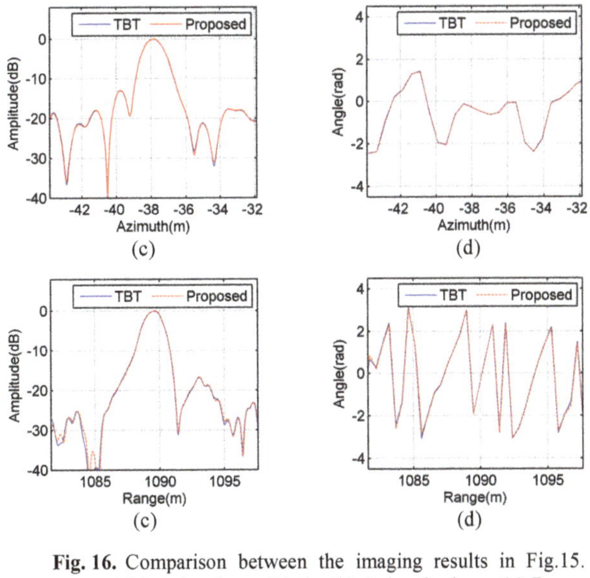

(c) (d)

(c) (d)

Fig. 16. Comparison between the imaging results in Fig.15. (a) Azimuth amplitude; (b) Azimuth phase; (c) Range amplitude; (d) Range phase.

In conclusion, the proposed raw signal generation method has high simulation precision and efficiency.

5. Conclusion

In the paper, a fast SAR raw signal generation method based on equivalent scatterer and subaperture processing is proposed. In this method, the subaperture processing and NUFFT are used to calculate the SAR system transfer function for the same subaperture pulses, then subaperture pulses' raw signal is generated by the convolution of the transmitted signal and system transfer function using FFT. It considers the radar's motion track to generate the precise raw signal for the real SAR. The utilization of the equivalent scatterer, subaperture processing and FFT (NUFFT) operation can greatly improve the efficiency of the SAR raw signal generation. The proposed method is verified by the simulation results, which has enriched and developed the SAR raw signal generation method.

However, to calculate the scattering coefficient of the equivalent scatterer for the same subaperture, the slant range approximation is used in the proposed simulation algorithm. Thus, it is needed to analyze the effect of the approximation error on the raw signal generation and imaging processing in the future research. In addition, with regard to the natural scene simulation, the complex data of the airborne SAR image is directly used as the target's backscattering coefficient. Thus, to generate more precise raw signal, the model and simulation of the nature scene should be investigated in the future research.

Acknowledgements

This work was supported by the National Natural Science Foundation of China under grant 61201329.

References

[1] CUMMING, I. G., WONG, F. H. *Digital Processing of Synthetic Aperture Radar Data: Algorithms and Implementation.* 1st ed. Norwood: Artech House, 2005.

[2] XIE, H., AN, D., HUANG, X., LI, X., ZHOU, Z. Fast factorised backprojection algorithm in elliptical polar coordinate for one-stationary bistatic very high frequency/ultrahigh frequency ultra wideband synthetic aperture radar with arbitrary motion. *IET Radar, Sonar and Navigation*, 2014, vol. 8, no. 8, p. 946–956, DOI: 10.1049/iet-rsn.2012.0350.

[3] XIE, H., AN, D., HUANG, X., ZHOU, Z. Fast time-domain imaging in elliptical polar coordinate for general bistatic VHF/UHF ultra-wideband SAR with arbitrary motion. *IEEE Journal of Selected Topics in Applied Earth Observations and Remote Sensing*, September 2014, DOI: 10.1109/JSTARS.2014.2347413.

[4] FRANCESCHETTI, G., MIGLIACCIO, M., RICCIO, D., SCHIRINZI, G. SARAS: A synthetic aperture radar (SAR) raw signal simulator. *IEEE Transactions on Geoscience and Remote Sensing*, 1992, vol. 30, no. 1, p. 110–123.

[5] CIMMINO, S., FRANCESCHETTI, G., IODICE, A., RICCIO, D., RUELLO, G. Efficient spotlight SAR raw signal simulation of

extended scenes. *IEEE Transactions on Geoscience and Remote Sensing*, 2003, vol. 41, no. 10, p. 2329–2337.

[6] QIU, X., HU, D., ZHOU, L., DING, C. A bistatic SAR raw data simulator based on inverse ω-k algorithm. *IEEE Transactions on Geoscience and Remote Sensing*, 2010, vol. 48, no. 3, p. 1540 to 1547.

[7] LIN, Y. C., SARABANDI, K. A Monte Carlo coherent scattering model for forest canopies using fractal-generated trees. *IEEE Transactions on Geoscience and Remote Sensing*, 1999, vol. 37, no. 1, p. 440–451.

[8] BROWN, C. G., SARABANDI, K., GILGENBACH, M. Physics-based simulation of high-resolution polarimetric SAR images of forested areas. In *Proceedings of IEEE International Geoscience and Remote Sensing Symposium*. Toronto (Canada), 2002, p. 466 to 468.

[9] FRANCESCHETTI, G., MIGLIACCIO, M., RICCIO, D. On ocean SAR raw signal simulation. *IEEE Transactions on Geoscience and Remote Sensing*, 1998, vol. 38, no. 1, p. 84–100.

[10] FRANCESCHETTI, G., IODICE, A., RICCIO, D. A canonical problem in electromagnetic backscattering from buildings. *IEEE Transactions on Geoscience and Remote Sensing*, 2002, vol. 40, no. 8, p. 1787–1801.

[11] FRANCESCHETTI, G., IODICE, A., RICCIO, D., RUELLO, G. SAR raw signal simulation for urban structures. *IEEE Transactions on Geoscience and Remote Sensing*, 2003, vol. 41, no. 9, p. 1986–1995.

[12] WANG, Z. L., XU, F., JIN, Y. Q., OGURA, H. A double Kirchhoff approximation for very rough surface scattering using the stochastic functional approach. *Radio Science*, 2005, vol. 40, no. 4, RS4011, DOI: 10.1029/2004RS003079.

[13] XU, F., JIN, Y. Q. Deorientation theory of polarimetric scattering targets and application to terrain surface classification. *IEEE Transactions on Geoscience and Remote Sensing*, 2005, vol. 43, no. 10, p. 2351–2364.

[14] XU, F., JIN, Y. Q. Imaging simulation of polarimetric synthetic aperture radar for comprehensive terrain scene using the mapping and projection algorithm. *IEEE Transactions on Geoscience and Remote Sensing*, 2006, vol. 44, no. 11, p. 3219–3234.

[15] YONGYAN, L. *SAR Image*. Harbin: Harbin Institute of Technology Press, 1999.

[16] FRANCESCHETTI, G., IODICE, A., RICCIO, D., RUELLO, G. A 2-D Fourier domain approach for spotlight SAR raw signal simulation of extended scenes. In *Proceedings of IEEE International Geoscience and Remote Sensing Symposium*. Toronto (Canada), 2002, p. 853–855.

[17] FRANCESCHETTI, G., IODICE, A., PERNA, S., RICCIO, D. Efficient simulation of airborne SAR raw data of extended scenes *IEEE Transactions on Geoscience and Remote Sensing*, 2006, vol. 44, no. 10, p. 2851–860.

[18] FRANCESCHETTI, G., IODICE, A., PERNA, S., RICCIO, D. SAR sensor trajectory deviations: Fourier domain formulation and extended scene simulation of raw signal. *IEEE Transactions on Geoscience and Remote Sensing*, 2006, vol. 44, no. 9, p. 2323 to 2334.

[19] FRANCESCHETTI, G., GUIDA, R., IODICE, A., RICCIO, D., RUELLO, G., STILLA, U. Simulation tools for interpretation of high resolution SAR images of urban areas. In *Proceedings of Urban Remote Sensing Joint Event*. Paris (France), 2007, p. 1–5.

[20] WANG, Y., ZHANG, Z., DENG, Y. Squint spotlight SAR raw signal simulation in the frequency domain using optical principles. *IEEE Transactions on Geoscience and Remote Sensing*, 2008, vol. 46, no. 8, p. 2208–2215.

[21] KHWAJA, A. S., FERRO-FAMIL, L., POTTIER, E. SAR raw data simulation using high precision focusing methods. In *Proceedings of European Radar Conference*. Paris (France), 2005, p. 33–36.

[22] KHWAJA, A. S., FERRO-FAMIL, L., POTTIER, E. Efficient SAR raw data generation for anisotropic urban scenes based on inverse processing. *IEEE Geoscience and Remote Sensing Letters*, 2009, vol. 6, no. 4, p. 757–761.

[23] HUANG, L., WANG, Z., ZHENG, T. A fast algorithm based on FFT used in simulation of SAR return wave signal. *Journal of Remote Sensing*, 2004, vol. 8, no. 2, p. 128–136.

[24] ZHANG, S., LONG, T., AENG, T., DING, Z. Space-borne synthetic aperture radar received data simulation based on airborne SAR image data. *Advances in Space Research*, 2008, vol. 41, no. 11, p. 181–182.

[25] WEN, L., ZHENG, T. A sub-aperture based SAR raw signal generation method. In *Proceedings of IET International Radar Conference*. Guilin (China), 2009, p. 89–92.

[26] ZHANG, S., ZHANG, W., KONG, L. SAR raw signal simulation based on sub-aperture processing. In *Proceedings of International Radar Conference*. Washington (USA), 2010, p. 569–572.

[27] LIU, Q. H., NGUYEN, N. An accurate algorithm for nonuniform fast Fourier transforms (NUFFT's). *IEEE Microwave and Guided Wave Letters*, 1998, vol. 8, no. 1, p. 18–20.

[28] ANDERSSON, F., MOSES, R., NATTERER, F. Fast Fourier methods for synthetic aperture radar imaging. *IEEE Transactions on Aerospace and Electronic Systems*, 2012, vol. 48, no. 1, p. 215 to 229.

About Authors ...

Hongtu XIE was born in Hunan, China. He received his B.Sc. from the Hunan University in 2008 and his M.Sc. from the National University of Defense Technology (NUDT) in 2010. His research interests include the low frequency ultra wideband (UWB) SAR imaging.

Daoxiang AN was born in Jilin, China. He received his B.Sc., M.Sc. and Ph.D. from the National University of Defense Technology in 2004, 2006 and 2011, respectively. He is currently a Lecturer with the NUDT. His research interests include the UWB SAR imaging.

Xiaotao HUANG was born in Hubei, China. He received his B.Sc. and Ph.D. from the National University of Defense Technology in 1990 and 1999, respectively. He is currently a Professor with the NUDT. His research interests include the radar theory and signal processing.

Zhimin ZHOU was born in Hebei, China. He received the B.Sc., M.Sc. and Ph.D. from the National University of Defense Technology in 1982, 1989, and 2002, respectively. He is currently a Professor with the NUDT. His research interest includes UWB radar system and signal processing.

First Order Sea Clutter Cross Section for HF Hybrid Sky-Surface Wave Radar

Yongpeng ZHU, Yinsheng WEI, Yajun LI

School of Electronics and Information Engineering, Harbin Institute of Technology, Harbin, 150001, China

zhuyp@hit.edu.cn, weiys@hit.edu.cn, liyajun1985happy@163.com

Abstract. *This paper presents a modified method to simulate the first order sea clutter cross section for high frequency (HF) hybrid sky-surface wave radar, based on the existent model applied in the bistatic HF surface wave radar. The modification focuses on the derivation of Bragg scattering frequency and the ionosphere dispersive impact on the clutter resolution cell. Meanwhile, an analytic expression to calculate the dispersive transfer function is derived on condition that the ionosphere is spherical stratified. Simulation results explicate the variance of the cross section after taking account of the influence triggered by the actual clutter resolution cell, and the spectral width of the first order sea clutter is defined so as to compare the difference. Eventually, experiment results are present to verify the rationality and validity of the proposed method.*

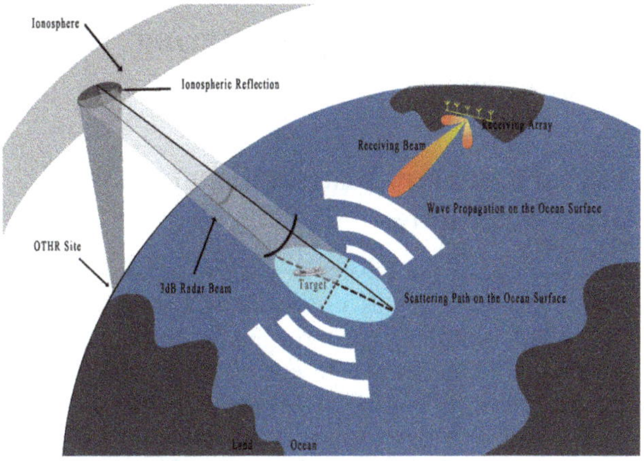

Fig. 1. The detection principle diagram of HF hybrid sky-surface wave radar.

Keywords

Clutter resolution cell, ionosphere dispersion, first order sea clutter cross section, HF hybrid sky-surface wave radar.

1. Introduction

The HF hybrid sky-surface wave radar is a novel radar configuration, which consists of the sky wave transmitting and surface wave receiving propagation path, and the basic schematic diagram of the detection principle is illustrated in Fig. 1. Such special hybrid travelling channel not only maintains the capacity of detecting targets over the horizon, but also makes the deployment of receiving antenna array much more flexible [1], [2].

However, the ocean surface, as the primary medium of ground wave diffraction, brings in the issue of sea clutter interference at the same time. Now that the sea clutter, especially the first order sea clutter, constructs the major detection background of targets with low Doppler frequency [3]-[5], therefore it is worthwhile to comprehend the characteristic of which so as to enhance the detection probability of targets [6]-[10]. Therefore, the radar cross section (RCS) model has been studied afterwards to simulate the first and second order RCS of the sea clutter applicable in the monostatic and bistatic HF surface wave radar (HFSWR) [11], [12]. However, given the hybrid propagation mode herein, the travelling path consists of the ionosphere reflection. Therefore, in order to explicate the ionospheric influence on the clutter cross section, we mainly focus on its influence on the clutter resolution cell A_c as a matter of fact that the normalized radar cross section denoted by the symbol σ^0 is given as $\sigma^0 = \sigma_c/A_c$ [13], where σ_c is the radar cross section of clutter occupying A_c. While, the effect on the clutter resolution cell imposed by ionosphere is an important problem that is to mention but a few. As far as the existing published analysis [14], [15], the ionospheric dispersive feature is regarded as the princeps factor leading to this. In order to explain the ionosphere dispersion, Lundborg [14] has derived an analytical expression of the ionosphere dispersive transfer function under the assumption that the ionosphere is plane stratified and the electron density profile obeys the parabolic model.

Besides, such new travelling path also introduces a number of new arguments such as the grazing angle and the lateral scattering angle etc. Therefore, a redefined formula to calculate the Bragg scattering frequency appears to be an issue worthy of study in the HF hybrid sky surface wave radar after taking account of these variable effects.

Given these key problems aforementioned, the paper is set out as follows. Section 2 focuses on the derivation of the Bragg scattering frequency accommodating to this hybrid propagation system. Section 3 explains the effect on the clutter resolution cell resulting from ionosphere dispersion property. In this section, we consider the case of spherical geometry gaining simplicity by neglecting the influence of the geomagnetic field and collisions. The expression of dispersive coefficient, a critical argument determining the dispersive transfer function, is derived in Appendix A, which is appropriate for a spherical stratified ionosphere and quasi-parabolic (Q-P) model. Section 4 simulates the radar cross section of the first order sea clutter after taking the influence of clutter resolution cell into consideration. Furthermore, experiment results are provided to verify the effectiveness of the simulation method. Eventually, our results are summarized in Section 5.

2. Derivation of Bragg Scattering Frequency

2.1 Spatial Geometry Distribution of the First Order Sea Clutter

HF sky-surface wave radar makes use of the sky wave radar station located inland as a transmitting station, monitoring the ocean area of interest through ionosphere reflection, and finally the backscattered echo is received by the shore-based receiver by means of diffraction on the ocean surface. A simplified propagation path can be seen in Fig. 2, where the symbol Tr and Rr indicate the position of transmitter and receiver, respectively. T is another ellipse focus relative to Rr.

As the time delay corresponds to the distance between the target and the receiver given the conventional radar signal processing, it is reasonable to take it as an independent variable to deduce the distribution of the first order sea clutter. Firstly, combined with the geometry relationship, we could obtain the following equations easily assuming that the ionosphere is not inclined, viz. $R_1 = R_2$

$$R_1^2 = h^2 + (R_t / 2)^2, \quad (1)$$

$$R_t^2 = D_0^2 + R^2 - 2D_0 R\cos\theta_r, \quad (2)$$

$$d = R_1 + R_2 + R = 2R_1 + R. \quad (3)$$

Based on (1) – (3), the target location can be expressed as

$$R = \frac{d^2 - 4h^2 - D_0^2}{2d - 2D_0\cos\theta_r}$$

where D_0 indicates the baseline distance between the transmitting and receiving stations; d is the group range which could be estimated by the time delay; R is the distance from the target to the receiving station; θ_r is the angle between R and the baseline; h represents the ionospheric reflection height; Δ_i is the grazing angle.

Afterwards, the group range d is postulated to be fixed, which is equivalent to make the time delay invariant and then the coordinates could be constructed as follows:

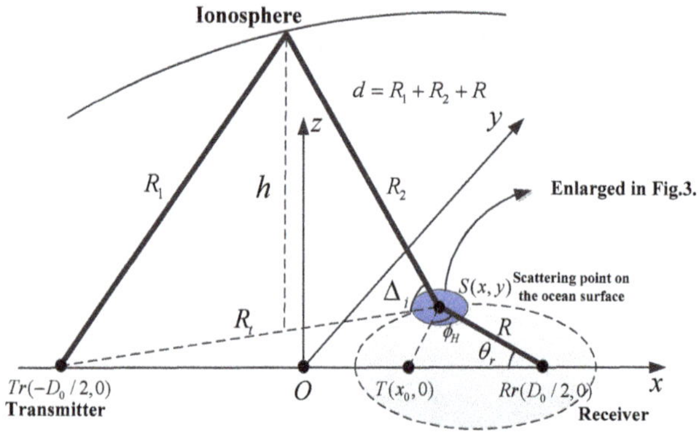

Fig. 2. Hybrid sky-surface propagation channel illustration (solid bold line) and the ellipse trajectory distribution of the sea clutter (dotted line) corresponding to a fixed time delay.

The baseline is x axis, the normal direction is y axis and the central point O is the origin; therefore, we have the following mathematical relationship

$$\begin{cases} (x + \frac{D_0}{2})^2 + y^2 = (2h\cot\Delta_i)^2 \\ (x - \frac{D_0}{2})^2 + y^2 = R^2 \end{cases}$$

Therefore, the coordinate (x, y) of the scattering point S on the ocean surface could be easily obtained by

$$\begin{cases} x = \dfrac{(2h\cot\Delta_i)^2 - R^2}{2D_0} \\ y = \pm\sqrt{(2h\cot\Delta_i)^2 - (x + D_0/2)^2} \end{cases} \quad (4)$$

Furthermore, based on (4), we figure out that the locus of S could also be described as an ellipse equation (the dashed line in Fig. 2.) with the focus $T(x_0, 0)$ and $Rr(D_0/2, 0)$.

$$\left(x - x_c\right)^2 / a^2 + y^2 / b^2 = 1, \quad (5)$$

$$a = \frac{(d^2 - 4h^2 - D_0^2)d}{2(d^2 - D_0^2)}, \quad c = \frac{(d^2 - 4h^2 - D_0^2)D_0}{2(d^2 - D_0^2)}, \quad (5a)$$

$$b = \sqrt{a^2 - c^2}, \qquad x_c = \frac{2h^2 D_0}{d^2 - D_0^2}. \quad (5b)$$

The corresponding parameters are obtained by (5a)–(5b), and the detailed derivation is not stated herein for the sake of the article length.

2.2 Derivation of the Sea Wave Vector Responsible for Bragg Scattering

Given the relationship between the incident wave vector and scattering wave vector illustrated in Fig. 3, the sea wave vector responsible for Bragg scattering could be regarded as the composition output of vectors. Furthermore,

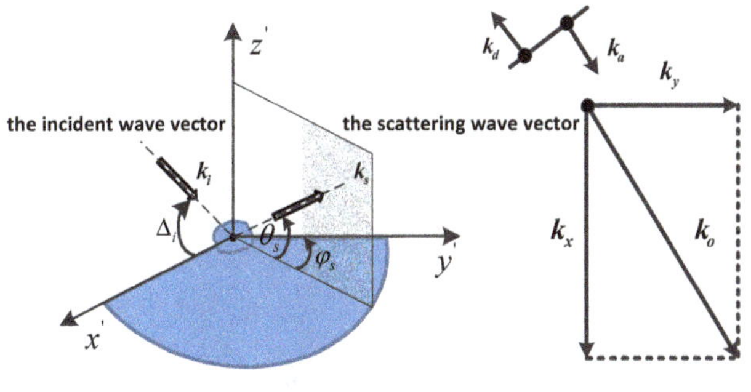

Fig. 3. Geometry for scattering from the patch on the ocean surface.

according to Fig. 3, the elementary scattering relationship could be expressed as

$$\boldsymbol{k}_x = \boldsymbol{k}_s \cos\theta_s \sin\varphi_s + \boldsymbol{k}_i \cos\Delta_i , \quad (6)$$

$$\boldsymbol{k}_y = \boldsymbol{k}_s \cos\theta_s \cos\varphi_s , \quad (7)$$

$$\boldsymbol{k}_o = \boldsymbol{k}_x + \boldsymbol{k}_y . \quad (8)$$

In Fig. 3, the ascending wave vector $\boldsymbol{k}_a$ and the descending wave vector $\boldsymbol{k}_d$ responsible for the positive and negative Bragg scattering frequency can be interpreted as in the same and opposite direction of $\boldsymbol{k}_o$. In addition, since the velocity v of a sea wave of length L is given by $v = \sqrt{gL/(2\pi)}$ [11] and the sea wave number k_0 is determined by $k_0 = 2\pi / L$, the intrinsic frequency for a particular sea wave can be obtained by (9), based on the assumption that the absolute value of the wave vector before and after scattering keeps equal, viz. $|\boldsymbol{k}_i| = |\boldsymbol{k}_s| = k$.

$$\begin{aligned}\omega_H &= 2\pi f_H = 2\pi v / L \\ &= \frac{2\pi}{L}\sqrt{\frac{gL}{2\pi}} = \sqrt{gk_0} \\ &= \sqrt{gk}\left(\cos^2\Delta_i + \cos^2\theta_s + 2\cos\theta_s \sin\varphi_s \cos\Delta_i\right)^{1/4}\end{aligned} \quad (9)$$

Actually, strong scattering happens once the sea wave intrinsic vibrating frequency resonates with the radar radio wave frequency. And the resonating frequency ω_H is regarded as in consistence with Bragg scattering frequency [11]. Furthermore, θ_S in Fig. 3 equals to zero since the radar radio wave received by the receiver is diffracted on the ocean surface. And compared with the geometry relationship in Fig. 2 and Fig. 3, it is easy to yield the result that $\varphi_s = \theta_r$. Therefore, equation (9) could be simplified as

$$\omega_H = \sqrt{gk}\left(\cos^2\Delta_i + 1 + 2\sin\theta_r \cos\Delta_i\right)^{1/4} . \quad (10)$$

Besides, allowing for the fact that the spatial geometry distribution of the first order sea clutter in bistatic HFSWR could be described by an ellipse [12], the result of which is also consistent with the conclusion derived in previous part. And the Bragg frequency in bistatic HF surface wave radar could be calculated by

$$\omega_B = \sqrt{2kg\cos(\phi/2)} \quad (11)$$

where ϕ indicates the bistatic angle.

Intuitively, we try to generate a similar expression with (11). Therefore, an approximate formula (12) is given by analogy. However, albeit with the unity in form, the calculation error still exists. Since the relative error between (10) and (12) is small enough to be ignored, (12) is reasonable to be treated as the formula calculating the Bragg scattering frequency for most circumstance.

$$\omega_H \approx \sqrt{2kg\cos(\phi_H / 2)} \quad (12)$$

where $\phi_H = 2\arctan(\frac{\sin\theta_r}{a/c - \cos\theta_r})$. The parameter a and c can be calculated through (5a) – (5b).

3. Analysis of Ionospheric Effect on the Clutter Resolution Cell

3.1 Effect of Ionosphere Dispersion on the Clutter Resolution Cell

It is well known that the clutter resolution cell is determined by the radar range resolution and azimuthal angle resolution, simultaneously. As far as the ionosphere influence is concerned, we emphasize on its distortion on the radar range resolution caused by the ionosphere dispersion profile. And the azimuthal angle resolution is not the focus of this work, since it is determined primarily by the aperture of the receiving array [13].

3.1.1 Analytical Expression of the Ionosphere Dispersion Transfer Function

Generally, the radar range resolution is usually determined by the waveform output of matching filter, which could be expressed as [15]

$$\rho(t) = \frac{1}{2\pi}\int M_r(j\omega)M_0^*(j\omega)\exp(j\omega t)d\omega . \quad (13)$$

According to (13), $M_r(j\omega)$ is the spectrum of the received signal and $M_0^*(j\omega)$ is the complex conjugate of the spectrum of the transmitted signal. On account of ionosphere dispersion, the received signal $m_r(t)$ should be written as the convolution output of the ionosphere dispersion transfer function $g(t)$ and $m(t)$ viz.

$$m_r(t) = \int_{-\infty}^{\infty} m(t-\tau)g(\tau)d\tau . \quad (14)$$

As for the dispersion transfer function, Lundborg [14] has yielded the ionospheric weighing function expressing the ionosphere dispersion causing the distortion of the pulse shape by (16), which forms a Fourier transform pair with (15).

	E Layer			F_1 Layer			F_2 Layer		
Parameters	$f_{cE}(MHz)$	$h_{mE}(km)$	$y_{mE}(km)$	$f_{cF_1}(MHz)$	$h_{mF_1}(km)$	$y_{mF_1}(km)$	$f_{cF_2}(MHz)$	$h_{mF_2}(km)$	$y_{mF_2}(km)$
	4.1	85	19	6	125	60	11	200	120
t_1	t_{1E}	0.10~1.03		$t_{1F_{1L}}$	0.17~3.11		$t_{1F_{2L}}$	0.20~6.14	
$\times 10^{-6} s$	—	—	—	$t_{1F_{1H}}$	18.30~24.56		$t_{1F_{2H}}$	17.58~24.56	

Tab. 1. Range of dispersive coefficient corresponding to each layer (the ground distance D = 2000 km).

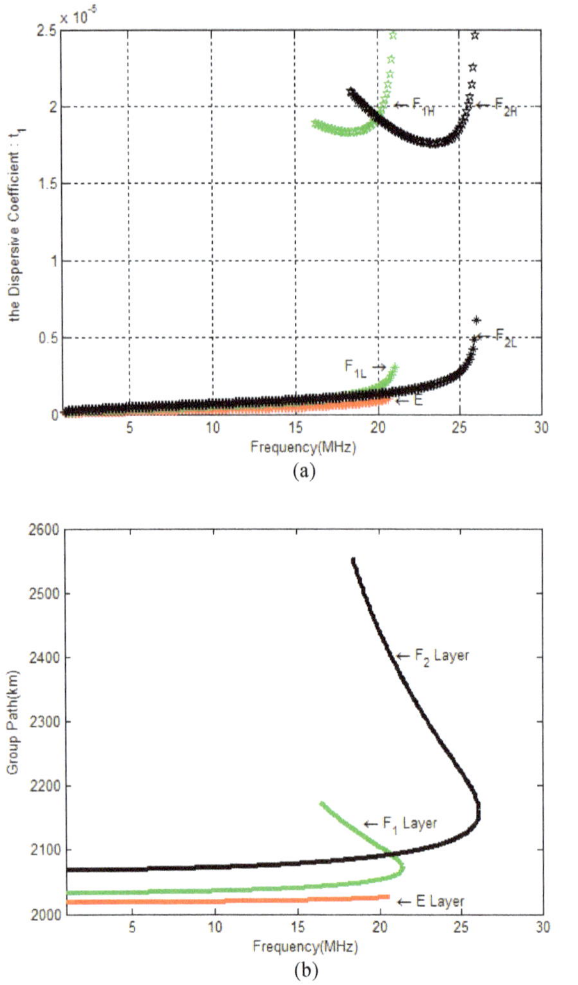

Fig. 4. Frequency variation of the (a) dispersive coefficient and (b) group path in a quasi-parabolic model for each ionosphere layer at the ground distance of 2000 km.

$$g(t) = t_1^{-1} \exp(-\pi i / 4 + \pi i t^2 / t_1^2), \tag{15}$$

$$G(\omega) = \mathcal{F}\left(g(t)\right) = \exp(-\pi i \omega^2 / f_1^2) \tag{16}$$

where t_1 is defined as the dispersive coefficient and $f_1 = 1 / t_1$ is called as the dispersive bandwidth. According to (16), it is easy to observe that the ionosphere dispersion function acts somewhat like a low pass filter in the frequency domain. And, as the unique parameter, the dispersive coefficient t_1 determines the scale of dispersive bandwidth essentially and therefore influences the extent of band pass filtering modulation on the signal envelope indirectly in accordance with (14).

Afterwards, an analytical expression to calculate t_1 under the assumption that the ionosphere is spherical stratified and the electron density of ionosphere obeys the quasi-parabolic model has been derived (see Appendix A). The reason why we choose such conditions is as follows:

- With regard of the detection over the horizon, one hop is usually more than hundreds of kilometers, making the assumption of plane stratified ionosphere structure untenable.
- Compared with the parabolic model adopted in [14], the quasi-parabolic model is an improved one with the introduction of a modified factor. Besides, it is much more appropriate to describe the ionospheric characteristic with the single layer structure.

Based on the formula derived in the Appendix A, Fig. 4 illustrates t_1 and the group path corresponding to E layer, F_1 layer and F_2 layer respectively versus different operating frequency. And the numerical calculation results are recorded in Tab. 1, where the first three parameters in each column indicate the critical frequency f_c, the distance between the bottom of ionosphere and the ground h_m and the ionosphere half thickness y_m respectively.

Furthermore, combined with the numerical calculation results, the property of the ionosphere dispersion could be boiled down to as follows: (1) the bandwidth for the low pass filter commonly ranges from hundreds of kilohertz to several megahertz; (2) In terms of F_1 and F_2 layers, as the transmitting frequency increases, there are generally two dispersive coefficient values corresponding to the high ray and low ray respectively. And the one with regard to the high ray exceeds that for the low ray; (3) As for each layer, the magnitude of dispersive coefficient value follows $F_2 > F_1 > E$.

3.1.2 Effect of Ionosphere Dispersion on Calculating the Area of Clutter Resolution Cell in Reality

According to Fig.4 (b), it is easy to observe that, as the frequency increases, there are generally two different values of group path, when the ground distance is fixed. And we call this the occurrence of multi-path in a single layer illustrated in Fig. 5.

One of the major influence triggered by multi-path propagation could be attributed to the pulse mixing due to the overlap of the convolution output of the high ray dispersion transfer function and the low ray one. Besides,

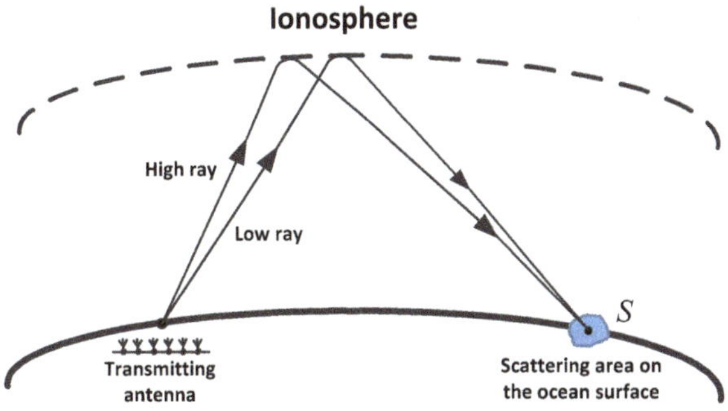

Fig. 5. A sketch map for the multi-path propagation including the high ray mode and low ray mode in a single layer.

combined with the filtering property of the ionosphere dispersive transfer function explicated in the previous section, the final signal $S_r(t)$ received by the receiver is a compound result imposed by the ionospheric band pass filtering and pulse mixing. And such modulating process could be expressed as

$$S_r(t) = \int e(t-\tau) g_L(\tau) d\tau + \int e(t-\tau) g_H(\tau) d\tau \,. \quad (17)$$

To explain this, Fig. 6(a) simulates the convolution output for a Gaussian narrow pulse reflecting from the F_1 layer. The relationship of bandwidth between $e(t)$, $g_H(t)$, and $g_L(t)$ satisfies as: $B_{wH} < B_{we} < B_{wL}$ (the subscript H and L indicate the high ray and low ray channel). Obviously, the echo reflecting from the low ray channel keeps nearly intact, while the envelope of the one corresponding to the high ray distorts seriously as B_{we} exceeds B_{wH} resulting in splitting and broadening of $e(t)$, which also deteriorates the envelope shape of final signal $S_r(t)$.

However, as for the HF radar, linear frequency modulated continuous wave (LFMCW) is usually implemented with a bandwidth of tens of kilohertz, which is far smaller than the ionosphere dispersive bandwidth. Therefore, the influence because of $g(t)$ filtering modulation would be nearly neglected, only remaining the envelop modulation impact due to pulse mixing. According to Fig. 6(b)–(c), the effect on the envelop modulation can be detected and confirmed both from the simulation and the experimental echo envelope.

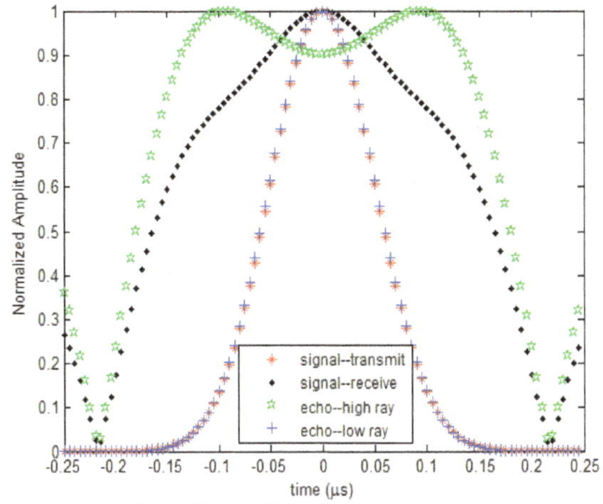

(a) The effect on Gaussian pulse propagation.

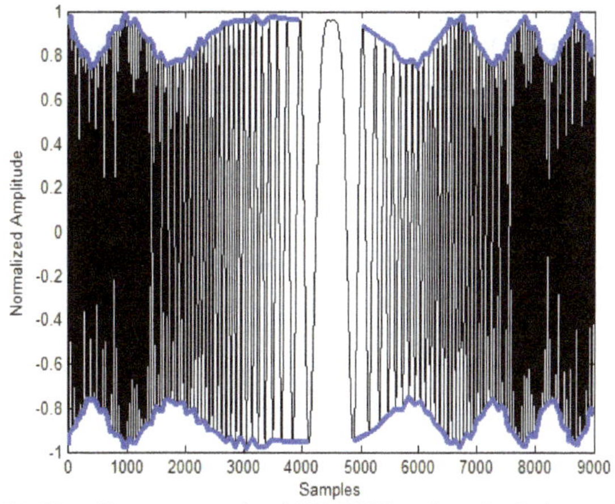

(b) The effect on narrow band LFMCW based on simulation result.

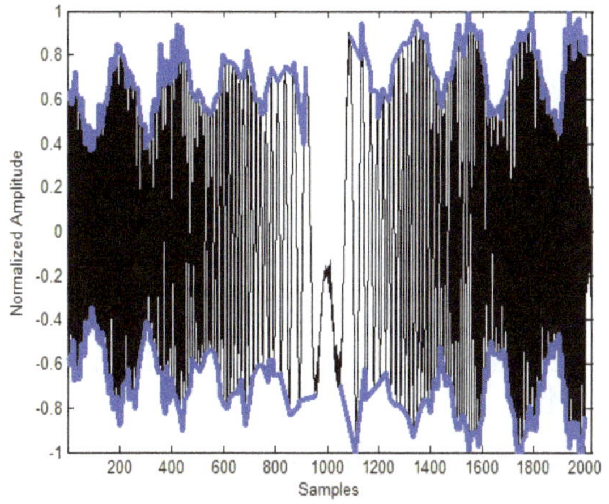

(c) The effect on narrow band LFMCW based on experiment result.

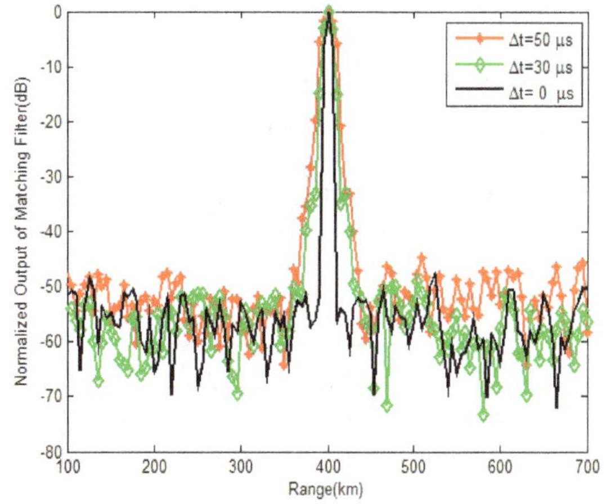

(d) Output of matching filter.

Fig. 6. The effect of ionosphere dispersion on pulse propagation and the output of matching filter.

Essentially, the envelop modulation would lead to the peak spreading of the output after range processing by (13). To illustrate this, Fig. 6(d) gives a number of results corresponding to different time delay Δt ($\Delta t = (P_H - P_L) / c_v$ indicates the time difference between the high ray and low ray path, and c_v is the velocity of light). Actually, the time interval between each discrete sample after range processing is $\tau = 1 / B_w$ (the derivation is given in Appendix B).

Based on which, it is reasonable to infer that as long as Δt is smaller than τ, the echoes scattered originally from two adjacent range bins would be folded into one, which is equivalent to the expansion of the clutter resolution cell area. This result is also in accordance with peak broadening of $\rho(t)$. Therefore, the area of such enlarged clutter resolution cell illuminated by the receiving beam with the azimuthal angle resolution $\Delta\theta$ at a distance of R could be calculated by

$$\Delta S = \frac{c_v(\tau + \Delta t)\Delta\theta R}{2}. \tag{18}$$

3.2 Limits of Ionosphere on the Clutter Resolution Cell

The ionosphere causes radar pulses with a certain bandwidth to spread in ground range due to the ionization profile [16]. If this spreading range exceeds the radar range resolution, then the performance is degraded. In addition, the azimuthal angle resolution is also restricted by the nature of fine-scale irregularities in the ionosphere plasma density [17]. On account of these, the ionosphere is bound to place a practical limit on the radar waveform bandwidth and azimuthal angle resolution in order to avoid worsening the radar detection performance. Recently, the formula to describe and calculate the influence quantitatively has been studied by Li [18]

$$\Delta B_{\lim} = \sqrt{\pi c_v \omega_{p0}^2 / (2\omega z_0 \sin(2\beta))}, \tag{19}$$

$$\Delta\theta_{\lim} = \frac{2\sqrt{2L\kappa_0 \langle n_1^2 \rangle (r_e \lambda \sec\beta)^2}}{k} \tag{20}$$

where r_e is the classical electron radius (2.8×10^{-15} m), $\langle n_1^2 \rangle$ is the mean square density of the plasma irregularities, L is the horizontal distance travelling in the ionosphere, κ_0 is the outer scale length of irregularities, ω_{p0} is the plasma frequency, z_0 is the distance between the ground and the bottom of ionosphere, β is the elevation angle, ω is the radar operating angular frequency, c_v is the velocity of light. Hence, based on (19)–(20), the area of the limited clutter resolution cell at a distance of R is given by

$$\Delta S_{\lim} = \frac{R\Delta\theta_{\lim} c_v}{2\Delta B_{\lim}} = \frac{Rc_v\sqrt{2L\kappa_0 \langle n_1^2 \rangle (r_e\lambda\sec\beta)^2}}{k\sqrt{\pi c_v \omega_{p0}^2/(2\omega z_0 \sin(2\beta))}} = \frac{2Rc_v}{k}\sqrt{\frac{L\kappa_0\langle n_1^2\rangle (r_e\lambda\sec\beta)^2 \omega z_0 \sin(2\beta)}{\pi c_v \omega_{p0}^2}} \tag{21}$$

In fact, the limit of ionosphere on the resolution cell discussed here mainly focuses on quantitatively calculating the area of the resolution cell, within which the backscattered sea clutter echo could be regarded as homogeneous and relevant so as to provide a reference on dividing the actual clutter resolution cell.

3.3 Comparison between $\Delta S_{\lim}$ and ΔS

Previously, the influence on the clutter resolution cell triggered by the ionosphere has been deeply discussed. Afterwards, we would put a stress on the comparison between the limit area $\Delta S_{\lim}$ and the one ΔS in reality. Tab. 2 provides a series of numeral results to explain this. Simulation parameters are selected as follows: the baseline $D_0 = 800$ km, $\omega_{p0}/(2\pi) = 5$ MHz, $z_0 = 130$ km, $L = 600$ km, $\langle n_1^2 \rangle = 10^{19}\text{m}^{-6}$, $\kappa_0 = 10^{-4}$/m, $B_w = 30$ kHz, the half thickness of the ionosphere $y_m = 60$ km and the distance between the earth center and the ionospheric layer corresponding to the maximum electron density $r_m = r_b + y_m = 6860$ km. For the sake of simplicity, we denote $\delta = \Delta S / \Delta S_{\lim}$.

According to Tab. 2, the relationship between $\Delta S_{\lim}$ and ΔS could be summed as: (1) the limit on the bandwidth and the azimuthal angle resolution is restricted with each other; (2) δ is generally proportional to the radar operating frequency, the time delay Δt and θ_r.

f (MHz)	$R = 200$ km, $\theta_r = 150°$						
	$\Delta B_{\lim}$ (kHz)	$\Delta\theta_{\lim}$ (°)	$\Delta S_{\lim}$ (km × km)	Δt (μs)	$\Delta B_{\lim} / B_w$	$\Delta\theta / \Delta\theta_{\lim}$	δ
5	262.9	7.3	14.6	0	8.8	1.4	12.0
8	205.3	2.9	7.4	0	6.8	3.5	23.7
13	155.2	1.2	4.1	0.2	5.1	8.8	42.1

(a) δ versus different operating frequency for a given detection clutter cell.

θ_r (°)	$d = 1200$ km, $f = 6$ MHz						
	$\Delta B_{\lim}$ (kHz)	$\Delta\theta_{\lim}$ (°)	$\Delta S_{\lim}$ (km × km)	Δt (μs)	$\Delta B_{\lim} / B_w$	$\Delta\theta / \Delta\theta_{\lim}$	δ
30	188.1	6.7	30.2	0	6.3	1.5	9.4
90	219.4	5.4	9.2	0	7.3	1.8	13.5
150	227.4	5.3	5.5	0	7.6	1.9	14.4

(b) δ versus different clutter cells for a given operating frequency.

Tab. 2. Comparison result between $\Delta S_{\lim}$ and ΔS.

4. First Order Sea Clutter Cross Section Simulation and Experiment Results

4.1 A Method to Simulate the First Order Sea Clutter Cross Section

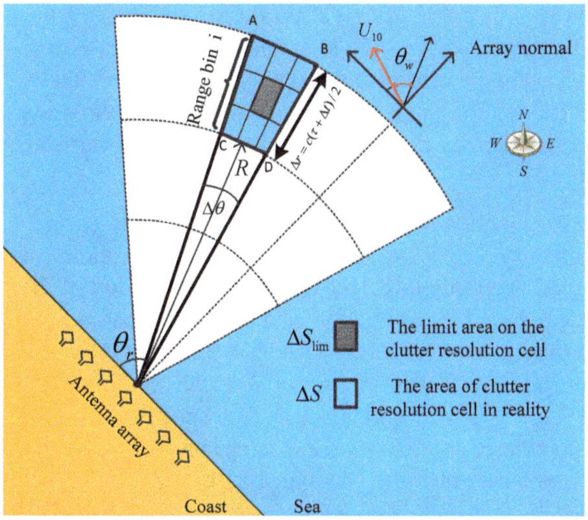

Fig. 7. Illustration for the relationship between ΔS_{lim} and ΔS.

The first order sea clutter cross section applied in the bistatic surface wave has been proposed by Walsh and Gill [12], [19]

$$\sigma_1(\omega_d) = 2^4 \pi k^2 \sum_{m=\pm 1} S(m\boldsymbol{K}) \frac{K_d^{5/2} \cos\phi_0}{\sqrt{g}} \Delta R \times Sa^2 \left[\frac{\Delta R}{2} \left(\frac{K_d}{\cos\phi_0} - 2k \right) \right] \tag{22}$$

where $K_d = \omega_d^2 / g$, ω_d is the Doppler angular frequency, $\phi_0 = \phi_H / 2$ is the half bistatic angle, ΔR indicates the range resolution, $Sa(\cdot)$ is the sinc function. The directional wave height spectrum $S(m\boldsymbol{K})$ considered here is the product of non-directional P-M spectrum and the standard form of the normalized directional distribution [12].

Given the following mathematic relationship in (23) and (24)

$$\Delta R \times Sa^2 \left[\frac{\Delta R}{2} \left(\frac{K_d}{\cos\phi_0} - 2k \right) \right] = 2\cos\phi_0 \times \frac{\Delta R}{2\cos\phi_0} \times Sa^2 \left[\frac{\Delta R}{2\cos\phi_0} (K_d - 2k\cos\phi_0) \right] \tag{23}$$

$$\lim_{M \to \infty} M Sa^2(Mx) = \pi \delta(x). \tag{24}$$

Equation (22) could be simplified as

$$\sigma_1(\omega_d) = 2^4 \pi k^2 \sum_{m=\pm 1} S(\mathrm{m}\boldsymbol{K}) \left[2kg\cos\phi_0 \right]^{5/2} \frac{\cos\phi_0}{\sqrt{g}} \times \frac{2\pi g \cos\phi_0}{2\sqrt{2kg\cos\phi_0}} \delta\left(\omega_d + m\sqrt{2kg\cos\phi_0} \right) = 2^6 \pi^2 k^4 \cos^4\phi_0 \times \sum_{m=\pm 1} S(\mathrm{m}\boldsymbol{K}) \delta\left(\omega_d + m\sqrt{2kg\cos\phi_0} \right) \tag{25}$$

Judging from the derivation process in [19], equation (25) is appropriate to be applied to calculate the radar cross section for the clutter within ΔS_{lim}, because the clutter distributed within ΔS_{lim} is equipped with the highest correlation. Hence, based on the analysis in Section 3, as well as the relationship illustrated in Fig. 7, a much more accurate model may be generated after taking ΔS_{lim} as the standard to divide ΔS. The dividing criterion refers to the relative scale of $B_w / \Delta B_{\text{lim}}$ and $\Delta\theta / \Delta\theta_{\text{lim}}$. By virtue of such manipulation, we suppose that the sea clutter distributed in each unit could be approximately treated as independent and identically distributed (i.i.d); thus, the mean cross section $\bar{\sigma}_1(\omega_d)$ corresponding to ΔS could be obtained by

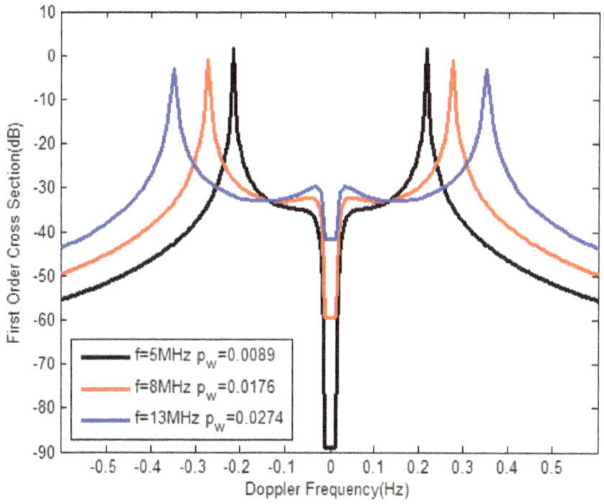

(a)

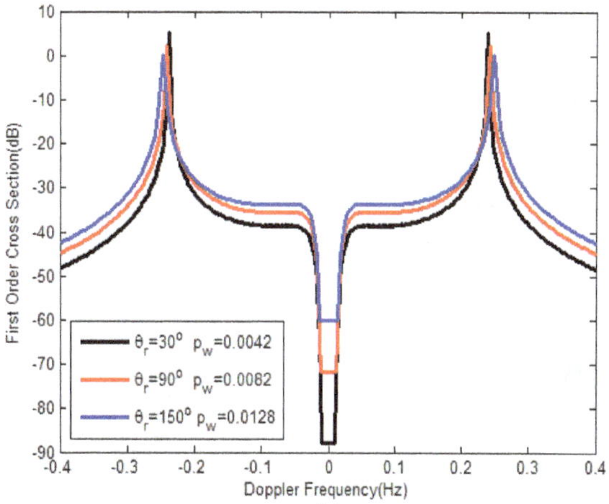

(b)

Fig. 8. The first order sea clutter cross section variance of operating frequency and θ_r in accordance with results of Tab. 2.

$$\bar{\sigma}_1(\omega_d) = \frac{\sum_{n=1}^{N} \sigma_1(\omega_d)_n \cdot \Delta S_{\lim}(n)}{\sum_{n=1}^{N} \Delta S_{\lim}(n)} \tag{26}$$

where N stands for the total number of divided clutter cells. Besides, in order to describe the spectral pattern quantitatively, the equivalent spectral width of the first order line is defined as

$$P_w = \int_{f_H-\Delta}^{f_H+\Delta} P(f)df \,/\, P_0 \tag{27}$$

where f_H is the Doppler frequency of the largest Bragg line. $P(f)$ is the spectral density and Δ is the theoretical Doppler shift of the Bragg line (Δ applied in this paper is equal to the radar Doppler resolution), P_0 indicates all the energy.

Furthermore, simulation has been conducted to explicate the variety of first order sea clutter cross section after taking account of the clutter division. The simulation parameters selected here are in consistent with the ones applied in Tab. 2. Combined with (25) and (26), Fig. 8 presents a straightforward variance of the first order cross section. Meanwhile, allowing for the fact that system timing effects as well as surface fluctuations and noise will, in practice, provide a smoothing effect. Therefore, the result in Fig. 8 has been simulated by convolving the cross section with a Hamming window in consistence with [19]. According to Fig. 8, the amplitude and spectral width are changing with different operating and detection parameters. Essentially, such phenomenon could be explained by the relative scale between $\Delta S_{\lim}$ and ΔS. That is to say, as long as ΔS exceeds the limit scale on the resolution cell, the clutter distributed in the area surpassing $\Delta S_{\lim}$ would be superimposed and overlapped. And for such a big area ΔS, the deteriorating correlation of the sea clutter is bound to result in spectral peak energy spreading to adjacent Doppler bins.

4.2 Simulation and Experiment Results

The experiment data was obtained from the newly-developed HF sky-surface wave experimental platform in HIT, which is illustrated in Fig. 9 (a)–(b). The basic operating parameters applied in this system are: the operating frequency $f = 15$ MHz, the baseline length $D_0 = 800$ km, the bandwidth $B_w = 40$ kHz, the pulse repetition period is 20 ms, coherent integration time (CIT) $CIT = 51.20$ s, the normal direction of the receiving array is 58° from north to west and the element number of the uniform linear array is eight. The reflection happens at F_2 layer. The sea state is 3 measured in Douglas offing state grade [13]. Fig.9 (c)–(d) give us an ad hoc image after taking the manipulation of range and Doppler processing. It is well to be spotted that the spectral pattern, especially the spectral width, varies with different receiving direction (θ_r) and range bins, therefore exhibiting a spatial variance property [7], [20].

In conjunction with the experiment result, the according simulation has been conducted to testify the validity of the method proposed in calculating the cross section of the first order sea clutter in Fig. 10. The simulation parameters applied herein are in accordance with those for the experiment and the calculation results of δ equal to 32, 65, 58 and 40 respectively. Simultaneously, the spectral width P_w is also calculated to contrast the discrepancy between each spectral pattern. According to Fig. 10, several straightforward conclusions could be yielded:

- Given a selected clutter cell, P_w is proportional to the array antenna receiving direction, indicating the occurrence of spectrum spreading. Besides, as for different range bins illuminated within the same beam, the clutter echo spectrum corresponding to the far range bin is sharpen in contrast with the ones from the range bin adjacent to the receiver. Albeit with the variance of θ_w (the definition is illustrated in Fig. 7) in each simulation, it merely affects the relative magnitude between the negative and positive Bragg peaks, but not the spectral width.
- The characteristic of the first order sea clutter spectral pattern is in accordance with the calculation result of δ. Such consistency also confirms the effectiveness and validity of clutter resolution cell division when calculating the actual sea clutter cross section.

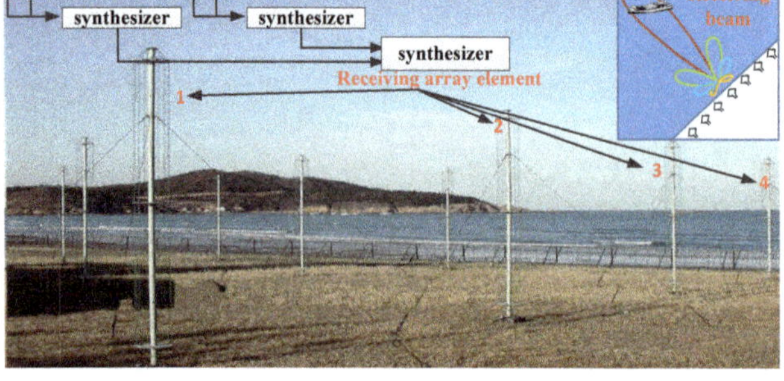

(a) An actual layout of radar receiving antenna array.

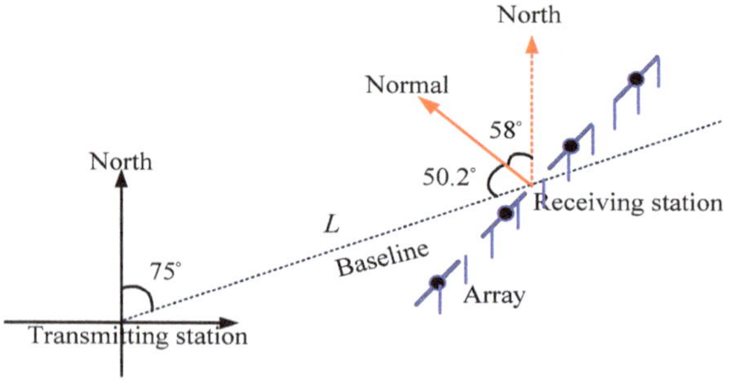

(b) The array geometric layout of the transmitting and receiving stations.

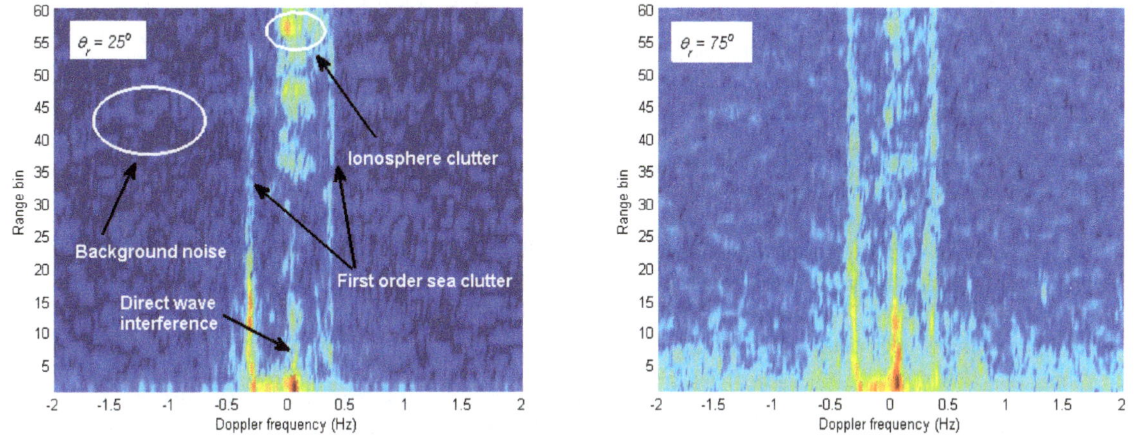

(c) The range and Doppler image with θ_r=25°. (d) The range and Doppler image with θ_r=75°.

Fig. 9. Illustration of the experimental platform and range-Doppler processing result of the experiment data versus different receiving direction θ_r.

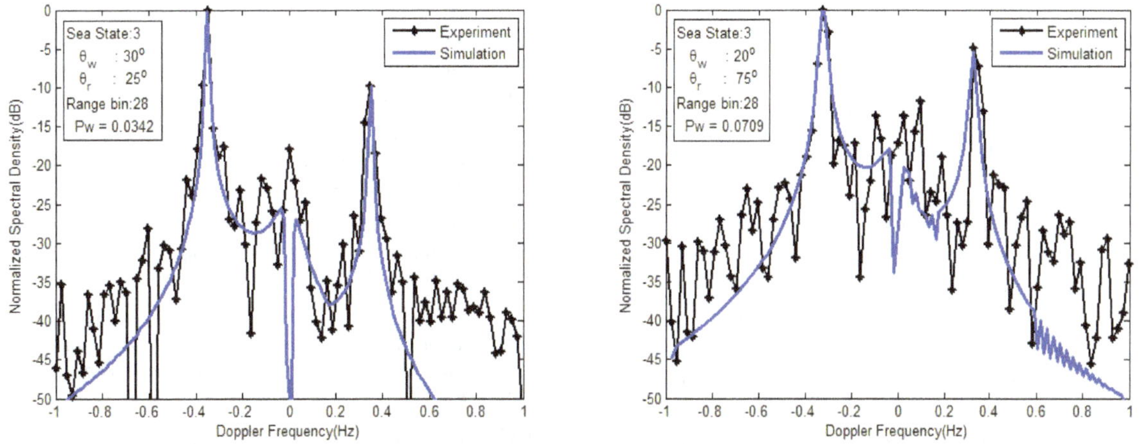

(a) Spectral pattern variance versus different receiving direction θ_r.

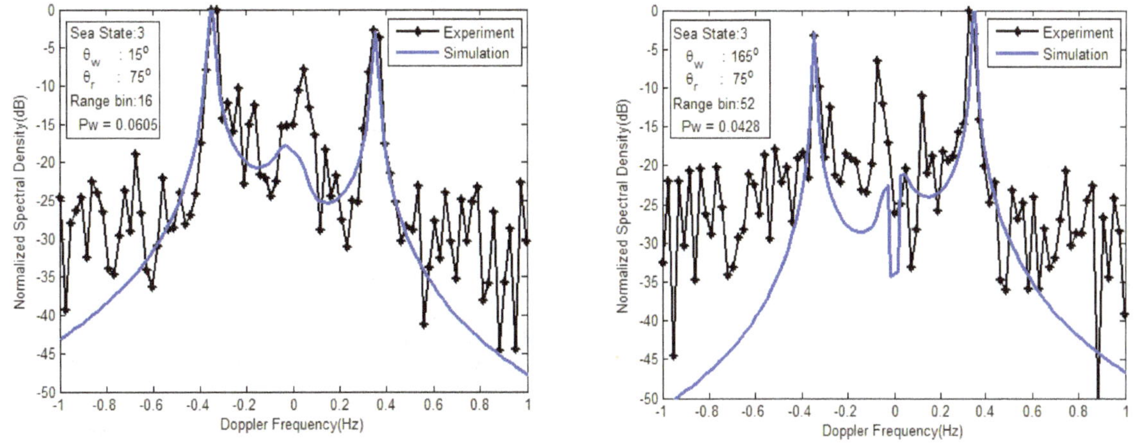

(b) Spectral pattern variance versus different range bins.

Fig. 10. Comparison between the simulation result and the normalized sea clutter spectral density obtained from the experiment data.

5. Conclusion

In this paper, with regard to the ad hoc hybrid sky-surface propagation mode, some elementary problems have been discussed and studied deeply and extensively.

The spatial geometry distribution of the first order sea clutter could be depicted as an ellipse for a fixed time delay under a relatively ideal condition. Based on which, the Bragg scattering frequency is derived. It can be recognized that the results are similar to those derived in the bistatic surface wave radar, but a redefinition is supposed to be conducted on some parameters, especially the bistatic angle.

Subsequently, we explicate how the ionosphere dispersion impacts on the clutter resolution cell. Specifically, we have demonstrated that the property of the dispersion transfer function acts somewhat like a band pass filter. According to this property, the ionosphere dispersion has put forward a clear restriction on the radar operating frequency as well as some other detection parameters so as to avoid the distortion on the transmitting signal and the occurrence of multi-path travelling. On top of this, the limit of ionosphere on the clutter resolution cell has been discussed so as to recognize how big the size of clutter resolution cell could be treated as a unit area, so that the traditional first order sea clutter cross section model derived by Walsh and Gill [12], [19] can be used.

Eventually, a modified simulation method for the first order sea clutter cross section is derived and discussed. Through the simulation, we have demonstrated how the spectral width spreads versus different δ. According to the compared result between experiment and simulation, the clutter cell division appears to be imperative when regarding the coherence between adjacent clutter cells. Therefore, in the actual detection, some ad hoc restriction should be put to minimize the area of the clutter resolution cell as much as possible so as to avoid the spreading of the clutter spectrum, which may necessitate a radically different radar system design.

Acknowledgements

This work was supported by the National Natural Science Foundation Project grant (No.61471144) to China. The authors appreciate the help from the Institute of Electronic Engineering Technology in HIT for providing the experimental data.

References

[1] MELYANOVSKI, P. A., TOURGENEV, I. S. Bistatic HF radar for oceanography applications with the use of both ground and space waves. *Telecommunications and Radio Engineering*, 1997, vol. 51, no. 2-3, p. 73–79.

[2] ZHAO, Z.X., WAN, X.R., ZHANG, D.L., CHENG, F. An experimental study of HF passive bistatic radar via hybrid sky-surface wave mode. *IEEE Transactions on Antennas and Propagation*, 2013, vol. 61, no. 1, p. 415–424.

[3] JIANG, W., DENG, W.B., YANG, Q. Analyses of sea clutter for HF over the horizon hybrid sky-surface wave radar. *Journal of Electronics & Information Technology*, 2011, vol. 33, no. 8, p. 1786–1791.

[4] LI, Y.J., WEI, Y.S. Analysis of first-order sea clutter spectrum characteristics for HF sky-surface wave radar. In *2013 International Conference on Radar*. Australia, 2013, p. 368–373.

[5] ANDRIĆ, M., BUJAKOVIĆ, D., BONDŽULIĆ, B., SIMIĆ, S., ZRNIĆ, B. Analysis of radar Doppler signature from human data. *Radioengineering*, 2014, vol. 23, no. 1, p. 11–19.

[6] LI, Y.J., WEI, Y.S., ZHANG, C., SHANG, C., TANG, X.D. Influence of ionosphere on SCR of HF hybrid sky-surface wave radar. In *IET International Radar Conference*. Xi'an (China), 2013, p. 1–6.

[7] LI, Y.J., WEI, Y.S., XU, R.Q., CHU, T.Q., WANG, Z.Q. Space-time characteristics and experimental analysis of broadening first-order sea clutter in HF hybrid sky-surface wave radar. *Radioengineering*, 2014, vol. 23, no. 3, p. 831–841.

[8] RIDDOLLS, R. J. Ship detection performance of a high frequency hybrid sky-surface wave radar. *Defense R&D Canada-Ottawa*, 2007, p. 1-42.

[9] ZHU, Y.P., SHANG, C., LI, Y.J., A FBLP based method for suppressing sea clutter in HFSWR. In *International Symposium on Antennas and Propagation*. Nanjing (China), 2013, p. 1090–1093.

[10] HOU, C.Y., KE, G., SHI, T.G., WANG, Y.X. Study on the detectability of the sky-surface wave hybrid radar. *Journal of Applied Mathematics*, 2014, p. 1-10.

[11] BARRICK, D. E. First-order theory and analysis of MF/HF/VHF scatter form the sea. *IEEE Transactions on Antennas and Propagation*, 1972, vol. 20, no. 1, p. 2–10.

[12] WALSH, J., DAWE, B. J. Development of a model for the first order bistatic ocean clutter radar cross section for ground wave radars. *Northern Radar Systems Limited contract report for the Defense Research Establishment Ottawa*, 1994, p. 1-74.

[13] BARTON, D. *Modern Radar Systems Analysis*. Norwood: Artech House, 1987.

[14] LUNDBORG, B. Pulse propagation through a plane stratified ionosphere. *Journal of Atmospheric and Terrestrial Physics*, 1990, vol. 52, no. 9, p. 759–770.

[15] KRETOV, N. V., RYSHKINA, T. Y., FEDOROVA, L. V. Dispersive distortions of transionospheric broadband VHF signals. *Radio Science*, 1992, vol. 27, no. 4, p. 491–495.

[16] RIDDOLLS, R. J. Limits on the detection of low-Doppler targets by a high frequency hybrid sky surface wave radar system. In *2008 IEEE Radar Conference*, Rome (Italy), 2008, p. 1–4.

[17] RIDDOLLS, R. J. Detection of aircraft by high frequency sky wave radar under auroral clutter-limited conditions. *Defense R&D Canada-Ottawa*, 2008, p. 1–38.

[18] LI, Y.J., WEI, Y.S., XU, R.Q. Influence of ionosphere on resolution cell of HF hybrid sky-surface wave radar. In *Antennas and Propagation Society International Symposium (APSURSI)*. Orlando (USA), 2013, p. 1028–1029.

[19] GILL, E. The scattering of high frequency electromagnetic radiation from the ocean surface: An analysis based on a bistatic ground wave radar configuration. *PhD Thesis*. Faculty of Engineering and Applied Science, Memorial University of Newfoundland, 1999.

[20] LI, Y. J., WEI, Y.S., XU, R.Q., SHANG, C. Simulation analysis and experimentation study on sea clutter spectrum for high-frequency hybrid sky-surface wave propagation mode. *IET Radar Sonar and Navigation*, 2014, p. 1-14.(in press, doi: 10.1049/iet-rsn.2013.0289)

[21] CROFT, T., HOOGASIAN, H. Exact ray calculations in a quasi-parabolic ionosphere with no magnetic field. *Radio Science*, 1968, vol. 3, no. 1, p. 69.

Appendix

A. Derivation of the Dispersive Coefficient for a Spherical Stratified Ionosphere

In this section, we shall derive the explicit expression of the dispersive coefficient t_1 under the assumption that the ionosphere is spherical stratified and the electron density of ionosphere submits to the quasi-parabolic model. The Q-P model, a modified one by a slight modification of the parabolic model, is introduced by Croft and Hoogasian [21]. As in the case of quasi-parabolic model, the quasi-parabolic layer of semi-thickness y_m can be modeled as

$$f_p^2 = \frac{f_c^2}{f^2}\left[1-\left(\frac{r-r_m}{y_m}\right)^2\left(\frac{r_b}{r}\right)\right] \quad r_b \le r \le \frac{r_m r_b}{r_b - y_m}$$

$$f_p^2 = 0 \qquad \text{Elsewhere}$$

Here the critical frequency is related to the maximum electron density N_m and wavenumber k by $f_c = \sqrt{kN_m}$ .r_b is the distance between the bottom of ionosphere and the earth center; and r_m indicates the distance between the earth center and the ionospheric layer corresponding to the maximum electron density. Actually, the ionization rises from zero at $r = r_b$ at the base of the ionosphere to a peak value N_m at $r = r_m = r_b + y_m$.

The expressions for the ground distance D and the group path P for the quasi-parabolic layer are

$$P = 2(1-F^2/A)r_b \sin - 2r_0 \sin\beta - (BF/2A^{3/2})\ln(U/V^2),$$

$$D = 2r_0\left\{(r-\beta) - \frac{Fr_0\cos\beta}{2\sqrt{C}}\times \left(\ln(U) - \ln\left(4C(F\sin\gamma + \frac{1}{r_b}\sqrt{C} + \frac{B}{2\sqrt{C}})^2\right)\right)\right\}$$

where $F = f/f_c$, $\gamma = \cos^{-1}\left[(r_0/r_b)\cos\beta\right]$, $U = B^2 - 4AC$, $A = F^2 - 1 + (r_b/y_m)^2$, $C = (r_b r_m / y_m)^2 - F^2 r_0^2 \cos^2\beta$, $V = 2Ar_b + B + 2r_b F A^{1/2}\sin\gamma$ and $B = -2r_m r_b^2 / y_m^2$.

In our analysis, the dispersive coefficient t_1 is primarily determined by the frequency derivative of the group path.

$$t_1 = \sqrt{\frac{1}{c_v}\cdot\frac{dP}{df}} = \sqrt{\frac{1}{c_v f_c}\left(\frac{\partial P}{\partial F} - \frac{\partial P}{\partial \beta}\frac{\partial D}{\partial F} / \frac{\partial D}{\partial \beta}\right)}$$

where

$$\frac{\partial D}{\partial F} = \frac{r_0^2\cos\beta}{C^{1/2}}\left[\left(\frac{F\dot{C}_F}{2C} - 1\right)\Phi + 4F\frac{\dot{A}_F C + A\dot{C}_F}{B^2 - 4AC} + 2F\frac{\dot{C}_F r_b F\sin\gamma + 2Cr_b\sin\gamma + 2\dot{C}_F C^{1/2}}{C^{1/2}\left(2C^{1/2}r_b F\sin\gamma + 2C + Br_b\right)}\right] \tag{A-1}$$

$$\frac{\partial D}{\partial \beta} = 2r_0(\dot{r}_\beta - 1) + \frac{Fr_0^2\cos\beta}{C^{1/2}}\left[(\frac{\dot{C}_\beta}{2C} + \tan\beta)\Phi + \frac{4A\dot{C}_\beta}{B^2 - 4AC} + 2\frac{\dot{C}_\beta r_b F\sin\gamma + 2\dot{r}_\beta C r_b F\cos\gamma + 2\dot{C}_\beta C^{1/2}}{C^{1/2}\left(2C^{1/2}r_b F\sin\gamma + 2C + Br_b\right)}\right] \tag{A-2}$$

$$\frac{\partial P}{\partial F} = \frac{2F^2 r_b\sin\gamma}{A^2}\dot{A}_F - \frac{4r_b\sin\gamma F}{A} - \left(\frac{BW}{2A^{3/2}} - \frac{3BFW}{4A^{5/2}}\dot{A}_F + \frac{BF}{2A^{3/2}}\dot{W}_F\right) \tag{A-3}$$

$$\frac{\partial P}{\partial \beta} = \left(2 - \frac{2F^2}{A}\right)r_b\dot{r}_\beta\cos\gamma r_b\dot{r}_\beta\cos\gamma - 2r_0\cos\beta - \frac{BF}{2A^{3/2}}\dot{W}_\beta \tag{A-4}$$

And the relevant parameters in (A-1) – (A-4) are

$$\dot{W}_F = -\frac{4\dot{A}_F C + 4A\dot{C}_F}{B^2 - 4AC} - 2\times\frac{2\dot{A}_F r_b + 2r_b A^{1/2}\sin\gamma + r_b F A^{-1/2}\dot{A}_F\sin\gamma}{V}$$

$$\dot{W}_\beta = -\left(4\frac{A\dot{C}_\beta}{B^2 - 4AC} + 2\frac{2r_b F A^{1/2}\dot{r}_\beta\cos\gamma}{V}\right),$$

$$\dot{A}_F = 2F,\ \dot{C}_F = -2Fr_0^2\cos^2\beta,\ \dot{r}_\beta = \frac{r_0\sin\beta}{\sqrt{r_b^2 - r_0^2\cos^2\beta}},$$

$$W = \ln\frac{U}{V^2} = \ln\frac{B^2 - 4AC}{\left(2Ar_b + B + 2r_b F A^{1/2}\sin\gamma\right)^2},$$

$$\dot{C}_\beta = F^2 r_0^2\sin 2\beta,\ \Phi = \ln\frac{\left(B^2 - 4AC\right)r_b^2}{\left(2C^{1/2}r_b F\sin\gamma + 2C + Br_b\right)^2}.$$

B. Derivation of Time Interval Corresponding to a Range Bin after Range Processing

We take the LFMCW as the emission signal and thus we have:

$$S_t(t) = \cos\left[2\pi\left(ft + K(t-nT)^2/2\right)\right].$$

The echo signal could be expressed as:

$$S_r(t) = \cos\left[2\pi(f(t-\tau_d) + K(t-nT-\tau_d)^2/2)\right]$$

where f is the radar operating frequency, $K = B_w/T$ is the modulation rate, $\tau_d = 2(R_0 - v_r t)/c_v$ is the time delay determined by the position of the target, T is the modulation period, $n = -\infty,\cdots,0,1,\cdots,+\infty$ is the index of modulation period. And the phase output $\psi_b(t)$ after mixing with the local oscillator is

$$\begin{aligned}\psi_b(t) &= \psi_t(t) - \psi_r(t)\\ &= 2\pi\left[ft + \frac{K}{2}(t-nT)^2 - f(t-\tau_d) - \frac{K}{2}(t-nT-\tau_d)^2\right]\\ &= 2\pi\left[f\tau_d + \frac{K}{2}\tau_d(2t - 2nT - \tau_d)\right]\\ &= 2\pi\left[f\cdot 2\frac{R_0 - v_r t}{c_v} + K\cdot 2\frac{R_0}{c_v}(t-nT) - \frac{K}{2}\tau_d^{\,2}\right]\\ &= 2\pi\left[\frac{2fR_0}{c_v} - f_d nT + (f_R - f_d)(t-nT) - \frac{K}{2}\tau_d^{\,2}\right]\end{aligned}$$

Here we just focus on the frequency f_R corresponding to the range of the target:

$$f_R = 2KR_0/c_v = Kt_d\ .$$

Hence, after the stretch processing, the time interval is:

$$\tau{=}\Delta t_d = \frac{1}{K}\Delta f_R = \frac{1}{K\times T} = \frac{1}{B_w}$$

where Δf_R is determined by the Doppler resolution.

About Authors...

Yongpeng ZHU was born in Heilongjiang, China. He received the M.S. degree from HIT in 2014. He is currently working on the Doctorate degree in Information and Communication Engineering from HIT. His research interests include the fields of signal processing and radar system simulation in HF OTH radar.

Yinsheng WEI was born in 1974. He received his M.S. and Ph.D. degrees in Communication and Information Systems from Harbin Institute of Technology (HIT) in 1998 and 2002, respectively. And then, he joined the Department of Electronics Engineering in HIT as a lecturer, and became a professor in 2011. He is a member of IEEE AES, and a senior member of CIE. His main researches include in radar signal processing and radar system analysis and simulation.

Yajun LI was born in 1983. He received the M.S. degree from Harbin Engineering University in Information and Communication Engineering in 2011. He is now a PhD student of School of Electronics and Information Engineering at Harbin Institute of Technology, China. His current research interests include space-time adaptive processing, suppression of sea clutter, and HF radar system simulation.

A Novel FastICA Method for the Reference-based Contrast Functions

Wei ZHAO[1], *Yimin WEI*[2], *Yuehong SHEN*[2], *Zhigang YUAN*[2], *Pengcheng XU*[1], *Wei JIAN*[2]

[1]College of Communications Engineering, PLA University of Science and Technology, Yudao Street, 210007 Nanjing, China
[2]Wireless Communications Lab of College of Communications Engineering, PLA University of Science and Technology, Yudao Street, 210007 Nanjing, China

gmajyie@126.com

Abstract. *This paper deals with the efficient optimization problem of Cumulant-based contrast criteria in the Blind Source Separation (BSS) framework, in which sources are retrieved by maximizing the Kurtosis contrast function. Combined with the recently proposed reference-based contrast schemes, a new fast fixed-point (FastICA) algorithm is proposed for the case of linear and instantaneous mixture. Due to its quadratic dependence on the number of searched parameters, the main advantage of this new method consists in the significant decrement of computational speed, which is particularly striking with large number of samples. The method is essentially similar to the classical algorithm based on the Kurtosis contrast function, but differs in the fact that the reference-based idea is utilized. The validity of this new method was demonstrated by simulations.*

Keywords

Blind source separation, FastICA, reference-based contrast functions, Kurtosis contrast function.

1. Introduction

For the latest decades, Blind Source Separation (BSS) has been applied in a wide variety of fields such as array processing, passive sonar, seismic exploration, speech processing, multi-user wireless communications, etc [1]. In the case of a linear multi-input/multi-output (MIMO) instantaneous system, BSS corresponds to Independent Component Analysis (ICA), which is now a well recognized concept. In this contribution, we mainly consider the efficient optimization issue of BSS in the FastICA framework, where statistically independent sources are linearly and instantaneously mixed.

In the linear MIMO systems, BSS has found interesting solutions through the optimization of so-called contrast functions [2], which are generally treated as separation criteria. Many separation criteria rely on higher-order statistics (e.g., the Kurtosis contrast function [3], [4]) or can be linked to higher-order statistics (e.g., the Constant Modulus contrast function [5]). These criteria are known to provide good results. Recently, some novel contrast schemes referred to as "reference-based" have been proposed in [6] and [7]. They are essentially the cross-statistics or cross-cumulants between the estimated outputs and reference signals [8]-[9]. These reference-based contrast functions have an appealing feature in common: the corresponding optimization algorithms are quadratic with respect to the searched parameters. First of all, we give a brief review of previous works on this subject.

- A maximization algorithm based on Singular Value Decomposition (SVD) has been proposed in [6] and [10], and it was shown to be significantly quicker than other maximization algorithms. However, the method often suffers from the need to have a good knowledge of the filter orders due to its sensitivity on a rank estimation [11].
- A gradient optimization method based on Kurtosis with reference signals introduced has been proposed in [12], which obtains an optimal step size and dose not require any rank estimation. Therefore, the drawback of the SVD-based method can be well overcome. But the reference signals involved in this method are fixed during the optimization process, which may lead to bad separation performance because of inappropriate initialization value of the corresponding reference signals.
- A similar gradient optimization algorithm based on Kurtosis maximization has been proposed in [13], in which the reference signals used in the separation process update after each one-dimensional optimization. So the separation quality of this method is better than that in [12].
- On the basis of the algorithms in [12] and [13], a new improvement method has been proposed in [11]. For this method, a tradeoff can be adjusted between performance and speed by introducing a new iterative update parameter. Moreover, the global convergence of this algorithm is proved in detail.

Inspired by [11] and [13], in this paper, we propose a new algorithm by considering the reference-based Kurtosis maximization in the FastICA optimization framework. The papers most directly linked to our approach are [11] on the one hand and [13] on the other hand. The former has provided a proof of stationary consistency of the reference-based contrast function, which is directly cited in our paper. The latter has proposed a gradient optimization algorithm with the reference signals updated, which contributes to the proposal of our method primarily. Besides the gradient optimization algorithms in [11]-[13], our proposed method provides another approach to an efficient optimization of reference-based contrast functions. To our knowledge, it has not been investigated yet despite its simplicity.

This paper is organized as follows. Section 2 describes the model and assumptions we consider in this paper. In Section 3, the separation criteria we use are presented. Our proposed algorithm can be found in Section 4. Simulation results are illustrated in Section 5 and Section 6 concludes this paper.

2. System Model and Assumptions

2.1 Instantaneous Mixture

We consider an observed M-dimensional $(M \geq 2)$ discrete-time signal, the nth sample of which is denoted by the column vector $\mathbf{x}(n)$ (where $n \geq 1$ holds implicitly in this whole paper). The observed signals result from a noise-free linear MIMO system, for which the input and output relationship is described as follows:

$$\mathbf{x}(\mathrm{n}) = \mathbf{As}(\mathrm{n}) \tag{1}$$

where A is a linear mixture matrix of $M \times N$, the elements of which are unknown constant. The N-dimensional $(N \geq 2)$ source vector $\mathbf{s}(\mathrm{n})$ is unknown and unobserved.

The objective of BSS is to recover source signals blindly only by using the observations. Similarly, we consider a linear separator, the output of which is described as:

$$\mathbf{y}(\mathrm{n}) = \mathbf{Wx}(\mathrm{n}) \tag{2}$$

where $\mathbf{W}$ is the separation matrix of $N \times M$. $\mathbf{y}(\mathrm{n})$ is the approximate estimation of $\mathbf{s}(\mathrm{n})$. We mention above that our proposed method presents a quadratic dependence of the searched parameters, where the "searched parameters" are the row vectors of $\mathbf{W}$.

2.2 Assumptions on the Sources

To be able to carry out the estimation blindly and successfully, we make some assumptions on the sources [7], which are shown as follows:

A1: For all i, the source sequence $s_i(n)$ is stationary, zero-mean and with unit variance.

A2: The source vectors $s_i(n), i \in \{1, \cdots, N\}$ are statistically mutually independent.

3. Separation Criteria

3.1 Notations

In order to describe the reference-based Kurtosis contrast function we utilize in this paper, we first introduce some notations. The Cumulant of a set of random variables is denoted by $Cum\{\bullet\}$. Note that we only consider the Cumulant of real-valued signals in this paper, even though the signals can be complex- or real-valued. The complex-valued signals case will be considered in our work later.

For any jointly stationary signals $y(n)$ and $z(n)$, we set

$$C\{y\} \stackrel{\Delta}{=} Cum\{y(n), y(n), y(n), y(n)\} = E\{y(n)^4\} - 3E\{y(n)^2\}^2 \tag{3}$$

$$\begin{aligned} C_z\{y\} &\stackrel{\Delta}{=} Cum(y(n), y(n), z(n), z(n)) \\ &= E\{y(n)^2 z(n)^2\} - E\{y(n)^2\}E\{z(n)^2\} - 2E^2\{y(n)z(n)\} \end{aligned} \tag{4}$$

where $E\{\bullet\}$ denotes the expectation value.

3.2 Reference Signals

Before introducing the reference-based contrast functions, we first introduce the corresponding "reference signals" we have mentioned above. Similarly to (2), we consider a separation matrix of $N \times M$ denoted by $\mathbf{V}$. The corresponding output can be denoted by

$$\mathbf{z}(\mathrm{n}) = \mathbf{Vx}(\mathrm{n}) \tag{5}$$

where the components of $\mathbf{z}(\mathrm{n})$ are the reference signals. Note that the reference signals have direct influence on the reference-based contrast function and their values do impact the optimization results, especially the initialization value.

As described in [11], the reference signals are artificially introduced in the algorithm for the purpose of facilitating the maximization of the contrast function. In [12], the reference signals are initialized arbitrarily and kept the same during whole optimization process. In [13], the reference signals are indirectly involved in the iterative optimization process. In other words, the reference signals update following the objective signals. More precisely, $\mathbf{V}$ updates following $\mathbf{W}$ in each loop iteration step. Then the separation quality of the algorithm in [13] is better than that in [12]. In [14] and [15], we have done some corresponding work to investigate the impact of reference signals, which is similar to [12]. Inspired by [13], we consider the case that the reference signals update circularly in this paper. Therefore, the performance of our algorithm in this paper is much better than those in [14] and [15].

3.3 Contrast Functions

Let us introduce the following criteria:

$$J(w) = \left| \frac{C\{y(n)\}}{E\left\{(y(n))^2\right\}^2} \right|^2, \tag{6}$$

$$I(w,v)=\left|\frac{C_z\{y(n)\}}{E\left\{(y(n))^2\right\}E\left\{(z(n))^2\right\}}\right|^2 \tag{7}$$

where J is the well-known Kurtosis contrast function, which has been proved to be a contrast function in [3] and [4]. I is the reference-based contrast function used in this paper, which has been proposed in [6] and the consistency of which to a stationary point has been proved in [11]. We mainly focus our attention on the efficient optimization of I in the FastICA framework, which leads to the proposal of our new algorithm.

Furthermore, besides the gradient algorithms in [11]-[13], our method provides another new approach to the efficient optimization of reference-based contrast functions. Recently, we have also done some work by introducing the reference signals to the Negentropy contrast criterion, based on which a family of more efficient and robust algorithms have been proposed such as the algorithm in [16].

4. Optimization Method

4.1 New Algorithm

We now introduce our new FastICA algorithm based on the reference-based Kurtosis contrast function. First, we give some definitions used in our method. As described in [11], ∇ denotes a gradient operator. ∇_1 and ∇_2 denote partial gradient operators with respect to the first and second parameters, respectively. More precisely, $\nabla J(w)$ is the vector composed of all partial derivatives of $J(w)$, whereas $\nabla_1 I(w,v)$ and $\nabla_2 I(w,v)$ are the vectors of partial derivatives of $I(w,v)$ with respect to w and v. We denote $\mathbf{W}$ and $\mathbf{V}$ by $\left(w^1,\ldots,w^N\right)^T$ and $\left(v^1,\ldots,v^N\right)^T$, where T means the transpose.

Combined with (4) and (7), we can get

$$\begin{aligned}I(w,v)&=\left|\frac{C_z\{y(n)\}}{E\left\{(y(n))^2\right\}E\left\{(z(n))^2\right\}}\right|^2\\&=\left|\frac{E[(w\mathbf{x}(n))^2(v\mathbf{x}(n))^2]-E[(w\mathbf{x}(n))^2]E[(v\mathbf{x}(n))^2]-2\{E[(w\mathbf{x}(n))(v\mathbf{x}(n))]\}^2}{E\left\{(w\mathbf{x}(n))^2\right\}E\left\{(v\mathbf{x}(n))^2\right\}}\right|^2.\end{aligned} \tag{8}$$

Because $\mathbf{x}(n)$ is prewhitened and $\mathbf{W}$ and $\mathbf{V}$ are normalized in our method, we can get $E\left\{(w\mathbf{x}(n))^2\right\}=E\left\{(v\mathbf{x}(n))^2\right\}=1$ and $ww^T=vv^T=1$. Then (8) can be reduced to

$$I(w,v)=\left|E[(w\mathbf{x}(n))^2(v\mathbf{x}(n))^2]-3w^2\right|^2. \tag{9}$$

Hence, the partial derivative of $I(w,v)$ with respect to w can be expressed as

$$\begin{aligned}\nabla_1 I(w,v)=\frac{\partial I(w,v)}{\partial w}&=4\left|E[(w\mathbf{x}(n))^2(v\mathbf{x}(n))^2]-3w^2\right|\\&(E[\mathbf{x}(n)(w\mathbf{x}(n))(v\mathbf{x}(n))^2]-3w).\end{aligned} \tag{10}$$

Our new proposed algorithm is shown as follows:

- Eliminate the mean value of $\mathbf{x}$ and prewhiten it.
- Initialize $\mathbf{W}$ and normalize it.
- $(M0)$ For $i=1,2,\cdots,N$ repeat $(M0)$

  Set $\mathbf{w}_0^i=\mathbf{w}_i$, $\mathbf{v}_0^i=\mathbf{w}_0^i$

  $(M0')$ For $k=0,1,\cdots,k_{\max}-1$ repeat $(M0')$

  - Set $\mathbf{d}_k=\nabla_1 I(\mathbf{w}_k^i,\mathbf{v}_k^i)$
  - $\mathbf{w}_{k+1}^i=\mathbf{d}_k$
  - Normalize $\mathbf{w}_{k+1}^i$
  - $\mathbf{w}_{k+1}^i=\mathbf{w}_{k+1}^i-\sum_{j=1}^{i-1}\mathbf{w}_j\mathbf{w}_j^T\mathbf{w}_i$
  - Renormalize $\mathbf{w}_{k+1}^i$
  - Set $\mathbf{v}_{k+1}^i=\mathbf{w}_{k+1}^i$
  - Renormalize $\mathbf{w}_{k+1}^i$
  - $\mathbf{w}_i=\mathbf{w}_{k_{\max}-1}^i$
- $\mathbf{y}=\mathbf{W}\mathbf{x}$

Here, $\mathbf{W}_0$ is the initializations of $\mathbf{W}$, which is chosen randomly. From the whole process, we can see the source signals are recovered one by one through each one-dimensional optimization in a deflationary manner. To prevent different one-dimensional optimization converging to the same maxima, a Gram-Schmidt-like decorrelation scheme is adopted as shown above.

4.2 Convergent Results

Because of the symmetry of $I(w,v)$, we can get the following relationship.

$$\begin{aligned}&I(w,v)=I(v,w),\\&\nabla_1 I(w,v)=\nabla_2 I(v,w),\\&\nabla J(w)=2\nabla_1 I(w,w)=2\nabla_2 I(w,w).\end{aligned} \tag{11}$$

From (11), we can see that, during one-dimensional optimization process, the reference-based contrast function is $I(w,v)$ with v fixed instead of $J(w)$. So the optimization algorithm is quadratic dependence on the searched parameters. To justify the convergence of the algorithm, besides the assumptions on source signals mentioned above, another assumption [11] is needed.

A3: The sources sequences $s_i(n), i\in\{1,\cdots,N\}$ are temporally independent and identically distributed (i.i.d.). Moreover, they have fourth-order Cumulants which are all of the same sign.

Now, we can give the following proposition [11] shown below:

Proposition 1: Assume that the sequences $w_k^i, i=1,\ldots,N$ are obtained according to the above algorithm with $k_{\max}$ infinite and all $w_k^i, i=1,\ldots,N$ are contained in a compat set. Then, under assumptions A1-A3, any convergent subsequence of $w_k^i, i=1,\ldots,N$ converges to points $(w^i)^*, i=1,\ldots,N$ respectively such that $\nabla J((w^i)^*)=0, i=1,\ldots,N$.

Separation Method		Average MSE				Median MSE				Average Execution Time (s)			
		Number of Samples				Number of Samples				Number of Samples			
		1000	5000	10000	30000	1000	5000	10000	30000	1000	5000	10000	30000
G-1	1st	0.0338	0.0123	0.0088	0.0052	0.0314	0.0125	0.0089	0.0051	3.793	10.743	18.934	53.401
	2nd	0.0365	0.0124	0.0088	0.0051	0.0346	0.0125	0.0089	0.0051				
	3rd	0.0367	0.0124	0.0088	0.0051	0.0356	0.0125	0.0089	0.0051				
G-2	1st	0.0364	0.0124	0.0088	0.0051	0.0338	0.0125	0.0089	0.0051	3.408	7.927	13.921	42.073
	2nd	0.0348	0.0124	0.0088	0.0051	0.0318	0.0125	0.0089	0.0051				
	3rd	0.0354	0.0123	0.0088	0.0051	0.0327	0.0124	0.0089	0.0051				
F-1	1st	0.0375	0.0123	0.0088	0.0051	0.0383	0.0124	0.0089	0.0051	0.687	3.095	6.069	16.389
	2nd	0.0352	0.0123	0.0087	0.0051	0.0355	0.0124	0.0089	0.0051				
	3rd	0.0361	0.0122	0.0086	0.0051	0.0370	0.0124	0.0084	0.0051				
F-2	1st	0.0370	0.0124	0.0088	0.0052	0.0361	0.0125	0.0089	0.0052	0.150	0.366	0.641	2.564
	2nd	0.0357	0.0124	0.0088	0.0051	0.0344	0.0125	0.0089	0.0051				
	3rd	0.0352	0.0124	0.0088	0.0051	0.0338	0.0125	0.0089	0.0051				

Tab. 1. MSE and execution time for different separation methods (100 Monte-Carlo runs).

Based on the assumptions A1-A3, the consistent convergence of *Proposition 1* can be proved by referring to *Proposition 1* in [11] and Zangwill's convergence theorem in [17].

5. Simulation Results

5.1 Experimental Data

In this section, we choose three speech signals as sources. Without loss of generality, we assume that the number of observations and sources are equal, i.e., $N = M = 3$. The iterative parameter $k_{\max} = 1000$. The initialization value of mixture matrix $\mathbf{W}_0$ is randomly chosen. And we implement each algorithm 1000 times independently and obtain the average value. Corresponding notations are described as follows:

G-1 denotes the classical gradient optimization algorithm based on the Kurtosis contrast function. However, the optimization step size is maximized to be optimal as presented in [11]-[13]. G-2 denotes the gradient optimization algorithm based on the reference-based Kurtosis contrast function proposed in [13], but differs in the fact that we apply it in the case of linear and instantaneous mixture in this paper. F-1 denotes the classical FastICA algorithm based on the Kurtosis. F-2 denotes our new FastICA algorithm in this paper.

Average MSE denotes the mean square estimation errors between sources and observations averaged over 1000 Monte-Carlo runs. Median MSE denotes the median mean square estimation errors between sources and observations averaged over 1000 Monte-Carlo runs. Average execution time denotes the execution time of recovering all sources averaged over 1000 Monte-Carlo runs.

1st, 2nd and 3rd denote the first, second and third recovered source signals, respectively. The experimental data in detail is shown in Tab. 1 on the top of this page.

5.2 Comments and Analysis

To compare the performance of above four algorithms clearly, the average execution time for samples from 1000 to 30000 is illustrated in Fig. 1.

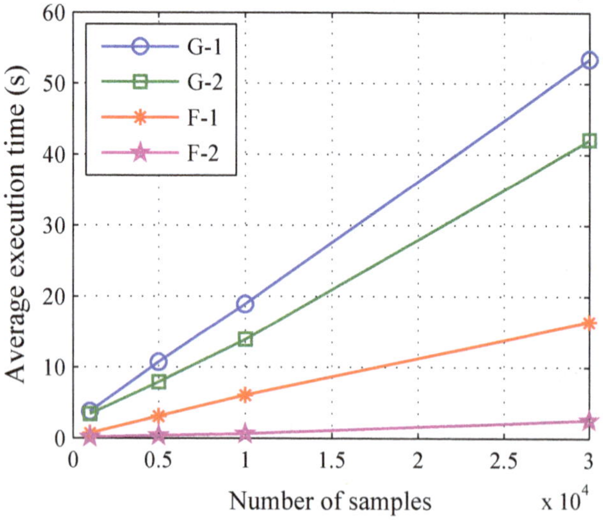

Fig. 1. Average execution time for samples from 1000 to 30000 (1000 Monte-Carlo runs).

Firstly, in the gradient optimization framework where G-1 and G-2 are considered, we can see that G-2 shows better performance than G-1 from Tab. 1 and Fig. 1. This has been confirmed and validated in [11] and [13], so detailed comparison and analysis are omitted here.

Secondly, in the FastICA optimization framework where F-1 and F-2 are considered, it can be clearly seen that

F-2 yields similar MSE value to F-1 from the corresponding rows in Tab. 1. This means our new algorithm F-2 performs well and can converge to the same stationary point. Moreover, the average MSE and median MSE value are very close to each other, which means that our method F-2 is very stable. However, we can obviously see that the computational time of F-2 is much less than that of F-1 with varying number of samples, which means our method is much quicker in terms of convergence speed. Furthermore, with an increasing number of samples, the advantage of our new method over F-1 is significantly apparent.

Finally, when the gradient and FastICA optimization schemes are considered together, the similar MSE value of G-1, G-2, F-1 and F-2 in Tab. 1 confirms the convergence performance for all of them. However, it can be observed clearly from Fig. 1 that our algorithm F-2 needs the least execution time among them. In other words, our proposed method F-2 is much quicker than F-1, and also much quicker than G-1 and G-2. It means that our method provides better performance than the corresponding classical algorithm on the one hand and than some other gradient optimization ones under same circumstances on the other hand.

6. Conclusion

A novel FastICA optimization algorithm based on the recently proposed reference-based Kurtosis contrast function is proposed in this paper. It is much more efficient than corresponding classical one in terms of computational speed. The performance of this new method is validated through simulations. Our future work includes the extension of our method to complex-valued signals and the application of our method to more complicated channel such as convolution mixture.

Acknowledgements

This work is supported by the NSF of Jiangsu Province of China under Grant No. BK2011117 and BK2012057, and by the National Natural Science Foundation of China under Grant No. 61172061 and 61201242.

References

[1] LAHAT, D., CARDOSO, J. F., MESSER, H. Second-order multidimensional ICA: Performance analysis. *IEEE Transactions on Signal Processing*, 2012, vol. 60, no. 9, p. 4598 - 4610.

[2] COMON, P., JUTTEN, C. (Eds.) *Handbook of Blind Source Separation, Independent Component Analysis and Applications.* New York, 2010.

[3] SIMON, C., LOUBATON, P., JUTTEN, C. Separation of a class of convolutive mixtures: A contrast function approach. *Signal Processing*, 2001, vol. 4, no. 81, p. 883 - 887.

[4] TUGNAIT, J. K. Identification and deconvolution of multichannel linear non-Gaussian processes using higher order statistics and inverse filter criteria. *IEEE Transactions on Signal Processing*, 1997, vol. 45, no. 3, p. 658 - 672.

[5] GODARD, D. N. Self-recovering equalization and carrier tracking in two-dimensional data communication systems. *IEEE Transactions on Communications*, 1980, vol. 28, no. 11, p. 1867 - 1875.

[6] CASTELLA, M., RHIOUI, S., MOREAU, E., PESQUET, J. C. Quadratic higher-order criteria for iterative blind separation of a MIMO convolutive mixture of sources. *IEEE Transactions on Signal Processing*, 2007, vol. 55, no. 1, p. 218 - 232.

[7] DUBROCA, R., DELUIGI, C., CASTELLA, M., MOREAU, E. A general algebraic algorithm for blind extraction of one source in a MIMO convolutive mixture. *IEEE Transactions on Signal Processing*, 2010, vol. 58, no. 5, p. 2484 - 2493.

[8] ADIB, A., MOREAU, E., ABOUTAJDINE, D. Source separation contrasts using a reference signal. *IEEE Signal Processing Letters*, 2004, vol. 11, no. 3, p. 312 - 315.

[9] CASTELLA, M., RHIOUI, S., MOREAU, E., PESQUET, J. C. Source separation by quadratic contrast functions: A blind approach based on any higher-order statistics. In *Proceedings of IEEE International Conference on Acoustics, Speech and Signal Processing.* Philadelphia (USA), 2005, vol. 3, p. III/569-III/572.

[10] KAWAMOTO, M., KOHNO, K., INOUYE, Y. Eigenvector algorithms incorporated with reference systems for solving blind deconvolution of MIMO-IIR linear systems. *IEEE Signal Processing Letters*, 2007, vol. 14, no. 12, p. 996 - 999.

[11] CASTELLA, M., MOREAU, E. New kurtosis optimization schemes for MISO equalization. *IEEE Transactions on Signal Processing*, 2012, vol. 60, no. 3, p. 1319 - 1330.

[12] CASTELLA, M., MOREAU, E. A new optimization method for reference-based quadratic contrast functions in a deflation scenario. In *Proceedings of IEEE International Conference on Acoustics, Speech and Signal Processing.* Taipei (Taiwan), 2009, p. 3161 - 3164.

[13] CASTELLA, M., MOREAU, E. A new method for kurtosis maximization and source separation. In *Proceedings of IEEE International Conference on Acoustics, Speech and Signal Processing.* Dallas (TX, USA), 2010, p. 2670 - 2673.

[14] ZHAO, W., YUE, H. S., WANG, J. G., YUAN, Z. G., JIAN, W. New methods for the efficient optimization of cumulant-based contrast functions. In *Proceedings of the 5th IET International Conference on Wireless, Mobile and Multimedia Networks.* Beijing (China), 2013, p. 345 - 349.

[15] ZHAO, W., YUE, H. S., WANG, J. G., YUAN, Z. G., JIAN, W. New kurtosis optimization algorithms for independent component analysis. In *Proceedings of the 4th IEEE International Conference on Information Science and Technology.* Shenzhen (China), 2014, p. 23 - 28.

[16] ZHAO, W., YUE, H. S., XU, P. C., WANG, J. G., YUAN, Z. G., WEI, Y. M., JIAN, W., LI, H. A new efficient reference-based negentropic algorithm for instantaneous ICA. In *Proceedings of the 3th International Conference on Convergence and its Application.* Seoul (Korea), 2014, p. 3 - 4.

[17] ZANGWILL, W. I. *Nonlinear Programming: A Unified Approach.* Englewood Cliffs (NJ, USA): Prentice-Hall, 1969.

21

On Amplify-and-Forward Relaying over Hyper-Rayleigh Fading Channels

Sajid H. ALVI, Shurjeel WYNE

Dept. of Electrical Engineering, COMSATS Institute of Information Technology, Islamabad, Pakistan

{sajid_hussain, shurjeel.wyne}@comsats.edu.pk.

Abstract. *Relayed transmission holds promise for the next generation of wireless communication systems due to the performance gains it can provide over non-cooperative systems. Recently hyper-Rayleigh fading, which represents fading conditions more severe than Rayleigh fading, has received attention in the context of many practical communication scenarios. Though power allocation for Amplify-and-Forward (AF) relaying networks has been studied in the literature, a theoretical analysis of the power allocation problem for hyper-Rayleigh fading channels is a novel contribution of this work. We develop an optimal power allocation (OPA) strategy for a dual-hop AF relaying network in which the relay-destination link experiences hyper-Rayleigh fading. A new closed-form expression for the average signal-to-noise ratio (SNR) at destination is derived and it is shown to provide a new upper-bound on the average SNR at destination, which outperforms a previously proposed upper-bound based on the well-known harmonic-geometric mean inequality. An OPA across the source and relay nodes, subject to a sum-power constraint, is proposed and it is shown to provide measurable performance gains in average SNR and SNR outage at the destination relative to the case of equal power allocation.*

Keywords

Relayed communications, hyper-Rayleigh fading, amplify-and-forward, power-allocation.

1. Introduction

Relayed transmission strategies are integral to the next generation of wireless communication systems due to the performance gains such as coverage extension and robustness to fading that these techniques can provide relative to non-cooperative systems [1], [2]. Amplify-and-Forward (AF) relaying, whereby the relay amplifies the received signal before re-transmitting it to the destination, has received considerable attention in the literature due to its low complexity of deployment, see for example [3], [4]. In variable gain AF schemes the instantaneous channel state information (CSI) of the previous hop is used to control the relay gain whereas in fixed/semi-blind AF relaying the relay gain is determined by channel statistics of the previous hop. Although the former scheme generally provides better diversity performance, the latter scheme trades-off the diversity performance with a lower complexity in the CSI estimation part and is therefore more attractive due to practical considerations [5].

When modeling land-mobile wireless communication channels, Rayleigh distributed fading is often considered to be the worst-case fading scenario [6], [7]. However, in recent years many published measurement campaigns have reported channels with fading more severe than Rayleigh or so-called hyper-Rayleigh fading; see for example [8] - [11] and references therein. These severe fading conditions have been observed in various scenarios of practical significance; in [8] the authors observed worse-than-Rayleigh fading in outdoor suburban measurements conducted at 1.5 GHz and they proposed to statistically model such fading with the Nakagami-m distribution with its fading severity parameter m taking values in the range $0.5 \leq m < 1$. The same authors also observed that the Weibull distribution with its shape factor less than 2 also provided an empirically best-fit in some cases. In [9] the authors performed indoor measurements at 2.4 GHz for two-way radio-frequency identification applications and used both the Nakagami-m and lognormal distributions to model the severe small-scale fading. In [10] measurements at 2.4 GHz within aircraft bodies were analyzed and the two-wave with diffuse power model, a physical wave model rather than a statistical approach, was proposed to model the observed hyper-Rayleigh fading. Finally, in [11] the authors reported vehicle-to-vehicle channel measurements in the 5 GHz band and observed that the weibull distribution provided the best-fit to worse-than-Rayleigh fading channel amplitudes. Despite this practical significance of hyper-Rayleigh fading scenarios a theoretical analysis of the power allocation problem for hyper-Rayleigh fading channels has not been performed previously even though power allocation for fading AF relaying networks has been widely studied in the literature, see for example [12], [13] and references therein. Through this work we aim to address this issue. In modeling hyper-Rayleigh fading we adopt the statistical approach of [8] and [9], which is to model the fading channel coefficient h as Nakagami-m distributed with its fading severity parameter m taking values in the range

$0.5 \leq m < 1$. The distribution of h under hyper-Rayleigh fading is then expressed as

$$f_h(h) = \frac{2}{\Gamma(m)} \left(\frac{m}{\Omega}\right)^m h^{2m-1} e^{-\frac{mh^2}{\Omega}}, \; 0.5 \leq m < 1 \quad (1)$$

where $\Gamma(\cdot)$ is the Gamma function [14, Eq. (8.310)], $\Omega = E[h^2]$, and $E[\cdot]$ denotes the statistical expectation operator. Our modeling approach is also motivated in part by the fact that the Nakagami-m distribution is widely considered for modeling AF relay channels, albeit with $m \geq 1$, and we aim to extend this work to include the practically significant case where $m < 1$, i.e., worse than Rayleigh fading.

Cooperative communications over Nakagami-m fading links has been extensively studied in the literature, see [15] - [18] and references therein. In [15] the authors have analyzed the outage probability and other statistical parameters for the relay network without diversity. In [16] and [17] the end-to-end SNR and the average symbol error probability have been analyzed for multi-hop communications for various fading distributions of the links. Furthermore, performance bounds for the multi-hop scenario are also proposed in the latter references. These works assume equal power allocation between the source and relay nodes. Optimal power allocation (OPA) for semi-blind AF relaying, subject to a sum-power constraint, has been studied in [19] for Rayleigh fading links, i.e., Nakagami-m fading with $m = 1$ and in [18] for Nakagami-m fading links with $m \geq 1$. This letter aims to extend these investigations to address OPA for dual-hop semi-blind AF relaying for the case where the relay-destination link experiences hyper-Rayleigh fading.

We develop a new closed form expression for the exact average SNR at the destination for a dual-hop semi-blind AF relaying system. The two hops are assumed to experience independent but not necessarily identically distributed Nakagami-m fading with arbitrary m values. We then develop an upper bound for the average SNR at the destination for the case where the relay-destination link experiences hyper-Rayleigh fading. As will be evident from the derivations presented in the sequel, our analysis does not impose any restriction on the m parameter of the Nakagami faded source-relay link, which may or may not experience hyper-Rayleigh fading, i.e., $m < 1$ or $m \geq 1$, respectively. The performance of the proposed upper bound is compared with a well-known upper bound proposed in [16], which is based on the harmonic-geometric mean inequality. We then propose an OPA strategy to increase the average SNR at destination by maximizing the proposed upper bound and compare its performance with OPA achieved by numerical maximization of the exact average SNR expression. An increase in the average SNR is desirable as it reduces the outage probability of the instantaneous SNR in a similar fashion to the outage reduction obtained by introducing a fading margin into the link budget [20]. The remainder of this paper is organized as follows. The system and channel model under consideration are introduced in Section 2. Section 3 provides analytical expressions for the exact average SNR at destination and its proposed upper bound. Section 4 contains analysis for the proposed power allocation. Section 5 contains the numerical and simulation results. Finally, concluding remarks are given in Section 6.

2. System and Channel Model

We consider a dual-hop network consisting of the source (s), relay (r), and destination (d) nodes. The s-d link is assumed to be non-existent or in a deep fade leading to a relayed network without cooperative diversity [1], [2], and [5]. Furthermore, the s, r nodes are assumed to transmit over orthogonal channels. Let h_{sr} and h_{rd} denote the independent flat-fading channel coefficients for the s-r and r-d links, respectively. The corresponding average channel power gains for these links are denoted by σ_{sr}^2 and σ_{rd}^2, respectively. We consider both h_{sr} and h_{rd} to be independent Nakagami-m fading with fading parameters m_1, and m_2, respectively. The instantaneous value of the equivalent SNR at the destination node can then be written as [21]

$$\gamma_{\text{eq}} = \frac{|A|^2 |h_{sr}|^2 |h_{rd}|^2 \rho}{|A|^2 |h_{rd}|^2 + 1} \quad (2)$$

where A is the fixed amplification factor provided by the semi-blind relay, $\rho = \frac{E_s}{N_o}$ is the SNR at the source node, E_s is the average transmit power at source, and N_o is the identical power spectral density of additive white Gaussian thermal noise present at all node inputs. Let the average transmit power at the relay, with power allocation parameter α, is $E_r = \alpha E_s$, then the relay's power gain can be expressed as [1]

$$|A|^2 = \frac{E_r}{E_s \sigma_{\text{sr}}^2 + N_0} = \frac{\alpha\rho}{\rho\sigma_{\text{sr}}^2 + 1}. \quad (3)$$

It is evident from the above relation that α can be used to optimize the link performance. Now substituting (3) into (2), the equivalent SNR for the relayed link can be expressed as

$$\gamma_{\text{eq}} = \frac{\alpha\rho^2 |h_{\text{sr}}|^2 |h_{\text{rd}}|^2}{1 + \rho\sigma_{\text{sr}}^2 + \alpha\rho |h_{\text{rd}}|^2}, \quad (4)$$

which can be equivalently written as

$$\gamma_{\text{eq}} = \frac{\rho |h_{\text{sr}}|^2 |h_{\text{rd}}|^2}{C + |h_{\text{rd}}|^2} \quad (5)$$

where $C = \frac{1+\rho\sigma_{sr}^2}{\alpha\rho}$.

3. End-to-End SNR

The SNR at destination when averaged over the channel fading can be written as

$$E\left[\gamma_{\text{eq}}\right] = E\left[\frac{\rho |h_{\text{sr}}|^2 |h_{\text{rd}}|^2}{C + |h_{\text{rd}}|^2}\right]. \quad (6)$$

Given that the channel gains $h_{\text{sr}}, h_{\text{rd}}$ are Nakagami-m distributed, it follows that the respective channel power gains $|h_{\mathrm{sr}}|^2$ and $|h_{\mathrm{rd}}|^2$ are Gamma distributed random variables [21]. We define $X = |h_{\mathrm{sr}}|^2$ and $Y = |h_{\mathrm{rd}}|^2$ as independent not necessarily identically-distributed Gamma random variables with the probability density function (pdf) for X written as [21]

$$f_X(x) = \frac{x^{a_X-1}e^{-\frac{x}{b_X}}}{\Gamma(a_X)b_X^{a_X}}.$$

The pdf for Y is defined similarly. The Gamma pdf parameters (a_X, b_X) for X can be related with parameters of the parent Nakagami-m distribution as $a_X = m_1$ and $b_X = \frac{\sigma_{\mathrm{sr}}^2}{m_1}$, and similarly for Y we have $a_Y = m_2$ and $b_Y = \frac{\sigma_{\mathrm{rd}}^2}{m_2}$. Then the average SNR for the relayed link can be written as

$$\begin{aligned} E\left[\gamma_{\mathrm{eq}}\right] &= E\left[\frac{\rho XY}{C+Y}\right] \\ &= \rho\int_{-\infty}^{\infty}\int_{-\infty}^{\infty} x\frac{y}{C+y}f_X(x)f_Y(y)\mathrm{d}x\mathrm{d}y \\ &= \rho\int_0^{\infty} x\frac{x^{a_X-1}e^{-\frac{x}{b_X}}}{\Gamma(a_X)b_X^{a_X}}\mathrm{d}x \\ &\quad\times\int_0^{\infty}\frac{y}{C+y}\frac{y^{a_Y-1}e^{-\frac{y}{b_Y}}}{\Gamma(a_Y)b_Y^{a_Y}}\mathrm{d}y \\ &= \rho a_X b_X \underbrace{\int_0^{\infty}\frac{y}{C+y}\frac{y^{a_Y-1}e^{-\frac{y}{b_Y}}}{\Gamma(a_Y)b_Y^{a_Y}}\mathrm{d}y}_{I_1}. \end{aligned} \tag{7}$$

The integral I_1 in (7) is further simplified by substitution and change of variables. After some manipulations, we obtain

$$I_1 = \frac{e^{C_1}}{\Gamma(a_Y)}\int_{C_1}^{\infty}\frac{(z-C_1)^{a_Y}e^{-z}}{z}dz \tag{8}$$

where the constant $C_1 = \frac{1+\rho\sigma_{\mathrm{sr}}^2}{\alpha\rho b_Y}$. Now using [14, Eq. (3.383.9)] and the identity [14, Eq. (8.331.1)],

$$\Gamma(x+1) = x\Gamma(x), \tag{9}$$

the integral I_1 can be expressed in closed form as

$$I_1 = a_Y e^{C_1}C_1^{a_Y}\Gamma(-a_Y, C_1) \tag{10}$$

where $\Gamma(a,x)$ is the upper incomplete Gamma function [14, Eq. (8.350.2)]. Substituting (10) into (7) we get

$$E\left[\gamma_{\mathrm{eq}}\right] = \rho a_X b_X a_Y e^{C_1}C_1^{a_Y}\Gamma(-a_Y, C_1). \tag{11}$$

The introduction of the incomplete Gamma function into (11) is novel in that it replaces the confluent hypergeometric function [14, Eq. (9.210.1)] conventionally used in SNR expressions for Nakagami-m faded relayed links, see for example [16], [18] and references therein. One may also observe from (11) that the average SNR at destination does not depend on m_1, which is canceled in the product $a_X b_X$. Therefore, our analysis is general in the sense that it is equally applicable whether only the r-d link is hyper-Rayleigh fading or additionally the s-r link is also hyper-Rayleigh fading. Now it can be shown that $\Gamma(-a,x)$ is upper-bounded as

$$\Gamma(-a,x) \leq \frac{e^{-x}x^{-a}}{x+a}2^{1-a}\Gamma(1+a),\ x>0. \tag{12}$$

Proof. See Appendix. □

Substituting (12) and C_1 into (11) and expressing the Gamma distribution parameters (a_X, b_X) and (a_Y, b_Y) in (11) in terms of their parent Nakagami-m parameters, the average SNR for the relayed link can be upper-bounded as,

$$E\left[\gamma_{\mathrm{eq}}\right] \leq \frac{\alpha\rho^2\sigma_{sr}^2\sigma_{rd}^2}{\rho\sigma_{sr}^2+\alpha\rho\sigma_{rd}^2+1}2^{1-m_2}\Gamma(1+m_2). \tag{13}$$

With a minor modification we obtain a more tractable upper-bound without significantly affecting the tightness as,

$$E\left[\gamma_{\mathrm{eq}}\right] < E\left[\gamma_{\mathrm{eq}}^{\mathrm{UB}}\right] \equiv \frac{\alpha\rho\sigma_{sr}^2\sigma_{rd}^2}{\sigma_{sr}^2+\alpha\sigma_{rd}^2}2^{1-m_2}\Gamma(1+m_2). \tag{14}$$

To compare the performance of the upper bound obtained in (14) we consider the upper-bound derived by [16] that is based on the well-known harmonic-geometric mean inequality of two positive numbers, which for the case under consideration, are the average SNRs of the two hops. The bound of [16] is valid for N-hop relayed communication over Nakagami-m fading links and, for the two-hop case ($N=2$) considered herein, leads to an alternative upper-bound expression for the average end-to-end SNR that can be expressed as,

$$E\left[\gamma_{\mathrm{eq}}^{\mathrm{UB\text{-}KG}}\right] = \frac{\Gamma\left(m_2+\frac{1}{2}\right)}{2\Gamma(m_2)\sqrt{m_2}}\times\rho\sigma_{sr}^2\sqrt{\frac{\alpha\rho\sigma_{rd}^2}{1+\rho\sigma_{sr}^2}}. \tag{15}$$

Proof. See Appendix. □

Now let $\rho_T = \frac{E_s+E_r}{N_o}$ denote the total transmit SNR. Then ρ can be written in terms of ρ_T and α as,

$$\rho = \frac{\rho_T}{(1+\alpha)}. \tag{16}$$

Substituting (16) into (14) allows the proposed upper bound for the average SNR to be written as

$$E\left[\gamma_{\mathrm{eq}}^{\mathrm{UB}}\right] = \frac{\rho_T}{1+\alpha}\left(\frac{\alpha\sigma_{\mathrm{sr}}^2\sigma_{\mathrm{rd}}^2}{\sigma_{\mathrm{sr}}^2+\alpha\sigma_{\mathrm{rd}}^2}\right)2^{1-m_2}\Gamma(1+m_2). \tag{17}$$

Similarly, inserting (16) into (15) leads to an alternative upper bound on the average SNR at destination that can be expressed as

$$E\left[\gamma_{\mathrm{eq}}^{\mathrm{UB\text{-}KG}}\right] = \frac{\rho_T}{1+\alpha}\left(G\sigma_{\mathrm{sr}}^2\sqrt{\frac{\alpha\rho_T\sigma_{\mathrm{rd}}^2}{1+\alpha+\rho_T\sigma_{\mathrm{sr}}^2}}\right) \tag{18}$$

where $G = \frac{\Gamma\left(m_2+\frac{1}{2}\right)}{2\Gamma(m_2)\sqrt{m_2}}$. Note that the distance-dependent pathloss effects are taken into account here by modeling the mean channel power gains as $\sigma_{\mathrm{sr}}^2 = d^{-n}$, and $\sigma_{\mathrm{rd}}^2 = (1-d)^{-n}$ for the s-r and r-d links, respectively, where n is the pathloss exponent.

4. Optimal Power Allocation

We now demonstrate that OPA, which maximizes the upper bound of the average SNR $\alpha = \alpha_1$, suffers no performance penalty relative to the OPA based on numerical maximization of the exact expression for average SNR $\alpha = \alpha_{\text{opt}}$. The former scheme has the advantage of being more tractable for mathematical analysis. Now from basic calculus the optimal value of α that maximizes (17) is found to be

$$\alpha_1 = \sqrt{\frac{\sigma_{sr}^2}{\sigma_{rd}^2}}. \tag{19}$$

It is apparent from the above formulation that this power allocation requires only the knowledge of the channel statistics. Substituting α_1 into (17), the upper bound of the average SNR can be written as

$$E\left[\gamma_{\text{eq}}^{\text{UB}}\right]_{\alpha=\alpha_1} = \rho_T \sigma_{\text{rd}}^2 \left(\frac{\sigma_{\text{sr}}^2}{\sigma_{\text{sr}}^2 + \sqrt{\sigma_{\text{sr}}^2 \sigma_{\text{rd}}^2}} \right)^2 2^{1-m_2}\Gamma(1+m_2). \tag{20}$$

4.1 Outage Analysis

The outage probability of the received SNR is one of the conventional metrics that is used to assess the performance of a wireless communication link. It is well-understood from the concept of fading-margin of a wireless link that increasing the average SNR value at destination reduces the outage probability of the instantaneous received SNR [20], [22]. In the previous section we have demonstrated the advantage of our proposed OPA strategy in providing an SNR gain at the destination; in this section we demonstrate that the same OPA strategy results in a reduced outage probability of the instantaneous received SNR, or conversely, a power saving when achieving a target outage. Given that the outage probability is inversely proportional to the average value of the received SNR, the proposed upper-bound maximizing power allocation strategy is shown to require smaller values of the total transmit SNR relative to equal power allocation to achieve the same outage probability. To this end, we begin by expressing the upper-bound from (14) as

$$E\left[\gamma_{\text{eq}}^{\text{UB}}\right] = \frac{\rho_T}{1+\alpha}\left(\frac{\alpha\sigma_{\text{sr}}^2}{\lambda+\alpha}\right) 2^{1-m_2}\Gamma(1+m_2) \tag{21}$$

where $\lambda = \sigma_{\text{sr}}^2/\sigma_{\text{rd}}^2$. Furthermore, let us denote the required total transmit SNR to achieve a target outage with optimal and equal power allocation as $\rho_T\,|_{\alpha=\alpha_1}$ and $\rho_T\,|_{\alpha=1}$, respectively. Then using (21) we can write the transmit SNR for optimal power allocation as

$$\rho_T\,|_{\alpha=\alpha_1} = \frac{(\lambda+\alpha_1)(1+\alpha_1)}{\alpha_1\sigma_{\text{sr}}^2 2^{1-m_2}\Gamma(1+m_2)} \times E\left[\gamma_{\text{eq}}^{\text{UB}}\right], \tag{22}$$

whereas for the equal power allocation case the transmit SNR can be expressed as

$$\rho_T\,|_{\alpha=1} = \frac{2(\lambda+1)}{\sigma_{\text{sr}}^2 2^{1-m_2}\Gamma(1+m_2)} \times E\left[\gamma_{\text{eq}}^{\text{UB}}\right]. \tag{23}$$

By rearranging (23) the upper bound average SNR can be written in terms of λ and $\rho_T\,|_{\alpha=1}$ as

$$E\left[\gamma_{\text{eq}}^{\text{UB}}\right] = \frac{\sigma_{\text{sr}}^2 2^{1-m_2}\Gamma(1+m_2)}{2(\lambda+1)}\rho_T\,|_{\alpha=1}. \tag{24}$$

Substituting (24) into (22), we finally get

$$\rho_T\,|_{\alpha=\alpha_1} = \frac{(\lambda+\alpha_1)(1+\alpha_1)}{2\alpha_1(\lambda+1)}\rho_T\,|_{\alpha=1}. \tag{25}$$

5. Numerical and Simulation Results

In this section we provide some numerical and simulation results in order to validate the accuracy of the proposed analytical results. Though the results of our theoretical analysis are valid for all practical values of the pathloss exponent n, for illustrative purposes values larger than 2, i.e., $n = 3$ and 4 are used. This choice is based on the heuristic reasoning that a cluttered environment, which results in small scale fading more severe than Rayleigh fading, is expected to exhibit a corresponding pathloss exponent larger than the free-space pathloss exponent of $n = 2$. For example in [9] the value $n = 4$ was reported for the 2-way channel and in general $n = 3,4$ or similar values are expected to be observed. For the results presented in this section, all simulations were carried out using the Matlab computational software, whereas the analytical plots were generated using both Matlab and Mathematica.

In Fig. 1, we plot the upper bound proposed in (17) against the normalized s-r link distance, $d = d_{\text{sr}}/d_{\text{sd}}$, where d_{ij} is the distance between nodes i and j. The exact SNR expression from (11) is also graphed in the same figure along with Monte Carlo simulations, using results of [23], which verify the correctness of our analysis. The upper bound from (18) is also plotted for comparison. From Fig. 1 one can observe that our proposed bound is significantly tighter than the upper bound of [16] for the case where the relay location approaches the source or destination location, i.e., the power imbalance between the average SNRs of the s-r and r-d links increases. When the relay is midway between these two nodes then our proposed upper-bound performs not significantly different from the bound of [16]. The looseness of the latter bound towards the link edges, as observed in the figure, can be attributed to the harmonic-geometric mean inequality, which loosens as the power imbalance between the average SNRs of the two hops increases.

In Fig. 2, the performance of the proposed power allocation strategy is demonstrated by plotting the SNR gain against the normalized distance d, where the SNR gain is defined as the quotient $E\left[\gamma_{\text{eq}}\right]/\rho_T$. Monte Carlo simulations are also shown in the figure to verify the correctness of the derived analytical expressions. One may observe from Fig. 2 that substituting $\alpha = \alpha_1$ into (11) provides the same SNR gain as that achieved by numerical optimization of (11). The crossovers seen between the two curves in the figure are

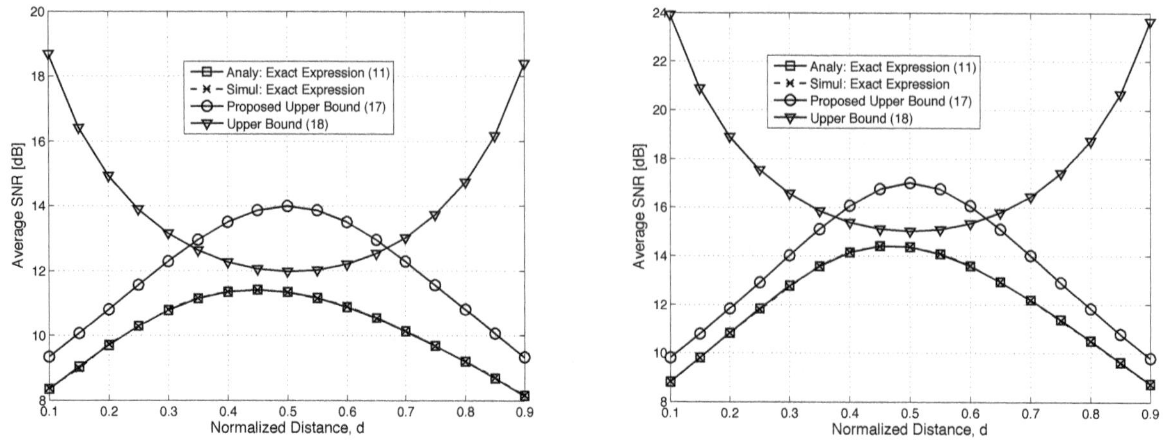

(a) pathloss exponent, n = 3. (b) pathloss exponent, n = 4.

Fig. 1. Comparison of exact average SNR with its upperbounds. Considered parameter values are: $(m_2, \alpha, \rho_{T,dB}) = (0.5, 1, 10)$.

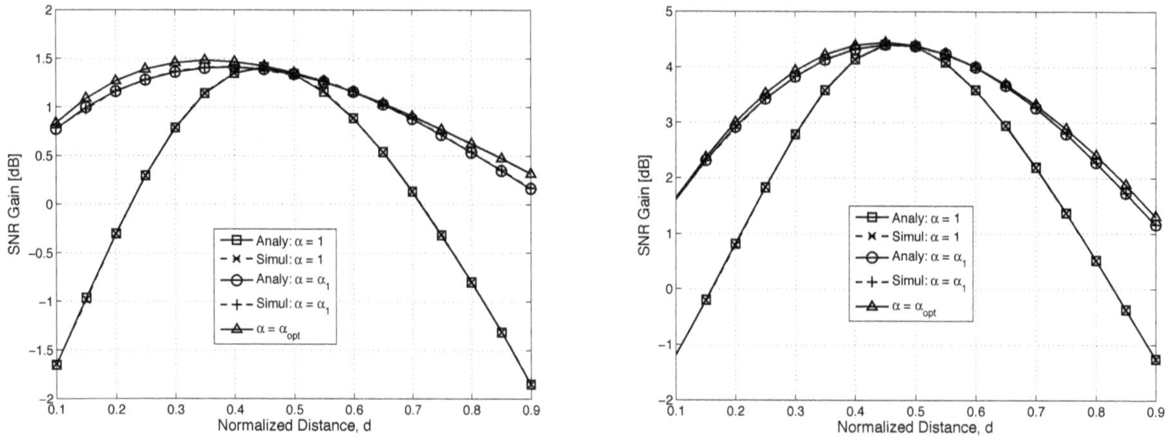

(a) pathloss exponent, n = 3. (b) pathloss exponent, n = 4.

Fig. 2. SNR gain as a function of normalized distance. Considered parameter values are: $(m_2, \rho_{T,dB}) = (0.5, 10)$.

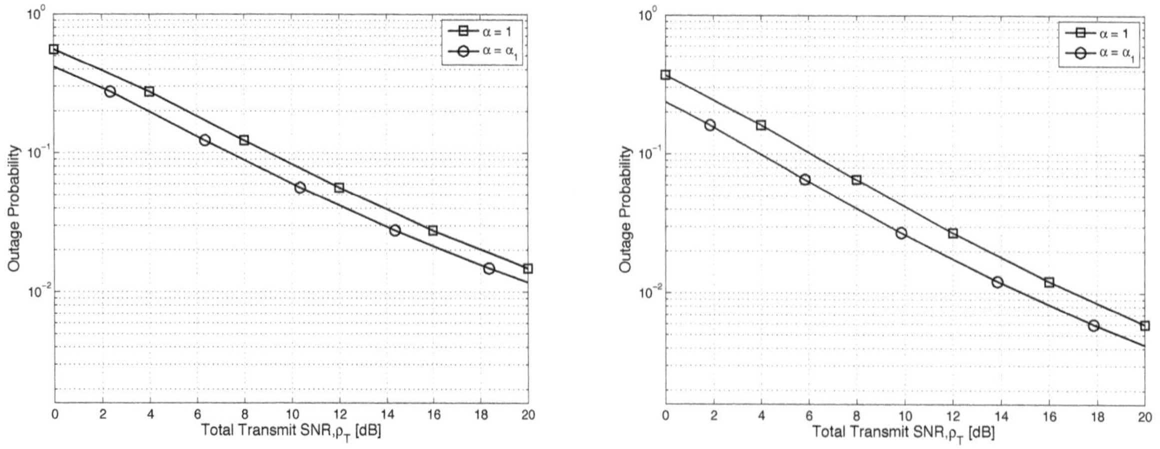

(a) pathloss exponent, n = 3. (b) pathloss exponent, n = 4.

Fig. 3. Outage probability as a function of total transmit SNR. Considered parameter values are: $(m_1, m_2, d, \gamma_{TH,dB}) = (1.2, 0.5, 0.75, 0)$.

due to the fact that we have optimized the upper bound rather than the exact average SNR expression, a fact which has also been observed in [19] for the case of Rayleigh fading.

In Fig. 3, the SNR outage performance of the proposed OPA scheme is shown in relation with the outage for equal power allocation, i.e., $\alpha = 1$. For the latter case, the SNR outage is plotted as a function of total transmit SNR by using the exact outage expression derived in [24, Eq. (20)] for dual-hop semi-blind AF relaying. This somewhat extensive outage expression, though straightforward to evaluate with computational software, is not reproduced here in the interest of brevity. Now to evaluate the outage performance for the $\alpha = \alpha_1$ case, (25) is used to compute the required total transmit SNR to achieve the same outage values as those previously achieved for the $\alpha = 1$ case. The threshold SNR for both outage curves is set to $\gamma_{TH} = 0$ dB without loss of generality. From Fig. 3 one may observe that the OPA strategy provides an SNR saving of around 2 dB over the equal power allocation case. Such savings in transmit power can be significant when taking into account the fact that the communication channel under consideration is subject to worse than Rayleigh fading. Furthermore, a similar range of values of the power savings have also been reported in [19] for power allocation based on the outage expression for the case of Rayleigh fading.

6. Conclusion

We have derived a new exact expression for the average SNR of a dual-hop semi-blind AF relaying network subject to Nakagami-m fading with arbitrary m. Additionally for the case when the r-d link is subject to worse-than-Rayleigh fading, we have proposed a new upper bound on the average SNR at destination, which has been shown to outperform a previously proposed bound that is based on the well-known harmonic-geometric mean inequality. We have also demonstrated that a power allocation strategy based on maximizing the proposed upper bound rather than the exact average SNR expression has no significant performance loss compared with the latter. The proposed OPA strategy has also been shown to reduce the outage probability relative to the case of equal power allocation. These results hold significance for relayed communication in all practical scenarios of interest, where hyper-Rayleigh fading can occur.

Appendix A: Proof of (12)

Using [14, Eq. (8.353.3)], $\Gamma(-a,x)$ can be expressed as

$$\Gamma(-a,x) = \frac{e^{-x}x^{-a}}{\Gamma(1+a)} \times \int_0^\infty \frac{e^{-t}t^a}{x+t}\mathrm{d}t, \quad Re(a) > -1, x > 0. \tag{26}$$

Multiplying (26) on the right by $(x+a)/(x+a)$ we obtain,

$$\Gamma(-a,x) = \frac{e^{-x}x^{-a}}{\Gamma(1+a)\,(x+a)} \int_0^\infty \frac{x+a}{x+t} e^{-t}t^a \mathrm{d}t. \tag{27}$$

Then using the gamma function inequality [25],

$$2^{a-1} \le \Gamma(1+a) \le 1, \; 0 < a \le 1. \tag{28}$$

$\Gamma(-a,x)$ can be upper-bounded as

$$\Gamma(-a,x) \le \frac{2^{1-a}e^{-x}x^{-a}}{x+a} \underbrace{\int_0^\infty \frac{x+a}{x+t} e^{-t}t^a \mathrm{d}t}_{I}. \tag{29}$$

The integral I appearing in (29) above can easily be shown to simplify to

$$I = \Gamma(1+a) + a \underbrace{\int_0^\infty \frac{t}{x+t} e^{-t}t^{a-1}\mathrm{d}t}_{I_1} - \underbrace{\int_0^\infty \frac{t}{x+t} e^{-t}t^a \mathrm{d}t}_{I_2} \tag{30}$$

where we have used the fact that $\Gamma(1+a) = \int_0^\infty e^{-t}t^a \mathrm{d}t$. We formulate an upper-bound for the integral I, by first establishing an upper-bound for the integral I_1. It can readily be observed from (30) that the integrand of I_1 satisfies the inequality

$$\frac{t}{x+t} e^{-t}t^{a-1} \le e^{-t}t^{a-1}, \; x > 0, \tag{31}$$

so that from basic calculus it follows that I_1 is upper-bounded as

$$I_1 = \int_0^\infty \frac{t}{x+t} e^{-t}t^{a-1}\mathrm{d}t \le \int_0^\infty e^{-t}t^{a-1}\mathrm{d}t, \; x > 0. \tag{32}$$

Now replacing I_1 in (30) with its upper-bound from (32) and using the Gamma function definition $\Gamma(a) = \int_0^\infty e^{-t}t^{a-1}\mathrm{d}t$ [14, Eq. (8.310.1)] together with (9), I is upper-bounded as

$$I \le 2\Gamma(1+a) - I_2, \; x > 0. \tag{33}$$

Using arguments similar to those for the upper-bound derivation for I_1 above, it can readily be shown that I_2 satisfies the inequality

$$I_2 \le \Gamma(1+a), \; x > 0. \tag{34}$$

Now using (34) in (33) we obtain the relation

$$I \le \Gamma(1+a), \; x > 0 \tag{35}$$

which when used in (29) leads to the inequality

$$\Gamma(-a,x) \le \frac{e^{-x}x^{-a}}{x+a} 2^{1-a}\Gamma(1+a), \; x > 0 \tag{36}$$

which is the desired result.

Appendix B: Proof of (15)

Using [16, Eq. (14)] and setting $k = 1$ for the first moment and $N = 2$ for the 2-hop case, the upper bound for the average SNR of the relayed link can be written as

$$E[S_2^{\text{UB-KG}}] = Z_2 \left[\left(\frac{\bar{\gamma}_1}{m_1} \right) \frac{\Gamma(m_1+1)}{\Gamma(m_1)} \right] \times \left[\left(\frac{\bar{\gamma}_2}{m_2} \right)^{1/2} \frac{\Gamma(m_2+1/2)}{\Gamma(m_2)} \right] \quad (37)$$

where $\bar{\gamma}_1 = \rho\sigma_{sr}^2$ and $\bar{\gamma}_2 = \alpha\rho\sigma_{rd}^2$ represent the average receive SNR for the source-relay and relay-destination links, respectively, and $\Gamma(a)$ is the Gamma function [14, Eq. (8.310)]. Furthermore, Z_2 is a constant dependent on the type of fixed gain used at the relay and is written as [16, Eq. (11)],

$$Z_2 = \frac{1}{2} \prod_{i=1}^{2} K_i^{-\frac{2-i}{2}} = \frac{1}{2\sqrt{K_1}}. \quad (38)$$

For the choice of relay gain given in (3), the constant K_1 in (38) is given by

$$K_1 = \frac{E_r}{A^2 N_o} = 1 + \bar{\gamma}_1. \quad (39)$$

Substituting (38), (39), and $\bar{\gamma}_1, \bar{\gamma}_2$ into (37) and using (9), we obtain (15) after simplification.

References

[1] LANEMAN, J. N., TSE, D. N. C., WORNELL, G. W. Cooperative diversity in wireless networks: Efficient protocols and outage behaviour. *IEEE Transactions on Information Theory*, 2004, vol. 50, no. 12, p. 3062 - 3080.

[2] PABST, R., WALKE, B. H. et al. Relay-based deployment concepts for wireless and mobile broadband radio. *IEEE Communications Magazine*, 2004, vol. 42, no. 9, p. 80 - 89.

[3] NABAR, R. U., BÖLCSKEI, H., KNEUBÜHLER, F. W. Fading relay channels: performance limits and space-time signal design. *IEEE Journal on Selected Areas in Communications*, 2004, vol. 22, no. 6, p. 1099 - 1109.

[4] LIU, K. J. R., SADEK, A. K., SU, W., KWASINSKI, A. *Cooperative Communications and Networking*. Cambridge (UK): Cambridge University Press, 2009.

[5] HASNA, M. O., ALOUINI, M.-S. A performance study of dual-hop transmissions with fixed gain relays. *IEEE Transactions on Wireless Communications*, 2004, vol. 3, no. 6, p. 1963 - 1968.

[6] FAN, Z., GUO, D., ZHANG, B. Outage probability and power allocation for two-way DF relay networks with relay selection. *Radioengineering*, 2012, vol. 21, no. 3, p. 795 - 801.

[7] POLAK, L., KRATOCHVIL, T. DVB-T and DVB-T2 performance in fixed terrestrial TV channels. In *Proceedings of International Conference on Telecommunications and Signal Processing (TSP)*. Prague (Czech Republic), 2012, p. 725 - 729.

[8] TANEDA, M. A., TAKADA, J., ARAKI, K. The problem of the fading model selection. *IEICE Transactions on Communications*, 2001, vol. E84-B, no. 3, p. 355 - 358.

[9] KIM, D., INGRAM, M. A., SMITH, W. W. Jr. Measurements of small-scale fading and path loss for long range RF tags. *IEEE Transactions on Antennas and Propagation*, 2003, vol. 51, no. 8, p. 1740 - 1749.

[10] FROLIK, J. On appropriate models for hyper-Rayleigh fading. *IEEE Transactions on Wireless Communications*, 2008, vol. 7, no. 12, p. 5202 - 5207.

[11] SEN, I., MATOLAK, D. W. Vehicle-vehicle channel models for the 5-GHz band. *IEEE Transactions on Intelligent Transportation Systems*, 2008, vol. 9, no. 2, p. 235 - 245.

[12] TABATABA, F. S., SADEGHI, P., PAKRAVAN, M. R. Outage probability and power allocation of amplify and forward relaying with channel estimation errors. *IEEE Transactions on Wireless Communications*, 2011, vol. 10, no. 1, p. 124 - 134.

[13] ZHANG, Y., MA, Y., TAFAZOLLI, R. Power allocation for bidirectional AF relaying over Rayleigh fading channels. *IEEE Communications Letters*, 2010, vol. 14, no. 2, p. 145 - 147.

[14] GRADSHTEYN, I. S., RYZHIK, I. M. *Table of Integrals, Series and Products*. 7th ed. Burlington (MA, USA): Academic Press, 2007.

[15] HASNA, M. O., ALOUINI, M.-S. Outage probability of multihop transmision over Nakagami fading channels. *IEEE Communications Letters*, 2003, vol. 7, no. 5, p. 216 - 218.

[16] KARAGIANNIDIS, G. K., ZOGAS, D. A., SAGIAS, N. C., TSIFTSIS, T. A., MATHIOPOULOS, P. T. Multihop communications with fixed-gain relays over generalized fading channels. In *Proceedings of IEEE Global Telecommunications Conference (GLOBECOM'04)*. Dallas (TX, USA), 2004, p. 36 - 40.

[17] KARAGIANNIDIS, G. K., TSIFTSIS, T. A., MALLIK, R. K. Bounds for multihop relayed communications in Nakagami-*m* fading. *IEEE Transactions on Communications*, 2006, vol. 54, no. 1, p. 18 - 22.

[18] ZHAI, C., LIU, J., ZHENG, L., CHEN, H. New power allocation schemes for AF cooperative communication over Nakagami-*m* fading channels. In *Proceedings of International Conference on Wireless Communications and Signal Processing (WCSP'09)*. Nanjing (China), 2009, p. 1 - 5.

[19] DENG, X., HAIMOVICH, A. M. Power allocation for cooperative relaying in wireless networks. *IEEE Communications Letters*, 2005, vol. 9, no. 11, p. 994 - 996.

[20] MOLISCH, A. F. *Wireless Communications*. 2nd ed. Chichester (UK): Wiley-IEEE Press, 2011.

[21] SIMON, M. K., ALOUINI, M.-S. *Digital Communications over Fading Channels*. 2nd ed. New York (USA): Wiley, 2005.

[22] GOLDSMITH, A. *Wireless Communications*. Cambridge (UK): Cambridge University Press, 2005.

[23] CAO, L., BEAULIEU, N. C. Simple efficient methods for generating independent and bivariate Nakagami-*m* fading envelope samples. *IEEE Transactions on Vehicular Technology*, 2007, vol. 56, no. 4, p. 1573-1579.

[24] XIA, M., XING, C., WU, Y.-C., AISSA, S. Exact performance analysis of dual-hop semi-blind AF relaying over arbitrary Nakagami-*m* fading channels. *IEEE Transactions on Wireless Communications*, 2011, vol. 10, no. 10, p. 3449 - 3459.

[25] LAFORGIA, A., NATALINI, P. On some inequalities for the Gamma function. *Advances in Dynamical Systems and Applications*, 2013, vol. 8, no. 2, p. 261 - 267.

About Authors ...

Sajid H. ALVI received his MSc and MPhil degrees in Electronics from Quaid-I-Azam University Islamabad, Pakistan in 2001 and 2006, respectively. Between 2001 and 2004, he was a faculty member at College of Electrical and Mechanical engineering, NUST, Islamabad, Pakistan. Since 2006 he is serving as a faculty member at COMSATS Institute of Information Technology (CIIT), Islamabad, Pakistan, where he is currently an assistant professor at the Department of Physics and pursuing his PhD at the Electrical Engineering Department of the same institute. Sajid Alvi's research interests are in cooperative communications and signal processing.

Shurjeel WYNE received his PhD from Lund University, Sweden in 2009. Between 2009 and 2010, he was a postdoctoral research fellow funded by the High-Speed Wireless Center at Lund University. Since 2010 he holds an Assistant Professorship at the Department of Electrical Engineering at CIIT, Islamabad, Pakistan. Dr. Wyne is a co-recipient of the best paper award of the Antennas and Propagation Track in the IEEE 77th Vehicular Technology Conference (VTC2013-Spring). His research interests are in wireless channel measurements and modeling, 60 GHz Communications, cooperative relay networks, and multi-antenna systems.

22

Performance Analysis of a Dual-Hop Cooperative Relay Network with Co-Channel Interference

Min LIN [1, 2], *Heng WEI* [3], *Jian OUYANG* [4], *Kang AN* [3], *Can YUAN* [3]

[1] PLA University of Science and Technology, Nanjing, China
[2] National Mobile Communications Research Laboratory, Southeast University, Nanjing, China
[3] College of Communication Engineering, PLA University of Science and Technology, Nanjing, China
[4] Inst. of Signal Processing and Transmission, Nanjing University of Posts and Telecommunications, Nanjing, China

linmin63@163.com, weiheng63@163.com, ouyangjian@njupt.edu.cn, ankang@nuaa.edu.cn, yuancanchina@gmail.com

Abstract. *This paper analyzes the performance of a dual-hop amplify-and-forward (AF) cooperative relay network in the presence of direct link between the source and destination and multiple co-channel interferences (CCIs) at the relay. Specifically, we derive the new analytical expressions for the moment generating function (MGF) of the output signal-to-interference-plus-noise ratio (SINR) and the average symbol error rate (ASER) of the relay network. Computer simulations are given to confirm the validity of the analytical results and show the effects of direct link and interference on the considered AF relay network.*

Keywords

Amplify-and-forward, cooperative relay network, co-channel interference, performance analysis.

1. Introduction

Recently, the application of relay technology has attracted many researchers as a low-cost strategy to increase the capacity, reliability and coverage of wireless networks without requiring extra power or bandwidth. (See [1], [2] and the citations therein for example.) Among the commonly used relay protocols, such as amplify-and-forward (AF), decode-and-forward (DF) and compress-and-forward (CF), AF scheme is of particular interest due to its simplicity. When only one relay is exploited to amplify the signal, the configuration is called as dual-hop relay network.

When the channel state information (CSI) is available, the performance of dual-hop AF relay networks has been well studied in a large body of open literatures [3]-[5]. In practice, however, co-channel interference (CCI) due to frequency reuse often severely degrades the performance of wireless systems. Motivated by this fact, the authors of [6] have investigated the effect of CCI on a variable gain AF relay network over Rayleigh fading channels. Moreover, the authors in [7] have analyzed the ergodic capacity of AF relay system with CCI present at both relay and destination over Nakagami-*m* fading channels. By considering the joint effects of CCI and imperfect CSI, the performance of a dual hop variable gain AF relaying with BF has been studied in [8], where the outage probability (OP), average symbol error rate (ASER) and asymptotic analysis at high SNR have been presented.

An extension of the previous works to the multi-antenna AF relaying has been presented in [9]-[11], respectively. In [9], the authors have analyzed the performance of multi-antenna AF relay systems with feedback delay at the source and CCI at the relay. By assuming that the statistical CSI is available at the relay, the analytical expressions for the OP, Ergodic capacity, ASER of a multi-antenna AF relay network have been obtained in [10]. Moreover, both the closed-form and asymptotic expressions of OP for multi-antenna AF relaying with have derived in [11], where the more general case with both the relay and destination corrupted by multiple co-channel interferers has been considered. However, in the aforementioned studies, the direct link between the source and destination is assumed to be unavailable due to heavy shading, which is not an actual model in many practical environments.

In this paper, unlike the previous related works, we conduct the performance of a dual-hop AF cooperative relay network, where direct link exists between the source and destination, and multiple CCIs corrupt the received signals at the relay. In particular, we derive the new theoretical expressions for the moment generating function (MGF) of the output signal-to-interference-plus-noise ratio (SINR) as well as the average symbol error rate (ASER) of the relay system. To the best of our knowledge, this is the first time such analytical results have been presented.

2. System Model

As shown in Fig. 1, we consider a dual-hop AF relay network, where a source S communicates with a destination D through a relay R. We assume that each node is equipped with a single antenna, and there exists a direct link between S and D. Here, all the wireless channels are subject to Rayleigh fading. The complete communication takes place in two time slots. In the first time slot, S broadcasts the signal to D and R through the fading channels h_0 and h_1, respectively. Meanwhile, R is also impaired

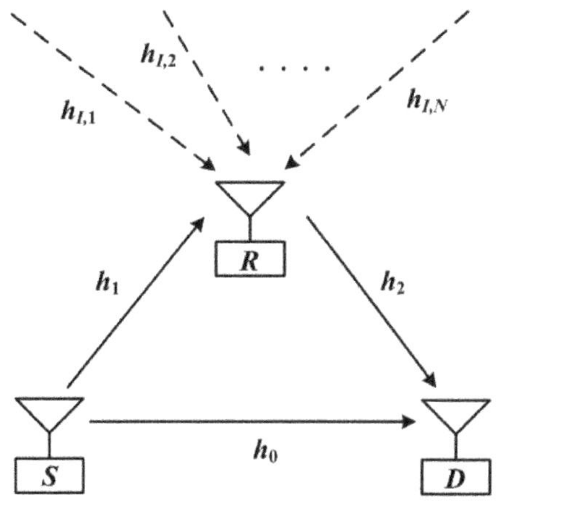

Fig. 1. System model of a dual-hop cooperative relay network with CCI.

by N CCIs through fading channel $h_{I,n}$ ($n = 1,2,\ldots,N$). As such, the received signals at D and R can be, respectively, expressed as

$$y_0(t) = \sqrt{P_S}\, h_0 x_S(t) + n_0(t), \tag{1}$$

$$y_1(t) = \sqrt{P_S}\, h_1 x_S(t) + \sum_{n=1}^{N} \sqrt{P_{I,n}}\, h_{I,n} x_{I,n}(t) + n_1(t) \tag{2}$$

where P_S is the transmit power at S, $P_{I,n}$ is the transmit power of nth-interference, $x_S(t)$ and $x_{I,n}(t)$ are the transmit signal from S and that from nth-interference, satisfying $\mathrm{E}[|x_S(t)|^2] = 1$ and $\mathrm{E}[|x_{I,n}(t)|^2] = 1$, $n_0(t)$ and $n_1(t)$ are the zero mean additive white Gaussian noises (AWGNs) at D and R, obeying $n_0(t) \sim \mathbb{N}_c(0,\sigma_0^2)$ and $n_1(t) \sim \mathbb{N}_c(0,\sigma_1^2)$, respectively, and E[.] denotes the expectation operator. In the second time slot, the received signal at R is amplified and forwarded to D through the fading channel h_2. Accordingly, the signal received at D can be written as

$$\begin{aligned} y_2(t) &= \sqrt{P_R}\, G_R h_2 y_1(t) + n_2(t) \\ &= \sqrt{P_S P_R}\, G_R h_1 h_2 x_S(t) + \sum_{n=1}^{N} \sqrt{P_{I,n} P_R}\, G_R h_{I,n} h_2 x_{I,n}(t) \\ &\quad + \sqrt{P_R}\, G_R h_2 n_1 + n_2(t) \end{aligned} \tag{3}$$

where P_R denotes the transmit power at R, $n_2(t)$ the AWGN at D with zero mean and variance σ_2^2. Furthermore, G_R is the amplify gain at R, given by [9]

$$G_R = \sqrt{\frac{1}{P_S |h_1|^2 + \sum_{n=1}^{N} P_{I,n} |h_{I,n}|^2 + \sigma_1^2}}\ . \tag{4}$$

Following the application of maximal ratio combining (MRC) at D, it is not difficult to find that the output SINR can be expressed as

$$\begin{aligned} \gamma &= \frac{P_S |h_0|^2}{\sigma_0^2} + \frac{P_S P_R G_R^2 |h_1|^2 |h_2|^2}{\sum_{n=1}^{N} P_{I,n} P_R G_R^2 |h_{I,n}|^2 |h_2|^2 + P_R G_R^2 |h_2|^2 \sigma_1^2 + \sigma_2^2} \\ &= \frac{P_S |h_0|^2}{\sigma_0^2} + \frac{\frac{P_S |h_1|^2}{\sigma_1^2} \frac{P_R |h_2|^2}{\sigma_2^2}}{\frac{P_S |h_1|^2}{\sigma_1^2} + \left(\frac{P_R |h_2|^2}{\sigma_2^2} + 1\right)\left(\sum_{n=1}^{N} \frac{P_{I,n} |h_{I,n}|^2}{\sigma_1^2} + 1\right)} \\ &= \gamma_0 + \frac{\gamma_1 \gamma_2}{\gamma_1 + (\gamma_2 + 1)(\gamma_3 + 1)} \\ &\cong \gamma_0 + \gamma_R \end{aligned} \tag{5}$$

where $\gamma_0 = |h_0|^2 \bar{\gamma}_0$, $\gamma_1 = |h_1|^2 \bar{\gamma}_1$, $\gamma_2 = |h_2|^2 \bar{\gamma}_2$ and $\gamma_3 = \sum_{n=1}^{N} |h_{I,n}|^2 \bar{\gamma}_{3,n}$ with $\bar{\gamma}_0 = P_S/\sigma_0^2$, $\bar{\gamma}_1 = P_S/\sigma_1^2$, $\bar{\gamma}_2 = P_R/\sigma_2^2$ and $\bar{\gamma}_{3,n} = P_{I,n}/\sigma_1^2$ being the average signal-to-noise ratios (SNRs). In (5), γ_0 and γ_R represent the output signal-to-noise ratio (SNR) of direct S-D link and the output SINR of the S-R-D relay link at D, respectively. Note that a similar expression has also been derived in [6, equation 5] for the case of ignoring direct link between S and D and noise at the relay, indicating that our work extends the study of previous works to a more general case.

3. Moment Generating Function

MGF is an important approach to evaluate the performance of wireless systems, which is defined as $M_r(s) = \mathrm{E}\left[e^{-s\gamma}\right]$. Clearly, the MGF of γ in (5) can be obtained as

$$\begin{aligned} M_r(s) &= \mathrm{E}\left[e^{-s(\gamma_0 + \gamma_R)}\right] = \mathrm{E}\left[e^{-s\gamma_0}\right]\mathrm{E}\left[e^{-s\gamma_R}\right] \\ &= M_{\gamma_0}(s) M_{\gamma_R}(s) \end{aligned} \tag{6}$$

where $M_{\gamma_0}(s)$ and $M_{\gamma_R}(s)$ are the MGF of γ_0 and γ_R, respectively. Since the *S-D* link undergoes Rayleigh fading, $|h_0|^2$ has a Chi-square distribution with 2 degrees of freedom and variance 1/2, the probability density function (PDF) is given by

$$f_{\gamma_0}(x) = \frac{1}{\bar{\gamma}_0} \exp\left(-\frac{x}{\bar{\gamma}_0}\right). \tag{7}$$

Thus, the MGF of γ_0 can be obtained as [4]

$$M_{\gamma_0}(s) = \mathrm{E}\left[e^{-s\gamma_0}\right] = \int_0^{\infty} e^{-sx} f_{\gamma_0}(x)\, dx = \frac{1}{1 + s\bar{\gamma}_0}\ . \tag{8}$$

To obtain the analytical MGF $M_{\gamma_R}(s)$ of γ_R, we first express the upper bound of γ_R as

$$\tilde{\gamma}_R = \min\left(\frac{\gamma_1}{\gamma_3 + 1}, \gamma_2\right) \cong \min(\gamma_{13}, \gamma_2) \tag{9}$$

for mathematical tractability. Thus, the MGF of γ_R can be approximated as

$$M_{\gamma_R}(s) \approx M_{\tilde{\gamma}_R}(s) = \mathrm{E}\left[e^{-s\tilde{\gamma}_R}\right] = \int_0^\infty e^{-sx} f_{\tilde{\gamma}_R}(x)dx = \int_0^\infty se^{-sx} F_{\tilde{\gamma}_R}(x)dx \quad (10)$$

where $F_{\tilde{\gamma}_R}(x)$ is the cumulative distribution function (CDF) of $\tilde{\gamma}_R$, given by

$$F_{\tilde{\gamma}_R}(x) = 1 - \left[1 - F_{\gamma_{13}}(x)\right]\left[1 - F_{\gamma_2}(x)\right] \quad (11)$$

with $F_{\gamma_{13}}(x)$ and $F_{\gamma_2}(x)$ being the CDF of γ_{13} and γ_2. Next, according to [12], the PDF of $\gamma_3 = \sum_{n=1}^{N} |h_{l,n}|^2 \overline{\gamma}_{3,n}$ can be written as

$$f_{\gamma_3}(x) = \sum_{i=1}^{t}\sum_{j=1}^{v_i} \frac{C_{i,j}}{(j-1)!(\overline{\gamma}_{3,i})^j} x^{j-1} \exp\left(-\frac{x}{\overline{\gamma}_{3,i}}\right) \quad (12)$$

with the coefficient $C_{i,j}$ given by

$$C_{i,j} = \frac{1}{(v_i - j)!\overline{\gamma}_{3,i}^{v_i - j}} \frac{\partial^{v_i - j}}{\partial s^{v_i - j}} \left[\prod_{k=1,k\neq i}^{t_n} \left(\frac{1}{1 + s\overline{\gamma}_{3,k}}\right)^{v_k}\right]\Bigg|_{s=-\overline{\gamma}_{3,i}^{-1}} \quad (13)$$

where v_i denotes the repeated times of $\overline{\gamma}_{3,i}$ satisfying $\sum_{i=1}^{t} v_i = N$. Consider the *S-R* link experiences Rayleigh fading, the CDF of γ_1 can be written as

$$F_{\gamma_1}(x) = 1 - \exp\left(-\frac{x}{\overline{\gamma}_1}\right). \quad (14)$$

Therefore, $F_{\gamma_{13}}(x)$ can be calculated as

$$\begin{aligned} F_{\gamma_{13}}(x) &= \Pr\left(\frac{\gamma_1}{\gamma_3 + 1} \leq x\right) = \int_0^\infty \int_0^{x(1+z)} f_{\gamma_1}(y) f_{\gamma_3}(z) dy dz \\ &= \int_0^\infty F_{\gamma_1}(x(1+z)) f_{\gamma_3}(z) dz \\ &= 1 - \exp\left(-\frac{x}{\overline{\gamma}_1}\right) \sum_{i=1}^{t}\sum_{j=1}^{v_i} \frac{C_{i,j}}{(j-1)!} \left(\frac{1}{\overline{\gamma}_{3,i}^j}\right)^j \\ &\quad \times \int_0^\infty z^{j-1} \exp\left(-\left(\frac{z}{\overline{\gamma}_1} + \frac{1}{\overline{\gamma}_{3,i}}\right)\right) dz. \end{aligned} \quad (15)$$

With the help of identity [13, equation 3.351.3], we have

$$F_{\gamma_{13}}(x) = 1 - \exp\left(-\frac{x}{\overline{\gamma}_1}\right) \sum_{i=1}^{t}\sum_{j=1}^{v_i} C_{i,j} \left(1 + \frac{\overline{\gamma}_{3,i}}{\overline{\gamma}_1}\right)^{-j}. \quad (16)$$

Meanwhile, the *R-D* link is subject to Rayleigh fading, one can obtain

$$F_{\gamma_2}(x) = 1 - \exp\left(-\frac{x}{\overline{\gamma}_2}\right). \quad (17)$$

Substituting (16) and (17) into (11), one can obtain the closed-form expression of $F_{\tilde{\gamma}_R}(x)$ as

$$F_{\tilde{\gamma}_R}(x) = 1 - \exp\left(-\left(\frac{1}{\overline{\gamma}_1} + \frac{1}{\overline{\gamma}_2}\right)x\right) \sum_{i=1}^{t}\sum_{j=1}^{v} C_{i,j} \left(1 + \frac{\overline{\gamma}_{3,i}}{\overline{\gamma}_1}\right)^{-j} \quad (18)$$

Finally, by plugging (18) into (10), after some algebraic manipulations, the MGF of γ_R can be approximately calculated as

$$\begin{aligned} M_{\gamma_R}(s) &\approx 1 - s\sum_{i=1}^{t}\sum_{j=1}^{v} C_{i,j} \int_0^\infty \left(1 + \frac{\overline{\gamma}_{3,i}}{\overline{\gamma}_1}\right)^{-j} \exp\left(-\left(s + \frac{1}{\overline{\gamma}_1} + \frac{1}{\overline{\gamma}_2}\right)x\right) dx \\ &= 1 - s\sum_{i=1}^{t}\sum_{j=1}^{v_i} C_{i,j} \frac{\overline{\gamma}_1}{\overline{\gamma}_{3,i}} U\left(1, 2-j; \left(s + \frac{1}{\overline{\gamma}_1} + \frac{1}{\overline{\gamma}_2}\right)\frac{\overline{\gamma}_1}{\overline{\gamma}_{3,i}}\right) \end{aligned} \quad (19)$$

where $U(a, b; x)$ denotes the confluent hypergeometric function [14]. In deriving (19), we have applied [14, equation 2.3.6.9]. Furthermore, using (8) and (19) into (6), one can readily obtain the approximate expression of $M_r(s)$, giving

$$\begin{aligned} M_r(s) \approx \frac{1}{1 + s\overline{\gamma}_0} - \frac{s}{1 + s\overline{\gamma}_0} \sum_{i=1}^{t}\sum_{j=1}^{v_i} C_{i,j} \frac{\overline{\gamma}_1}{\overline{\gamma}_{3,i}} \\ \times U\left(1, 2-j; \left(s + \frac{1}{\overline{\gamma}_1} + \frac{1}{\overline{\gamma}_2}\right)\frac{\overline{\gamma}_1}{\overline{\gamma}_{3,i}}\right). \end{aligned} \quad (20)$$

4. Average Symbol Error Rate

The ASER of wireless communication systems with various modulation formants over fading channels can be expressed as [15]

$$P_s = \int_0^\theta aM_\gamma\left(\frac{b}{\sin^2\phi}\right)d\phi \quad (21)$$

where a, b and θ are the modulation specific constants. However, to the best of our knowledge, due to the integral, the closed-form solution is mathematically unavailable by directly employing the above formula. To overcome this, we provide an approximate yet accurate alternative method to evaluate the ASER.

For the case of *M*-ary phase-shift keying (*M*-PSK) modulation signal, we have

$$\begin{aligned} P_{M\text{-PSK}} &= \frac{1}{\pi}\int_0^{(M-1)\pi/M} M_\gamma\left(\frac{\sin^2(\pi/M)}{\sin^2\phi}\right)d\phi \\ &= \underbrace{\frac{1}{\pi}\int_0^{\pi/2} M_\gamma\left(\frac{\sin^2(\pi/M)}{\sin^2\phi}\right)d\phi}_{I_1} + \underbrace{\frac{1}{\pi}\int_{\pi/2}^{(M-1)\pi/M} M_\gamma\left(\frac{\sin^2(\pi/M)}{\sin^2\phi}\right)d\phi}_{I_2} \end{aligned} \quad (22)$$

By using the definition of MGF, L_1 can be calculated as

$$\begin{aligned} L_1 &= \frac{1}{\pi}\int_0^{\pi/2}\int_0^{\infty}\exp\left(-\frac{x\sin^2(\pi/M)}{\sin^2\phi}\right)f_\gamma(x)\,dx\,d\phi \\ &\approx \int_0^{\infty}\left(\frac{1}{12}\exp\left(-x\sin^2\left(\frac{\pi}{M}\right)\right)\right. \\ &\quad \left.+\frac{1}{4}\exp\left(-\frac{4x\sin^2(\pi/M)}{3}\right)\right)f_\gamma(x)\,dx \\ &= \frac{1}{12}M_\gamma\left(\sin^2\left(\frac{\pi}{M}\right)\right)+\frac{1}{4}M_\gamma\left(\frac{4}{3}\sin^2\left(\frac{\pi}{M}\right)\right). \quad (23) \end{aligned}$$

In deriving (23), we have applied the following approximation [16]

$$\frac{2}{\pi}\int_0^{\frac{\pi}{2}}\exp\left(-\frac{x}{\sin^2\theta}\right)d\theta \approx \frac{1}{6}\exp(-x)+\frac{1}{2}\exp\left(-\frac{4}{3}x\right). \quad (24)$$

Similarly, with the help of identity [16]

$$\begin{aligned} \frac{1}{\pi}\int_{\frac{\pi}{2}}^{\frac{(M-1)}{M}\pi}\exp\left(-\frac{x}{\sin^2\theta}\right)d\theta &\approx \left(\frac{(M-1)}{2M}-\frac{1}{4}\right) \\ &\times\left(\exp(-x)+\exp\left(-\frac{x}{\sin^2((M-1)\pi/M)}\right)\right) \quad (25) \end{aligned}$$

L_2 can be approximately obtained as

$$\begin{aligned} L_2 &= \frac{1}{\pi}\int_{\pi/2}^{(M-1)\pi/M}\int_0^{\infty}\exp\left(-\frac{x\sin^2(\pi/M)}{\sin^2\phi}\right)f_\gamma(x)\,dx\,d\phi \\ &\approx \int_0^{\infty}\left(\left(\frac{M-1}{2M}-\frac{1}{4}\right)\left(\exp\left(-x\sin^2\left(\frac{\pi}{M}\right)\right)\right.\right. \\ &\quad \left.\left.+\exp\left(-\frac{x\sin^2(\pi/M)}{\sin^2((M-1)\pi/M)}\right)\right)\right)f_\gamma(x)\,dx \\ &= \left(\frac{M-1}{2M}-\frac{1}{4}\right)M_\gamma\left(\sin^2\left(\frac{\pi}{M}\right)\right) \\ &\quad +\left(\frac{M-1}{2M}-\frac{1}{4}\right)M_\gamma\left(\frac{\sin^2(\pi/M)}{\sin^2((M-1)\pi/M)}\right). \quad (26) \end{aligned}$$

Furthermore, the approximate ASER expression for M-PSK modulation signal can be written as

$$\begin{aligned} P_{M\text{-PSK}} &\approx \left(\frac{M-1}{2M}-\frac{1}{4}\right)M_\gamma\left(\frac{\sin^2(\pi/M)}{\sin^2((M-1)\pi/M)}\right) \\ &\quad +\left(\frac{M-1}{2M}-\frac{1}{6}\right)M_\gamma\left(\sin^2\left(\frac{\pi}{M}\right)\right)+\frac{1}{4}M_\gamma\left(\frac{4}{3}\sin^2\left(\frac{\pi}{M}\right)\right). \quad (27) \end{aligned}$$

In a similar manner, the ASERs of M-ary Pulse Amplitude Modulation (M-PAM) signals can be calculated as

$$\begin{aligned} P_{M\text{-PAM}} &= 2\left(\frac{M-1}{\pi M}\right)\int_0^{\frac{\pi}{2}}M_\gamma\left(\frac{3}{(M^2-1)\sin^2\phi}\right)d\phi \\ &= 2\left(\frac{M-1}{\pi M}\right)\int_0^{\frac{\pi}{2}}\int_0^{\infty}\exp\left(-\frac{3x}{(M^2-1)\sin^2\phi}\right)f_\gamma(x)\,dx\,d\phi \\ &\approx \left(\frac{M-1}{\pi M}\right)\int_0^{\infty}\left(\frac{1}{3}\exp\left(-\frac{3x}{M^2-1}\right)+\exp\left(-\frac{4x}{3(M^2-1)}\right)\right)f_\gamma(x)\,dx \\ &= \frac{M-1}{M}\left(\frac{1}{6}M_\gamma\left(\frac{3}{M^2-1}\right)+\frac{1}{2}M_\gamma\left(\frac{4}{M^2-1}\right)\right). \quad (28) \end{aligned}$$

As for M-ary Quadrature Amplitude Modulation (M-QAM) signals, it can be considered as two independent $\sqrt{M}$ -PAM signals with its ASER being expressed as

$$\begin{aligned} P_{M\text{-QAM}} &= \underbrace{\frac{4}{\pi}\left(1-\frac{1}{\sqrt{M}}\right)\int_0^{\pi/2}M_\gamma\left(\frac{3}{(M-1)\sin^2\phi}\right)d\phi}_{L_3} \\ &\quad \underbrace{-\frac{4}{\pi}\left(1-\frac{1}{\sqrt{M}}\right)^2\int_0^{\pi/4}M_\gamma\left(\frac{3}{(M-1)\sin^2\phi}\right)d\phi}_{L_4}. \quad (29) \end{aligned}$$

Following a similar manner to derive (23), L_3 can be obtained as

$$L_3 \approx \left(1-\frac{1}{\sqrt{M}}\right)\left(\frac{1}{3}M_\gamma\left(\frac{3}{M-1}\right)+M_\gamma\left(\frac{4}{M-1}\right)\right). \quad (30)$$

In addition, by applying an approximate expression [3], we have

$$L_4 \approx \frac{1}{6}\left(1-\frac{1}{\sqrt{M}}\right)^2\left(3M_\gamma\left(\frac{6}{M-1}\right)+M_\gamma\left(\frac{3}{M-1}\right)\right). \quad (31)$$

Therefore, an approximate expression of $P_{M\text{-QAM}}(\gamma)$ can be calculated as

$$\begin{aligned} P_{M\text{-QAM}}(\gamma) &\approx \left(1-\frac{1}{\sqrt{M}}\right)\left(\frac{1}{3}M_\gamma\left(\frac{3}{M-1}\right)+M_\gamma\left(\frac{4}{M-1}\right)\right) \\ &\quad +\frac{1}{6}\left(1-\frac{1}{\sqrt{M}}\right)^2\left(3M_\gamma\left(\frac{6}{M-1}\right)+M_\gamma\left(\frac{3}{M-1}\right)\right). \quad (32) \end{aligned}$$

5. Numerical Results

In this section, we confirm the validity of the derived analytical results through comparison with Monte Carlo simulations, and investigate the impacts of the direct link and CCIs on the performance of the considered cooperative relay network. Here we assume that R is corrupted by 3 CCIs with the SNRs denoted as $\bar{\gamma}_{I,1}$, $\bar{\gamma}_{I,2}=0.6\bar{\gamma}_{I,1}$ and $\bar{\gamma}_{I,3}=0.8\bar{\gamma}_{I,1}$. In all the plots, we define $\bar{\gamma}_1=\bar{\gamma}_2=\bar{\gamma}$,

$\eta = \overline{\gamma}/\overline{\gamma}_0$ and $\rho = \overline{\gamma}/\overline{\gamma}_{I,1}$, where the cases of no direct link and no interferences are denoted as $\eta = \infty$ dB and $\rho = \infty$ dB.

Fig. 2, Fig. 3 and Fig. 4 show, respectively, the ASER versus $\overline{\gamma}$ in terms of QPSK, 8PAM and 16QAM modulation for different η. As we see, since the approximate method is used to calculate the MGF of the output SINR in (10), there exist a little deviation at low SNR, the good match between the analytical results and the Monte Carlo experiments in general can still be obtained. The scenarios confirm the validity of the derived analytical expressions. . In addition, it can also be found that the ASER performance is improved with the increase of direct *S-D* link strength. This is because the direct link can enhance the output SINR.

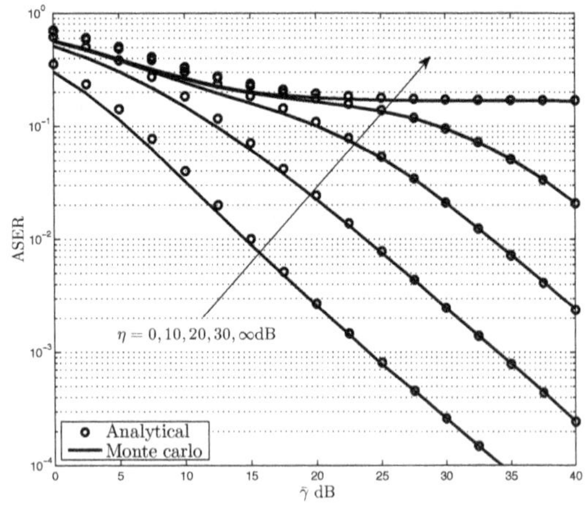

Fig. 2. ASER versus average SNR for QPSK with $N = 3$ and $\rho = 10$ dB.

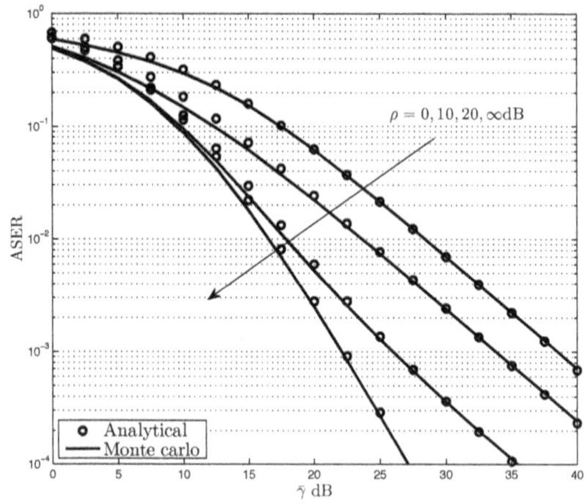

Fig. 5. ASER versus average SNR for QPSK with $N = 3$ and $\eta = 10$ dB.

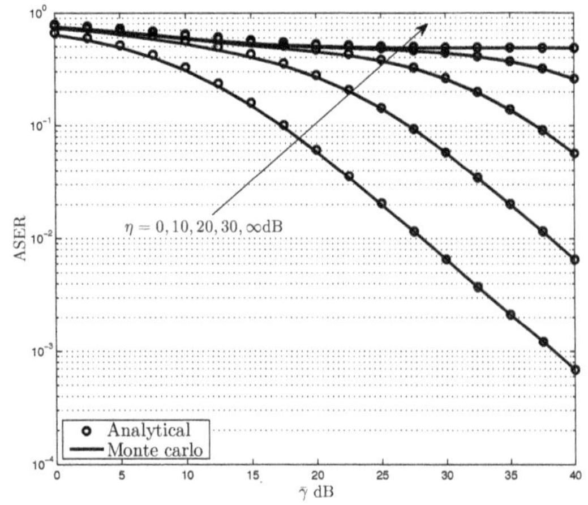

Fig. 3. ASER versus average SNR for 8PAM with $N = 3$ and $\rho = 10$ dB.

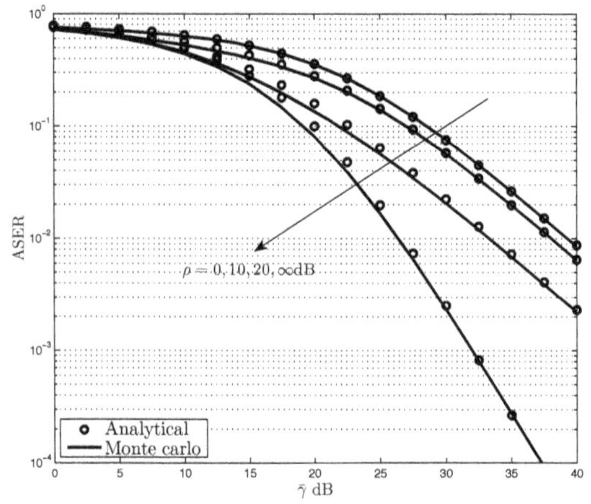

Fig. 6. ASER versus average SNR for 8PAM with $N = 3$ and $\eta = 10$ dB.

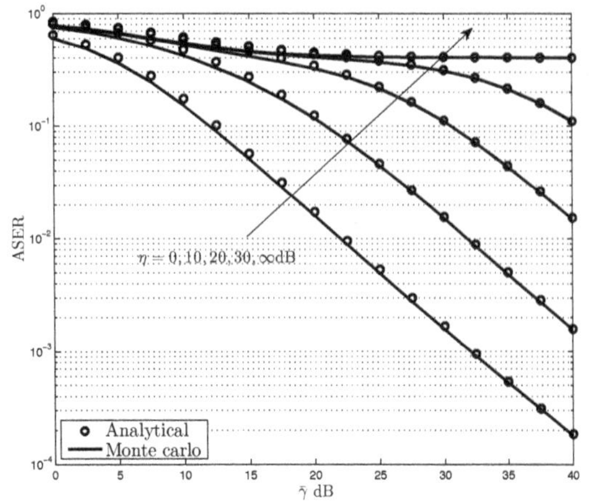

Fig. 4. ASER versus average SNR for 16QAM with $N = 3$ and $\rho = 10$ dB.

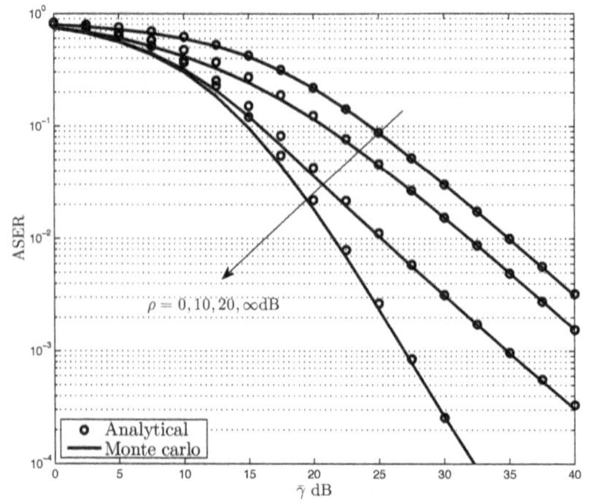

Fig. 7. ASER versus average SNR for 16QAM with $N = 3$ and $\eta = 10$ dB.

Fig. 5, Fig. 6 and Fig. 7 depict the ASERs of QPSK, 8PAM and 16QAM modulations against SNR for various ρ. As we expect, at middle and high SNR regimes, a well agreement between the analytical results and Monte Carlo experiments is observed. Furthermore, it can also be seen that the ASERs of the network decreases with the increase of ρ. This is because when ρ is increased, the CCIs are degraded, resulting in the increase of output instantaneous SNR at the destination.

6. Conclusions

In this paper, a dual-hop AF cooperative relay network has been investigated in the presence of *S-D* direct link and multiple CCIs at the relay. In particular, we have derived the new analytical expressions for the MGF of the output SINR as well as the ASER of the system. Furthermore, we have also provided numerical results to demonstrate the validity of the theoretical analysis, and examined the effects of direct link and interference on the considered AF relay network. It should be pointed out the extension of the presented method to the more complex networks, such as more than one nodes having multi-antenna and/or multiple relays are exploited to amplify the signals, is our further work.

Acknowledgements

This work is supported by the National Natural Science Foundation of China (61271255), the Natural Science Foundation of Jiangsu Province (BK20131068), and the Open Research Fund of National Mobile Communications Research Laboratory in Southeast University (2012D15).

References

[1] KAKITANI, M., BRANTE, G., SOUZA, R. Energy efficiency analysis of a two-dimensional cooperative wireless sensor network with relay selection. *Radioengineering*, 2013, vol. 22, no. 2, p. 549 to 557.

[2] FAN, Z. J., GUO, D. X., ZHANG, B. N. Outage probability and power allocation for two-way DF relay networks with relay selection. *Radioengineering*, 2012, vol. 21, no. 3, p. 795–801.

[3] LI, M., LIN, M., YU, Q., ZHU, W.-P., DONG, L. Optimal beamformer design for two hop MIMO AF relay networks over Rayleigh fading channels. *IEEE Journal on Selected Areas in Communications*, 2012, vol. 30, no. 8, p. 1402–1414.

[4] OUYANG, J., LIN, M., ZHUANG, Y. Performance analysis of cooperative relay networks over asymmetric fading channels. *Electronics Letters*, 2012, vol. 48, no. 21, p.1370–1371.

[5] LI, M., AN, K., OUYANG, J., WEI, H., et al. Effect of feedback delay on dual-hop fixed gain relay networks over mixed fading channels. *Transactions on Emerging Telecommunications Technologies*, 2014, vol. 25, no. 10, p. 1045–1055.

[6] SURAWEERA, H. A., HARI, K. G., NALLANATHAN, A. Performance analysis of two hop amplify-and-forward systems with interference at the relay. *IEEE Communications Letters*, 2010, vol. 14, no. 8, p. 692–694.

[7] TRIGUI, I., AFFES, S., STEPHENNE, A. On the ergodic capacity of amplify-and-forward relay channels with interference in Nakagami-m fading. *IEEE Transactions on Communications*, 2013, vol. 61, no. 8, p. 3136–3145.

[8] AN, K., LIN, M., OUYANG, J., WEI, H. Beamforming in dual-hop AF relaying with imperfect CSI and co-channel interference. *Wireless Personal Communication*, 2014, vol. 78, no. 2, p. 1187 to 1197.

[9] PHAN, H., DUONG, T. Q., ELKASHLAN, M., ZEPERNICK, H.-J. Beamforming amplify-and-forward relay networks with feedback delay and interference. *IEEE Signal Processing Letters*, 2012, vol. 19, no. 1, p. 16–19.

[10] AN, K., LIN, M., OUYANG, J., WEI, H. Performance analysis of beamforming in two-hop AF relay networks with antenna correlation and interference. *AEU-International Journal of Electronics and Communications*, 2014, vol. 68, no. 7, p. 587–594.

[11] DING, H., HE, C., JIANG, L.-G. Performance analysis of fixed gain MIMO relay systems in presence of co-channel interference. *IEEE Communications Letters*, 2012, vol. 16, no. 7, p. 1133–1136.

[12] CUI, X. W., FENG, Z. M. Lower capacity bound for MIMO correlated fading channels with keyhole. *IEEE Communications Letters*, 2004, vol. 8, p. 500–502.

[13] GRADSHTEYN, I. S., RYZHIK, I. M. *Table of Integrals, Series, and Products*. 7th ed. Academic Press, 2007.

[14] PRUDNIKOV, A. P., BRYCHKOV, Y. A., MARICHEV, O. I. *Integrals and Series*. Vol. 1, 1st ed. New York (USA): Gordon and Breach Science Publishers, 1986.

[15] SIMON, M. K., ALOUINI, M. S. *Digital Communication over Fading Channels*. 2nd ed. Wiley, 2005.

[16] MCKAY, M. R., ZANELLA, A., COLLINGS, I. B., et al. Error probability and SINR analysis of optimum combining in Rician fading. *IEEE Transactions on Communications*, 2009, vol. 57, no. 3, p. 676–687.

About Authors ...

Min LIN was born in Zhejiang Province, China, in 1972. He received the B.S. degree from the National University of Defense Technology, the M.S. degree from Nanjing Institute of Communication Engineering, and Ph.D. degree from the Southeast University in 1993, 2000, and 2008, respectively, all in Electrical Engineering. His current research interests include wireless communications, array signal processing and antenna technology. He is the author or coauthor of more than 70 papers, and holds 4 patents. Dr. Lin is a Member of IEEE and Senior Member of Chinese Institute of Electronics (CIE).

Heng WEI (corresponding author) was born in Jiangsu Province, China, in 1989. He received the B.S. degree from the College of Electronic and Information Engineering, Nanjing University of Aeronautics and Astronautics in 2012. He is currently working towards the M.S. degree in Electronic Engineering, from the PLA University of Science and Technology. His research interests are MIMO technique and wireless relay communications.

Jian OUYANG was born in Jiangsu Province, China, in 1983. He received the B.S., M.S. and Ph.D. degrees from the College of Computer Science and Technology, Nanjing University of Aeronautics and Astronautics in 2007, 2010, and 2014 respectively. Since July 2014, he has been with the Institute of Signal Processing and Transmission, Nanjing University of Posts and Telecommunications (NUPT). His current research interests include wireless cooperative networks.

Kang AN was born in Xinjiang Uygur autonomous region, China, in 1989. He received the B.S. degree from the College of Electronic and Information Engineering, Nanjing University of Aeronautics and Astronautics in 2011, and M.S. degree in Electronic Engineering, from the College of Communications Engineering, PLA University of Science and Technology in 2014. He is currently working towards the Ph.D. degree at the College of Communications Engineering, PLA University of Science and Technology. His research interests include hybrid satellite-terrestrial network, cooperative communications, cognitive radio and array signal processing.

Can YUAN was born in Anhui Province, China, in 1991. He received the B.S. degree from the College of Science, Nanjing University of Aeronautics and Astronautics in 2013. He is currently working towards the M.S. degree in Communication Engineering at the College of Communications Engineering, PLA University of Science and Technology. His research interests are array signal processing and wireless cooperative communications.

Permissions

All chapters in this book were first published in Radioengineering, by Spolecnost pro radioelektronicke inzenyrstvi; hereby published with permission under the Creative Commons Attribution License or equivalent. Every chapter published in this book has been scrutinized by our experts. Their significance has been extensively debated. The topics covered herein carry significant findings which will fuel the growth of the discipline. They may even be implemented as practical applications or may be referred to as a beginning point for another development.

The contributors of this book come from diverse backgrounds, making this book a truly international effort. This book will bring forth new frontiers with its revolutionizing research information and detailed analysis of the nascent developments around the world.

We would like to thank all the contributing authors for lending their expertise to make the book truly unique. They have played a crucial role in the development of this book. Without their invaluable contributions this book wouldn't have been possible. They have made vital efforts to compile up to date information on the varied aspects of this subject to make this book a valuable addition to the collection of many professionals and students.

This book was conceptualized with the vision of imparting up-to-date information and advanced data in this field. To ensure the same, a matchless editorial board was set up. Every individual on the board went through rigorous rounds of assessment to prove their worth. After which they invested a large part of their time researching and compiling the most relevant data for our readers.

The editorial board has been involved in producing this book since its inception. They have spent rigorous hours researching and exploring the diverse topics which have resulted in the successful publishing of this book. They have passed on their knowledge of decades through this book. To expedite this challenging task, the publisher supported the team at every step. A small team of assistant editors was also appointed to further simplify the editing procedure and attain best results for the readers.

Apart from the editorial board, the designing team has also invested a significant amount of their time in understanding the subject and creating the most relevant covers. They scrutinized every image to scout for the most suitable representation of the subject and create an appropriate cover for the book.

The publishing team has been an ardent support to the editorial, designing and production team. Their endless efforts to recruit the best for this project, has resulted in the accomplishment of this book. They are a veteran in the field of academics and their pool of knowledge is as vast as their experience in printing. Their expertise and guidance has proved useful at every step. Their uncompromising quality standards have made this book an exceptional effort. Their encouragement from time to time has been an inspiration for everyone.

The publisher and the editorial board hope that this book will prove to be a valuable piece of knowledge for researchers, students, practitioners and scholars across the globe.

List of Contributors

Zhiyuan ZHAO
Institute of Communications Engineering, PLA University of Science and Technology, Nanjing 210007, China
Nanjing Telecommunication Technology Institute, Nanjing 210007, China

Jiang CHEN
Nanjing Telecommunication Technology Institute, Nanjing 210007, China

Lin YANG
Nanjing Telecommunication Technology Institute, Nanjing 210007, China

Kunhe CHEN
Nanjing Telecommunication Technology Institute, Nanjing 210007, China

Yee Hui LEE
School of Electrical and Electronic Engineering, Nanyang Technological University, 50 Nanyang Avenue, Singapore 639798, Singapore

Feng DONG
School of Electrical and Electronic Engineering, Nanyang Technological University, 50 Nanyang Avenue, Singapore 639798, Singapore

Yu Song MENG
National Metrology Centre, Agency for Science, Technology and Research (A*STAR) 1 Science Park Drive, Singapore 118221, Singapore

Yajun LI
School of Electronics and Information Engineering, Harbin Institute of Technology, Harbin, 150001, China

Yinsheng WEI
School of Electronics and Information Engineering, Harbin Institute of Technology, Harbin, 150001, China

Rongqing XU
School of Electronics and Information Engineering, Harbin Institute of Technology, Harbin, 150001, China

Tianqi CHU
School of Electronics and Information Engineering, Harbin Institute of Technology, Harbin, 150001, China

Zhuoqun WANG
School of Electronics and Information Engineering, Harbin Institute of Technology, Harbin, 150001, China

Yongping SONG
College of Electronics Science and Engineering, National University of Defense Technology, Changsha, 410073, China

Tian JIN
College of Electronics Science and Engineering, National University of Defense Technology, Changsha, 410073, China

Biying LU
College of Electronics Science and Engineering, National University of Defense Technology, Changsha, 410073, China

Jun HU
College of Electronics Science and Engineering, National University of Defense Technology, Changsha, 410073, China

Zhimin ZHOU
College of Electronics Science and Engineering, National University of Defense Technology, Changsha, 410073, China

Peio LÓPEZ ITURRI
Dept. of Electrical and Electronic Engineering, Public Univ. of Navarre, Campus de Arrosadía, 31006, Pamplona, Spain

Leire AZPILICUETA
Dept. of Electrical and Electronic Engineering, Public Univ. of Navarre, Campus de Arrosadía, 31006, Pamplona, Spain

Juan Antonio NAZABAL
Dept. of Electrical and Electronic Engineering, Public Univ. of Navarre, Campus de Arrosadía, 31006, Pamplona, Spain

Carlos FERNÁNDEZ-VALDIVIELSO
Dept. of Electrical and Electronic Engineering, Public Univ. of Navarre, Campus de Arrosadía, 31006, Pamplona, Spain

Jesús SORET
Dept. of Electrical and Electronic Engineering, University of Valencia, Burjassot, Valencia, Spain

Francisco FALCONE
Dept. of Electrical and Electronic Engineering, Public Univ. of Navarre, Campus de Arrosadía, 31006, Pamplona, Spain

Fatih GENÇ
CU OPEN Lab, Dept. of Electronic and Communication Engineering, Ç ankaya University,Yukarıyurtçu Mahallesi Mimar Sinan Caddesi No:4, 06790, Etimesgut, Ankara, Turkey

M. Anıl REŞAT
Dept. of Electrical and Electronics Engineering, Yildırım Beyazıt University, Ç ankırı Caddesi Ç içek sok. No:3 , Altındăg, Ankara, Turkey

Asuman SAVAŞÇ IHABEŞ
Dept. of Electrical and Electronics Engineering, Nuh Naci Yazgan University, Erkilet Dere Mah., Kocasinan, Kayseri/Turkey

Özgür ERTUĞ
Telecommunication and Signal Processing Laboratory, Dept. of Electrical and Electronics Engineering, Gazi University, Yükselis¸ Sk. No:5, Maltepe, Ankara/Turkey

Min HUANG
The State Key Laboratory of Integrated Service Networks, Xidian University, Xiˇ ˙ Zan, 710071, P.R.China

Bingbing LI
The State Key Laboratory of Integrated Service Networks, Xidian University, Xiˇ ˙ Zan, 710071, P.R.China

Bing LI
State Key Lab. of ISN, Xidian University, Xi'an, 710071, P. R. China

Baoming BAI
State Key Lab. of ISN, Xidian University, Xi'an, 710071, P. R. China
Science and Technology on Information Transmission and Dissemination in Communication Networks Laboratory, Shijiazhuang, 050002, P. R. China

Mohammad Zahidul H. BHUIYAN
Dept. of Navigation and Positioning, Finnish Geodetic Institute, Kirkkonummi, Finland

Heidi KUUSNIEMI
Dept. of Navigation and Positioning, Finnish Geodetic Institute, Kirkkonummi, Finland

Stefan SÖDERHOLM
Dept. of Navigation and Positioning, Finnish Geodetic Institute, Kirkkonummi, Finland

Esa AIROS
Defence Forces Technical Research Centre, Riihimäki, Finland

Daniel ARBET
Institute of Electronics and Photonics, Slovak University of Technology, Ilkovičova 3, 812 19 Bratislava, Slovakia

Gabriel NAGY
Institute of Electronics and Photonics, Slovak University of Technology, Ilkovičova 3, 812 19 Bratislava, Slovakia

Viera STOPJAKOV Á
Institute of Electronics and Photonics, Slovak University of Technology, Ilkovičova 3, 812 19 Bratislava, Slovakia

Martin KOVÁ Č
Institute of Electronics and Photonics, Slovak University of Technology, Ilkovičova 3, 812 19 Bratislava, Slovakia

Nikola STOJANOVIĆ
University of Niš, Faculty of Electronics, A. Medvedeva 14, 18000 Niš, Serbia

Negovan STAMENKOVIĆ
University of Prishtina, Faculty of Natural Science and Mathematics, 28220 K. Mitrovica, Serbia

Vidosav STOJANOVI´C
University of Niš, Faculty of Electronics, A. Medvedeva 14, 18000 Niš, Serbia

Simon KUPCHAN
Dept. of Electrical and Electronic Engineering, Ariel University of Samaria, Ariel 40700, Israel

Monika PINCHAS
Dept. of Electrical and Electronic Engineering, Ariel University of Samaria, Ariel 40700, Israel

Ivan PRUDYUS
Institute of Telecommunications, Radioelectronics and Electronic Devices, Lviv Polytechnic National University, St. Bandery st., 12, 79013 Lviv, Ukraine

Valeriy OBORZHYTSKYY
Institute of Telecommunications, Radioelectronics and Electronic Devices, Lviv Polytechnic National University, St. Bandery st., 12, 79013 Lviv, Ukraine

Chi-Un LEI
Department of Electrical and Electronic Engineering, The University of Hong Kong, Hong Kong

Sezai Alper TEKİN
Dept. of Industrial Design Engineering, Erciyes University, 38039, Kayseri, Turkey

Hamdi ERCAN
Dept. of Avionics, Erciyes University, 38039, Kayseri, Turkey

Mustafa ALÇI
Dept. of Electrical and Electronics Engineering, Erciyes University, 38039, Kayseri, Turkey

Richa SRIVASTAVA
Dept. of Electronics and Communication Engineering, NSIT, New Delhi, India

Maneesha GUPTA
Dept. of Electronics and Communication Engineering, NSIT, New Delhi, India

Urvashi SINGH
Dept. of Electronics and Communication Engineering, NSIT, New Delhi, India

Jorge TORRES
Dept. of Telecommunication and Telematic, CUJAE University, 114 Street, 11901, La Habana, Cuba

Fidel HERNANDEZ
Dept. of Telecommunication and Electronics, UPR University, Marti 270, Pinar del Río, Cuba

Joachim HABERMANN
Department for ICT, Electrical Engineering & Mechatronics, TH Mittelhessen University of Applied Sciences,D-61169 Friedberg, Germany

Hongtu XIE
College of Electronic Science and Engineering, National University of Defense Technology, Changsha, Hunan, P. R. China, 410073

Daoxiang AN
College of Electronic Science and Engineering, National University of Defense Technology, Changsha, Hunan, P. R. China, 410073

Xiaotao HUANG
College of Electronic Science and Engineering, National University of Defense Technology, Changsha, Hunan, P. R. China, 410073

Zhimin ZHOU
College of Electronic Science and Engineering, National University of Defense Technology, Changsha, Hunan, P. R. China, 410073

Yongpeng ZHU
School of Electronics and Information Engineering, Harbin Institute of Technology, Harbin, 150001, China

Yinsheng WEI
School of Electronics and Information Engineering, Harbin Institute of Technology, Harbin, 150001, China

Yajun LI
School of Electronics and Information Engineering, Harbin Institute of Technology, Harbin, 150001, China

Wei ZHAO
College of Communications Engineering, PLA University of Science and Technology, Yudao Street, 210007 Nanjing, China

Yimin WEI
Wireless Communications Lab of College of Communications Engineering, PLA University of Science and Technology, Yudao Street, 210007 Nanjing, China

Yuehong SHEN
Wireless Communications Lab of College of Communications Engineering, PLA University of Science and Technology, Yudao Street, 210007 Nanjing, China

Zhigang YUAN
Wireless Communications Lab of College of Communications Engineering, PLA University of Science and Technology, Yudao Street, 210007 Nanjing, China

Pengcheng XU
College of Communications Engineering, PLA University of Science and Technology, Yudao Street, 210007 Nanjing, China

Wei JIAN
Wireless Communications Lab of College of Communications Engineering, PLA University of Science and Technology, Yudao Street, 210007 Nanjing, China

Sajid H. ALVI
Dept. of Electrical Engineering, COMSATS Institute of Information Technology, Islamabad, Pakistan

Shurjeel WYNE
Dept. of Electrical Engineering, COMSATS Institute of Information Technology, Islamabad, Pakistan

Min LIN
PLA University of Science and Technology, Nanjing, China
National Mobile Communications Research Laboratory, Southeast University, Nanjing, China

Heng WEI
College of Communication Engineering, PLA University of Science and Technology, Nanjing, China

Jian OUYANG
Inst. of Signal Processing and Transmission, Nanjing University of Posts and Telecommunications, Nanjing, China

Kang AN
College of Communication Engineering, PLA University of Science and Technology, Nanjing, China

Can YUAN
College of Communication Engineering, PLA University of Science and Technology, Nanjing, China

www.ingramcontent.com/pod-product-compliance
Lightning Source LLC
Chambersburg PA
CBHW050442200326
41458CB00014B/5041